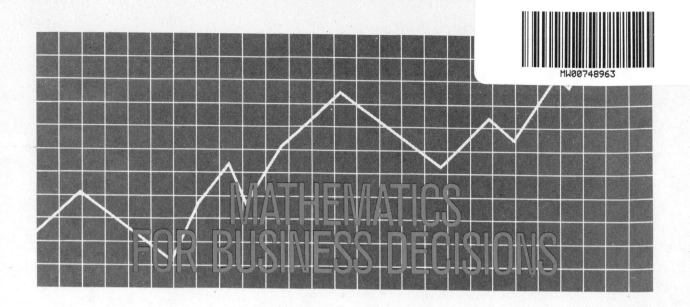

# MATHEMATICS FOR BUSINESS DECISIONS

## SECOND EDITION

### DAVID R. PETERSON
*Iowa Lakes Community College*
*Estherville, Iowa*

### KATHLEEN N. MILLER, Ed.D.
*Daytona Beach Community College*
*Daytona Beach, Florida*

**GLENCOE**

McGraw-Hill

New York, New York
Columbus, Ohio
Woodland Hills, California
Peoria, Illinois

Sponsoring Editors: Gary Milo and Mark B. Moscowitz
Editing Supervisor: Suzette André
Design and Art Supervisor: Caryl Valerie Spinka
Production Supervisor: Catherine Bokman

Text Designer: Suzanne Bennett & Associates
Cover Designer: John Keithley & Associates

**Library of Congress Cataloging-in-Publication Data**

Peterson, David R.
  Mathematics for business decisions.

  1. Business mathematics.  I. Miller, Kathleen N.  II. Title.
HF5691.P48  1989        512.1        88-37287
ISBN 0-07-049630-7

*Mathematics for Business Decisions,* Second Edition

Imprint 2000
Copyright © 1989 by Glencoe/McGraw-Hill. All rights reserved. Copyright © 1989, 1983 by
McGraw-Hill, Inc. All rights reserved. Printed in the United States of America. Except as permitted
under the United States Copyright Act of 1976, no part of this publication may be reproduced or distributed in any form or by any means, or stored in a database or retrieval system, without the prior
written permission of the publisher.

Send all inquiries to:
Glencoe/McGraw-Hill
8787 Orion Place
Columbus, OH 43240


 9 10 11 12 13 14 15  009  03 02 01 00

ISBN 0-07-049630-7

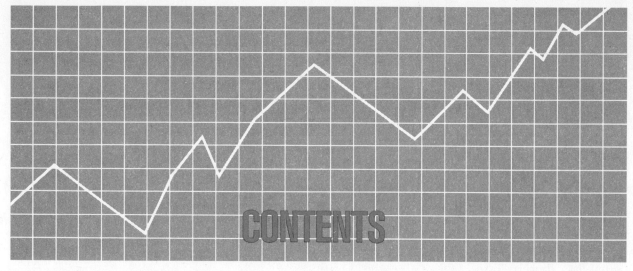

# CONTENTS

PREFACE

Anyone who pursues a business career will find his or her daily activities filled with various business mathematics calculations. These might include preparing a bank reconciliation, calculating payroll, ordering merchandise, pricing goods, selling merchandise, calculating interest on loans or credit sales, determining percent relationships, and a wide range of other calculations.

*Mathematics for Business Decisions,* Second Edition, is designed to teach prospective business employees, managers, and owners how to make these commonly occurring business calculations and how to apply these calculations to realistic business situations.

## TEXTBOOK FORMAT

The textbook is divided into 15 chapters and 3 appendixes. Chapters 1 and 2 provide instructional review of the basic mathematics fundamentals of addition, subtraction, multiplication, and division of whole numbers, decimals, and fractions. Chapters 3 to 15 provide instruction on how to perform many business mathematics calculations. The appendixes provide instruction on how to use a hand-held calculator, a realistic simulation, and additional case problems.

### CHAPTER CONTENT

The chapters in *Mathematics for Business Decisions,* Second Edition, are arranged in a logical sequence which parallels the order in which a businessperson might encounter the topics. Each chapter is divided into several lessons. Each lesson presents a topic that is studied in smaller segments, or subtopics. Features found in each chapter are as follows.

### LEARNING OBJECTIVES

Each chapter starts with a list of learning objectives which identify calculations the student should be able to perform after studying the chapter.

## CHAPTER INTRODUCTIONS

Each chapter introduction provides an overview of the topic presented in that chapter. This might serve as a springboard for class discussion to broaden the students' understanding of the concept to be studied in that chapter.

## INSTRUCTIONAL MATERIAL

Each calculation is described and explained before the calculation is performed. This provides the student with an understanding of why the calculation is performed and how the result is used in business practice.

## EXAMPLES AND SOLUTIONS

An example and step-by-step solution follow the instructional material. These demonstrate exactly how the calculation is to be performed and provide a handy reference for students.

## PRACTICE PROBLEMS

After the instructional material, example, and solution for each topic are practice problems. Answers for the practice problems are provided at the bottom of the page, upside down, so that students can get immediate feedback on their performance.

## EXERCISES

At the end of each lesson are exercises, presented in a format similar to that used for practice problems. These can be completed in class or assigned as homework.

## BUSINESS APPLICATIONS

Word problems, called business applications, follow the exercises at the end of each lesson. These require the student to apply to business situations what has been learned in that lesson. The business applications can be completed in class or assigned as homework.

## CHAPTER REVIEWS

A chapter review, covering all calculations presented in the chapter, is provided at the end of each chapter as reinforcement of the principles learned.

## CUMULATIVE REVIEWS

Cumulative reviews appear after Chapters 5, 9, 12, and 15. These provide a short review and refresher of topics covered so far in the book.

**ELECTRONIC CALCULATORS**

Instructional material on how to use a hand-held calculator is presented in Appendix A. Calculator features such as memory and percent are explained in a step-by-step format. Examples and solutions show how to use the calculator, and practice problems are provided for students to develop their calculator skills.

**THE HIT FACTORY: A BUSINESS SIMULATION**

This realistic simulation, presented in Appendix B, projects the student into the role of an employee for the Hit Factory, a professional recording company. Each activity (problem) in the simulation is not dependent upon any other activity. Therefore, even if one or more chapters of the textbook is not covered, the simulation can still be used by eliminating those topics from the simulation. The simulation is described fully in the teaching-suggestions portion of the instructor's manual.

**CASE PROBLEMS**

Additional word problems are provided in Appendix C for a variety of uses, including extra practice, makeup work, and extra credit. Each case problem identifies the lesson(s) in which the principle involved was discussed.

## TRANSPARENCIES

Fifty-three transparencies are available that provide worked-out solutions to complicated problems, blank forms, tables, and charts to facilitate teaching and learning.

## INSTRUCTOR'S MANUAL AND KEY

The *Instructor's Manual and Key* provides many aids to make instruction more meaningful and easier. The following features are provided.

**TEACHING SUGGESTIONS**

For each chapter, potential trouble spots that students might encounter are identified, and suggestions for helping students through these difficult areas are made.

Another feature of the teaching suggestions is additional background information for each chapter which might be shared in class discussion. Alternative terminology to that used in the textbook, additional facts, comments on business practices, and other types of information are provided.

**SOLUTION KEY**

Solutions to all calculations are provided. Step-by-step procedures are shown for all word problems.

**TESTS**

Two pretests are provided on the basic mathematics principles covered in Chapters 1 and 2. Also provided are four tests for each chapter and two final tests.

## MICROCOMPUTER TEST BANK

All the tests provided for this textbook are available on computer disk. This allows the instructor to easily and quickly combine items from various tests to make new test forms.

# ACKNOWLEDGMENTS

The authors are grateful to the following reviewers who supplied valuable expertise during the development of this edition:

Barbara Blickensderfer, Community College of Philadelphia, Philadelphia, Pennsylvania

James Cox, Lane Community College, Eugene, Oregon

E. Joseph Dorzweiler, Longview Community College, Lees Summit, Missouri

Ann Fleming, King's College, Charlotte, North Carolina

Darwin Grimm, Iowa Lakes Community College, Estherville, Iowa

Jean Hara, Honolulu Community College, Honolulu, Hawaii

Ron Hickman, CPA, Estherville, Iowa

Carol Johnson, Business Mathematics Instructor, Mineola, New York

Richard Lietzau, Inver Hills Community College, Inver Grove Heights, Minnesota

Jones Mayberry, Wake Technical Community College, Raleigh, North Carolina

Rosalie Morgan, Delaware County Community College, Media, Pennsylvania

Harry Munns, Daytona Beach Community College, Dayton Beach, Florida

Clara Nelson, Central Piedmont Community College, Charlotte, North Carolina

Anne Peterson, Iowa Lakes Community College, Estherville, Iowa

Mildred Polisky, Business Education Consultant, Milwaukee, Wisconsin

Nancy Schendel, Iowa Lakes Community College, Estherville, Iowa

John Senn, IGL Audio, Spirit Lake, Iowa

Doreen Stern, Business Education Consultant, Yardley, Pennsylvania

David Wheaton, Jerra Technical College, Freemont, Ohio

Richard H. Wirth, Business Mathematics Instructor, Cleveland, Ohio

**David R. Peterson**
**Kathleen N. Miller**

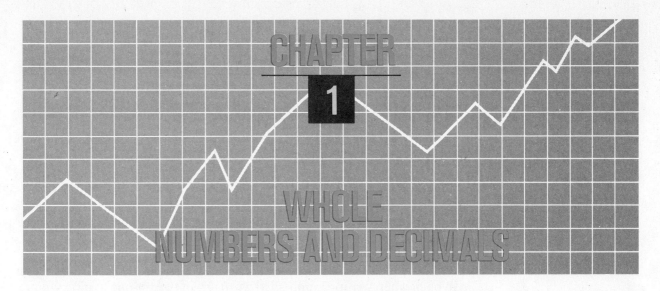

# CHAPTER 1

# WHOLE NUMBERS AND DECIMALS

## LEARNING OBJECTIVES

**1.** Identify place value from units place through millions place.

**2.** Round numbers.

**3.** Develop speed and accuracy when adding and subtracting vertically and horizontally.

**4.** Verify addition and subtraction problems.

**5.** Utilize the four-step method to solve word problems.

**6.** Develop speed and accuracy when multiplying and dividing numbers.

**7.** Verify multiplication and division problems.

Although computers and calculators are used to solve many business problems, businesspersons still must have a working knowledge of basic math and problem-solving skills. This chapter will present a review of the fundamental operations of addition, subtraction, multiplication, and division, emphasizing speed and accuracy when calculating.

## LESSON 1-1    ADDITION

**READING AND WRITING NUMBERS**

The most common number system, the **decimal** number system, uses 10 digits: 1, 2, 3, 4, 5, 6, 7, 8, 9, 0. The value of each digit in a number is determined by its position in relationship to the decimal point. Each position has a value that is 10 times greater than the position immediately to its right. Numbers to the left of the decimal point are **whole numbers;** numbers to the right of the decimal point are **decimal numbers.**

Numbers are read from left to right. Use the word "and" to indicate the decimal point. To facilitate reading numbers, commas are inserted in whole numbers after every third digit starting at the decimal point. The place value chart on page 2 shows the value of each place from trillions place to trillionths place.

| Whole Numbers | | | | | | | | | | | | | (Decimal point) | Decimals | | | | | | | | | | | |
|---|---|---|---|---|---|---|---|---|---|---|---|---|---|---|---|---|---|---|---|---|---|---|---|---|---|
| Trillions | Hundred billions | Ten billions | Billions | Hundred millions | Ten millions | Millions | Hundred thousands | Ten thousands | Thousands | Hundreds | Tens | Units | | Tenths | Hundredths | Thousandths | Ten thousandths | Hundred thousandths | Millionths | Ten millionths | Hundred millionths | Billionths | Ten billionths | Hundred billionths | Trillionths |

5 , 7 8 2 , 3 9 0 . 1 7 4 2

Numbers are written as figures and as words on documents such as checks and legal forms. Write a number in words exactly as it is read. Use a hyphen when writing numbers in words from twenty-one to ninety-nine. The number 5,782,390.1742 in the above example is read, "five million, seven hundred eighty-two thousand, three hundred ninety and one thousand, seven hundred forty-two ten thousandths."

| EXAMPLES | SOLUTIONS |
|---|---|
| 25,672.15 | Twenty-five thousand, six hundred seventy-two and fifteen hundredths |
| 0.2681 | Two thousand, six hundred eighty-one ten thousandths |
| 27.841 | Twenty-seven and eight hundred forty-one thousandths |
| $592.43 | Five hundred ninety-two dollars and forty-three cents |

## PRACTICE PROBLEMS

Write in word form.

1. 36.18
2. $1,425.67
3. 371.9
4. 0.245
5. $14,256.75

## ROUNDING NUMBERS

Rounded numbers are used in business when only an approximate amount or a "ballpark" figure is needed. A number can be rounded to any desired number of places. Money amounts are usually rounded to the nearest cent or the nearest dollar. Use the following steps when rounding numbers:

1. Identify the digit in the place being rounded. This digit is called the **round-off digit.**
2. If the digit to the right of the round-off digit is 5 or greater, add 1 to the round-off digit. If it is less than 5, the round-off digit remains the same.
3. Change all the digits to the right of the round-off digit to zero.

ANSWERS FOR PRACTICE PROBLEMS

(1) Thirty-six and eighteen hundredths
(2) One thousand, four hundred twenty-five dollars and sixty-seven cents
(3) Three hundred seventy-one and nine tenths
(4) Two hundred forty-five thousandths
(5) Fourteen thousand two hundred fifty-six dollars and seventy-five cents

## EXAMPLE 1

Round 34,867 to the nearest hundred.

### SOLUTION

34,867 ── Hundreds place

6 is greater than 5

34,867 rounded to the nearest hundred is 34,900.

## EXAMPLE 2

Round $157.934 to the nearest cent.

### SOLUTION

$157.934 ── Hundredths place

4 is less than 5

$157.934 rounded to the nearest cent is $157.93 or $157.934 ≈ $157.93.

The symbol ≈ means the answer is rounded.

### PRACTICE PROBLEMS

Round to the indicated place.

1. Round 4,865 to the nearest hundred.
2. Round 7,986 to the nearest ten.
3. Round 32.04538 to the nearest thousandth.
4. Round 46.1497 to the nearest tenth.
5. Round $845.525 to the nearest dollar.

## ADDITION

Addition is used in everyday business activities such as calculating total sales, finding the number of items in stock, or totaling a bank deposit. In an addition problem, the numbers being added are called the **addends.** The answer is called the **sum** or **total.** When adding whole numbers with two or more digits, align the numbers starting with the digits in the units place. The decimal point is assumed to be at the far right when adding whole numbers.

### EXAMPLE

34 + 139 + 7 =

### SOLUTION

$$
\begin{array}{r}
34 \\
139 \\
+\ \ 7 \\
\hline
180
\end{array}
$$
Addends

180   Sum or total

ANSWERS FOR PRACTICE PROBLEMS: (1) 4,900  (2) 7,990  (3) 32.045  (4) 46.1  (5) $846.00

To increase speed and accuracy when adding, use the following combinations of two and three numbers that total 10. First add the combinations totaling 10; then add the remaining numbers to obtain the sum.

| 1 | 2 | 3 | 4 | 5 | 1 | 1 | 1 | 1 | 2 | 2 | 3 | 4 |
|---|---|---|---|---|---|---|---|---|---|---|---|---|
| 9 | 8 | 7 | 6 | 5 | 1 | 2 | 3 | 4 | 2 | 3 | 3 | 4 |
|   |   |   |   |   | 8 | 7 | 6 | 5 | 6 | 5 | 4 | 2 |

---

**EXAMPLE**

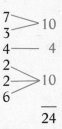

```
7
3 }10
4 — 4
2
2 }10
6
  ──
  24
```

---

When adding decimals, align the decimal points and add as though you were adding whole numbers. If the addends have a varying number of decimal places, add one or more zeros to the right of the last digit so that each addend has the same number of decimal places. Placing zero to the right of the decimal point does not change the value of the number. For example, 1.4, 1.40, and 1.400 have the same value.

---

**EXAMPLE**

1,375.3 + 860.28 + 9.049 =

**SOLUTION**

```
  1,375.300
    860.280
+     9.049
  ─────────
  2,244.629
```

---

Check or verify addition by adding in the reverse order. If the sum at the top is the same as the sum at the bottom, the addition should be correct.

---

**EXAMPLE**

Carrie Davis totaled the following check amounts on a deposit slip. Verify the total.

```
$1,875.26
   860.97
    36.12
─────────
$2,772.35
```

**SOLUTION**

Add in reverse order to verify.

```
$2,772.35 ↑
  1,875.26
    860.97
     36.12
─────────
$2,772.35   Addition is correct
```

---

Numerical information is recorded horizontally on many business documents such as sales reports, payroll registers, and inventory records. When adding numbers horizontally, add the digits farthest to the right, then add the digits second from the right, and so on. When necessary, remember to carry before adding the next set of digits.

---

**EXAMPLE**

32 + 15 + 98 =

**SOLUTION**

**Step 1**  3$^2$ + 1$^5$ + 9$^8$ =  $^5$

**Step 2**  $^1$$^3$2 + $^1$5 + $^9$8 = $^{14}$5

---

Horizontal addition can be verified automatically on many business forms because the figures are often added both horizontally and vertically. This is called *crossfooting.* The sum of the horizontal totals and the sum of the vertical totals should be equal. This sum is called the *grand total.*

---

**EXAMPLE**

Total the vertical columns and the horizontal rows on the following sales report. Find the grand total.

**SOLUTION**

| | A | B | C | D | E |
|---|---|---|---|---|---|
| 1 | | QUARTERLY SALES REPORT | | | |
| 2 | | | | | |
| 3 | Salesperson | January | February | March | Totals |
| 4 | | | | | |
| 5 | J. Randall | $41,628.00 | $31,891.00 | $29,954.00 | $103,473.00 |
| 6 | T. Schwartz | 22,904.00 | 31,009.00 | 24,560.00 | 78,473.00 |
| 7 | Y. Taylor | 18,979.00 | 21,259.00 | 19,702.00 | 59,940.00 |
| 8 | I. Veller | 33,100.00 | 26,983.00 | 32,600.00 | 92,683.00 |
| 9 | | | | | |
| 10 | Totals | $116,611.00 | $111,142.00 | $106,816.00 | $334,569.00 |
| 11 | | | | | |
| 12 | | | | | |
| 13 | | | | | |
| 14 | | | | | |
| 15 | | | | | |
| 16 | | | | | |
| 17 | | | | | |
| 18 | | | | | |

## PRACTICE PROBLEMS

Add using combinations of 10 where possible. Verify sums.

| | 1. 5,386 | 2. $316.85 | 3. 5.286 | 4. $721.86 |
|---|---|---|---|---|
| | 704 | 57.80 | 17.059 | 3.14 |
| | 6,913 | 970.15 | .682 | 82.19 |
| | 8,117 | 3.24 | 3.052 | .48 |
| | 825 | 684.13 | 45.226 | 310.60 |

Rewrite in vertical form; then add. Verify sums.

**5.** 3.56 + 29 + 148 + 0.307 + 69.2 =
**6.** 1,520 + 372.14 + 29.976 + 534.318 + 4.8 =
**7.** 146.82 + 5.7 + 3.91 + 8.16 + 5.2 =
**8.** 15,862.10 + 38.526 + 6.89 + 0.583 + 1.81 =

Find the horizontal totals, vertical totals, and grand total.

**9.**  186 +  36 +  19 =
        23 + 149 + 137 =
       204 +  61 +  56 = _____
         +     +     = _____

**10.** $27.51 + $16.83 + $12.08 =
         59.60 +  73.92 +  46.18 =
         46.09 +  17.08 +  24.32 = _____
           +       +       = _____

## SOLVING WORD PROBLEMS

In business, many problems are presented in sentence form called **word problems.** The following four-step method will make word problems easier to solve.

**Step 1** READ    Read the problem carefully.
**Step 2** PLAN    Determine what is known and what needs to be found. Choose a strategy for solving the problem.
**Step 3** DO      Perform the calculation.
**Step 4** CHECK   Estimate to see if your answer is reasonable and logical.

Estimating the answer to a business math problem can help avoid careless mistakes and increase the chances of obtaining an accurate answer. To estimate answers, round the numbers to the desired place value and perform the calculation. Compare the estimated answer with the actual answer to see if the answer is reasonable.

---

## EXAMPLE

Michelle Hovey went shopping and spent the following amounts: $12.85,

---

$55.45, $23.98, $99.95, and $71.16. The store clerk charged Michelle $263.39 for her purchases. Round the amounts to the nearest ten dollars and estimate the sum to see if the total is reasonable.

## SOLUTION

| Actual Amounts | Rounded Amounts | |
|---|---|---|
| $12.85 | $ 10 | |
| 55.45 | 60 | |
| 23.98 | 20 | |
| 99.95 | 100 | |
| 71.16 | 70 | |
| | $260 | Estimated sum |

Watch for the key words "plus," "sum," "combined," "total," and "added to," which indicate an addition problem.

## EXAMPLE

Christopher Davis worked 3 days last week and had daily earnings as follows: Monday, $52.25; Tuesday, $63.75; and Wednesday, $72.45. What were his total earnings for the week?

## SOLUTION

| Step 1 | READ | The key phrase is "total earnings," so add. | | | |
|---|---|---|---|---|---|
| Step 2 | PLAN | Monday's Earnings + | Tuesday's Earnings + | Wednesday's Earnings = | Total Earnings |
| Step 3 | DO | $52.25 + | $63.75 + | $72.45 = | $188.45 |
| Step 4 | CHECK | $50.00 + | $60.00 + | $70.00 = | $180.00 Estimate |

## PRACTICE PROBLEMS

1. Round to the nearest dollar; then estimate.

| $15.12 | $25.10 |
|---|---|
| 9.86 | 39.60 |
| 12.05 | 22.95 |
| 3.95 | 17.80 |
| 8.08 | 24.05 |

2. Round to the nearest hundred; then estimate.

| 695 | 387.6 |
|---|---|
| 409 | 2,105.2 |
| 821 | 791.5 |
| 97 | 427.6 |
| 298 | 1,907.2 |

Solve using the four-step method.

3. Nettleton Furniture had furniture on display costing $12,376. Additional furniture was added to the display floor costing $1,574, $3,690, and $1,816. What was the total cost of the furniture on display?

4. Ken Wiggins works in a factory where he strings tennis racquets. In 5 days last week, Ken strung 128, 124, 132, 126, and 133 racquets. How many racquets did Ken string last week?

ANSWERS FOR
PRACTICE PROBLEMS

(1) $49; $130   (2) 2,300; 5,600   (3) $12,376 + 1,574 + 3,690 + 1,816 = $19,456   (4) 128 + 124 + 132 + 126 + 133 = 643

## LESSON 1-1     EXERCISES

Write in word form.

**1.** 3,609 _____

**2.** $276.18 _____

**3.** 70,349.762 _____

_____

**4.** 0.16 _____

**5.** $50,005.95 _____

**6.** $37.98 _____

Round to the indicated place.

| | Nearest Tenth | Nearest Hundred | Nearest Thousand |
|---|---|---|---|
| **7.** 6,218.094 | _____ | _____ | _____ |
| **8.** 1,496.55 | _____ | _____ | _____ |
| **9.** 17,249.329 | _____ | _____ | _____ |
| **10.** 4,724.94 | _____ | _____ | _____ |
| **11.** 81,478.96 | _____ | _____ | _____ |

Add.

| **12.** 76 | **13.** 286 | **14.** 479 | **15.** 3,506 | **16.** 45.89 |
|---|---|---|---|---|
| 19 | 173 | 65 | 2,814 | 1.06 |
| 43 | 405 | 302 | 82 | 3.14 |
| 27 | 619 | 8 | 975 | 50.23 |
| 48 | 381 | 91 | 4,006 | .47 |

| **17.** 27.15 | **18.** $86.05 | **19.** $126.19 | **20.** $4,386.42 | **21.** $586.92 |
|---|---|---|---|---|
| 3.8 | 9.43 | 5.60 | 90.17 | 47.80 |
| 9.62 | 62.20 | 3.42 | 820.34 | 816.74 |
| 145.00 | 14.67 | 85.00 | 6,921.65 | 3.98 |
| 1.07 | 8.89 | .08 | 709.48 | 47.06 |

| **22.** 6,497 | **23.** $586.13 | **24.** 6.825 | **25.** 681.27 | **26.** 3,743.01 |
|---|---|---|---|---|
| 805 | 8.75 | 62.254 | 4.13 | 459.92 |
| 7,824 | 510.79 | 2.503 | 91.28 | 198.13 |
| 9,228 | 4.23 | .286 | .84 | 826.14 |
| 936 | 314.86 | 95.071 | 601.30 | 389.62 |

Rewrite in vertical form, and then add. Verify sums.

**27.** 36.02 + 1.99 + 0.24 + 0.183 = _____

**28.** 120.9 + 92.23 + 8.342 + 42.78 = _____

**29.** 301.41 + 423.62 + 902.051 + 84.143 = _____

**30.** 10.204 + 195.35 + 142.8013 = _____

**31.** Find the vertical, horizontal, and grand totals for the following report for the Alkett Company.

| | A | B | C | D | E |
|---|---|---|---|---|---|
| 1 | | COMMISSION | EARNED | | |
| 2 | | | | | |
| 3 | Salesperson | January | February | March | Totals |
| 4 | | | | | |
| 5 | R. Cranshaw | $1,216.21 | $1,273.10 | $1,549.89 | $4,039.20 |
| 6 | T. Folley | 1,409.87 | 1,575.80 | 1,411.10 | 4,396.77 |
| 7 | Q. Issy | 1,242.38 | 1,254.54 | 1,148.96 | 3,645.88 |
| 8 | E. Noel | 1,417.31 | 1,328.24 | 1,308.49 | 4,054.04 |
| 9 | G. Reely | 998.50 | 1,023.94 | 1,253.15 | 3,275.59 |
| 10 | | | | | |
| 11 | Totals | $6,284.27 | $6,455.62 | $6,671.59 | $19,411.48 |
| 12 | | | | | |
| 13 | | | | | |
| 14 | | | | | |
| 15 | | | | | |
| 16 | | | | | |
| 17 | | | | | |
| 18 | | | | | |

## BUSINESS APPLICATIONS

Solve the following word problems using the four-step method.

**32.** Mel Holbrook made sales of $246.80, $321.76, $108.14, $482.12, and $95.07 yesterday. What were his total sales for the day?

_____

**33.** Karen went shopping and made purchases of the following amounts: $6.98, $22.15, $38.20, $109.89, and $14.79. Estimate to the nearest dollar Karen's total purchase.

_____

**10** ■ LESSON 1-1

**34.** The Hitmaker Record Shoppe had tapes in stock costing $86,214.95. The next day, the store received shipments of tapes costing $3,580.15 and $8,972.49. What was the total cost of tapes in stock?

_____

**35.** Last month Camden Company had the following sales by department: clothing, $47,386; cosmetics, $16,284; jewelry, $25,698; luggage, $4,281; and shoes, $8,425. What was Camden's total sales for the month?

_____

**36.** Jean Claussen has the following monthly expenses: rent, $695; food, $220; utilities, $90; car, $190; loan payments, $275; and entertainment, $225. Round each amount to the nearest hundred dollars and estimate Jean's total monthly expenses.

_____

**37.** The Pacesetters' Bicycle Club went on a 5-day bicycle ride covering the following number of miles each day: 30.73, 37.16, 35.09, 42.6, and 39.41. What is the total miles club members rode?

_____

**38.** Melinda Hanover ordered the following garments: suit, $185.59; blouse, $54.79; shoes, $62.50; and dress, $109.95. If sales tax and shipping charges of $32.24 were added to her bill, what is the total cost of her order?

_____

**39.** Last year Bill Adams earned $375.50 interest on his certificate of deposit, $186.70 interest on his savings account, $228.50 interest on his mutual fund, and $424.80 interest on his IRA. What was the total amount of interest Bill earned from his investments?

_____

**40.** Four women form a partnership to buy their own bookstore. Each contributes the following amount of money: Kathy, $26,523; Mary, $22,634; Kim, $17,389; and Denise, $28,957. Round each woman's contribution to the nearest hundred and estimate the total amount of money contributed by the women.

_____

**41.** The Ellis Tire Mart has a balance of $150 in its office safe. The morning deposit into the safe was $456.78 and the afternoon deposit was $1,009.48. What was the total amount in the safe at the end of the day?

_____

Subtraction is used daily in business to find the balance in an account, the amount of items left in stock, and other important amounts. Subtraction is the process of deducting from one number called the **minuend** another number called the **subtrahend.** The result is called the **difference.**

| | |
|---|---|
| 175 | Minuend |
| − 80 | Subtrahend |
| 95 | Difference |

When subtracting, it is important to align decimals so that units are above units, tens are above tens, and so on. Begin subtracting digits in the right column and move left. When the digit in the subtrahend is larger than the digit in the minuend, borrow as shown in the following example.

### EXAMPLE

$28.09 − 15.25 =

### SOLUTION

| Step 1 | Step 2 | Step 3 | Step 4 |
|---|---|---|---|
| $28.09 | $28.09 | $28.09 | $28.09 |
| 15.25 | 15.25 | 15.25 | 15.25 |
| 4 | 84 | 2.84 | $12.84 |

To verify the accuracy of subtraction, add the difference to the subtrahend. The result should equal the minuend.

Subtraction

$28.09
− 15.25
$12.84

Subtraction is correct

Verification

$12.84
+ 15.25
$28.09

When numbers are written horizontally on business forms such as ledger accounts or payroll reports, learn to subtract without rewriting. To subtract horizontally, first subtract the digits farthest to the right and move to the left. Borrowing is done the same way as it is when subtracting vertically.

### PRACTICE PROBLEMS

Subtract and verify answers.

1. $486.94
   271.62

2. 5,682
   3,234

3. $578.20
   95.68

4. 5.069
   0.872

Subtract horizontally and verify answers.

5. 768 − 532 =     6. 1,295 − 383 =     7. $2,075 − 1,239 =

## SOLVING WORD PROBLEMS

When solving word problems involving subtraction, use the four-step method: read, plan, do, and check. Check the answer by rounding the minuend and the subtrahend and subtracting the rounded numbers. Watch for the key words "difference," "remainder," "subtract," "less," "minus," and "decrease," which indicate a subtraction problem.

### EXAMPLE

Carlos Fernandez purchased a car for $15,725. Two years later he sold the car for $11,850. What was the difference between the purchase price and the selling price of the car?

### SOLUTION

| Step 1 | READ | The key word is "difference," so subtract. |
|--------|------|--------------------------------------------|
| Step 2 | PLAN | Purchase Price − Selling Price = Difference |
| Step 3 | DO   | $15,725 − $11,850 = $3,875 |
| Step 4 | CHECK | $16,000 − $12,000 = $4,000 Estimate |

### PRACTICE PROBLEMS

Round to the nearest thousand, and then estimate the answer.

1. 15,780 − 6,921 =          2. 189,200 − 23,840 =

Solve using the four-step method.

3. The original price of a dining room table and chairs was $395.99. It is now selling for $225.59. What is the amount of the price reduction?
4. Dr. Langly paid $1,942.50 for car repairs. His insurance company reimbursed him for $1,154.43. How much did he actually pay for car repairs?

## LESSON 1-2    EXERCISES

Subtract and verify answers.

| 1. | $568.92<br>− 489.16 | 2. | $7,287.16<br>− 908.49 | 3. | $305.09<br>− 286.83 | 4. | $27,506.88<br>− 19,873.05 | 5. | $18,305.62<br>− 9,650.85 |
|---|---|---|---|---|---|---|---|---|---|

| 6. | 53,589<br>−47,394 | 7. | 16,905<br>−14,816 | 8. | 25,004<br>−18,436 | 9. | 17,805<br>− 9,056 | 10. | 38,145<br>−26,875 |
|---|---|---|---|---|---|---|---|---|---|

| 11. | 0.875<br>−0.500 | 12. | 18.005<br>− 4.970 | 13. | 456.125<br>− 97.850 | 14. | 92.050<br>−48.175 | 15. | 3,562.18<br>− 496.79 |
|---|---|---|---|---|---|---|---|---|---|

| 16. | $4,206.25<br>− 798.19 | 17. | $86,582<br>− 78,008 | 18. | $345,869<br>− 108,267 | 19. | $7,621.84<br>− 928.73 | 20. | $20,056.93<br>− 16,724.82 |
|---|---|---|---|---|---|---|---|---|---|

Estimate by rounding to the nearest hundred and subtracting.

**21.** $1,316.12 − 805.68 = _____

**22.** $806.95 − 592.07 = _____

**23.** 18,065 − 12,750 = _____

**24.** 21,958 − 1,316 = _____

**25.** Find the weekly net pay for the following employees of Leans Electronics, Inc.

| Employee | Gross Pay | − | Deductions | = | Net Pay |
|---|---|---|---|---|---|
| Y. Heller | $432.50 | − | $173.00 | = | _____ |
| I. Kelkner | 620.75 | − | 286.20 | = | _____ |
| E. Pompey | 564.90 | − | 253.80 | = | _____ |
| A. Quency | 502.30 | − | 220.60 | = | _____ |
| L. Tully | 659.40 | − | 290.73 | = | _____ |

## BUSINESS APPLICATIONS

Solve using the four-step method.

**26.** On September 5, Julie Lamoni had $315.80 in her savings account. During September, she withdrew $56 and $95 from her account. How much is in the account after these withdrawals?

**27.** The amount of Dan Simpson's paycheck was $1,652.60. He made the following payments from his paycheck: rent, $550.75; food, $225.15; utilities, $112.98; auto expenses, $174.90; clothing, $162.59; and entertainment, $227.32. He put the remaining amount in his savings account. How much did Dan put in his savings account?

_____

**28.** Salesperson Judy Johnsrude used a company car for a sales trip to another city. The odometer's reading was 28,308 when she left and 28,673 when she returned. How many miles did Judy travel?

_____

**29.** Sterling Department Store is selling a set of two matching lamps for $140.50. If the lamps ordinarily sell for $75.30 each, how much can be saved by buying two lamps as a set?

_____

**30.** The Oakland Business Supply House sells a computer for $1,159, a printer for $395, and a software package for $219.35. J & W Computer Trends sells the same computer for $997, the printer for $452, and the software package for $246.50. Which store offers a better buy for the three items?

_____

Multiplication is a basic math operation used in business to calculate such amounts as interest on loans, discounts on purchases, insurance rates, and gross pay. Multiplication is a rapid method of adding repeated addends. For example, it is faster to multiply $6 \times 4$ than to add $6 + 6 + 6 + 6$. The numbers being multiplied are called *factors;* the answer is called the *product.*

$$
\begin{array}{r}
6 \\
\times\ 4 \\
\hline
24
\end{array}
$$

 6 } Factors
 24 Product

When multiplying by numbers containing more than one digit, partial products are calculated. Write each partial product one place to the left of the previous partial product. Add the partial products to find the final product.

## EXAMPLE

$348 \times 125 =$

## SOLUTION

$$
\begin{array}{r}
348 \\
\times 125 \\
\hline
1740 \\
696 \\
348 \\
\hline
43,500
\end{array}
$$

1740 First partial product ($348 \times 5$)
696 Second partial product ($348 \times 2$)
348 Third partial product ($348 \times 1$)
43,500 Product

When multiplying by a factor containing zeros in the middle, place a zero in the partial product directly below the zero in the factor.

## EXAMPLE

$\$892 \times 407 =$

## SOLUTION

**Step 1** Find the first partial product.

$$
\begin{array}{r}
\$892 \\
\times\ 407 \\
\hline
6244
\end{array}
$$

**Step 2** Write a zero directly below the zero in the factor. Then, find the next partial product and add.

$$
\begin{array}{r}
\$892 \\
\times\ 407 \\
\hline
6\ 244 \\
356\ 80 \\
\hline
\$363,044
\end{array}
$$

Verify a multiplication computation by reversing the factors and multiplying. If the products are the same, the multiplication should be correct.

## EXAMPLE

| Multiplication | | Verification |
|---|---|---|
| 82 | Multiplication | 12 |
| ×12 | is correct | ×82 |
| 984 ← | | → 984 |

### PRACTICE PROBLEMS

Multiply and verify answers.

| **1.** 918 | **2.** 36 | **3.** 179 | **4.** 1,372 | **5.** 2,843 |
|---|---|---|---|---|
| × 9 | ×17 | × 43 | × 105 | × 307 |

## MULTIPLYING NUMBERS ENDING IN ZEROS

A number that ends with one or more zeros is called a **multiple of 10.** When multiplying factors ending in zero, first multiply the digits to the left of the ending zeros. Then, add to the product the total number of zeros in both factors.

### EXAMPLE

$2,600 \times 70 =$

### SOLUTION

| 2,600 | 2 zeros |
|---|---|
| × 70 | 1 zero |
| 182,000 | 3 zeros |

### PRACTICE PROBLEMS

Multiply.

| **1.** 340 | **2.** 1,700 | **3.** 180 | **4.** 9,300 | **5.** 4,670 |
|---|---|---|---|---|
| × 30 | × 500 | ×400 | ×3,200 | × 600 |

## MULTIPLYING DECIMALS

Multiply decimal numbers as you would whole numbers. The number of decimal places in the product is equal to the total number of decimal places in the factors. When placing the decimal point in the product, begin counting from the right. Add zeros to the left of the product if necessary.

### EXAMPLE 1

$\$61.09 \times 3.2 =$

### SOLUTION

| $61.09 | 2 decimal places |
|---|---|
| × 3.2 | 1 decimal place |
| 12 218 | |
| 183 27 | |
| $195.488 | 3 decimal places |

## EXAMPLE 2

$0.216 \times 0.43 =$

## SOLUTION

$$\begin{array}{r} \$0.216 \quad \text{3 decimals} \\ \times \quad 0.43 \quad \text{2 decimals} \\ \hline 648 \\ 864 \\ \hline \$0.09288 \quad \text{5 decimals (add zero to left)} \end{array}$$

## PRACTICE PROBLEMS

Multiply and verify answers.

| 1. | 475 | 2. | 35.15 | 3. | 2.876 | 4. | 85.02 | 5. | 319.79 |
|---|---|---|---|---|---|---|---|---|---|
| | $\times$ 3.5 | | $\times$ 5.9 | | $\times$ 9.4 | | $\times$ 3.78 | | $\times$ 6.39 |

## MULTIPLYING BY A POWER OF 10

The numbers 10, 100, 1,000, 10,000, and so on are called **powers of 10.** When multiplying by powers of 10, simply move the decimal point to the right in the product as many places as there are zeros in the power of 10. Add zeros to the right when necessary.

## EXAMPLES

$5.94 \times 10 = 59.4$

$45.349 \times 100 = 4,534.9$

$\$85 \times 1,000 = \$85,000$   (Add 3 zeros)

$\$35.52 \times 10,000 = \$355,200$   (Add 2 zeros)

## PRACTICE PROBLEMS

Multiply.

1. $68 \times 10 =$
2. $5.65 \times 1,000 =$
3. $\$36.85 \times 100 =$
4. $0.5672 \times 10 =$
5. $35.1 \times 1,000 =$

## SOLVING WORD PROBLEMS

When solving word problems involving multiplication, follow the four-step method: read, plan, do, check. A small error in a multiplication computation can result in a large error in the product. Estimating products will usually detect large errors.

## EXAMPLE

The Round-Rib Restaurant was billed $4,176 for the purchase of 18 tables selling for $232 each. Estimate the total price of the tables to see if the bill is reasonable.

ANSWERS FOR PRACTICE PROBLEMS

**Bottom: (1)** 680   **(2)** 5,650   **(3)** $3,685   **(4)** 5.672   **(5)** 35,100

**Top: (1)** 1,662.5   **(2)** 207.385   **(3)** 27.0344   **(4)** 321.3756   **(5)** 2,043.4581

## SOLUTION

| Actual Amounts | Estimated Amounts |
|---|---|
| $232 | $200 |
| × 18 | × 20 |
| | $4,000  Estimated total |

When solving word problems, watch for the key words "times," "product," or "fractional part of," which indicate multiplication.

## EXAMPLE

Last year Terri Tyson deposited $823.20 in her savings account. This year Terri was able to deposit 3 times as much. How much money did she deposit this year?

## SOLUTION

| Step 1 | READ | The key word is "times," so multiply. |
|---|---|---|
| Step 2 | PLAN | Last Year's Savings × 3 = This Year's Savings |
| Step 3 | DO | $823.20    × 3 =    $2,469.60 |
| Step 4 | CHECK | $800    × 3 =    $2,400  Estimate |

## PRACTICE PROBLEMS

Round factors to the first digit, and then estimate products.

1. $1,340
   ×  382

2. 821
   × 42

3. 208
   × 74

4. 4,002
   ×  37

5. 8,940
   × 450

Solve using the four-step method.

6. Ed Wallstein earned $2,150 a month last year. What were Ed's total earnings for the year?

7. The cost of manufacturing a children's swing set is $48.92. How much will it cost to manufacture 196 sets?

## LESSON 1-3    EXERCISES

Multiply.

1.  58
    × 7

2.  173
    × 15

3.  905
    × 29

4.  314
    × 18

5.  2,079
    × 35

6.  4,516
    × 80

7.  3,093
    × 12

8.  4,619
    × 305

9.  51,265
    × 208

10.  4,207
     × 45

11.  $486.99
     × 4.65

12.  7053.7
     × 204

13.  2,316
     × 3.09

14.  $605.55
     × 1.16

15.  $21.89
     × 143

16.  3,240.6
     × 2005

17.  $29.67
     × .108

18.  $5,005
     × 1.25

19.  $3.287
     × 0.25

20.  0.0718
     × 0.362

**21.** $0.825 \times 10 =$

**22.** $420 \times 40 =$

**23.** $240 \times 200 =$

**24.** $1.298 \times 10{,}000 =$

**25.** $3.62 \times 1{,}000 =$

**26.** $\$532.19 \times 100 =$

**27.** $1{,}200 \times 30 =$

**28.** $560 \times 250 =$

**29.** $1{,}720 \times 50 =$

**30.** $14{,}500 \times 60 =$

## BUSINESS APPLICATIONS

Solve using the four-step method.

**31.** The clerk-typist at Insurance World works 37 hours a week for $8.54 an hour. What is her weekly pay?

_____

**32.** Auburn Manufacturing Company has 16 employees. Last year each employee worked 250 days and earned $90 per day. What is the total amount Auburn paid to its 16 employees for the year?

_____

**33.** Newell Candy Manufacturing Company purchased 37 sacks of sugar weighing 50 pounds each and 28 sacks of sugar weighing 25 pounds each. How many pounds of sugar did Newell buy?

_____

**34.** Last week Denise Hatfield worked 40 hours at her regular job and earned $7.50 per hour. She also worked 15 hours at a part-time job and earned $5.75 per hour. What was Denise's total earnings for the week?

_____

**35.** Lakeview Grocery purchased 18 cases of oranges containing 56 oranges each. Of these, 14 oranges were spoiled and had to be thrown away. How many oranges did Lakeview have available to sell?

_____

**36.** Econo Rental rented to customers four garden tractors for 6 hours each at $10.50 per hour. How much did Econo's earn from the rentals?

_____

**37.** Minerva's Fashions sold 12 dresses at $76.50 each and 26 dresses at $89.95 each. What was the total amount received from the sale of these dresses?

_____

**38.** A new car can be purchased for $500 down and 36 monthly payments of $255. What is the total cost of the car?

_____

**39.** The Starlight Convention Center sponsored a concert featuring the Metropolitan Symphony Orchestra. Four thousand, seven hundred sixty-nine tickets were sold at $12.50 each. How much did Starlight have left after paying the symphony $22,500?

_____

**40.** Roberta Favis manages a 100-unit apartment complex. Twelve of the units rent for $410 each, 37 rent for $610 each, 42 rent for $685 each, and the remaining rent for $715 each. What is the total monthly rent Roberta collects from the 100 apartment units?

_____

Division is used in business to calculate such amounts as the average of a group of numbers or the cost of a single item sold in bulk. Division is the process of determining how many times one number is contained in another. The **divisor** is the number doing the dividing, the **dividend** is the number being divided, and the **quotient** is the answer. The division problem, 240 divided by 15, can be written in any one of the following ways:

Dividend ÷ Divisor = Quotient
  240   ÷   15   =   16

$$\frac{\text{Dividend}}{\text{Divisor}} \qquad \frac{240}{15} = 16 \quad \text{Quotient}$$

$$\text{Divisor} \quad 15\overline{)240} \quad \begin{matrix} 16 & \text{Quotient} \\ & \text{Dividend} \end{matrix}$$

When calculating long division, use the following four steps: divide, multiply, subtract, bring down. Repeat these steps until you can no longer bring down a digit from the dividend.

**EXAMPLE**

$285 \div 12 =$

**SOLUTION**

| | | | | |
|---|---|---|---|---|
| **Step 1** | DIVIDE | $28 \div 12$ | $12\overline{)285}$ with $2$ | $12\overline{)285}$ with $23$ |
| **Step 2** | MULTIPLY | $12 \times 2$ | $24$ | $24$ |
| **Step 3** | SUBTRACT | $28 - 24$ | $45$ | $45$ |
| **Step 4** | BRING DOWN | Bring down the 5 | | $36$ |
| **Step 5** | REPEAT | Repeat the first four steps. | | $9$ Remainder |

The remainder can be expressed as a whole number, as a decimal, or as a fraction by placing the remainder over the divisor.

```
   23 r9          23.75           23 9/12
12)285         12)285.00       12)285
   24             24
   45             45
   36             36
    9             90
                  84
                  60
                  60
```

Verify the accuracy of division by multiplying the divisor by the quotient and adding the remainder, if any. The result should equal the dividend.

## EXAMPLE

Division                                    Verification

$$
\begin{array}{r}
35 \text{ r}19 \\
25\overline{)894} \\
75 \\
\hline
144 \\
125 \\
\hline
19
\end{array}
$$

Division is correct

$$
\begin{array}{r}
25 \quad \text{Divisor} \\
\times 35 \quad \text{Quotient} \\
\hline
125 \\
75 \\
\hline
875 \\
19 \quad \text{Remainder} \\
\hline
894 \quad \text{Dividend}
\end{array}
$$

## PRACTICE PROBLEMS

Divide and verify answers. Express remainders as whole numbers.

**1.** $17\overline{)256}$   **2.** $18\overline{)305}$   **3.** $104\overline{)1,478}$   **4.** $25\overline{)865}$   **5.** $384\overline{)34,060}$

**DIVIDING DECIMALS**

When dividing a decimal by a whole number, place the decimal point in the quotient directly above the decimal point in the dividend.

## EXAMPLE

$6\overline{)553.50}$

## SOLUTION

$$
\begin{array}{r}
92.25 \\
6\overline{)553.50}
\end{array}
$$
Align decimal points

If there is a remainder, add zeros in the dividend until no remainder is found or the desired number of decimal places is calculated in the quotient.

## EXAMPLE

Divide and round to the nearest thousandth.

$86\overline{)500}$

## SOLUTION

$$
\begin{array}{r}
5.8139 \approx 5.814 \\
86\overline{)500.0000} \\
430 \\
\hline
70\ 0 \\
68\ 8 \\
\hline
1\ 20 \\
86 \\
\hline
340 \\
258 \\
\hline
820 \\
774 \\
\hline
\end{array}
$$

When dividing by a decimal, first convert the divisor to a whole number by moving the decimal point to the right. Then, move the decimal point in the dividend the same number of places to the right and divide. Add zeros to the right of the dividend if necessary. Moving the decimal point the same number of places in the divisor and the dividend does not change the value of the quotient.

**EXAMPLE**

$9.25\overline{)310.8}$

**SOLUTION**

```
              33.6
9 25)310 80.0
      277 5
       33 30
       27 75
        5 550
        5 550
```

**PRACTICE PROBLEMS**

Divide and verify answers. Round to the nearest thousandth.

1. $26\overline{)\$928.72}$     2. $27\overline{)416}$     3. $35\overline{)1,359}$     4. $63\overline{)927.45}$

5. $105\overline{)705.86}$     6. $3.6\overline{)358.12}$     7. $9.18\overline{)87.14}$     8. $6.55\overline{)506.8}$

9. $8.5\overline{)44.12}$     10. $60.5\overline{)300.25}$

**DIVIDING BY NUMBERS ENDING IN ZERO**

A shortcut can be used when a divisor contains ending zeros. Drop the zeros in the divisor, and move the decimal point in the dividend to the left the same number of places as the number of zeros dropped.

**EXAMPLES**

$400\overline{)182,000}$     $30\overline{)1,926}$

**SOLUTIONS**

```
    455.00              64.2
4)1,820.00          3)192.6
```

**PRACTICE PROBLEMS**

Divide using the shortcut. Round to the nearest thousandth.

1. $40\overline{)39.15}$     2. $600\overline{)3,009}$     3. $120\overline{)25,000}$

4. $200\overline{)981.4}$     5. $30\overline{)45.22}$

ANSWERS FOR
PRACTICE PROBLEMS

**Bottom: (1)** 0.979   **(2)** 5.015   **(3)** 208.333   **(4)** 4.907   **(5)** 1.507

**Top: (1)** $35.72   **(2)** 15.407   **(3)** 38.829   **(4)** 14.721   **(5)** 6.722   **(6)** 99.478
**(7)** 9.492   **(8)** 77.374   **(9)** 5.191   **(10)** 4.963

Division by a power of 10 can be done rapidly using a shortcut. To obtain the quotient, move the decimal point in the dividend as many places to the left as there are zeros in the divisor. Add zeros to the left when necessary.

### EXAMPLES

$8{,}903.2 \div 10 = 890.32$

$7.45 \div 100 = 0.0745$ (add one zero)

$0.26 \div 1{,}000 = 0.00026$ (add two zeros)

### PRACTICE PROBLEMS

Divide using the shortcut. Round to the nearest thousandth.

1. $100\overline{)2.82}$
2. $1{,}000\overline{)45{,}732.5}$
3. $10\overline{)23.9}$
4. $100\overline{)147.85}$
5. $10{,}000\overline{)33932.2}$

## SOLVING WORD PROBLEMS

When solving word problems involving division, follow the four-step procedure: read, plan, do, check. Watch for the key words "divided by," "average," or "quotient," which indicate a division problem.

### EXAMPLE

The profits from the High Profit Investment firm are divided 11 ways among the owners. If the profits from last year totaled $32,232.45, how much did each owner receive? Round to the nearest cent.

### SOLUTION

| | | |
|---|---|---|
| **Step 1** READ | Key word is "divided," so divide. | |
| **Step 2** PLAN | Year's Profit ÷ Owners = Profit per Owner | |
| **Step 3** DO | $32{,}232.45 \div 11 = \$2{,}930.22$ | |
| **Step 4** CHECK | $\$32{,}000 \div 10 = \$3{,}200$ Estimate | |

### PRACTICE PROBLEMS

Solve using the four-step method.

1. Michelle Oates operates a hair styling salon and charges $12.50 for a haircut. Yesterday, she earned $187.50 from haircuts. How many haircuts did she give?
2. When the price of gold was $426.20 per ounce, Joseph Waller purchased $4,048.90 worth of gold. How many ounces did he buy?

## LESSON 1-4    EXERCISES

Divide. Express remainders as whole numbers.

**1.** $15\overline{)57}$           **2.** $16\overline{)145}$           **3.** $45\overline{)1,315}$

**4.** $215\overline{)24,615}$        **5.** $52\overline{)6,638}$        **6.** $15\overline{)5,155}$

**7.** $18\overline{)27,313}$        **8.** $21\overline{)24,473}$        **9.** $46\overline{)22,447}$

Divide. Round to the nearest hundredth.

**10.** $82\overline{)105.3}$        **11.** $55\overline{)125.8}$        **12.** $\overline{)98.76}$

      

**13.** $24\overline{)42.68}$

**14.** $35\overline{)305.6}$

**15.** $1.1\overline{)98}$

**16.** $30.1\overline{)114}$

**17.** $1.2\overline{)356}$

**18.** $3.41\overline{)5.26}$

Divide using a shortcut. Round to the nearest thousandth.

**19.** $40\overline{)261,300}$

**20.** $1,400\overline{)64,700}$

**21.** $300\overline{)455.3}$

**22.** $20\overline{)62.2}$

**23.** $100\overline{)45.28}$

**24.** $10\overline{)1.72}$

**25.** $1{,}000\overline{)3.56}$        **26.** $10\overline{)1.528}$        **27.** $100\overline{)875}$

## BUSINESS APPLICATIONS

Solve using the four-step method. Round to the nearest cent.

**28.** Alexander Insurance Company mailed 500 letters to its customers. The total cost for the mailing consisted of envelopes, $15; printing of brochures, $240; and postage, $195. What was the cost per letter mailed?

_____

**29.** Nelson pays $1,364 per year for health insurance, $884.42 per year for car insurance, and $326.95 per year for homeowners' insurance. How much total insurance does he pay each month if he pays each on a monthly basis?

_____

**30.** At the Farmers' Highway Market, 100 pounds of potatoes can be purchased for $97. A 5-pound sack of potatoes can be purchased at the same price per pound. How much will 5 pounds of potatoes cost?

_____

**31.** In a typical month, Stacy Langman's expenses include: rent, $580; utilities, $85; food, $225; car, $366; and telephone, $55. In a 30-day month, what are Stacy's daily expenses?

_____

**32.** Calvin Mason drove 2,442.6 miles on a lecture tour. He used a total of 118 gallons of gasoline. How many miles per gallon did his automobile get?

_____

**33.** Growth Industries leased a 2,575 square foot office for $37,337.50 for the year. What was the annual cost per square foot?

_____

**34.** The 23 members of the Franklin High School Business Club raffled off a television set to raise money to attend the national convention. The television set cost $375. Total raffle ticket sales were $1,219. After paying for the television set, the money is to be split equally among the club members. How much will each club member receive?

_____

**35.** At Video Connection, Arrow brand video cassette tapes usually sell for $8.98. As a special promotion, three Arrow tapes can be purchased for $20.65. How much can a customer save per tape by buying at the special price?

_____

**36.** LDP Industries purchased 46 square yards of carpet at a total cost of $862.50. What was the cost per square yard?

_____

**37.** Leonard drove 333.3 miles on 12.9 gallons of gas, and Rachel drove 381.5 miles on 15.7 gallons of gas. Whose car gets the better gas mileage per gallon? Round to the nearest tenth.

_____

## CHAPTER 1 REVIEW

Write out in words.

**1.** 3,507 = _____

Add.

**2.** $1,436.29
278.14
2,050.76
392.41

**3.** 25,306.18
41,914.49
382.06
5,705.21

**4.** 27 + 18 + 43 + 60 =

**5.** 12 + 20 + 68 + 91 =

Round numbers.

**6.** 46.0586 rounded to nearest hundredth = _____

**7.** 8,659.625 rounded to the nearest hundred = _____

Subtract.

**8.** $15,307.28
− 9,618.49

**9.** 689.305
− 91.653

**10.** 724 − 469 =

**11.** 902 − 374 =

Multiply. Round answers to two decimal places.

**12.** 516
× 24

**13.** 306
× 19

**14.** 32.19
× 3.5

**15.** $780.25
× 6.125

**16.** $426.50 × 100 =

**17.** 905.875 × 10 =

**18.** 230 × 20 =

**19.** 1,200 × 30 =

Divide. Express answers as a whole number and a remainder.

**20.** 1,470 ÷ 24 =

**21.** 2,385 ÷ 56 =

Divide. Express answers as a whole number and a decimal. Round to two decimal places.

**22.** 1,280 ÷ 42 =

**23.** 867 ÷ 60 =

## BUSINESS APPLICATIONS

**24.** Picture Perfect purchased 50 picture frames for $104.50. How much did they pay for each frame?

_____

**25.** Jamey Stone purchased a camera for $219.95, film for $3.79, a camera case for $37.24, and a photo album for $12.49. If Jamey gave the clerk $300, how much change did he receive?

_____

**26.** Tru-Quality Hardware sold 15 gallons of paint for $9.29 a gallon, 9 gallons of paint for $11.49 a gallon, 22 gallons of paint for $15.35 a gallon, and 11 gallons of paint for $12.95 a gallon. What was the total amount received for the paint sold?

_____

**27.** Gregory Fashion Place is having a "buy one, get one free" sale. Cheryl purchased two sweaters for $59.95, which originally sell for $34 each. How much money did Cheryl save on the purchase?

_____

**28.** The Business Club purchased sweatshirts for $11.19 each. They sold 529 for $19.95 each. How much money is left after the club pays for the sweatshirts?

_____

# CHAPTER 2

# FRACTIONS

## LEARNING OBJECTIVES

1. Identify proper and improper fractions.
2. Raise fractions to higher terms.
3. Reduce fractions to lower terms.
4. Convert improper fractions to mixed numbers and mixed numbers to improper fractions.
5. Add and subtract fractions and mixed numbers.
6. Multiply and divide fractions and mixed numbers.
7. Multiply using cancellations.

Although fractions are not as widely used as whole numbers and decimals, they are used in many important business calculations. For example, a financial investor may calculate the cost of stock with a price of $37\frac{1}{8}$, or a store clerk may calculate the total cost of 200 items, each priced at $\frac{1}{6}$ of a dollar. In this chapter, you will learn how to perform calculations involving fractions and how to convert fractions to decimals.

## LESSON 2-1    EQUIVALENT FRACTIONS

**TYPES OF FRACTIONS**

A fraction indicates a part of a whole or unit. For example, the rectangle below represents one whole unit. There are six equal parts in the unit. The fraction $\frac{5}{6}$ is represented by the 5 shaded parts of the unit. The bottom number, called the **denominator,** identifies the number of equal parts the whole has been divided into. The top number, called the **numerator,** identifies the number of parts being considered. The numerator and denominator are called the **terms** of the fraction.

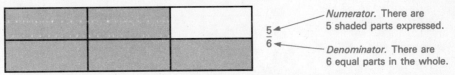

$\frac{5}{6}$

*Numerator.* There are 5 shaded parts expressed.

*Denominator.* There are 6 equal parts in the whole.

A *proper fraction* is used to express a value that is less than one whole unit. The numerator is always less than the denominator in a proper fraction. The following are examples of proper fractions:

$\frac{5}{8}$  $\frac{2}{3}$  $\frac{7}{8}$  $\frac{12}{32}$

LESSON 2-1  ■  **35**

An *improper fraction* is equal to one, such as $\frac{5}{5}$, or greater than one, such as $\frac{9}{6}$. The numerator is always equal to or greater than the denominator. The following are examples of improper fractions:

$\frac{8}{8}$   $\frac{5}{2}$   $\frac{16}{12}$   $\frac{7}{5}$

A *mixed number* also is used to express a value that is more than one whole unit. A mixed number consists of a whole number and a fraction, such as $1\frac{1}{2}$, $3\frac{2}{3}$, and $12\frac{4}{5}$.

## PRACTICE PROBLEMS

**1.** Circle the proper fractions.

$\frac{13}{13}$   $\frac{6}{8}$   $1\frac{2}{3}$   $\frac{1}{6}$   $\frac{12}{24}$   $\frac{9}{20}$   $\frac{9}{5}$

**2.** Circle the improper fractions.

$2\frac{7}{8}$   $\frac{8}{6}$   $2\frac{7}{12}$   $12\frac{2}{3}$   $\frac{12}{12}$   $\frac{10}{8}$   $\frac{14}{15}$

**3.** Circle the mixed numbers.

$\frac{9}{9}$   $4\frac{7}{8}$   $\frac{4}{5}$   $32\frac{1}{2}$   $\frac{16}{20}$   $7\frac{6}{14}$   $\frac{3}{5}$

## RAISING FRACTIONS TO HIGHER TERMS

All fractions being added or subtracted must have the same denominator. Therefore, fractions with different denominators, called **unlike fractions,** must be converted to equivalent values with the same denominator, called a **common denominator.** Fractions that represent the same value are called *equivalent fractions.* The shaded area in the following illustration shows that $\frac{1}{2}$, $\frac{2}{4}$, $\frac{3}{6}$, and $\frac{4}{8}$ are equivalent fractions.

$\frac{1}{2}$

$\frac{2}{4}$

*Equivalent fractions.*
All have the same value.

$\frac{3}{6}$

$\frac{4}{8}$

To find an equivalent fraction, multiply or divide the numerator and denominator by the same number. In the above illustration, the fraction $\frac{1}{2}$ is raised to higher terms by multiplying the numerator and denominator by 4. The fraction $\frac{4}{8}$ is reduced to lowest terms by dividing the numerator and denominator by 4.

## EXAMPLES

$\frac{1 \times 4}{2 \times 4} = \frac{4}{8}$        $\frac{4 \div 4}{8 \div 4} = \frac{1}{2}$

To raise a fraction to higher terms when the new denominator is already given, use the following steps:

## EXAMPLE

$\frac{3}{4} = \frac{?}{28}$

## SOLUTION

**Step 1** Divide the new denominator by the original denominator.

$28 \div 4 = 7$

**Step 2** Multiply the result by the original numerator to find the new numerator.

$\frac{3 \times 7}{4 \times 7} = \frac{21}{28}$

## PRACTICE PROBLEMS

Find the missing numerators.

1. $\frac{2}{3} = \frac{}{24}$    2. $\frac{1}{4} = \frac{}{16}$    3. $\frac{7}{8} = \frac{}{32}$    4. $\frac{4}{9} = \frac{}{81}$

5. $\frac{4}{6} = \frac{}{108}$    6. $\frac{12}{13} = \frac{}{65}$    7. $\frac{5}{6} = \frac{}{72}$    8. $\frac{3}{8} = \frac{}{96}$

**REDUCING FRACTIONS TO LOWEST TERMS**

Since it is easier to work with smaller numbers, answers to any operations with fractions should be reduced to lowest terms. To reduce a fraction to lowest terms, find the greatest number that can be divided evenly into both the numerator and the denominator. This number is called the **greatest common divisor.** A fraction is in lowest terms when no number except 1 can be divided evenly into the numerator and the denominator.

## EXAMPLE 1                                 SOLUTION

Reduce $\frac{12}{16}$ to lowest terms.          4 is the greatest common divisor.

$\frac{12}{16} \div \frac{4}{4} = \frac{3}{4}$

## EXAMPLE 2                                 SOLUTION

Reduce $\frac{9}{27}$ to lowest terms.           9 is the greatest common divisor.

$\frac{9}{27} \div \frac{9}{9} = \frac{1}{3}$

Always try the smaller term as the greatest common divisor as shown in the above example. If the greatest common divisor cannot be easily determined, it may be necessary to divide more than once to reduce the fraction to lowest terms.

## EXAMPLE 1                                 SOLUTION

Reduce $\frac{36}{42}$ to lowest terms.          $\frac{36}{42} \div \frac{2}{2} = \frac{18}{21} \div \frac{3}{3} = \frac{6}{7}$

## EXAMPLE 2                                 SOLUTION

Reduce $\frac{30}{48}$ to lowest terms.          $\frac{30}{48} \div \frac{3}{3} = \frac{10}{16} \div \frac{2}{2} = \frac{5}{8}$

ANSWERS FOR PRACTICE PROBLEMS

(1) $\frac{16}{24}$    (2) $\frac{4}{16}$    (3) $\frac{28}{32}$    (4) $\frac{36}{81}$    (5) $\frac{72}{108}$    (6) $\frac{60}{65}$    (7) $\frac{60}{72}$    (8) $\frac{36}{96}$

Reduce to lowest terms.

**1.** $\frac{3}{9} =$      **2.** $\frac{10}{12} =$      **3.** $\frac{9}{36} =$      **4.** $\frac{18}{20} =$

**5.** $\frac{10}{15} =$      **6.** $\frac{8}{12} =$      **7.** $\frac{14}{16} =$      **8.** $\frac{22}{24} =$

## CONVERTING MIXED NUMBERS TO IMPROPER FRACTIONS

In most cases, it is more convenient to compute fractions when they are in the form of a proper or an improper fraction. To convert a mixed number to an improper fraction, multiply the denominator by the whole number and add the product to the numerator. The denominator remains the same. Whole numbers can be expressed as improper fractions by placing the whole number over one.

| EXAMPLE | SOLUTION |
|---|---|
| Convert $3\frac{5}{8}$ to an improper fraction. | $\dfrac{(8 \times 3) + 5}{8} = \dfrac{24 + 5}{8} = \dfrac{29}{8}$ |

### PRACTICE PROBLEMS

Convert to improper fractions.

**1.** $2\frac{1}{4} =$      **2.** $4\frac{1}{2} =$      **3.** $7\frac{1}{3} =$      **4.** $8\frac{2}{3} =$

**5.** $8 =$      **6.** $9\frac{1}{4} =$      **7.** $10\frac{2}{3} =$      **8.** $10 =$

## CONVERTING IMPROPER FRACTIONS TO WHOLE OR MIXED NUMBERS

Although mixed numbers are easier to compute if they are converted to improper fractions, the final answer is usually converted back to a mixed number. An improper fraction is converted to a whole or mixed number by dividing the numerator by the denominator. The remainder, if any, is expressed as a fraction and reduced to lowest terms.

### EXAMPLES

Convert $\frac{20}{6}$ and $\frac{105}{3}$ to whole or mixed numbers. Reduce to lowest terms.

### SOLUTIONS

$$\frac{20}{6} = 6\overline{)\begin{array}{l} 3\frac{2}{6} = 3\frac{1}{3} \\ 20 \\ \underline{18} \\ 2 \end{array}} \qquad \frac{105}{3} = 3\overline{)105} = 35$$

### PRACTICE PROBLEMS

Convert to whole or mixed numbers. Reduce to lowest terms.

**1.** $\frac{9}{2} =$      **2.** $\frac{14}{3} =$      **3.** $\frac{16}{4} =$      **4.** $\frac{25}{4} =$

**5.** $\frac{33}{6} =$      **6.** $\frac{76}{8} =$      **7.** $\frac{17}{3} =$      **8.** $\frac{39}{3} =$

ANSWERS FOR
PRACTICE PROBLEMS

**Bottom:** (1) $4\frac{1}{2}$ (2) $4\frac{2}{3}$ (3) 4 (4) $6\frac{1}{4}$ (5) $5\frac{1}{2}$ (6) $9\frac{1}{2}$ (7) $5\frac{2}{3}$ (8) 13

**Middle:** (1) $\frac{9}{4}$ (2) $\frac{9}{2}$ (3) $\frac{22}{3}$ (4) $\frac{26}{3}$ (5) $\frac{8}{1}$ (6) $\frac{37}{4}$ (7) $\frac{32}{3}$ (8) $\frac{10}{1}$

**Top:** (1) $\frac{1}{3}$ (2) $\frac{5}{6}$ (3) $\frac{1}{4}$ (4) $\frac{9}{10}$ (5) $\frac{2}{3}$ (6) $\frac{2}{3}$ (7) $\frac{7}{8}$ (8) $\frac{11}{12}$

**38** ■ LESSON 2-1

## LESSON 2-1    EXERCISES

Raise the following fractions to higher terms.

**1.** $\frac{4}{5} = \frac{}{30}$

**2.** $\frac{2}{3} = \frac{}{15}$

**3.** $\frac{3}{8} = \frac{}{48}$

**4.** $\frac{6}{7} = \frac{}{49}$

**5.** $\frac{11}{12} = \frac{}{144}$

**6.** $\frac{8}{9} = \frac{}{108}$

**7.** $\frac{14}{15} = \frac{}{105}$

**8.** $\frac{1}{2} = \frac{}{56}$

Reduce the following fractions to lowest terms.

**9.** $\frac{3}{12}$

**10.** $\frac{8}{56}$

**11.** $\frac{9}{12}$

**12.** $\frac{10}{24}$

**13.** $\frac{20}{32}$

**14.** $\frac{9}{15}$

**15.** $\frac{9}{48}$

**16.** $\frac{12}{15}$

Convert the following mixed numbers to improper fractions.

**17.** $1\frac{5}{12}$

**18.** $3\frac{1}{2}$

**19.** $6\frac{3}{4}$

**20.** $6\frac{2}{3}$

**21.** $7\frac{3}{8}$

**22.** $4\frac{7}{8}$

**23.** $12\frac{2}{3}$

**24.** $9\frac{5}{6}$

Convert the following improper fractions to whole or mixed numbers.
Reduce to lowest terms.

**25.** $\frac{8}{3}$

**26.** $\frac{9}{4}$

**27.** $\frac{13}{5}$

**28.** $\frac{26}{6}$

**29.** $\frac{45}{8}$

**30.** $\frac{39}{6}$

**31.** $\frac{144}{8}$

**32.** $\frac{63}{9}$

## BUSINESS APPLICATIONS

Reduce answers to lowest terms.

**33.** Max Starsky purchased a 48-ounce box of detergent. Max used 12 ounces to wash his clothes. Express as a fraction the portion of the box Max used.

**34.** Ellen McDay went shopping with $20. She spent all but $4. Express as a fraction the amount of money Ellen spent of that she took along shopping.

_____

**35.** Margaret is baking a cake. The recipe calls for $2\frac{1}{4}$ cups of sugar. Margaret has only a quarter-cup size measuring cup. How many quarter cups of sugar should she add to the cake mix?

_____

**36.** Germain Janssen ran $3\frac{1}{2}$ miles. How many eighths of a mile is this?

_____

**ADDING AND SUBTRACTING LIKE FRACTIONS**

Fractions that have the same denominator or a common denominator are called *like fractions,* such as $\frac{1}{8}$, $\frac{3}{8}$, and $\frac{7}{8}$. When adding or subtracting like fractions, add or subtract the numerators and place the answer over the common denominator. If necessary, reduce the answer to lowest terms or change the improper fraction to a mixed number.

| EXAMPLE 1 | SOLUTION |
|---|---|
| $\frac{2}{6} + \frac{5}{6} + \frac{3}{6} + \frac{4}{6} =$ | $\frac{2+5+3+4}{6} = \frac{14}{6} = 2\frac{2}{6} = 2\frac{1}{3}$ |

| EXAMPLE 2 | SOLUTION |
|---|---|
| $\frac{7}{8} - \frac{1}{8} =$ | $\frac{7-1}{8} = \frac{6}{8} = \frac{3}{4}$ |

**PRACTICE PROBLEMS**

**1.** $\frac{1}{5} + \frac{3}{5} + \frac{2}{5} + \frac{4}{5} =$     **2.** $\frac{3}{4} + \frac{1}{4} + \frac{1}{4} + \frac{2}{4} =$

**3.** $\frac{7}{8} + \frac{6}{8} + \frac{1}{8} + \frac{3}{8} =$     **4.** $\frac{8}{9} - \frac{5}{9} =$

**5.** $\frac{12}{18} - \frac{9}{18} =$     **6.** $\frac{18}{3} - \frac{12}{3} =$

**FINDING THE LEAST COMMON DENOMINATOR**

*Unlike fractions* are fractions that have different denominators, such as $\frac{1}{4}$, $\frac{3}{8}$, and $\frac{4}{7}$. Before adding or subtracting unlike fractions, the fractions must be converted to their equivalent values with a common denominator. Since a set of fractions has many common denominators, it is easier to compute with the smallest or *least common denominator* (LCD). The LCD is the smallest denominator that can be divided evenly by all the denominators in the series.

In some cases, the largest given denominator can be used as the LCD. In the following example, the largest denominator, 20, can be used as the LCD because the other denominators, 4 and 5, can divide evenly into 20 ($20 \div 4 = 5$ and $20 \div 5 = 4$). After finding the LCD, raise each fraction to higher terms.

**EXAMPLE**

Find the LCD and raise to higher terms with the LCD as denominator: $\frac{3}{4}$, $\frac{2}{5}$, $\frac{1}{20}$

**SOLUTION**

$\frac{3 \times 5}{4 \times 5} = \frac{15}{20}$          $\frac{2 \times 4}{5 \times 4} = \frac{8}{20}$          $\frac{1 \times 1}{20 \times 1} = \frac{1}{20}$

If none of the denominators can be used as a common denominator, the LCD can be determined by inspection. For example, the LCD for the fractions $\frac{1}{6}$ and $\frac{3}{4}$ is 12 since the denominators, 6 and 4, can divide evenly into 12 ($12 \div 6 = 2$ and $12 \div 4 = 3$).

---

### EXAMPLE

Find the LCD and raise to higher terms with the LCD as denominator:
$\frac{1}{6}$, $\frac{3}{4}$

### SOLUTION

$\frac{1 \times 2}{6 \times 2} = \frac{2}{12}$ $\qquad$ $\frac{3 \times 3}{4 \times 3} = \frac{9}{12}$

---

### PRACTICE PROBLEMS

Find the LCD and raise to higher terms with the LCD as denominator.

| **1.** $\frac{1}{3}$ | **2.** $\frac{1}{2}$ | **3.** $\frac{7}{9}$ | **4.** $\frac{1}{2}$ | **5.** $\frac{4}{5}$ |
|---|---|---|---|---|
| $\frac{3}{4}$ | $\frac{3}{8}$ | $\frac{1}{3}$ | $\frac{2}{3}$ | $\frac{3}{4}$ |
| $\frac{5}{6}$ | $\frac{3}{4}$ | $\frac{5}{6}$ | $\frac{1}{5}$ | $\frac{9}{10}$ |

When the LCD cannot be easily determined by inspection, it can be calculated by using the **prime factor method.** The prime factor method uses prime numbers to find the LCD. A **prime number** is only divisible by itself and 1. Examples of prime numbers are 2, 3, 5, 7, 11, 13, 17, and 19. The number 4 is not a prime number since it is also divisible by 2. Use the following steps to find the LCD using the prime factor method.

---

### EXAMPLE

Find the LCD for $\frac{3}{4}$, $\frac{5}{6}$, $\frac{3}{12}$, and $\frac{5}{15}$ using the prime factor method. Raise to higher terms with the LCD as denominator.

### SOLUTION

**Step 1** Place the denominators in a horizontal row. Divide the denominators by a prime number that divides evenly into as many of the denominators as possible. In this case, 2, since it divides 4, 6, and 12.

$2\overline{)4\ \ 6\ \ 12\ \ 15}$

**Step 2** Write the quotients and any numbers not evenly divisible by the prime number directly above the dividends.

$\begin{array}{c} 2\ \ 3\ \ \ 6\ \ 15 \\ \hline 2\overline{)4\ \ 6\ \ 12\ \ 15} \end{array}$

**Step 3** Divide by another prime number common to as many numbers as possible in the second row. In this case, 3, since it divides 3, 6, and 15. Write the quotients and any undivided numbers directly above the dividends.

$\begin{array}{c} 2\ \ 1\ \ \ 2\ \ \ 5 \\ \hline 3\overline{)2\ \ 3\ \ \ 6\ \ 15} \\ \hline 2\overline{)4\ \ 6\ \ 12\ \ 15} \end{array}$

---

**Step 4** Continue dividing by a prime number until all quotients across the top row are 1.

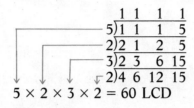

```
      1  1   1   1
   5) 1  1   1   5
   2) 2  1   2   5
   3) 2  3   6  15
   2) 4  6  12  15
```

**Step 5** Multiply the prime number divisors to obtain the LCD.

```
           1  1   1   1
        5) 1  1   1   5
        2) 2  1   2   5
        3) 2  3   6  15
        2) 4  6  12  15
```

$5 \times 2 \times 3 \times 2 = 60$ LCD

**Step 6** Convert each fraction to its equivalent value with the LCD as denominator.

$$\frac{3}{4} \times \frac{15}{15} = \frac{45}{60}$$

$$\frac{5}{6} \times \frac{10}{10} = \frac{50}{60}$$

$$\frac{3}{12} \times \frac{5}{5} = \frac{15}{60}$$

$$\frac{5}{15} \times \frac{4}{4} = \frac{20}{60}$$

## PRACTICE PROBLEMS

Find the LCD using the prime factor method. Raise to higher terms with the LCD as denominator.

| 1. $\frac{1}{8}$ | 2. $\frac{1}{5}$ | 3. $\frac{3}{12}$ | 4. $\frac{1}{3}$ |
|---|---|---|---|
| $\frac{2}{3}$ | $\frac{2}{3}$ | $\frac{1}{6}$ | $\frac{4}{5}$ |
| $\frac{3}{12}$ | $\frac{2}{15}$ | $\frac{3}{4}$ | $\frac{3}{4}$ |
| $\frac{1}{6}$ | $\frac{1}{9}$ | $\frac{1}{18}$ | $\frac{5}{12}$ |

ANSWERS FOR
PRACTICE PROBLEMS

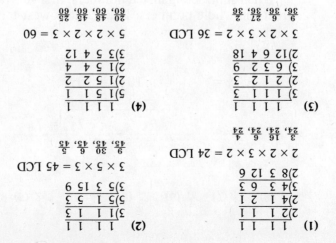

(1)
```
   1  1  1  1
2) 2  1  1  1
2) 4  1  2  1
3) 8  3  4  6
2) 8  3  12  6
```
$2 \times 2 \times 3 \times 2 = 24$ LCD

$\frac{3}{24}, \frac{16}{24}, \frac{6}{24}, \frac{4}{24}$

(2)
```
   1  1  1  1
3) 3  1  1  1
5) 9  1  5  3
   9  3  15  3
```
$3 \times 5 \times 3 = 45$ LCD

$\frac{5}{45}, \frac{30}{45}, \frac{9}{45}, \frac{45}{45}$

(3)
```
   1  1  1  1
3) 3  1  1  1
2) 3  1  2  3
3) 6  3  2  9
2) 18  6  4  12
```
$2 \times 3 \times 3 \times 2 = 36$ LCD

$\frac{36}{36}, \frac{27}{36}, \frac{6}{36}, \frac{9}{36}$

(4)
```
   1  1  1  1
5) 1  5  1  1
2) 2  5  1  2
2) 4  5  1  2
3) 12  5  4  3
```
$5 \times 2 \times 2 \times 3 = 60$

$\frac{20}{60}, \frac{48}{60}, \frac{45}{60}, \frac{25}{60}$

Before adding or subtracting unlike fractions, convert the fractions to equivalent values with a common denominator. Then, add or subtract the numerators and place the answer over the common denominator. If the answer is an improper fraction, convert to a mixed number and reduce to lowest terms if necessary.

### EXAMPLE 1

$$\frac{3}{4} + \frac{2}{3} + \frac{5}{6} =$$

SOLUTION

$$\frac{3 \times 3}{4 \times 3} = \frac{9}{12}$$
$$\frac{2 \times 4}{3 \times 4} = \frac{8}{12}$$
$$+\frac{5 \times 2}{6 \times 2} = \frac{10}{12}$$
$$\frac{27}{12} = 2\frac{3}{12} = 2\frac{1}{4}$$

### EXAMPLE 2

$$\frac{6}{8} - \frac{1}{3} =$$

SOLUTION

$$\frac{6 \times 3}{8 \times 3} = \frac{18}{24}$$
$$-\frac{1 \times 8}{3 \times 8} = \frac{8}{24}$$
$$\frac{10}{24} = \frac{5}{12}$$

### PRACTICE PROBLEMS

Add or subtract. Reduce to lowest terms.

1.  $\frac{5}{6}$
    $\frac{8}{12}$
    $+\frac{2}{9}$

2.  $\frac{3}{4}$
    $\frac{2}{3}$
    $+\frac{5}{6}$

3.  $\frac{5}{6}$
    $\frac{7}{8}$
    $+\frac{3}{4}$

4.  $\frac{2}{10}$
    $\frac{1}{8}$
    $+\frac{3}{5}$

5.  $\frac{8}{9}$
    $-\frac{1}{4}$

6.  $\frac{18}{30}$
    $-\frac{1}{2}$

7.  $\frac{5}{6}$
    $-\frac{1}{5}$

8.  $\frac{7}{8}$
    $-\frac{5}{6}$

When adding mixed numbers, first add the fractions, then add the whole numbers. After the fractions are added, the sum may be an improper fraction. In this case, convert the improper fraction to a mixed number; then add the whole numbers. Reduce to lowest terms if necessary.

### EXAMPLE

$$2\frac{2}{3} + 6\frac{5}{9} =$$

### SOLUTION

$$2\frac{2}{3} = 2\frac{6}{9}$$
$$+6\frac{5}{9} = 6\frac{5}{9}$$
$$8\frac{11}{9} = 8 + 1\frac{2}{9} = 9\frac{2}{9}$$

When subtracting mixed numbers, first subtract the fractions, then subtract the whole numbers. Reduce to lowest terms if necessary.

### EXAMPLE

$$10\frac{5}{12} - 3\frac{1}{4} =$$

### SOLUTION

$$10\frac{5}{12} = 10\frac{5}{12}$$
$$- 3\frac{1}{4} = 3\frac{3}{12}$$
$$7\frac{2}{12} = 7\frac{1}{6}$$

ANSWERS FOR PRACTICE PROBLEMS (upside down): (1) $1\frac{11}{18}$ (2) $2\frac{1}{4}$ (3) $2\frac{11}{24}$ (4) $\frac{37}{40}$ (5) $\frac{23}{36}$ (6) $\frac{1}{10}$ (7) $\frac{19}{30}$ (8) $\frac{1}{24}$

If the fraction in the subtrahend is larger than the fraction in the minuend, borrow a 1 from the whole number and add it to the fraction. Change the 1 to a fraction with the same denominator as the fraction it is being added to.

**EXAMPLE**

$18\frac{1}{4} - 5\frac{7}{8} =$

**SOLUTION**

$$18\frac{1}{4} = 18\frac{2}{8} = 17 + \frac{8}{8} + \frac{2}{8} = \quad 17\frac{10}{8}$$
$$-\ 5\frac{7}{8} = \quad 5\frac{7}{8} = \qquad\qquad -\ 5\ \frac{7}{8}$$
$$\overline{\qquad\qquad\qquad\qquad\qquad\quad 12\ \frac{3}{8}}$$

When subtracting a mixed number from a whole number, borrow a 1 from the whole number and convert the 1 to a fraction; then subtract.

**EXAMPLE**

$16 - 5\frac{5}{8} =$

**SOLUTION**

$$16\ \ = 15\frac{8}{8}$$
$$-\ 5\frac{5}{8} = \ \ 5\frac{5}{8}$$
$$\overline{\qquad\qquad 10\frac{3}{8}}$$

**PRACTICE PROBLEMS**

Add or subtract. Reduce to lowest terms.

1. $12\frac{1}{3}$  
    $8\frac{1}{2}$  
   $+\ 4\frac{5}{12}$

2. $3\frac{1}{4}$  
    $7\frac{1}{2}$  
   $+1\frac{1}{8}$

3. $4\frac{1}{6}$  
    $3\frac{1}{8}$  
   $+2\frac{1}{2}$

4. $3\frac{1}{9}$  
    $6\frac{1}{4}$  
   $+5\frac{2}{3}$

5. $15\frac{5}{9}$  
   $-\ 4\frac{1}{3}$

6. $35$  
   $-\ 3\frac{3}{4}$

7. $18\frac{1}{2}$  
   $-\ 5\frac{7}{8}$

8. $30\frac{1}{3}$  
   $-16\frac{4}{5}$

## LESSON 2-2     EXERCISES

Find the LCD for the following series of fractions. Convert each fraction to its equivalent with the LCD as its denominator.

**1.** $\frac{3}{18}, \frac{1}{24}, \frac{2}{36}$                 **2.** $\frac{2}{12}, \frac{1}{15}, \frac{3}{30}$

**3.** $\frac{1}{6}, \frac{2}{9}, \frac{2}{5}$                 **4.** $\frac{2}{30}, \frac{1}{18}, \frac{4}{45}$

Add. Reduce to lowest terms.

**5.**
$$\begin{array}{r} \frac{1}{4} \\ \frac{3}{4} \\ \frac{2}{4} \\ +\frac{3}{4} \end{array}$$

**6.**
$$\begin{array}{r} \frac{3}{8} \\ \frac{1}{8} \\ \frac{7}{8} \\ +\frac{3}{8} \end{array}$$

**7.**
$$\begin{array}{r} \frac{1}{6} \\ \frac{2}{6} \\ \frac{5}{6} \\ +\frac{3}{6} \end{array}$$

**8.**
$$\begin{array}{r} \frac{3}{4} \\ \frac{7}{18} \\ \frac{1}{2} \\ +\frac{4}{9} \end{array}$$

**9.**
$$\begin{array}{r} \frac{1}{4} \\ \frac{1}{3} \\ \frac{1}{12} \\ +\frac{5}{6} \end{array}$$

**10.**
$$\begin{array}{r} \frac{3}{4} \\ \frac{1}{2} \\ +\frac{5}{16} \end{array}$$

**11.**
$$\begin{array}{r} \frac{4}{9} \\ \frac{5}{8} \\ \frac{1}{6} \\ +\frac{5}{12} \end{array}$$

**12.**
$$\begin{array}{r} 13\frac{1}{2} \\ 24\frac{1}{3} \\ +33\frac{5}{6} \end{array}$$

**13.**
$$\begin{array}{r} 19\frac{1}{4} \\ 23\frac{5}{8} \\ +24\frac{1}{2} \end{array}$$

**14.**
$$\begin{array}{r} 44\frac{1}{6} \\ 11\frac{2}{3} \\ +104\frac{6}{9} \end{array}$$

**15.**
$$\begin{array}{r} 15\frac{7}{9} \\ 31\frac{2}{7} \\ 33\frac{1}{6} \end{array}$$

**16.**
$$\begin{array}{r} 43\frac{2}{5} \\ 18\frac{1}{10} \\ + 7\frac{2}{7} \end{array}$$

Subtract. Reduce to lowest terms.

**17.**
$$\begin{array}{r} \frac{9}{14} \\ -\frac{7}{14} \end{array}$$

**18.**
$$\begin{array}{r} \frac{7}{8} \\ -\frac{1}{8} \end{array}$$

**19.**
$$\begin{array}{r} \frac{5}{8} \\ -\frac{1}{3} \end{array}$$

        

**20.**  $\frac{2}{3}$
$-\frac{2}{5}$

**21.**  $7$
$-5\frac{1}{8}$

**22.**  $12\frac{3}{4}$
$-\ 5\frac{1}{2}$

**23.**  $28\frac{1}{4}$
$-15\frac{7}{8}$

**24.**  $12\frac{1}{6}$
$-\ 4\frac{5}{8}$

**25.**  $15\frac{2}{3}$
$-13\frac{1}{4}$

**26.**  $\frac{21}{25}$
$-\frac{4}{15}$

**27.**  $32$
$-14\frac{5}{6}$

**28.**  $101\frac{2}{9}$
$-\ 74\frac{3}{18}$

## BUSINESS APPLICATIONS

Reduce to lowest terms.

**29.** Diane purchased a piece of cloth measuring $62\frac{1}{4}$ inches. She needs 57 inches of the material for a skirt she is sewing. How many inches of the cloth will be left over?

**30.** Sherri Cline purchased a nut mix consisting of $\frac{3}{4}$ pound peanuts, $\frac{1}{8}$ pound almonds, $\frac{5}{6}$ pound cashews, and $\frac{1}{2}$ pound walnuts. How many pounds does the mixture weigh?

**31.** It is $32\frac{6}{10}$ miles between Alyson's home and her office. She drove $13\frac{1}{2}$ miles on her way to work before stopping for gas. How many more miles is it to her office?

**32.** Last week Gary Hilgers worked $6\frac{3}{4}$ hours as a cashier, $12\frac{1}{8}$ hours as a salesperson, and $8\frac{1}{2}$ hours stocking shelves. How many hours did Gary work?

**33.** Ralph Wilson put $20\frac{1}{2}$ gallons of gas in his delivery truck before beginning work in the morning. At the end of the day, he had $6\frac{3}{5}$ gallons of gas left. How many gallons of gas did he use during the day?

**34.** The company exhibit center is $9\frac{2}{3}$ yards on each of two sides, and $6\frac{1}{6}$ yards on each of the other two sides. What is the distance around the exhibit center?

**35.** Stock in the LDT Corporation opened at $62\frac{5}{8}$ today on the New York Stock Exchange and closed at $70\frac{1}{8}$. What was the amount of increase?

**36.** The Elton Construction Company painted three office complexes. They used $18\frac{2}{5}$ gallons of paint on the first office complex, $11\frac{3}{4}$ gallons of paint on the second office complex, and $21\frac{3}{5}$ gallons on the third office complex. If the company bought 55 gallons of paint to paint the complexes, how many gallons of paint were left?

_____

**37.** Tom Krueger pays $\frac{1}{3}$ of his income for rent, $\frac{1}{4}$ for food, $\frac{1}{6}$ for clothing, and $\frac{1}{5}$ for utilities. What fraction of his income is left after paying his expenses?

_____

**38.** Carla Hammer purchased a piece of canvas 108 inches long. She cut pieces of the following lengths in inches to be used for her paintings: $26\frac{1}{2}$, $32\frac{1}{4}$, and $40\frac{3}{8}$. How long is the unused piece of canvas that remains?

_____

**MULTIPLYING FRACTIONS**

To multiply fractions, first multiply the numerators, and then multiply the denominators. Reduce products to lowest terms if necessary.

### EXAMPLE

$$\frac{3}{4} \times \frac{2}{3} = \frac{6}{12} = \frac{1}{2}$$

Convert whole numbers and mixed numbers to improper fractions before multiplying.

### EXAMPLES

$$5 \times 2\frac{5}{6} = \frac{5}{1} \times \frac{17}{6} = \frac{85}{6} = 14\frac{1}{6}$$

$$2\frac{1}{4} \times 1\frac{3}{8} = \frac{9}{4} \times \frac{11}{8} = \frac{99}{32} = 3\frac{3}{32}$$

*Cancellation* is the process of dividing any numerator and any denominator by the same number. Canceling before multiplying simplifies the calculation. The denominator and numerator to be canceled do not have to be in the same fraction. Also, more than one cancellation may be performed in a problem. After canceling, multiply the remaining terms.

### EXAMPLE 1

$$\frac{16}{18} \times \frac{9}{20} =$$

**SOLUTION**

Divide 16 and 20 by 4; divide 18 and 9 by 9.

$$\frac{\overset{4}{\cancel{16}}}{\underset{2}{\cancel{18}}} \times \frac{\overset{1}{\cancel{9}}}{\underset{5}{\cancel{20}}} = \frac{4}{2} \times \frac{1}{5} = \frac{4}{10} = \frac{2}{5}$$

### EXAMPLE 2

$$\frac{4}{5} \times \frac{10}{12} \times 9 =$$

**SOLUTION**

Divide 4 and 12 by 4; divide 5 and 10 by 5; divide 9 and 3 by 3.

$$\frac{\overset{1}{\cancel{4}}}{\cancel{5}} \times \frac{\overset{2}{\cancel{10}}}{\underset{\underset{1}{\cancel{3}}}{\cancel{12}}} \times \frac{\overset{3}{\cancel{9}}}{1} = \frac{1}{1} \times \frac{2}{1} \times \frac{3}{1} = \frac{6}{1} = 6$$

### PRACTICE PROBLEMS

Multiply. Use cancellation where possible. Reduce to lowest terms.

1. $\frac{6}{8} \times \frac{4}{9} =$        2. $36 \times \frac{5}{6} =$        3. $\frac{7}{8} \times \frac{2}{14} =$

4. $\frac{32}{35} \times \frac{7}{16} =$        5. $\frac{7}{8} \times \frac{2}{3} \times \frac{3}{7} =$        6. $6\frac{3}{4} \times 3\frac{1}{9} =$

7. $\frac{6}{8} \times 8\frac{2}{3} =$        8. $\frac{3}{4} \times 3 =$        9. $6\frac{4}{6} \times 4\frac{3}{8} =$

**(1)** $\frac{1}{3}$  **(2)** 30  **(3)** $\frac{1}{8}$  **(4)** $\frac{2}{5}$  **(5)** $\frac{1}{4}$  **(6)** 21  **(7)** $6\frac{1}{2}$  **(8)** $2\frac{1}{4}$  **(9)** $29\frac{5}{6}$

## DIVIDING FRACTIONS

When dividing fractions, invert (turn upside down) the divisor and multiply. The divisor is the second fraction. The inverted fraction is called a **reciprocal** of the fraction. Use cancellation whenever possible before multiplying.

### EXAMPLE

$$\frac{8}{9} \div \frac{2}{3} =$$

### SOLUTION

Invert the divisor and multiply. Use cancellation.

$$\frac{8}{9} \div \frac{2}{3} = \frac{8}{9} \times \frac{3}{2} = \frac{\overset{4}{\cancel{8}}}{\underset{3}{\cancel{9}}} \times \frac{\overset{1}{\cancel{3}}}{\underset{1}{\cancel{2}}} = \frac{4}{3} \times \frac{1}{1} = \frac{4}{3} = 1\frac{1}{3}$$

Convert whole numbers and mixed numbers to improper fractions before inverting and multiplying.

### EXAMPLES

$$2\frac{2}{3} \div 6 = \frac{8}{3} \div \frac{6}{1} = \frac{8}{3} \times \frac{1}{6} = \frac{\overset{4}{\cancel{8}}}{3} \times \frac{1}{\underset{3}{\cancel{6}}} = \frac{4}{3} \times \frac{1}{3} = \frac{4}{9}$$

$$2\frac{2}{5} \div 3\frac{1}{3} = \frac{12}{5} \div \frac{10}{3} = \frac{12}{5} \times \frac{3}{10} = \frac{\overset{6}{\cancel{12}}}{5} \times \frac{3}{\underset{5}{\cancel{10}}} = \frac{6}{5} \times \frac{3}{5} = \frac{18}{25}$$

### PRACTICE PROBLEMS

Divide. Cancel where possible. Reduce to lowest terms.

1. $\frac{5}{6} \div \frac{15}{16} =$      2. $\frac{8}{9} \div \frac{2}{3} =$      3. $\frac{18}{20} \div \frac{3}{4} =$

4. $3 \div \frac{1}{8} =$      5. $10\frac{1}{2} \div \frac{3}{4} =$      6. $4\frac{1}{2} \div 6\frac{1}{4} =$

7. $12\frac{2}{3} \div 4 =$      8. $5\frac{3}{4} \div \frac{1}{2} =$      9. $\frac{9}{14} \div \frac{6}{7} =$

Multiply. Reduce to lowest terms.

**1.** $\frac{4}{5} \times \frac{1}{2} =$

**2.** $\frac{8}{9} \times \frac{3}{4} =$

**3.** $\frac{12}{15} \times \frac{5}{6} =$

**4.** $\frac{3}{8} \times \frac{4}{5} =$

**5.** $\frac{9}{10} \times \frac{12}{15} =$

**6.** $7 \times \frac{1}{4} =$

**7.** $8 \times \frac{3}{7} =$

**8.** $2\frac{2}{3} \times 9 =$

**9.** $16 \times 3\frac{1}{4} =$

**10.** $\frac{7}{8} \times 12 =$

**11.** $4\frac{2}{3} \times 6\frac{3}{4} =$

**12.** $9\frac{1}{3} \times \frac{3}{7} =$

**13.** $2\frac{1}{2} \times \frac{4}{5} =$

**14.** $4\frac{7}{8} \times 6\frac{2}{3} =$

**15.** $5\frac{1}{5} \times 3\frac{3}{4} =$

**16.** $\frac{3}{4} \times \frac{7}{8} \times \frac{8}{9} =$

**17.** $2\frac{2}{3} \times \frac{1}{4} \times \frac{5}{6} =$

**18.** $3\frac{1}{8} \times \frac{4}{5} \times \frac{3}{6} =$

**19.** $6\frac{2}{3} \times 5 \times \frac{1}{4} \times 3 =$

**20.** $7\frac{1}{2} \times \frac{3}{5} \times 6 \times 2 =$

Divide. Reduce to lowest terms.

**21.** $\frac{3}{4} \div \frac{3}{18} =$

**22.** $\frac{5}{6} \div \frac{15}{24} =$

**23.** $\frac{8}{27} \div \frac{2}{3} =$

**24.** $\frac{11}{12} \div \frac{1}{6} =$

**25.** $\frac{1}{2} \div \frac{3}{10} =$

**26.** $5 \div \frac{3}{4} =$

**27.** $\frac{8}{9} \div 6 =$

**28.** $15\frac{1}{3} \div 8 =$

**29.** $12 \div 5\frac{2}{14} =$

**30.** $10\frac{1}{8} \div 27 =$

**31.** $6\frac{9}{24} \div 2\frac{1}{4} =$

**32.** $7\frac{2}{9} \div \frac{13}{15} =$

**33.** $9\frac{2}{8} \div 3\frac{3}{8} =$

**34.** $\frac{7}{8} \div 3\frac{2}{4} =$

**35.** $3\frac{4}{32} \div 6\frac{2}{8} =$

**36.** $3\frac{7}{16} \div \frac{11}{12} =$

**37.** $6\frac{1}{15} \div \frac{7}{27} =$

**38.** $20 \div 3\frac{1}{15} =$

**39.** $3\frac{5}{24} \div 8\frac{1}{4} =$

**40.** $13\frac{3}{6} \div \frac{9}{24} =$

## BUSINESS APPLICATIONS

**41.** The Martin Toy Manufacturing Company makes Halloween costumes. Each costume requires $2\frac{5}{8}$ yards of material. How many yards of material are required to make 24 costumes?

_____

**42.** The Pancake Hut serves approximately $2\frac{1}{3}$ cups of coffee to each customer every morning. If they served 123 people yesterday morning, how many cups of coffee did they serve?

_____

**43.** Mary and Scott Rajohn purchased a home for $95,000. They were required to make a down payment of $\frac{1}{5}$ of the purchase price. What was the amount of their down payment?

_____

**54** ■ LESSON 2-3

**44.** Roger worked 9 hours on Friday and 3 hours on Saturday assembling swing sets for Creative Play, Inc. He spent about $\frac{3}{4}$ of an hour assembling each set. Approximately how many swing sets did he assemble?

_____

**45.** The Holt Land Development Company bought 235 acres to divide and sell for single-family lots. If each lot will be $1\frac{2}{3}$ acres, how many lots will there be to sell?

_____

**46.** Dr. Gerald Marvin paid $1,178 for stocks costing $73\frac{5}{8}$ per share and $3,690 for stocks costing $51\frac{1}{4}$ per share. How many shares did Dr. Marvin buy?

_____

**47.** Bellview Interiors received a shipment of 135 lamps. The owner returned $\frac{1}{5}$ of the lamps because of breakage. The owner expects to sell $\frac{3}{4}$ of the remaining lamps by the end of the week. How many lamps will remain unsold?

_____

**48.** The Fresh-Pop Company purchased 136 bushels of popping corn to be repackaged in smaller containers for resale. Each smaller container holds $\frac{2}{23}$ of a bushel. How many smaller containers can be packaged from the 136 bushels?

**49.** Mel Johnson planted $\frac{3}{4}$ of his garden with the following crops: $1\frac{1}{2}$ acres strawberries, $\frac{3}{4}$ acre raspberries, $\frac{1}{4}$ acre watermelons, and 1 acre cantelope. How many acres are in Mel's garden?

**CONVERTING FRACTIONS TO DECIMALS**

To convert a fraction to a decimal, divide the numerator by the denominator. If the division does not come out even, add zeros in the dividend and round to the desired place. Use the same procedure to convert a mixed number to a mixed decimal; the whole number remains the same.

---

### EXAMPLE

Convert to a decimal. Round to the nearest thousandth.

### SOLUTION

Fraction $\longrightarrow$ Decimal

$$\frac{1}{6} \;=\; 6\overline{)1.0000} \;\;\approx\;\; 0.167 \quad (.1666)$$

$$\frac{7}{8} \;=\; 8\overline{)7.000} \;\;=\;\; 0.875 \quad (.875)$$

$$5\tfrac{1}{3} \;=\; 5 + 3\overline{)1.0000} \;\approx\; 5.333 \quad (.3333)$$

---

Sometimes it is necessary or more convenient to perform a calculation with a fraction's decimal equivalent.

---

### EXAMPLE

$8\tfrac{1}{21} + 3\tfrac{1}{13} =$

### SOLUTION

Convert mixed numbers to decimal numbers, round to the nearest thousandth, and then add.

Fraction $\longrightarrow$ Decimal

$$8\tfrac{1}{21} \;=\; 8 + 21\overline{)1.0000} \;\approx\;\; 8.048 \quad (.0476)$$

$$+3\tfrac{1}{3} \;=\; 3 + 3\overline{)1.0000} \;\approx\; +3.333 \quad (.3333)$$

$$\overline{\phantom{xxxxxxxxxxxxxxxx}11.381}$$

---

### PRACTICE PROBLEMS

Convert to decimals. Round to the nearest thousandth.

**1.** $\frac{2}{9} =$     **2.** $\frac{11}{12} =$     **3.** $\frac{5}{6} =$     **4.** $\frac{5}{8} =$

**5.** $3\frac{6}{7} =$     **6.** $2\frac{2}{3} =$     **7.** $\frac{14}{15} =$     **8.** $\frac{4}{5} =$

Convert to decimal equivalents and round to the nearest thousandth. Add or subtract.

**9.**   $7\frac{1}{2}$     **10.**   $15\frac{1}{3}$     **11.**   $7\frac{6}{10}$     **12.**   $20\frac{1}{13}$
    $+8\frac{1}{2}$         $+\,7\frac{1}{8}$         $-4\frac{1}{7}$         $-18\frac{2}{14}$

---

**ANSWERS FOR PRACTICE PROBLEMS**

**(1)** 0.222   **(2)** 0.917   **(3)** 0.833   **(4)** 0.625   **(5)** 3.857   **(6)** 2.667   **(7)** 0.933   **(8)** 0.800   **(9)** 16   **(10)** 22.458   **(11)** 3.457   **(12)** 1.934

## CONVERTING DECIMALS TO FRACTIONS

To convert a decimal to a fraction, drop the decimal point and place that number in the numerator of the fraction. The denominator is 1 followed by as many zeros as there are decimal places in the decimal. Reduce the fraction to lowest terms if necessary. Use the same procedure to convert a mixed decimal to a mixed number. The whole number remains the same.

### EXAMPLE 1

Convert 0.375 to a fraction. Reduce to lowest terms.

**SOLUTION**

Decimal $\longrightarrow$ Fraction

$$\underbrace{0.375}_{\text{3 decimal places}} = \underbrace{\tfrac{375}{1000}}_{\text{3 zeros}} = \tfrac{3}{8}$$

### EXAMPLE 2

Convert 2.75 to a mixed number. Reduce to lowest terms.

**SOLUTION**

Decimal $\longrightarrow$ Fraction

$$\underbrace{2.75}_{\text{2 decimal places}} = 2\underbrace{\tfrac{75}{100}}_{\text{2 zeros}} = 2\tfrac{3}{4}$$

### PRACTICE PROBLEMS

Convert to a decimal or mixed number. Reduce to lowest terms.

**1.** 0.25 =        **2.** 0.4 =        **3.** 0.375 =        **4.** 0.1 =
**5.** 4.125 =        **6.** 0.5 =        **7.** 5.75 =        **8.** 9.625 =

The following chart shows some of the more common decimal equivalents used in business:

### DECIMAL EQUIVALENT CHART

| | | |
|---|---|---|
| $\frac{1}{2} = 0.5$ | $\frac{4}{5} = 0.8$ | $\frac{1}{9} = 0.1111$ |
| $\frac{1}{4} = 0.25$ | $\frac{1}{6} = 0.1667$ | $\frac{1}{16} = 0.0625$ |
| $\frac{3}{4} = 0.75$ | $\frac{1}{7} = 0.1429$ | $\frac{3}{16} = 0.1875$ |
| $\frac{1}{3} = 0.3333$ | $\frac{5}{6} = 0.8333$ | $\frac{5}{16} = 0.3125$ |
| $\frac{2}{3} = 0.6667$ | $\frac{1}{8} = 0.125$ | $\frac{7}{16} = 0.4375$ |
| $\frac{1}{5} = 0.2$ | $\frac{3}{8} = 0.375$ | $\frac{9}{16} = 0.5625$ |
| $\frac{2}{5} = 0.4$ | $\frac{5}{8} = 0.625$ | $\frac{11}{16} = 0.6875$ |
| $\frac{3}{5} = 0.6$ | $\frac{7}{8} = 0.875$ | $\frac{13}{16} = 0.8125$ |

ANSWERS FOR
PRACTICE PROBLEMS

(1) $\frac{1}{4}$  (2) $\frac{2}{5}$  (3) $\frac{3}{8}$  (4) $\frac{1}{10}$  (5) $4\frac{1}{8}$  (6) $\frac{1}{2}$  (7) $5\frac{3}{4}$  (8) $9\frac{5}{8}$

## LESSON 2-4     EXERCISES

Convert the following fractions to decimals. Round to the nearest thousandth.

**1.** $\frac{3}{5}$                    **2.** $\frac{1}{8}$                    **3.** $\frac{1}{6}$

**4.** $\frac{11}{12}$                    **5.** $\frac{4}{7}$                    **6.** $\frac{7}{18}$

**7.** $4\frac{6}{11}$                    **8.** $7\frac{1}{10}$                    **9.** $3\frac{5}{6}$

**10.** $\frac{27}{52}$                    **11.** $\frac{18}{20}$                    **12.** $\frac{50}{99}$

Convert the following decimals to fractions. Reduce to lowest terms.

**13.** 0.625          **14.** 0.333          **15.** 0.35

**16.** 6.4          **17.** 12.75          **18.** 2.15

**19.** 0.45          **20.** 0.075          **21.** 0.65

Convert to decimal equivalents and add or subtract. Round fractions to the nearest thousandth.

**22.**   $16\frac{1}{4}$          **23.**   $21\frac{3}{20}$          **24.**   $8$          **25.**   $30\frac{5}{6}$
   $+17\frac{3}{8}$                      $-6\frac{1}{2}$                $-\frac{1}{9}$                $-20\frac{1}{3}$

LESSON 2-4  ■  **59**

26. $\frac{7}{8}$
    $-\frac{1}{3}$

27. $45\frac{1}{8}$
    $-36\frac{1}{4}$

28. $5\frac{1}{2}$
    $-5\frac{1}{3}$

29. $10\frac{4}{10}$
    $-\ 6\frac{2}{5}$

## BUSINESS APPLICATIONS

Round to the nearest cent.

30. Alan Jenssen purchased $4\frac{1}{8}$ pounds of steak at $4.12 per pound. How much did the steak cost?

_____

31. Last year Rachel Baker received a bonus of $440.37. Rachel's employer informed her that this year's bonus will be $1\frac{2}{5}$ of last year's bonus. What will Rachel's bonus be this year?

_____

32. The Northern Hardware store sells brackets for $37\frac{1}{2}$ cents each. How much will 128 brackets cost?

_____

33. Gemini Electric has 24 digital clock radios for sale. Gemini paid $14.80 each for $\frac{1}{4}$ of the radios, $16.50 each for $\frac{1}{8}$ of the radios, $22.50 each for $\frac{1}{2}$ of the radios, and paid $35 each for the remaining radios. What was the total cost of all the radios?

_____

34. The cost of producing one widgit is $0.375. How many widgits can be produced for $483.75? (Convert to fractions before calculating.)

_____

**60** ■ LESSON 2-4

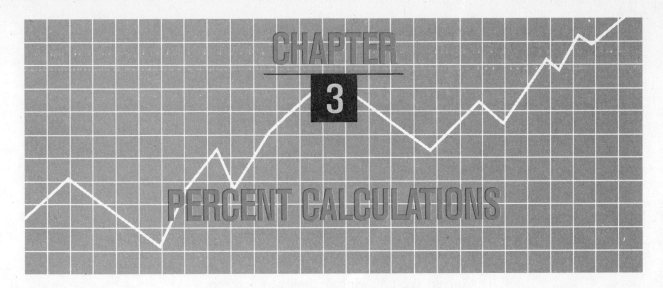

## CHAPTER 3

## PERCENT CALCULATIONS

### LEARNING OBJECTIVES

**1.** Convert decimals, fractions, and percents.

**2.** Calculate the base, part, and rate using the percent formula.

**3.** Calculate the percent of increase and decrease.

Managers can learn a good deal about the performance of their business by analyzing and comparing the dollar amounts of various items. Often, however, a clearer picture of business performance can be obtained by comparing the relationship of one amount to another. One way in which relationships are commonly expressed is as *percents.* In this chapter, you will learn how to make various percent calculations and also how to use percents to analyze amounts and to compare one amount to another.

## LESSON 3-1    CONVERTING DECIMALS, FRACTIONS, AND PERCENTS

**CONVERTING DECIMALS AND FRACTIONS TO PERCENTS**

Percents show the relationship of a number to a whole or base of 100. Percent means "out of 100." For example, 30% means "30 parts out of a total of 100 parts." Percents, decimals, and fractions all represent parts of a whole. For example, the percent 30% is also expressed as the decimal 0.30 and the fraction $\frac{30}{100}$. Percents can be more than 100%. For example, 130% means that more than one unit has been divided into 100 equal parts, and 130 parts are being considered.

All decimals and fractions can be expressed as percents. To convert a decimal to a percent, move the decimal point two places to the right and add a percent sign (%). Insert zeros to the right as needed. If the decimal has more than two digits, the resulting percent will include decimal parts of a whole percent.

## EXAMPLE

Convert the following decimals to percents by moving the decimal point two places to the right and adding a percent sign.

## SOLUTION

| Decimal ——————→ | | Percent |
|---|---|---|
| 0.35 | 0.35 | 35% |
| 0.8 | 0.80 | 80% — Add zero |
| 0.1375 | 0.1375 | 13.75% — Decimal part of percent |
| 2.5 | 2.50 | 250% — Add zero |

## PRACTICE PROBLEMS

Convert the following decimals to percents.

**1.** 1.05    **2.** 0.08    **3.** 0.62    **4.** 0.875    **5.** 0.7
**6.** 0.625    **7.** 3.4    **8.** 0.37    **9.** 1.057    **10.** 0.74

To convert a fraction to a percent, first convert the fraction to its decimal equivalent by dividing the numerator by the denominator. Then, convert the decimal to a percent. Use the same procedure when converting a mixed number, such as $2\frac{5}{8}$, to a percent; the whole number remains the same.

## EXAMPLE

Convert the following fractions to percents. Round percents to the nearest tenth.

## SOLUTION

| Fraction ——————————→ | | Decimal ——————→ | Percent |
|---|---|---|---|
| $\frac{2}{5}$ | $5\overline{)2.0}$ .4 | 0.4 | 40% |
| $\frac{1}{3}$ | $3\overline{)1.0000}$ .3333 | 0.3333 | 33.3% |
| $1\frac{1}{5}$ | $1 + 5\overline{)1.0}$ .2 | 1.2 | 120% |

## PRACTICE PROBLEMS

Convert the following fractions to percents. Round percents to the nearest tenth.

**1.** $\frac{7}{8}$    **2.** $\frac{1}{8}$    **3.** $2\frac{2}{3}$    **4.** $1\frac{1}{6}$    **5.** $\frac{3}{8}$
**6.** $\frac{3}{4}$    **7.** $\frac{3}{5}$    **8.** $\frac{2}{9}$    **9.** $1\frac{4}{7}$    **10.** $\frac{9}{14}$

ANSWERS FOR
PRACTICE PROBLEMS

**Top:** (1) 105%   (2) 8%   (3) 62%   (4) 87.5%   (5) 70%   (6) 62.5%
(7) 340%   (8) 37%   (9) 105.7%   (10) 74%

**Bottom:** (1) 0.875 = 87.5%   (2) 0.125 = 12.5%   (3) 2.6666 = 266.7%
(4) 1.1666 = 116.7%   (5) 0.375 = 37.5%   (6) 0.75 = 75%   (7) 0.6 = 60%
(8) 0.2222 = 22.2%   (9) 1.5714 = 157.1%   (10) 0.6428 = 64.3%

## CONVERTING PERCENTS TO DECIMALS AND FRACTIONS

To convert a percent to a decimal, move the decimal point two places to the left and drop the percent sign. Insert zeros to the left if necessary.

### EXAMPLE

Convert the following percents to decimals.

### SOLUTION

| Percent | ⟶ | Decimal |
|---------|----|---------|
| 23% | 23% | 0.23 |
| 5% | 05% | 0.05 |
| 3.7% | 03.7% | 0.037 |

A fractional percent such as $\frac{1}{4}\%$ is less than 1%. To change a fractional percent to a decimal, first change the fraction to a decimal. Then, move the decimal point two places to the left and drop the percent sign. Use the same procedure when the percent is a mixed number, such as $5\frac{3}{4}\%$; the whole number remains the same. Add zeros to the left if necessary.

### EXAMPLE

Convert the following percents to decimals.

### SOLUTION

Change fraction to decimal before moving decimal point to the left

| Percent | ⟶ | Decimal |
|---------|----|---------|
| $\frac{1}{4}\%$ | 00.25% | 0.0025 |
| $5\frac{3}{4}\%$ | 05.75% | 0.0575 |

### PRACTICE PROBLEMS

Convert the following percents to decimals.

1. 40%  **2.** 37.5%  **3.** 70%  **4.** 155%  **5.** $1\frac{1}{2}\%$
6. 4%  **7.** $\frac{1}{8}\%$  **8.** 0.62%  **9.** 100%  **10.** $20\frac{1}{4}\%$

To convert a percent to a fraction, first convert the percent to a decimal; then, convert the decimal to a fraction. Reduce to lowest terms.

### EXAMPLE

Convert the following percents to fractions.

### SOLUTION

| Percent | ⟶ Decimal | ⟶ Fraction |
|---------|-----------|------------|
| 36% | 0.36 | $\frac{36}{100} = \frac{9}{25}$ |
| $37\frac{1}{2}\%$ | 0.375 | $\frac{375}{1000} = \frac{3}{8}$ |
| 112% | 1.12 | $1\frac{12}{100} = 1\frac{3}{25}$ |

ANSWERS FOR PRACTICE PROBLEMS

**(1)** 0.4 **(2)** 0.375 **(3)** 0.7 **(4)** 1.55 **(5)** 0.015 **(6)** 0.04 **(7)** 0.00125 **(8)** 0.0062 **(9)** 1 **(10)** 0.2025

Convert the following percents to fractions. Reduce to lowest terms.

**1.** $12\frac{1}{2}\%$    **2.** 48%    **3.** 205%    **4.** 4%    **5.** $62\frac{1}{2}\%$

**6.** 6.25%    **7.** 115%    **8.** 80%    **9.** 22.5%    **10.** 45%

**EXPRESSING RATIOS AS FRACTIONS, DECIMALS, AND PERCENTS**

Business managers routinely analyze their business operations by comparing amounts such as income, costs, and expenses. *Ratios* are often used to compare one amount to another and the results from one month or year to other similar time periods. A ratio compares a part of an amount to a whole amount or a part to a part. A ratio can be expressed as a fraction, decimal, or percent. For example, the ratio of 1 part to 5 parts can be expressed as $\frac{1}{5}$, 0.20, or 20%.

DIFFERENT EXPRESSIONS OF RATIOS

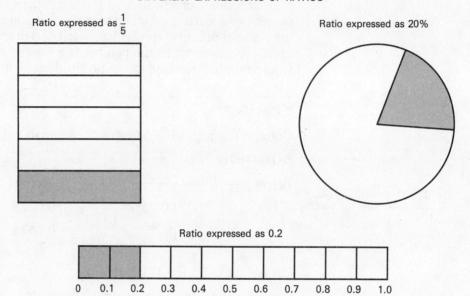

Ratio expressed as $\frac{1}{5}$

Ratio expressed as 20%

Ratio expressed as 0.2

0   0.1   0.2   0.3   0.4   0.5   0.6   0.7   0.8   0.9   1.0

**EXAMPLE**

In July Cutty's Boat Rental had labor expenses of $4,000 and rent expenses of $1,200. Total operating expenses were $16,000. What ratio is the labor expense to the total operating expenses for July? What is the ratio of rent expenses to labor expenses? Express the ratio as a fraction, decimal, and percent.

**SOLUTION**

**Step 1** Find the ratio of labor expenses to total operating expenses.

$$\frac{\text{Part}}{\text{Whole}} = \frac{\text{Labor Expense}}{\text{Total Operating Expenses}} = \frac{\$4,000}{\$16,000} = 0.25 \text{ or } 25\% \text{ or } \frac{1}{4}$$

**Step 2** Find the ratio of rent expense to labor expense.

$$\frac{\text{Part}}{\text{Part}} = \frac{\text{Rent Expense}}{\text{Labor Expense}} = \frac{\$1,200}{\$4,000} = 0.30 \text{ or } 30\% \text{ or } \frac{3}{10}$$

ANSWERS FOR PRACTICE PROBLEMS

**(10)** $\frac{9}{20}$

**(1)** $\frac{1}{8}$   **(2)** $\frac{12}{25}$   **(3)** $2\frac{1}{20}$   **(4)** $\frac{1}{25}$   **(5)** $\frac{5}{8}$   **(6)** $\frac{1}{16}$   **(7)** $1\frac{3}{20}$   **(8)** $\frac{4}{5}$   **(9)** $\frac{9}{40}$

## PRACTICE PROBLEMS

Find the ratio of each expense to the total operating expenses. Express ratios as fractions and percents.

| | Expense | Amount | Ratio Fraction | Percent |
|---|---|---|---|---|
| 1. | Fuel | $1,800 | _____ | _____ |
| 2. | Insurance | 1,400 | _____ | _____ |
| 3. | Labor | 4,000 | _____ | _____ |
| 4. | Rent | 2,000 | _____ | _____ |
| 5. | Repairs | 800 | _____ | _____ |
| | Total operating expenses | $10,000 | | |

## LESSON 3-1     EXERCISES

Convert the following decimals to percents.

**1.** 0.05 =          **2.** 7 =          **3.** 1.25 =          **4.** 0.0004 =

**5.** 0.009 =          **6.** 0.85 =          **7.** 0.1832 =          **8.** 3.04 =

**9.** 1.304 =          **10.** 2.062 =          **11.** 0.945 =          **12.** 0.8 =

Convert the following fractions to percents. Round percents to the nearest tenth.

**13.** $\frac{2}{3}$ =          **14.** $\frac{5}{8}$ =          **15.** $\frac{1}{3}$ =          **16.** $4\frac{1}{8}$ =

**17.** $1\frac{2}{9}$ =          **18.** $\frac{7}{15}$ =          **19.** $\frac{3}{5}$ =          **20.** $\frac{1}{12}$ =

**21.** $\frac{1}{25}$ =          **22.** $\frac{2}{13}$ =          **23.** $\frac{1}{50}$ =          **24.** $1\frac{9}{20}$ =

Convert the following percents to decimals.

**25.** $62\frac{1}{2}\%$ =          **26.** 125% =          **27.** 645% =          **28.** $2\frac{3}{4}\%$ =

**29.** $\frac{5}{8}\%$ =          **30.** 200% =          **31.** 0.7% =          **32.** 0.25% =

**33.** 83% =          **34.** 7.5% =          **35.** 27% =          **36.** 15% =

Convert the following percents to fractions. Reduce to lowest terms.

**37.** 50% =          **38.** 25% =          **39.** $37\frac{1}{2}\%$ =          **40.** 40% =

**41.** $87\frac{1}{2}\%$ =          **42.** 12% =          **43.** $2\frac{1}{2}\%$ =          **44.** 160% =

**45.** 250% =          **46.** 0.05% =          **47.** 12.5% =          **48.** 237.5% =

Find the missing fraction, decimal, or percent equivalents for each of the following.

|      | Fraction | Decimal | Percent |
|------|----------|---------|---------|
| **49.** | _____ | 0.02 | 2% |
| **50.** | $\frac{1}{4}$ | _____ | 25% |
| **51.** | $\frac{2}{25}$ | 0.08 | _____ |
| **52.** | $\frac{1}{5}$ | 0.20 | _____ |
| **53.** | _____ | 0.625 | $62\frac{1}{2}\%$ |
| **54.** | $\frac{3}{4}$ | _____ | 75% |
| **55.** | $\frac{3}{5}$ | 0.60 | _____ |
| **56.** | $\frac{1}{8}$ | _____ | $12\frac{1}{2}\%$ |
| **57.** | $\frac{3}{8}$ | _____ | $37\frac{1}{2}\%$ |
| **58.** | _____ | 0.40 | 40% |

Listed on top of page 70 are the sales of children's clothing and the total sales for Kakeldy's Clothing Store for 10 consecutive weeks. For each week, express the ratio of children's clothing sales to total sales as both a fraction and as a percent. Round percents to the nearest tenth.

| Week of | Children's Clothing Sales | Total Sales | Ratios | |
|---|---|---|---|---|
| | | | Fraction | Percent |
| 59. October 3 | $ 3,319 | $16,595 | _____ | _____ |
| 60. October 10 | 4,250 | 17,000 | _____ | _____ |
| 61. October 17 | 4,380 | 13,140 | _____ | _____ |
| 62. October 24 | 2,250 | 13,500 | _____ | _____ |
| 63. October 31 | 4,800 | 10,800 | _____ | _____ |
| 64. November 7 | 3,985 | 15,940 | _____ | _____ |
| 65. November 14 | 6,600 | 17,600 | _____ | _____ |
| 66. November 21 | 4,350 | 14,500 | _____ | _____ |
| 67. November 28 | 8,900 | 22,250 | _____ | _____ |
| 68. December 5 | 15,820 | 25,312 | _____ | _____ |

## BUSINESS APPLICATIONS

69. During May the total sales in women's wear for Leonard's Apparel, Inc., was $112,500. Total sales for the year were $1,687,500. Express the ratio of sales for May to the total sales for the year as a fraction. Reduce to lowest terms.

70. The salary expense for the Tillman Pool Company totaled $4,258 for the month of August. The yearly salary expense totaled $44,709. Express the ratio of the salary expenses for August to the yearly salary expenses as a decimal. Round to the nearest thousandth.

71. Kuow Manufacturing Company's total telephone expense for last month was $646.05. Of this, $430.70 was for long-distance calls. Express the ratio of long-distance calls to the total telephone expense as a fraction.

**72.** Last month Broadway Sewing Center had fabric sales of $14,500 and machine sales of $105,600. Find the ratio of fabric sales to total sales for the month. Express as decimal. Round to the nearest thousandth.

_____

**73.** Camper Country employs 210 part-time employees and 290 full-time employees. Find (a) the ratio of part-time employees to full-time employees and (b) the ratio of part-time employees to total employees. Express ratios as percents.

(a) _____

(b) _____

**74.** Bettye Verow's monthly consulting expenses include the following: commuting, $679; parking, $126; lodging, $704; and clothing, $362. Her monthly income from consulting is $7,484. What is the ratio of Bettye's total monthly expenses to her monthly income? Express the ratio as a fraction.

_____

**75.** Adrand Telecommunications Company's monthly traveling expenses are $12,624. Their total operating expenses are $265,104. Find the ratio of traveling expenses to total expenses. Express the ratio as a decimal. Round to the nearest thousandth.

_____

**76.** Last month Barlow Investments' total income was $24,975. Of this, $16,650 was earned from the sale of residential properties. What is the ratio of income from the sale of residential properties to total income? Express the ratio as a fraction.

_____

**77.** Last month Steven Wiley contacted 270 clients. Of these contacts Steven sold his product to 162 clients. What is the ratio of total sales to total clients contacted? Express the ratio (a) as a fraction and (b) as a percent.

**(a)** _____

**(b)** _____

**78.** Last month National Beauty Supply House had monthly sales to beauty salons of $78,945. Monthly sales to individual cosmetologists totaled $34,560. Find the ratio (a) of individual sales to salon sales and (b) salon sales to total monthly sales. Express ratios as decimals. Round to the nearest thousandth.

**(a)** _____

**(b)** _____

### IDENTIFYING THE PART, BASE, AND RATE

Percents also are used in business and in our daily lives to show the relationship that exists among two or more numbers. For example, percents are used to determine a down payment on an automobile, sales tax on purchases, profit on sales, and interest on savings accounts.

In any percent calculation, there are three elements, the base, the part, and the rate. The **base** is the whole amount; it is the original amount to which other amounts are compared.

The **part** is an amount that is a portion of the base. It is expressed in the same unit of measure as the base, usually in dollars.

The **rate** is a percent indicating the relationship of the part to the base. The rate is almost always followed by a percent sign (%).

The following examples show the relationship of the base, part, and rate in many business problems.

| Part | Rate | Base |
|------|------|------|
| Sales tax of $6,000 | is 6% | of sales of $100,000. |
| Raise of $1,200 | is 11% | of last year's salary of $10,909. |
| Reduced price of $45 | is 65% | of the retail price of $69.23. |
| Interest of $260 | is 8% | of savings of $3,250. |
| Clothing sales of $10,500 | is __15__ % | of total sales of $72,300. |

### EXAMPLE

Paulette Perkins had total sales of $25,000 last week. She earned a $1,250 commission, which is 5% of her total sales. Identify the base, part, and rate.

### SOLUTION

**Step 1**  The base is $25,000 ("whole amount").
**Step 2**  The part is $1,250 ("portion" of the total sales).
**Step 3**  The rate is 5% (shows the relationship between the total sales and the commission).

### PRACTICE PROBLEMS

Identify the base, part, and rate.

1. Rondell purchased a coat for $235. A $5\frac{1}{2}$% sales tax of $12.93 was added to the purchase price.

   base = _____        part = _____        rate = _____

2. Tom Underwood answered 54 of the 60 questions on the test correctly. He received a 90% grade on his test.

   base = _____        part = _____        rate = _____

3. Anne Loewen paid a down payment of $10,050, which represents 15% of the purchase price of $67,000 for a condo.

   base = _____        part = _____        rate = _____

ANSWERS FOR PRACTICE PROBLEMS

(1) $235, $12.93, $5\frac{1}{2}$%    (2) 60, 54, 90%    (3) $67,000, $10,050, 15%

If any two of these elements are known, the third can be found using the percent formula, **Part = Base × Rate.** The following illustration is an easy guide to help you calculate the base, rate, or part. Simply cover the letter which represents the element you are trying to find. The remaining letters in the illustration reveal the formula.

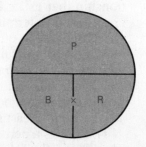

To find the part when the base and the rate are given, multiply the base times the rate. Change the rate to a decimal before multiplying. The circle below illustrates the formula for finding the part.

Part = Base × Rate

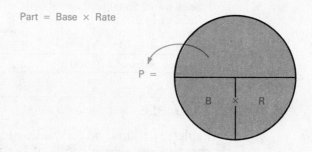

P =

### EXAMPLE

Karen Dial's current annual salary is $22,000. She will receive an 8% annual salary raise. How much of a salary raise will Karen receive?

### SOLUTION

**Step 1**   Identify the given elements and the missing element.
The base is $22,000 (the annual salary).
The rate is 8% (rate is expressed as a percent).
The part is missing (the annual raise).

**Step 2**   Choose the formula that will find the missing element and solve the problem.

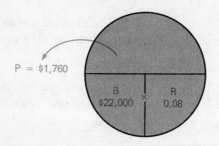

P = $1,760

$$P \ = \ B \ \times \ R$$
$$\$1{,}760 = \$22{,}000 \times 0.08$$

## PRACTICE PROBLEMS

Solve using the percent formula. Round to the nearest tenth.

**1.** $7\frac{1}{4}$% of 7,800 =     **2.** 15% of 890 =
**3.** 120.5% of 740 =     **4.** 1.8% of 400 =

**5.** Gerald Robinson saves 7% of his monthly income of $2,660. How much does Gerald save each month?
**6.** Reva Desmore earns $452 a week. Her employer deducts 32% of her earnings for city, state, and federal taxes. What is the total amount of taxes deducted each week?

**INCREASE AND DECREASE PROBLEMS— FINDING THE PART**

When an amount increases or decreases over a period of time, the amount and percent of change are often analyzed and compared. The original amount is always the base and the new amount is the part.

When finding the part in an increase or decrease problem, use the formula, **Part = Base × Rate**. If the amount increases, the part is greater than the base and the rate of the increased amount is 100% plus the rate of increase. For example, the enrollment at Western College increased by 23% from 10,500 to 12,915. The original amount of 10,500 is the base, the new amount of 12,915 is the part, and the rate of the increased amount is 123%.

| This year's enrollment (123%) |
|---|
| Last year's enrollment (100%) + Increase (23%) |

### EXAMPLE

Five years ago Hanneco, Inc., purchased a warehouse for $80,000. Now the value of the building is estimated to have increased by 25% over the purchase price. What is the estimated current value of the warehouse?

### SOLUTION

**Step 1** Identify the given elements and the missing element.
The base is $80,000 (original purchase price).
The rate is 125% (original purchase price of 100% plus a 25% increase).
The part is missing (current value of the warehouse).

**Step 2** Choose the formula that will find the missing element and solve the problem.

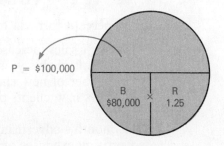

$$P = B \times R$$
$$\$100,000 = \$80,000 \times 1.25$$

Part is greater than base

**(1)** 565.5   **(2)** 133.5   **(3)** 891.7   **(4)** 7.2   **(5)** $186.20   **(6)** $144.64

LESSON 3-2 ■ **75**

If an amount decreases, the part is less than the base and the rate of the decreased amount is 100% minus the rate of decrease. For example, sales decreased by 25% from $200,000 to $150,000. The original sales of $200,000 is the base, the new sales of $150,000 is the part, and the new sales rate is 75%.

| New sales (75%) |
| Original sales (100%) − Decrease (25%) |

## EXAMPLE

Last year the Crowly Company had an advertising budget of $14,700. This year the advertising budget was reduced by 10%. What is the amount of this year's advertising budget?

### SOLUTION

**Step 1**  Identify the given elements and the missing element.
The base is $14,700 (last year's budget).
The rate is 90% (last year's budget of 100% minus a 10% decrease).
The part is missing (this year's budget).

**Step 2**  Choose the formula that will find the missing element and solve the problem.

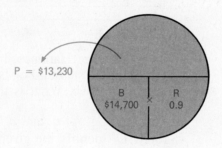

P = $13,230

$$P = B \times R$$
$$\$13,230 = \$14,700 \times 0.9$$

Part is less than the base

## PRACTICE PROBLEMS

Use the percent formula to solve the following problems.

1. This year's sales increased 40% over last year's sales of $110,900. Find the amount of this year's sales.
2. The number of new clients this year decreased by 12% compared to last year's new clients of 125. What is the number of new clients this year?
3. This month's advertising expenses are 75% greater than last month's advertising expenses of $12,100. Find the amount of this month's expenses.
4. On Tuesday the manager of the produce department of a local grocery store had 110 heads of lettuce. On Thursday, the amount of lettuce was reduced by 60%. How many heads of lettuce were left?

ANSWERS FOR
PRACTICE PROBLEMS

(1) $155,260   (2) 110   (3) $21,175   (4) 44

**76** ■ LESSON 3-2

## LESSON 3-2    EXERCISES

Identify the base, part, and rate in each of the following statements.

1. Arleen's raise of $3,916 is an increase of 10% over her last year's salary of $39,160.

   base =                      part =                      rate =

2. The Western Auto Parts Company earned 9% interest of $11,970 on the company's investment of $133,000

   base =                      part =                      rate =

3. The fire damage to the Orange County Warehouse was $52,010. This amount is 35% of the estimated value of the warehouse of $148,600.

   base =                      part =                      rate =

4. Property Investment, Inc., purchased apartments for $500,900 and hopes to receive 10% more or $50,090 more when they sell.

   base =                      part =                      rate =

5. The Hamletons invest 12% of their annual income of $56,700. This amounts to $6,804 a year.

   base =                      part =                      rate =

Solve using the percent formula. Round to the nearest hundredth.

6. $12\frac{1}{2}\%$ of 320 =

7. 125% of 400 =

8. $1\frac{3}{4}\%$ of 180 =

9. 65% of 600 =

10. 10.5% of 60 =

11. 8% of $480.50 =

12. 5% of $35,500 =

13. 106% of $38,250 =

14. 300% of $89.56 =

15. $27\frac{1}{4}\%$ of 4,004 =

16. 60% of 150 =

17. 15% of $140.56 =

18. 135% of $5,890 =

19. 8% of 340 =

20. 20% of $350 =

21. $\frac{1}{4}\%$ of 2,350 =

The percents that total expenses are of sales are shown below for the first half of the year for the McNally Company. Find the total expenses for each month. Round to the nearest hundred dollar.

| Month | Sales | Percent Total Expenses Are of Sales | Total Expenses |
|---|---|---|---|
| 22. January | $20,900 | 80.8% | _____ |
| 23. February | 24,500 | 96.2% | _____ |
| 24. March | 25,000 | 87.5% | _____ |
| 25. April | 25,900 | 92.5% | _____ |
| 26. May | 26,300 | 83.7% | _____ |
| 27. June | 24,800 | 95.6% | _____ |

## BUSINESS APPLICATIONS

Solve the following using the percent formula. Round to the nearest hundredth.

**28.** Sales for the Oscar Radilly Company for the first quarter of the year were $63,900. Operating expenses for the same period of time were 83.7% of sales. Find the operating expenses.

_____

**29.** The weekly sales goal for Carla Andrews was $4,250. She reached 105.5% of her goal. What were her total sales for the week?

_____

**30.** Last year's sales at the Foxwood Rental, Inc., were $234,900. This year sales were 17.5% higher than last year. What are this year's sales? Round to the nearest dollar.

_____

**31.** Mike Ridout answered correctly 90% of the 150 questions on the state life insurance exam. How many questions did Mike answer correctly?

_____

**32.** A set of golf clubs regularly sells for $475 at Jaydee's Sport Center. The clubs were reduced in price by 30% for a sale. What is the amount of the price reduction?

_____

**33.** Production at the Southtex Fiberglass Plant last year was 83,560 units. This year production is down 8%. How many units were produced this year? Round to the nearest whole unit.

_____

**34.** Yesterday, Davis Manufacturing Company made 15,800 plastic telephones. Of these, 1.5% had minor defects. How many telephones were defective?

_____

**35.** The manager at the Amber Light Grill conducted a poll and found that 85% of Saturday night's 320 patrons were repeat customers. How many of Saturday night's customers were repeat customers?

_____

**36.** Curtis Morley took two of his business clients to lunch. The bill was $68.64. Curtis wants to leave the waiter a tip equal to 15% of the total bill. How much of a tip should he leave?

_____

**37.** This year the Tyler Tee Building Company reduced its overhead expenses by $17\frac{3}{4}$%. If last year's overhead expenses were $345,720, what are this year's overhead expenses?

_____

**38.** The maintenance department of the Harkness Manufacturing Company purchased a lawn tractor and mower for $3,673. The sales tax of $6\frac{1}{2}$% was added to the purchase price. How much sales tax was added?

_____

**39.** Rasmus Company purchased an intercom system costing $840 on a plan where they could pay 15% of the purchase price and pay the remaining amount later. Find (a) the amount first paid and (b) the remaining amount to be paid.

(a) _____

(b) _____

**40.** The Mavis Supply Company paid income tax of $12,456 last year. This year the income tax paid by the Mavis Company increased by 11%. How much income tax did they pay this year?

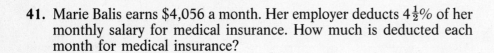

_____

**41.** Marie Balis earns $4,056 a month. Her employer deducts $4\frac{1}{2}$% of her monthly salary for medical insurance. How much is deducted each month for medical insurance?

_____

**42.** During the month of October, sales were $13,492. Sales in November were 16% higher than in October. Sales in December were 20% higher than sales in November. Find total sales for (a) November and (b) December.

**(a)**  _____

**(b)** _____

**USING THE PERCENT FORMULA TO FIND THE BASE**

If the part and the rate are known, the base can be found by dividing the part by the rate. The circle below illustrates the formula for finding the base.

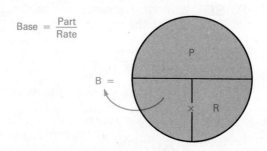

$$Base = \frac{Part}{Rate}$$

### EXAMPLE

At the Crowly Company, 20% of the employees take their vacations during the month of May. If 125 employees were on vacation in May, what is the total number of employees at Crowly?

### SOLUTION

**Step 1**   Identify the given elements and the missing element.
The part is 125 (employees on vacation).
The rate is 20%.
The base is missing (the total number of employees).

**Step 2**   Choose the formula that will find the missing element and solve the problem.

$$B = \frac{P}{R} = \frac{125}{0.2} = 625 \text{ Total number of employees}$$

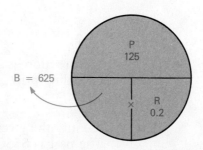

### PRACTICE PROBLEMS

Use the percent formula to solve the following. Round to the nearest hundredth.

1. $2,160 is 12% of _____
2. 1,020 is $8\frac{1}{2}$% of _____
3. 2,160 is 6.75% of _____
4. $140.75 is 10% of _____

**5.** Find the amount of total sales if the $6\frac{1}{2}\%$ sales tax collected by the Super Service Center is $1,290.

**6.** The First Savings and Loan building is insured for 80% of its value. If it is insured for $1,250,000, what is the value of the building?

## INCREASE AND DECREASE PROBLEMS— FINDING THE BASE

When finding the base in an increase or decrease problem, use the formula, **Base = Part ÷ Rate.** When an amount increases, the base is less than the part, and the rate is 100% plus the rate of increase. When an amount decreases, the base is greater than the part, and the rate is 100% minus the rate of decrease.

### EXAMPLE 1

Betty Matuska's annual salary for this year is $27,000, which is an 8% increase over her annual salary for last year. What was Betty's annual salary last year?

### SOLUTION

**Step 1**   Identify the given elements and the missing element.
The part is $27,000 (this year's salary).
The rate is 108% (last year's salary of 100% plus an 8% increase).
The base is missing (last year's salary).

**Step 2**   Choose the formula that will find the missing element and solve the problem.

$$B = \frac{P}{R} = \frac{\$27{,}000}{1.08} = \$25{,}000 \text{ Last year's salary}$$

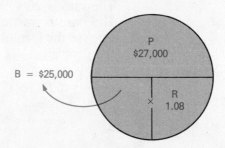

### EXAMPLE 2

Jerad's Furniture Store is selling a sleeper sofa for $350, which is 45% off the original selling price. What is the original selling price?

### SOLUTION

**Step 1**   Identify the given elements and the missing element.
The part is $350 (the reduced price).
The rate is 55% (the original selling price of 100% minus the decrease of 45%).
The base is missing (the original selling price).

**Step 2**   Choose the formula that will find the missing element and solve the problem.

$$B = \frac{P}{R} = \frac{\$350}{0.55} = \$636.363 \approx \$636.36 \text{ Original Selling Price}$$

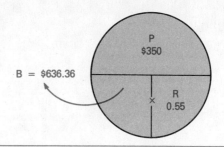

## PRACTICE PROBLEMS

Solve using the percent formula.

1. Student enrollment grew to 14,000 students, which is a 25% increase over last semester. Find the student enrollment for last semester.
2. John Dowdel had sales of $230,000 this month, which was a 19% decrease from last month. What were John's sales last month? Round to the nearest thousand.
3. The cost of building a new Blue Diamond home is $70,000, which is an increase of 27% over the cost 2 years ago. What was the cost of building a Blue Diamond home 2 years ago? Round to the nearest thousand.
4. This month's sales for the Fashion Trend shop are 30% higher than last month's sales. If this month's sales are $15,400, what were last month's sales? Round to the nearest dollar.
5. Maria Lopez sold her stocks for $1,890, which is a 35% decrease in the price she originally paid for them. How much did Maria originally pay for the stocks? Round to the nearest cent.

ANSWERS FOR
PRACTICE PROBLEMS

(1) 11,200  (2) $284,000  (3) $55,000  (4) $11,846  (5) $2,907.69

## LESSON 3-3    EXERCISES

Use the percent formula to solve the following problems. Round amounts to the nearest hundredth.

1. $2,205 is $10\frac{1}{2}$% of _____

2. 900 is 34% of _____

3. _____ is 6.75% of $32,000

4. _____ is 8% of $1,200

5. 9,360 is 156% of _____

6. _____ is 29% of 4,483

7. 510 is 6% of _____

8. $14.20 is 0.8% of _____

9. _____ is 1.7% of 3,482.35

10. $1,080 is 108% of _____

11. $115 is 62% of _____

12. _____ is 293% of 308.26

According to a report in a consumer magazine, a person should spend a certain percent of his or her monthly income on housing. Shown below is the monthly cost of housing and the percent that housing costs should be of total monthly income. Find the total income required in each case. Round to the nearest dollar.

| | Monthly Cost of Housing | Percent Housing Cost Is of Total Monthly Income | Total Monthly Income |
|---|---|---|---|
| 13. | $ 800 | 35% | _____ |
| 14. | 900 | 34% | _____ |
| 15. | 1,000 | 33% | _____ |
| 16. | 1,100 | 32% | _____ |
| 17. | 1,200 | 30% | _____ |
| 18. | 1,300 | 29% | _____ |

## BUSINESS APPLICATIONS

Solve the following using the percent formula. Round to the nearest hundredth.

19. Carrie Eike received a $2,160 pay raise, which was $6\frac{3}{4}$% of her salary from last year. Find (a) Carrie's salary last year and (b) Carrie's salary this year.

(a) _____

(b) _____

**20.** Chisco Stork received $3,230 when he sold his coin collection. He received 56% more than what he paid for the collection 6 years ago. How much did he pay for the collection?

_____

**21.** Jim Watts missed 28 questions on the state investment securities exam. This was 20% of the questions asked. How many questions were on the exam?

_____

**22.** Dr. Jarvis Davis contributes 4.25% of his annual income to a retirement annuity. If his salary is $280,300 a year, how much is his yearly annuity contribution?

_____

**23.** Carl McNutt receives 12% of the cost of each craft he sells for Sailcraft, Inc. Carl sold a craft and received $1,925. What was the purchase price of the craft?

_____

**24.** The Cole Brothers Produce Company received 560 crates of vegetables, of which 5% were destroyed due to spoilage. How many crates were destroyed due to spoilage?

_____

**25.** Habhab Tarib sold his company for $1,340,000, which was an increase of $24\frac{1}{4}$% over the purchase price of the company. How much did Habhab originally pay for the company?

_____

**26.** The local teachers' union states that of the 975 teachers in the school district, 60% belong to the teachers' union. Therefore, how many teachers are in the union?

_____

**27.** Business consultant Margaret Peta says that advertising expense should be 8% of sales for a company the size of Bandow, Inc. Last year, advertising expense was $22,948. According to Ms. Peta's guideline, what should sales have been last year?

_____

**28.** Sherri Canton spends $352.35 each month on commuting and parking costs, which is 9% of her total monthly income. What is Sherri's total monthly income?

_____

**29.** Roger Douglas withdrew 75% of his savings to use as a down payment for a new house. After withdrawing the money, Roger had a balance of $4,500 in his account. What was the balance in his account before he withdrew the money?

_____

**30.** After cost-cutting measures, Echo Company's monthly operating expenses were reduced by 20% from last year's operating expense. Currently, monthly expenses are $17,600. What were monthly operating expenses last year?

_____

**31.** Five years ago, Clark Auto Body Company purchased an antique auto to use in promotions. The auto is now worth $22,340, which is an increase of 62% over the purchase price 5 years ago. How much did Clark pay for the auto? Round to the nearest dollar.

_____

**32.** Michelle, Ltd., a clothing store, decreased the price of a fur coat by $270. This was a decrease of 23% of the original price. What was the coat's original price?

_____

**33.** Pamela Schole received a commission of $1,493. If this amount represented 28% of her sales, what were her sales?

_____

**34.** The Tollman Company has 21% of their total investments in real estate. If their real estate is worth $289,400, what is the total worth of their investments? Round to the nearest dollar.

_____

**35.** A recent survey stated that 55% of the residents had college educations. If 180,000 residents of the city have college educations, how many people live in the city? Round to the nearest whole number.

_____

**36.** A new manufacturing machine promises to produce 20% more units per hour than the old model. If 280 units per hour were produced on the old model, how many units are expected to be produced per hour on the new model?

_____

**USING THE PERCENT FORMULA TO FIND THE RATE**

To find the rate when the part and the base are given, divide the part by the base. The circle below illustrates the formula for finding the rate. Percents are usually rounded to the nearest tenth.

$$\text{Rate} = \frac{\text{Part}}{\text{Base}}$$

R =

P
B  ×

---

### EXAMPLE 1

Last month Kelloway Company had total expenses of $45,000. Of the total expenses, $17,000, were for wages. What percent of the total expenses are wages?

**SOLUTION**

**Step 1**  Identify the given elements and the missing element.
The base is $45,000 (the total expenses).
The part is $17,000 (the wages).
The rate is missing.

**Step 2**  Choose the formula that will find the missing element and solve the problem.

$$R = \frac{P}{B} = \frac{\$17,000}{\$45,000} = 0.3777 \approx 37.8\% \text{ Wage expense}$$

Convert to a percent rounded to the nearest tenth

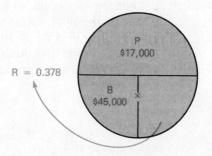

R = 0.378

P
$17,000

B
$45,000  ×

### EXAMPLE 2

Opel Manufacturing built 4,000 preconstructed panels and shipped 2,400 of them. What percent of the panels built still remain to be shipped?

**SOLUTION**

**Step 1**  Identify the given elements and the missing element.
The base is 4,000 (total number of panels).
The part is 1,600 (4,000 − 2,400 = 1,600 remaining panels).
The rate is missing (remaining panels).

**Step 2** Choose the formula that will find the missing element and solve the problem.

$$R = \frac{P}{B} = \frac{1,600}{4,000} = 0.4 \text{ or } 40\% \text{ Remaining panels}$$

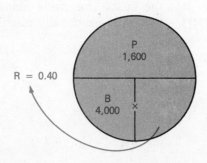

## PRACTICE PROBLEMS

Solve using the percent formula. Round percents to the nearest tenth.

1. 20 is what percent of 70?
2. 29 is what percent of 30?
3. 52 is what percent of 39?
4. 15 is what percent of 112?
5. $153 is what percent of $90?
6. $322 is what percent of $1,200

7. The Retail Marketing Club spent $275 of the $527 in their account on travel expenses for the state convention. What percent of the money in their account did they spend on travel expenses?
8. Elmer Witmire purchased $543 worth of merchandise and charged $296.50 of it using his credit card. What percent of his purchases did he charge on his credit card?
9. Mr. and Mrs. Jolly save $418 per month from their monthly income of $4,180. What percent of their monthly income do they save?
10. Of the 1,675 customers who purchased items from the Card & Party Store last month, 623 were returning customers. What percent of the customers were returning customers?

## FINDING THE PERCENT OF INCREASE OR DECREASE

A manager may want to know the percent of increase in sales from one month to the next, or a customer may want to know the percent of decrease in the cost of an item on sale. To find the percent of increase or decrease, divide the amount of increase or decrease by the original amount. The original amount is always the base. Use the formula, **Rate = Part / Base**.

$$\frac{\text{Rate of Increase}}{\text{or Decrease } (R)} = \frac{\text{Amount of Increase or Decrease } (P)}{\text{Original Amount } (B)}$$

### EXAMPLE 1

This year's sales for the Hemmingway Book Club were $1,300,000. Last year's sales were $1,000,000. What was the percent of increase in sales?

**90** ■ LESSON 3-4

**SOLUTION**

**Step 1** Find the amount of increase.

New Amount − Original Amount = Amount of Increase
  $1,300,000 − $1,000,000 = $300,000

**Step 2** Identify the given element and the missing element.
The base is $1,000,000 (original amount).
The part is $300,000 (amount of increase).
The rate is missing (percent of increase).

**Step 3** Choose the formula that will find the missing element; then solve the problem.

$$R = \frac{P}{B} = \frac{\$300,000}{\$1,000,000} = 0.3 \text{ or } 30\% \text{ Increase in sales}$$

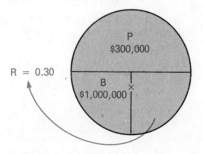

**EXAMPLE 2**

Membership in the Small Business Association this year is 544. Last year the membership was 620. What is the percent of decrease in membership?

**SOLUTION**

**Step 1** Find the amount of decrease.

Original Amount − New Amount = Amount of Decrease
    620 − 544 = 76

**Step 2** Identify the given elements and the missing element.
The base is 620 (original amount).
The part is 76 (amount of decrease).
The rate is missing (percent of decrease).

**Step 3** Choose the formula that will find the missing element and solve the problem.

$$R = \frac{P}{B} = \frac{76}{620} = 0.1225 \text{ or } 12.3\% \text{ Decrease in membership}$$

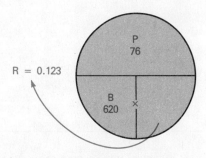

## PRACTICE EXERCISES

Solve using the percent formula. Round percents to the nearest tenth.

1. This year utility expenses for the Bi-Tex Company are $12,400 compared to utility expenses of $11,100 last year. What is the rate of increase in the company's utility expenses?
2. What is the rate of decrease in value of inventory if this year's inventory is valued at $113,700 and last year's inventory was valued at $115,850?
3. Gordon & Sons purchased an office building for $954,600 and later sold it for $1,025,000. What was the percent of increase made on selling the building?
4. A personal computer was reduced for sale from $1,295 to $825. What was the rate of decrease on the original selling price?
5. The Personal Companion Mail Order Company increased their selling territory from 32 states to 47 states. What was the percent of increase?

## LESSON 3-4     EXERCISES

Use the percent formula to solve the following. Round percents to the nearest tenth and amounts to the nearest hundredth.

1. $5,220 is _____% of $4,640

2. _____ is 43.5% of 72

3. $246 is _____% of $100

4. 14 is 13.1% of _____

5. $32.20 is _____% of $670.30

6. 156 is _____% of 185.5

7. 2,670 is 89% of _____

8. _____ is 1% of $985

9. 80 is _____% of 2,000

10. 79 is _____% of 50

11. $89.45 is _____% of $230

12. 600 is 60% of _____

Find the percent credit card sales are of total sales for the following months for the Patio Restaurant. Round percents to the nearest tenth.

| Month | Credit Card Sales | Total Sales | Percent Credit Card Sales Are of Total Sales |
|---|---|---|---|
| 13. January | $15,000 | $40,000 | _____ |
| 14. February | 12,000 | 36,000 | _____ |
| 15. March | 11,000 | 44,000 | _____ |
| 16. April | 26,000 | 52,000 | _____ |
| 17. May | 21,000 | 45,000 | _____ |
| 18. June | 20,000 | 46,000 | _____ |

Find the amount and percent of increase for each of the following expenses from year 19X1 to 19X2. Show decrease amounts and percents in parentheses. Round percents to the nearest tenth.

| Expenses | 19X1 | 19X2 | Amount of Increase (or Decrease) | Rate of Increase (or Decrease) |
|---|---|---|---|---|
| 19. Advertising | $ 216,240 | $ 195,670 | _____ | _____ |
| 20. Vehicles | 21,780 | 27,500 | _____ | _____ |
| 21. Bad debts | 14,290 | 10,930 | _____ | _____ |
| 22. Depreciation | 278,320 | 289,340 | _____ | _____ |
| 23. Labor | 670,340 | 703,390 | _____ | _____ |
| 24. Rent | 62,593 | 73,450 | _____ | _____ |
| 25. Telephone | 18,935 | 15,980 | _____ | _____ |
| 26. Utilities | 19,476 | 22,833 | _____ | _____ |
| 27. Miscellaneous | 10,950 | 8,459 | _____ | _____ |
| 28. Total expenses | 1,312,924 | 1,347,552 | _____ | _____ |

## BUSINESS APPLICATIONS

Solve the following using the percent formula. Round amounts to the nearest hundredth and percents to the nearest tenth.

**29.** Rent expense for the Jarvell Advertising Company for the month of October was $1,450. What percent of the total operating expenses of $6,040 is the rent expense?

_____

**30.** Six years ago the Albee Corporation had 114 employees. The Albee Corporation now has 258 employees. What is the percent of increase in employees?

_____

**31.** Cunningham Jewelry had total sales of $29,840 last month. Watches accounted for $10,320 of the total sales. What percent of total sales is watch sales?

_____

**32.** Last year the Lowerys rented their apartment for $895 a month. The rent has now increased to $1,030 a month. What is the percent of increase in the rent?

_____

**33.** Mae Lietz had actual sales of $45,290 last month. Her sales goal for the month was $35,600. What percent of the sales goal are the actual sales?

_____

**34.** Sycamore Investment Syndicate purchased a retail store complex for $365,500. The estimated value of the retail complex has increased to $450,000. What is the percent of increase in the value of the complex?

_____

**35.** Karen Patterson earns $32,650 a year. Her sister Sally earns $46,330 a year. Karen's earnings are what percent of Sally's earnings?

_____

**36.** The Baltimore Bearing Company increased its production by 24% over the previous year after installing new machinery. If this year's production was 3,489 units, what was last year's production? Round to the nearest whole unit.

_____

**37.** The fund-raising goal of the community fund for last year was $125,000. The actual amount raised was $123,500. What percent of the goal was reached?

_____

**38.** Of the 1,750 students who took the college placement test, 60% passed. Of the 60% who passed, 40% enrolled in college. How many students enrolled in college?

_____

**39.** Joy Draimer assembled 1,138 of the 1,150 units to be assembled. What percent of the total units did she assemble?

_____

**40.** On Tuesday gold sold for $430.75 per ounce. On Friday gold sold for $415.60 per ounce. What is the percent of decrease in price from Tuesday to Friday?

**41.** In January the value of inventory for the Tallyho Company was $11,600. In February the value of inventory was $12,400. The value of the February inventory was what percent of the value of the January inventory?

**42.** Ralph and Joanne budgeted $675 for air fare, $455 for hotel accommodations, and $532 for meals for their vacation. Ralph and Joanne find that 25% of their travel budget remains for shopping and entertainment. How much remains for shopping and entertainment?

**43.** The regular retail price for a car stereo at Stanley's Auto Accessories is $999. On sale, the stereo sells for $749. The sales price is what percent of the retail price?

**44.** Last year the price of an airline ticket from Memphis to New York City was $286.50. This year the same flight costs $254.70. What is the percent of decrease in the cost of the flight?

# CHAPTER 4

# PAYROLL

## LEARNING OBJECTIVES

**1.** Calculate gross pay using the hourly wage, piecework, salary, and commission methods.

**2.** Determine payroll deductions to calculate net pay.

**3.** Complete a payroll register.

A payroll is a list prepared by an employer showing names of employees and the amounts to be paid to each for work they have performed.

There are different methods used to calculate employee earnings including wages, piecework, salary, commission, and combinations of these plans. In this chapter you will learn how to calculate an employee's total earnings using all these methods. You will also learn how to determine employees' take-home net pay after calculating income taxes and other deductions.

```
                        Karman - Gordan, Inc.

                        Weekly Payroll Register

For the Week Ending: October 17, 19--
###############################################################################
                       GROSS  EARNINGS                      DEDUCTIONS

                                                      FED.
                 HOURLY            OVER    GROSS   FICA  INC.          TOTAL   NET
EMPLOYEE  HOURS  RATE     REG.     TIME    PAY     TAX   TAX    OTHER  DED.    PAY
---------------------------------------------------------------------------------------
T. Cass    40    $10.75  $430.00           $430.00 $32.29 $47.00 $5.50  $84.79  $345.21
J. Kesi    43    $9.65   $386.00  $43.43   $429.43 $32.25 $45.00        $77.25  $352.18
M. Ortez   40    $11.15  $446.00           $446.00 $33.49 $54.00 $10.00 $97.49  $348.51
O. Parks   45    $10.25  $410.00  $76.88   $486.88 $36.56 $60.00 $8.00  $104.56 $382.32
L. Ryan    38    $9.60   $364.80           $364.80 $27.40 $31.00 $5.50  $63.90  $300.90
D. Scott   41    $10.40  $416.00  $15.60   $431.60 $32.41 $41.00        $73.41  $358.19
J. Valen   40    $10.15  $406.00           $406.00 $30.49 $48.00 $10.00 $88.49  $317.51
```

Today, many businesses use a computer spreadsheet like that shown above to maintain the payroll.

Factory workers, office personnel, and retail store clerks are usually paid a specific dollar amount per hour worked. In some production industries, workers are paid per acceptable item produced. In this lesson, you will be taught these two methods of paying these employees.

Total amount earned by any employee is called **gross pay.** Gross pay, however, is not the amount paid to them. Deductions may be subtracted from gross pay to determine **net pay,** the amount actually received by the employee.

**WAGES**

The earnings of employees paid at an hourly rate are called **wages.** Employees are paid a **regular rate** for a predetermined number of **regular hours** worked per week, such as 40 or $37\frac{1}{2}$. Hours worked beyond the regular hours are called **overtime hours** and are paid at an **overtime rate.** Usually, the overtime rate is $1\frac{1}{2}$ times the regular rate and called **time and a half.** Payment for work performed on Sundays and holidays is usually at a higher rate, often double the regular hourly rate and called **double time.** Only the final answer is rounded. To find the **regular pay,** multiply the regular hours worked by the regular rate. To find the **overtime pay,** multiply the overtime hours by the regular rate by the overtime factor. For example the overtime factor for time and a half is $1\frac{1}{2}$ or 1.5. The gross pay is regular pay plus the overtime pay. If there is no overtime, the gross pay is equal to the regular pay.

### EXAMPLE 1

Beth Hastings, a clothing store cashier, earns $6.35 per hour and works regular hours of 40 hours per week with no overtime. What is her gross pay for the week?

### SOLUTION

Regular Hours × Regular Rate = Gross Pay ← Regular pay = gross pay
    40        ×     $6.35    =  $254.00

### EXAMPLE 2

Gregg Allen, a production worker at Paxton Electronics, is paid $9.75 per hour and time and a half for hours worked in excess of 40 hours in a week. Last week, Gregg worked 43 hours. What was his gross pay?

### SOLUTION

**Step 1**  Calculate the regular pay.

Regular Hours × Regular Rate = Regular Pay
    40        ×     $9.75    =  $390.00

**Step 2**  Calculate the overtime pay.

Overtime Hours × Regular Rate × Overtime Factor = Overtime Pay
      3         ×     $9.75    ×       1.5        = $43.875 ≈ $43.88

**Step 3**   Calculate gross pay.

Regular Pay + Overtime Pay = Gross Pay
  $390.00   +   $43.88   = $433.88

## EXAMPLE 3

Mary Higgins is paid $5.75 an hour for a regular 40-hour week and time and a half for hours worked over 40. She worked the following hours from Monday to Friday respectively: $10\frac{1}{2}$, $8\frac{3}{4}$, $9\frac{1}{4}$, 11, 10 hours. Find her gross pay.

## SOLUTION

**Step 1**   Calculate total hours worked.

$10\frac{1}{2} + 8\frac{3}{4} + 9\frac{1}{4} + 11 + 10 = 49\frac{1}{2}$

**Step 2**   Calculate the regular pay.

Regular Hours × Regular Rate = Regular Pay
    40        ×     $5.75    =   $230.00

**Step 3**   Calculate the overtime pay.

$49\frac{1}{2} - 40 = 9\frac{1}{2} = 9.5$ ⟶

| | | | Overtime | | |
|---|---|---|---|---|---|
| Overtime Hours | × | Regular Rate | × | Factor | = Overtime Pay |
| 9.5 | × | $5.75 | × | 1.5 | = $81.9375 ≈ $81.94 |

**Step 4**   Calculate gross pay.

Regular Pay + Overtime Pay = Gross Pay
  $230.00   +   $81.94    =  $311.94

---

Certain industries pay time and a half for any hours worked over eight hours in any one day, regardless of how many hours are worked in a week. This is called the ***overtime-hours-per-day method.***

As you can see in the following example, the overtime-hours-per-day method may pay employees more than they may have earned if the other overtime pay method was used. Industry-wide tradition usually determines which of the overtime methods is used.

---

## EXAMPLE

Bill Wheeler worked the following hours last week. He is paid a regular rate of $10.25 an hour for a regular 8-hour day. The overtime rate is time and a half on the overtime-hours-per-day method. Find his gross pay for the week.

| Day | Hours Worked | Regular Hours | Overtime Hours |
|---|---|---|---|
| Monday | $8\frac{1}{2}$ | | $\frac{1}{2}$ |
| Tuesday | 5 | | 0 |
| Wednesday | 8 | 8 | 0 |
| Thursday | 10 | 8 | 2 |
| Friday | 9 | 8 | 1 |
| Total | $40\frac{1}{2}$ | 37 | $3\frac{1}{2}$ |

## SOLUTION

**Step 1** Calculate the regular pay.

Regular Hours × Regular Rate = Regular Pay
    37     ×     $10.25     =     $379.25

**Step 2** Calculate the overtime pay.

Overtime    Regular    Overtime      Overtime
  Hours  ×   Rate  ×  Factor  =     Pay
   3.5   × $10.25 ×   1.5   = $53.8125 ≈ $53.81

**Step 3** Calculate gross pay.

Regular Pay + Overtime Pay = Gross Pay
   $379.25   +     $53.81     =   $433.06

## PRACTICE PROBLEMS

Using the overtime-hours-per-day method, find the regular hours, overtime hours, regular pay, overtime pay, and gross pay for the following employees. Each employee is paid a regular rate of $10.50.

| | Hours Worked | | | | | Regular Hours | Overtime Hours | Regular Pay | Overtime Pay | Gross Pay |
|---|---|---|---|---|---|---|---|---|---|---|
| Employee | M | T | W | Th | F | | | | | |
| **1.** D. Gore | 9 | 6 | 8 | 10 | 8 | 38 | 3 | 399 | 47.25 | 446.25 |
| **2.** T. Graham | 7½ | 9 | 11 | 5 | 8½ | 36.5 | 4.5 | 383.25 | 70.88 | 454.13 |
| **3.** C. Harthief | 8 | 8 | 10½ | 9 | 7 | 39 | 3.5 | | | |
| **4.** T. Reisis | 6 | 9 | 7 | 9 | 10 | | | | | |
| **5.** T. Stewart | 8 | 7½ | 8½ | 7 | 9 | | | | | |

Besides these two methods of paying overtime, some companies pay **double time** for weekend and holiday shifts. Industries that operate on a twenty-four-hour-a-day basis often pay **shift differentials** to attract workers to less desirable hours. All-night restaurants and public utilities, for instance, traditionally offer shift differential for the swing shift (4 p.m. to midnight) and the graveyard shift (midnight to 8 a.m.).

Employees who work staggered hours, that is, work only during peak periods of operation, are paid a **split-shift premium.** Such employees may work four hours on, four off, and four on again.

Instead of overtime pay, employees sometimes are offered **compensatory time** (comp. time). Such employees are given time off their regular work hours as compensation for overtime hours worked. Comp. time is usually 1½ hours for every 1 hour of overtime worked.

ANSWERS FOR
PRACTICE PROBLEMS

| | Regular Hours | Overtime Hours | Regular Pay | Overtime Pay | Gross Pay |
|---|---|---|---|---|---|
| **(1)** | 38 | 3 | $399.00 | $47.25 | $446.25 |
| **(2)** | 36½ | 4½ | 383.25 | 70.88 | 454.13 |
| **(3)** | 39 | 3½ | 409.50 | 55.13 | 464.63 |
| **(4)** | 37 | 4 | 388.50 | 63.00 | 451.50 |
| **(5)** | 38½ | 1½ | 404.25 | 23.63 | 427.88 |

## PRACTICE PROBLEMS

Calculate the gross pay for each employee. Overtime is $1\frac{1}{2}$ times the regular rate for hours worked in excess of 40 hours in a week. Round gross pay to the nearest cent.

| Employee | Hours | Regular Rate | Regular Pay | Overtime Pay | Gross Pay |
|---|---|---|---|---|---|
| 1. Tim Manly | 40 | $8.75 | | | |
| 2. Donna Smith | 41 | 9.00 | 360 | 13.50 | |
| 3. John Roushe | 38 | 7.50 | | | |
| 4. Jay Anderson | 42 | 8.75 | | | |
| 5. Tammy Lassiter | 45 | 9.25 | | | |

6. Carol Duncan receives $5.10 per hour and time and a half for all time worked over 8 hours on any day. Last week her hours were as follows: Monday, $8\frac{1}{2}$; Tuesday, 8; Wednesday, $9\frac{3}{4}$; Thursday, $5\frac{1}{2}$; and Friday, $7\frac{1}{2}$ hours. Find her regular pay, overtime pay, and gross pay for the week.

## STRAIGHT PIECEWORK PLAN

Some production workers are paid a specific rate for each piece produced. This is known as the *piecework plan.* It is important to note that employees are paid only for units produced that are of acceptable quality.

Under the straight piecework plan, the worker is paid the same rate for each item produced. Multiply the number of items produced by the rate per item to find gross pay.

Number of Items $\times$ Rate per Item = Gross Pay

### EXAMPLE

Angie Kline receives $0.80 for each lamp she assembles at Lamplighter Company. Yesterday she assembled 67 lamps. What was her gross pay?

### SOLUTION

Number Produced $\times$ Rate = Gross Pay
67 $\times$ $0.80 = $53.60

| | Regular Pay | Overtime Pay | Gross Pay |
|---|---|---|---|
| ANSWERS FOR PRACTICE PROBLEMS | | | |
| (1) | $350.00 | $ 0 | $350.00 |
| (2) | 360.00 | 13.50 | 373.50 |
| (3) | 285.00 | 0 | 285.00 |
| (4) | 350.00 | 26.25 | 376.25 |
| (5) | 370.00 | 69.38 | 439.38 |
| (6) | 188.70 | 17.21 | 205.91 |

Determine the gross pay for each of these employees using the straight piecework plan.

| Employee | Number Produced | Rate per Item | Gross Pay |
|---|---|---|---|
| 1. Ernie Freeman | 637 | $0.10 | _____ |
| 2. Mavis VanDaalen | 518 | 0.17 | _____ |
| 3. Joan Bach | 76 | 0.92 | _____ |
| 4. Carla Allen | 183 | 0.41 | _____ |
| 5. Dale Bates | 103 | 0.73 | _____ |

## DIFFERENTIAL PIECEWORK PLAN

As an incentive to encourage high production, some businesses pay employees on a *differential piecework plan.* Under this plan, employees are paid an increasingly higher rate as the number of items produced increases. To find the gross pay using this method, find the amount earned at each rate, then find the sum of the amounts.

### EXAMPLE

Using the scale shown below, Rally, Inc., pays employees who assemble midget race car instrument panels on a differential piecework plan. On Tuesday Madeline Ames assembled 41 acceptable panels. What was her gross pay for the day?

#### DIFFERENTIAL PIECEWORK SCALE

| Daily Items Produced | Rate |
|---|---|
| 1–10 | $1.25 |
| 11–20 | 1.50 |
| 21–30 | 1.75 |
| Over 30 | 2.00 |

### SOLUTION

Determine number produced at each rate and calculate gross pay.

Number Produced × Differential Rate = Amount Earned

| Number Produced | | Differential Rate | | Amount Earned |
|---|---|---|---|---|
| 10 | × | $1.25 | = | $12.50 |
| 10 | × | 1.50 | = | 15.00 |
| 10 | × | 1.75 | = | 17.50 |
| 11 | × | 2.00 | = | 22.00 |

Sum of amounts is
$67.00 ← the gross pay.

(1) $63.70   (2) $88.06   (3) $69.92   (4) $75.03   (5) $75.19

I See you
Avatar

## PRACTICE PROBLEMS

Find each employee's gross pay using the differential piecework scale shown below.

**DIFFERENTIAL PIECEWORK SCALE**

| Items Produced | Rate |
|---|---|
| 1–50 | $0.11 |
| 51–100 | 0.13 |
| 101–150 | 0.17 |
| 151–200 | 0.20 |
| Over 200 | 0.25 |

| Employee | Number Produced | Gross Pay |
|---|---|---|
| 1. Jean Bixby | 230 | 38 |
| 2. George Wills | 214 | _____ |
| 3. Coleen Orr | 187 | _____ |
| 4. Steve Speer | 280 | _____ |
| 5. Marion Sadler | 310 | _____ |
| 6. Carl Davis | 415 | _____ |
| 7. Dale Evers | 256 | _____ |
| 8. Kathy Jacobs | 440 | _____ |
| 9. Barb Hoien | 325 | _____ |
| 10. Cindy Grems | 385 | _____ |

## LESSON 4-1    EXERCISES

Calculate the regular pay, overtime pay, and gross pay for each employee. Overtime is $1\frac{1}{2}$ times the regular rate for hours worked in excess of 40 hours in a week.

| Employee | Hours | Regular Rate | Regular Pay | Overtime Pay | Gross Pay |
|---|---|---|---|---|---|
| 1. Shelly Davis | 40 | $ 8.50 | 340 | 0 | 340 |
| 2. Al Everett | 42 | 8.95 | 358 | 26.85 | 384.85 |
| 3. John Frank | 38 | 12.50 | 475 | 0 | 475 |
| 4. Jan Griffin | 43 | 12.35 | 494 | 55.58 | 549.58 |
| 5. Bill Hakes | 45 | 9.65 | 386 | 72.38 | 458.38 |

Using the overtime-hours-per-day method, find regular hours, overtime hours, regular pay, overtime pay, and gross pay for the following employees. Each employee is paid a regular rate of $11.50.

| Employee | M | T | W | Th | F | Regular Hours | Overtime Hours | Regular Pay | Overtime Pay | Gross Pay |
|---|---|---|---|---|---|---|---|---|---|---|
| 6. Tom Adams | 9 | 6 | 9 | 9 | 8 | 38 | 3 | 437 | 51.75 | $488.75 |
| 7. Joanne Bates | 7 | 8 | $10\frac{1}{2}$ | $9\frac{1}{2}$ | 8 | 39 | 4 | 448.50 | 69 | $517.50 |
| 8. Don Carlyle | 8 | 9 | $7\frac{1}{2}$ | 10 | 8 | 39.5 | 3 | 454.25 | 51.75 | $506.00 |

Determine the gross pay for each of the following employees using the straight piecework plan.

| Employee | Number Produced | Rate per Item | Gross Pay |
|---|---|---|---|
| 9. Donna Gates | 185 | $0.55 | 101.75 |
| 10. Beth Hanson | 1,560 | 0.06 | 93.60 |
| 11. Carl Jackson | 346 | 0.42 | 145.32 |
| 12. Diane Learner | 75 | 0.90 | 67.50 |
| 13. Ted Madden | 390 | 0.26 | 101.40 |

Find each employee's gross pay using the differential piecework scale shown below.

**DIFFERENTIAL PIECEWORK SCALE**

| Daily Items Produced | Rate |
|---|---|
| 1–100 | $0.08 |
| 101–200 | 0.10 |
| 201–300 | 0.12 |
| Over 300 | 0.16 |

| Employee | Number Produced | Gross Pay |
|----------|-----------------|-----------|
| **14.** Kay Aljets | 518 | _____ |
| **15.** Jill Bonjovi | 624 | _____ |
| **16.** Sid Clark | 410 | _____ |
| **17.** Bev Dees | 486 | _____ |
| **18.** Paul Eenen | 285 | _____ |

## BUSINESS APPLICATIONS

**19.** Judy Wilson earns $8 per hour and is paid time and a half for hours worked over 40 in a week. Last week Judy worked 48 hours. What percent of Judy's gross pay for the week came from overtime?

$40 \times 8 = 320$      $320 + 96 = 416$

$8 \times 8 \times 1.5 = 96$

$8 \times 1.5 = 12$

$96 \div 416 = 0.23076$

 23.08

**20.** Last week Fred Dalton worked the following hours: Monday, $9\frac{1}{2}$; Tuesday, 6; Wednesday, 8; Thursday, 10; and Friday, 11. His regular rate of pay is $9.35 an hour for a regular 8-hour day. For overtime, he is paid time and a half on the overtime-hours-per-day method. Find his gross pay for the week.

$38 \times 9.35 = 355.30$

$6.5 \times 9.35 \times 1.5 = 91.16$

$355.30$
$+ 91.16$
$=$

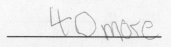 $=$   446.46

**21.** Bill Roede is paid $15 for each swing set he assembles at Johnson Hardware. Last week Bill assembled 20 swing sets in a 40-hour week. How much more or less was Bill's gross pay than if he had been paid $6.50 per hour?

$15 \times 20 = 300$

$6.50 \times 40 = 260$

$300 - 260 = 40$

_____ 40 more

**22.** Krista Mergen is paid $2.20 for every blouse she sews at LaRue Clothing Manufacturing Company. Last week in 40 hours Krista sewed 115 blouses. What was Krista's average gross pay per hour for the week?

$$253 \div 40 = 6.33$$

$^\$6.33$ per hour

**23.** Yesterday Carla Michaels attached 192 emblems to sailboats, for which she was paid a straight piecework rate of $0.41 each. How much more or less would she have earned if the following differential piecework scale had been used?

$$0.41 \times 192 = 78.72$$

**DIFFERENTIAL PIECEWORK SCALE**

| Items Produced | Rate |
| --- | --- |
| 1–50 | $0.25 |
| 51–100 | 0.30 |
| 101–150 | 0.40 |
| Over 150 | 0.60 |

$$50 \times 0.25 = 12.50$$
$$50 \times 0.30 = 15$$
$$50 \times 0.40 = 20$$
$$42 \times 0.60 = 25.20 + = 72.70$$

PW 72.70
hw 78.72

$^\$6.02$ Less

**24.** Dean Edwards is paid a regular rate of $12.30 an hour for an 8-hour day. His overtime rate is time and a half paid on the overtime-hours-per-day method. For the previous week, his hours worked were as follows: Monday, 6; Tuesday, 7; Wednesday, $8\frac{1}{2}$; Thursday, 9; and Friday, 10. Find his gross pay for the week.

Rh) $37 \times 12.30 = 455.10$
Oh) $3.5 \times 12.30 \times 1.5 = 64.58$

519.68

519.68

**25.** George Henderson is paid an overtime rate of time and a half using the overtime-hours-per-day method. On Monday he worked 9 hours; on Tuesday, 10 hours; on Wednesday, 8 hours, on Thursday, 7 hours; and on Friday, $9\frac{1}{2}$ hours. His regular rate of pay is $8.35 an hour for a regular 8-hour day. Find his weekly gross pay.

$39 \times 8.35 = 325.65$

$4.5 \times 8.35 \times 1.5 = 56.36 +$

$382.01$

_382.01_

**26.** Johnston Fishing Equipment Company pays Carla Evers $6.50 per hour for a 40-hour week tying fishing lures. Johnston is considering paying Carla on a piecework basis, paying $0.80 for every lure she ties. How many fishing lures must Carla tie in a week to equal her current gross pay?

$40 \times 6.50 = 260$

$260 \div 0.80 = 325$

_325 lures_

**27.** Phillips Company has a special direct mail project that will require assembling, folding, and inserting in envelopes 60,000 mail pieces. Through time trials, Phillips's cost accountant has determined that a person should be able to prepare four mailings per minute. Phillips's goal is to pay the temporary workers hired to assemble the mailings at a piecework rate that will equal earnings of $6.00 per hour. (a) Based on the cost accountant's estimates, how much should Phillips pay for each mailing piece? (b) What will be Phillips's total cost for processing this mailing? (c) Approximately how many hours will it take to prepare the mailing?

(a) _____

(b) _____

(c) _____

Managers, supervisors, and other professionals are generally paid a flat amount per month or year. Salespeople often are paid a percent of the dollar amount of their sales. Still other managers and salespeople are paid on some combination or variation of these two methods.

**SALARY**

A *salary* is a predetermined, set amount paid to an employee. Ordinarily, a salary is stated as an annual amount and is paid in equal pay periods. *Pay periods* can be weekly, biweekly, semimonthly, or monthly. To find the amount of salary paid each pay period, divide the annual salary by the number of pay periods per year. This amount is the gross pay for that particular pay period. The table below shows the number of pay periods per year.

Annual Salary   Pay Periods   Gross Pay per
 (Gross Pay)  ÷  per Year  =  Pay Period

Salary per   Pay Periods   Annual
Pay Period ×  per Year  =  Salary

| Pay Period | Paychecks per Year |
|---|---|
| Monthly | 12 |
| Semimonthly (twice per month) | 24 |
| Biweekly (every 2 weeks) | 26 |
| Weekly | 52 |

---

**EXAMPLE 1**

Sheila Mason receives an annual salary of $22,200 per year. What is her gross pay per month?

**SOLUTION**

               Pay Periods      Gross Pay
Annual Salary ÷  per Year  = per Pay Period
   $22,200   ÷     12    =    $1,850

**EXAMPLE 2**

Allan Price receives a biweekly salary of $1,250. What is Allan's annual gross pay?

**SOLUTION**

Biweekly   Pay Periods   Annual
 Salary  ×  per Year  =  Salary
 $1,250  ×     26    = $32,500

---

Annual sales for supervisory and management personnel at Gateway Public Relations, Inc., are shown below. Convert each annual salary to the amount earned for other pay periods. Round amounts to the nearest dollar.

| | Employee | Annual Salary | Monthly Salary | Semi-monthly Salary | Biweekly Salary | Weekly Salary |
|---|---|---|---|---|---|---|
| 1. | David McGovern | $ 38,520 | _____ | _____ | _____ | _____ |
| 2. | Betty Sims | 42,000 | _____ | _____ | _____ | _____ |
| 3. | Alex Johnson | 48,000 | _____ | _____ | _____ | _____ |
| 4. | George Willis | 66,000 | _____ | _____ | _____ | _____ |
| 5. | Paula Nobel | 120,000 | _____ | _____ | _____ | _____ |

Convert each of the following to annual salaries.

| | Employee | Pay Period | Salary | Annual Salary |
|---|---|---|---|---|
| 6. | Shelly Crees | Weekly | $ 315.50 | _____ |
| 7. | Shawn Budach | Biweekly | 814.20 | _____ |
| 8. | Terri Lohman | Monthly | 2,180.50 | _____ |
| 9. | Ali Hilburn | Semimonthly | 1,975.25 | _____ |
| 10. | Dee Andrews | Weekly | 605.00 | _____ |

**SALARY PLUS BONUS**

Often those in management who have direct responsibility for sales volume or the company's overall profitability receive a **bonus,** an additional amount based on performance. This bonus, calculated as a percent of sales made or profit earned above a certain amount, is called a **quota.** The gross pay is the sum of the salary plus the bonus.

Salary + Bonus = Gross Pay

---

**ANSWERS FOR PRACTICE PROBLEMS**

| | Monthly Salary | Semimonthly Salary | Biweekly Salary | Weekly Salary |
|---|---|---|---|---|
| (1) | $ 3,210 | $1,605 | $1,482 | $ 741 |
| (2) | 3,500 | 1,750 | 1,615 | 808 |
| (3) | 4,000 | 2,000 | 1,846 | 923 |
| (4) | 5,500 | 2,750 | 2,538 | 1,269 |
| (5) | 10,000 | 5,000 | 4,615 | 2,308 |

(6) $16,406.00  (7) $21,169.20  (8) $26,166.00  (9) $47,406.00
(10) $31,460.00

## EXAMPLE

The sales manager at Entertainment Age, Inc., Harley Comstock, receives an annual salary of $50,000 and a bonus of 1% on all sales over $20,000,000. Last year sales were $21,540,375. What was his gross pay for the year?

## SOLUTION

**Step 1**  Calculate sales above quota.

Total Sales − Quota = Sales Above Quota

$21,540,375 − $20,000,000 = $1,540,375

**Step 2**  Calculate the bonus.

Sales Above Quota × Bonus Rate = Bonus

$1,540,375 × 0.01 = $15,403.75

**Step 3**  Calculate gross pay.

Salary + Bonus = Gross Pay

$50,000.00 + $15,403.75 = $65,403.75

## PRACTICE PROBLEMS

Calculate the bonus and gross pay for each of the following district managers who are paid a salary and a bonus on sales above their quota.

| District | Salary | Sales Quota | Actual Sales | Bonus Rate | Bonus | Gross Pay |
|---|---|---|---|---|---|---|
| 1. Spokane | $25,000 | $4,000,000 | $4,940,300 | 2% | _____ | _____ |
| 2. Trenton | 22,500 | 3,700,000 | 4,100,976 | 2% | _____ | _____ |
| 3. Memphis | 27,500 | 4,100,000 | 3,875,631 | 2% | _____ | _____ |
| 4. St. Paul | 25,000 | 4,000,000 | 4,000,000 | 2% | _____ | _____ |
| 5. Eugene | 30,000 | 4,200,000 | 4,760,428 | 2% | _____ | _____ |

## COMMISSION

Salespersons are often paid on a *commission* basis, whereby earnings are calculated as a percent of sales made. Some salespeople are paid strictly by commission, while others receive a base salary plus commission.

Commissions based on sales motivate salespeople to sell at maximum levels since the more one sells, the more one earns.

ANSWERS FOR
PRACTICE PROBLEMS

| | Bonus | Gross Pay |
|---|---|---|
| (1) | $18,806.00 | $43,806.00 |
| (2) | 8,019.52 | 30,519.52 |
| (3) | 0 | 27,500.00 |
| (4) | 0 | 25,000.00 |
| (5) | 11,208.56 | 41,208.56 |

## STRAIGHT COMMISSION

Under a *straight-commission plan*, gross pay is based entirely upon the amount of sales. To find the gross pay, the amount of sales is multiplied by the *commission rate*, which is usually stated as a percent. In retail operations, customer returns and allowances (discounts, etc.) usually are deducted from total sales to find *net sales* before commissions are determined.

Sales × Commission Rate = Commission (Gross Pay)

### EXAMPLE 1

Last month Jerry Abbott's sales were $90,565, on which he received a commission of $3\frac{1}{4}$ percent. What was Jerry's gross pay? Round to the nearest cent.

### SOLUTION

Sales × Commission Rate = Gross Pay

$90,565 × 0.0325 = $2,943.362 ≈ $2,943.36

### EXAMPLE 2

During March Ann Garland had total sales of $33,180, with returns and allowances of $3,800. Her rate of commission is 13%. Find Ann's gross pay.

### SOLUTION

**Step 1** Deduct returns and allowances from total sales.

Total Sales − Returns and Allowances = Net Sales

$33,180 − $3,800 = $29,380

**Step 2** Find the gross pay.

Net Sales × Commission Rate = Gross Pay

$29,380 × 0.13 = $3,819.40

## PRACTICE PROBLEMS

Calculate the monthly gross pay for each salesperson. Round to the nearest cent.

| | Salesperson | Sales | Commission Rate | Gross Pay |
|---|---|---|---|---|
| 1. | Leroy Barret | $125,642 | 3.0% | _____ |
| 2. | Stan Towers | 46,904 | 2.5% | _____ |
| 3. | Sam Freeland | 80,018 | 2.5% | _____ |

| | Salesperson | Sales | Returns and Allowances | Commission Rate | Gross Pay |
|---|---|---|---|---|---|
| 4. | Joyce Baker | $ 79,407 | $ 3,484 | 3.0% | _____ |
| 5. | Jessie Germain | 156,137 | 12,396 | 2.2% | _____ |

## GRADUATED COMMISSION

A *graduated commission scale* provides an increasingly higher commission rate as sales volume increases. With the graduated commission scale shown below for Lakeshore Furniture, a salesperson receives an 8% commission on the first $50,000 of sales, a 10% commission on the next $50,000 of sales, a 12% commission on the next $50,000 of sales, and a 15% commission on all sales above $150,000. To find gross pay, add the amount of commission earned at each commission rate. Companies use this system to encourage sales people to put forth a good effort and to reward those salespeople with high productivity.

### LAKESHORE FURNITURE: GRADUATED COMMISSION SCALE

| Level | Sales | Commission Rate |
|-------|-------|-----------------|
| 1 | Up to $ 50,000 | 8% |
| 2 | 50,001– 100,000 | 10% |
| 3 | 100,001– 150,000 | 12% |
| 4 | Over 150,000 | 15% |

### EXAMPLE

Kathy Galvin is a salesperson for Lakeshore Furniture. She is paid on the graduated commission scale shown above. Last year Kathy's sales were $220,365. What was Kathy's annual gross pay?

### SOLUTION

Find the commission earned at each level.

| Level | Sales | × | Commission Rate | = | Commission Earned |
|-------|-------|---|------|---|-------------------|
| 1 | $50,000 | × | 0.08 | = | $ 4,000.00 |
| 2 | 50,000 | × | 0.10 | = | 5,000.00 |
| 3 | 50,000 | × | 0.12 | = | 6,000.00 |
| 4 | 70,365 | × | 0.15 | = | 10,554.75 |

$25,554.75 Gross pay

### PRACTICE PROBLEMS

Use the graduated commission scale shown above for Lakeshore Furniture to calculate the gross pay earned by each of the following salespeople.

| Salesperson | Sales | Gross Pay |
|-------------|-------|-----------|
| 1. Garry Hammond | $212,631 | _____ |
| 2. Bev Madson | 251,096 | _____ |
| 3. Gloria Trent | 187,448 | _____ |
| 4. Sam Bixby | 110,316 | _____ |
| 5. Pat Elwood | 347,874 | _____ |

ANSWERS FOR
PRACTICE PROBLEMS   (1) $24,394.65   (2) $30,164.40   (3) $20,617.20   (4) $10,237.92   (5) $44,681.10

## SALARY PLUS COMMISSION

Under the **salary-plus-commission** arrangement, the salesperson receives a guaranteed salary plus a commission on total sales or on sales over a specified amount. The base salary assures employees of a regular income, and the commission is an incentive to produce more sales.

Sometimes base salary is considered a loan against future commissions. In such situations, salary is called a **draw.** To calculate gross pay for a given pay period, first calculate the commission, then subtract the draw.

### EXAMPLE 1

Marilyn Winthers works as a sales representative at Maxim Carpeting. She receives a base salary of $1,000 a month plus a commission of 3 percent on total sales. Last month her sales amounted to $25,316. What was her gross pay?

### SOLUTION

**Step 1** Calculate the commission.

Sales $\times$ Commission Rate = Commission
$25,316 $\times$ 0.03 = $759.48

**Step 2** Calculate the gross pay.

Base Salary + Commission = Gross Pay
$1,000 + $759.48 = $1,759.48

### EXAMPLE 2

Kate Garvin produced April sales totaling $57,120. She receives a 6% commission against a monthly draw of $1,500. Calculate her gross pay after the drawing account is subtracted.

### SOLUTION

**Step 1** Calculate the total commissions.

Total Sales $\times$ Commission Rate = Total Commission
$57,120 $\times$ 0.06 = $3,427.20

**Step 2** Deduct draw from total commission.

Total Commission $-$ Draw = Gross Pay
$3,427.20 $-$ $1,500 = $1,927.20

### PRACTICE PROBLEMS

Find the gross pay for each salesperson. Round to the nearest cent.

| Employee | Salary | Sales | Commission Rate | Commission | Gross Pay |
|---|---|---|---|---|---|
| 1. Darren Milner | Base $ 800 | $ 7,412 | 3.0% | _____ | _____ |
| 2. Shelly Huber | Draw 1,200 | 12,905 | 17.5% | _____ | _____ |

## LESSON 4-2     EXERCISES

Convert each of the annual salaries below to the amount earned for other pay periods. Round to the nearest dollar.

| | Annual Salary | Monthly Salary | Semimonthly Salary | Biweekly Salary | Weekly Salary |
|---|---|---|---|---|---|
| 1. | $ 30,000 | _____ | _____ | _____ | _____ |
| 2. | 38,000 | _____ | _____ | _____ | _____ |
| 3. | 50,000 | _____ | _____ | _____ | _____ |
| 4. | 60,000 | _____ | _____ | _____ | _____ |
| 5. | 100,000 | _____ | _____ | _____ | _____ |

Convert the following to annual salaries.

| | Pay Period | Salary | Annual Salary |
|---|---|---|---|
| 6. | Monthly | $2,400 | _____ |
| 7. | Weekly | 485 | _____ |
| 8. | Semimonthly | 1,650 | _____ |
| 9. | Biweekly | 1,800 | _____ |
| 10. | Weekly | 710 | _____ |

Calculate the bonus and gross pay for each of the following sales managers, who are paid a salary and a bonus on sales above their quota.

| | Sales Manager | Salary | Sales Quota | Actual Sales | Bonus Rate | Bonus | Gross Pay |
|---|---|---|---|---|---|---|---|
| 11. | T. Lipka | $30,000 | $600,000 | $742,165 | 5% | _____ | _____ |
| 12. | M. Cooke | 40,000 | 500,000 | 741,924 | 4% | _____ | _____ |
| 13. | E. Jensen | 45,000 | 900,000 | 852,316 | 3% | _____ | _____ |
| 14. | M. Wohlman | 38,000 | 450,000 | 712,352 | 3% | _____ | _____ |
| 15. | T. Seagrem | 42,000 | 550,000 | 817,058 | 4% | _____ | _____ |

Calculate the gross pay for each of the following salespeople, who are paid on a straight commission basis.

| | Salesperson | Monthly Sales | Commission Rate | Monthly Gross Pay |
|---|---|---|---|---|
| 16. | A. LaDue | $215,415 | 2.5% | _____ |
| 17. | K. Sommers | 46,924 | 7.4% | _____ |
| 18. | T. Vlasic | 15,515 | 8.2% | _____ |
| 19. | R. Slade | 156,290 | 6.5% | _____ |
| 20. | B. Newcomb | 34,018 | 3.2% | _____ |

Use the graduated commission scale on page 118 to calculate the gross pay for the salespeople shown.

## GRADUATED COMMISSION SCALE

| Sales | Commission Rate |
|---|---|
| Up to $20,000 | 3% |
| 20,001–40,000 | 4% |
| Over 40,000 | 5% |

| | Salesperson | Sales | Gross Pay |
|---|---|---|---|
| 21. | T. Hess | $25,375 | _____ |
| 22. | L. Mangren | 18,208 | _____ |
| 23. | P. Gomez | 41,912 | _____ |
| 24. | M. Barnes | 51,206 | _____ |
| 25. | J. Ming | 65,385 | _____ |

Find the commission and gross monthly pay for each salesperson, who receives a guaranteed salary plus a commission on all sales made. Round to the nearest cent.

| | Employee | Base Salary | Sales | Commission Rate | Commission | Gross Pay |
|---|---|---|---|---|---|---|
| 26. | B. Sanow | $1,200 | $ 8,215 | 3.5% | _____ | _____ |
| 27. | M. Rice | 1,500 | 5,180 | 4.0% | _____ | _____ |
| 28. | D. Hart | 1,000 | 12,982 | 6.5% | _____ | _____ |
| 29. | T. Sumar | 2,000 | 9,465 | 5.5% | _____ | _____ |
| 30. | J. Rosen | 1,800 | 15,308 | 8.0% | _____ | _____ |

Find the commission and gross pay for each salesperson, who receives a draw plus a commission on all sales made. Round to the nearest cent.

| | Employee | Draw | Sales | Commission Rate | Commission | Gross Pay |
|---|---|---|---|---|---|---|
| 31. | G. Marlowe | $1,600 | $16,509 | 12.0% | _____ | _____ |
| 32. | S. Cohan | 2,200 | 24,813 | 11.4% | _____ | _____ |
| 33. | M. Hacher | 1,300 | 42,191 | 13.1% | _____ | _____ |
| 34. | R. Lowy | 900 | 65,395 | 10.5% | _____ | _____ |
| 35. | T. Piper | 1,400 | 33,764 | 11.9% | _____ | _____ |

## BUSINESS APPLICATIONS

36. Last year Sam Freeland earned $18,500 as a production worker in Milburn Company's factory. This year he switched to the sales staff, on which he is paid a straight commission. If he made sales of $312,500 and his commission rate is 8%, how much more or less did he earn this year than last?

**37.** Arlo Jenkins receives a base salary of $800 per month and a commission of 1.5% on all sales he makes. Last month he made sales of $11,385. What was his gross pay for the month? Round to the nearest cent.

_____

**38.** Nancy Juhl has a monthly draw of $1,200 at Picasso Set Products, and 3.6% on all sales. During May her total sales were $55,308. Find her (a) commission and (b) gross pay.

**(a)** _____

**(b)** _____

**39.** For the month of October, Larry Hill, a hardware salesperson for Plumbers Anonymous, had sales of $125,971, with a 2% commission rate on all sales. His draw is $2,000. Determine Larry's (a) commission and (b) gross pay.

**(a)** _____

**(b)** _____

**40.** Roger Ridout, a lighting fixture salesperson, receives a monthly draw of $1,400 and a 15% commission on all sales. Total sales for April were $10,521. Compute his (a) commission and (b) gross pay.

**(a)** _____

**(b)** _____

**41.** Donna DeLaCroce receives a $30,000 annual salary plus a bonus of $2\frac{1}{4}$% of sales made above a quota of $3,000,000. Last year sales were $3,824,625. What was Donna's gross pay for the year? Round to the nearest cent.

_____

**42.** Mary Unger is a real estate salesperson who receives a 3.5% commission on all houses she sells. Yesterday she sold a $147,500 house. What was Mary's commission from this sale?

_____

**43.** Neil Glenwood is paid on a graduated commission scale that pays 6% for sales of $1 to $10,000; 8% for sales of $10,001 to $25,000; and 10% for sales over $25,000. Last month Neil's sales were $37,586. What was Neil's gross pay for the month?

_____

**44.** Mike Devine earns a salary of $1,400 per month. Last year he also earned $6,145 in commissions. What percent of Mike's annual gross pay was from commissions earned? Round the percent off at two places.

_____

**45.** Marge Generis was offered a job at Central City Furniture at a semimonthly salary of $850. Gracious Living Furniture offered her a biweekly salary of $775. How much more or less will Marge's annual salary be if she takes the job at Central City Furniture?

_____

**46.** Last year district sales manager Dave Pasterini earned a salary of $28,000 and received a bonus of 2% on sales of $310,585 above his quota. What percent of his gross pay for the year resulted from his bonus? Round your answer off to two decimal places.

_____

After calculating an employee's gross pay, the employer must determine various deductions before calculating the employee's net pay. Deductions may be required, such as withholding taxes, or voluntary, such as insurance or savings. Several types of deductions are discussed in this lesson.

### SOCIAL SECURITY TAX (FICA)

The Federal Insurance Contributions Act (FICA) tax, commonly known as *social security,* is currently levied at the rate of 7.51% of a maximum amount of $45,000 paid to an employee in a calendar year. The employers must from their own funds match the amount of FICA tax withheld from an employee's pay and remit both to the government. Self-employed individuals must pay a higher rate than an employee, currently 13.02% on the first $45,000 earned, since there is no matching contribution by an employer.

The first step in calculating the FICA tax to be withheld from an employee's pay is to determine how much of the current pay period's gross pay is taxable. For instance, if the maximum taxable earnings is $45,000 and an employee's previous total gross pay for the year is $44,200, only $800 of the current pay period's gross pay is FICA taxable ($45,000 − $44,200 = $800). Taxable earnings are multiplied times the 7.51% FICA tax rate to find the FICA tax.

While social security tax rates and maximum taxable earnings amounts change periodically, the method of calculation is always the same. Refer to the Circular E, Employer's Tax Guide, for the FICA rate for a specific year. Circular E is a publication of the Federal Government.

---

### EXAMPLE 1

Jane Hirsch's gross pay for this pay period is $1,500. Her previous gross pay this year was $12,000. The FICA tax is 7.51% of the first $45,000 earned in a calendar year. How much should be withheld from Jane's gross pay for FICA taxes?

**SOLUTION**

**Step 1**  Determine how much of the $1,500 is taxable.

| Previous Gross Pay | + | Current Gross Pay | = | Total Gross Pay to Date |
|---|---|---|---|---|
| $12,000 | + | $1,500 | = | $13,500 |

Since Jane's yearly gross salary to date is less than $45,000, her entire gross pay of $1,500 is taxable.

**Step 2**  Calculate the FICA tax.

| Taxable Earnings | × | FICA Tax Rate | = | FICA Tax |
|---|---|---|---|---|
| $1,500 | × | 0.0751 | = | $112.65 |

### EXAMPLE 2

John Olson's gross pay for this pay period is $4,800. His previous gross pay this year was $42,000. The FICA tax is 7.51% of the first $45,000 earned in a calendar year. How much should be withheld for FICA taxes?

## SOLUTION

**Step 1** Determine how much of the $4,800 is taxable.

| Previous Gross Pay | + | Current Gross Pay | = | Total Gross Pay to Date |
|---|---|---|---|---|
| $42,000 | + | $4,800 | = | $46,800 |

| Maximum Taxable Earnings | − | Previous Gross Pay | = | Taxable Earnings |
|---|---|---|---|---|
| $45,000 | − | $42,000 | = | $3,000 |

**Step 2** Calculate the FICA tax.

| Taxable Earnings | × | FICA Tax Rate | = | FICA Tax |
|---|---|---|---|---|
| $3,000 | × | 0.0751 | = | $225.30 |

## EXAMPLE 3

James Vincent is a self-employed carpenter. Last year his total earnings were $54,790. Find his social security tax for the year.

## SOLUTION

| Taxable Earnings | × | Self-Employment Tax Rate | = | Self-Employment Tax |
|---|---|---|---|---|
| $45,000 | × | 0.1302 | = | $5,859.00 |

## PRACTICE PROBLEMS

Determine the taxable income and the amount of FICA tax to be withheld from each of the following employee's gross pay this pay period. The FICA tax is 7.51% of the first $45,000 earned in a calendar year.

| Employee | Previous Gross Pay This Year | Gross Pay This Pay Period | Taxable Income | FICA Tax |
|---|---|---|---|---|
| 1. S. Brown | $20,000 | $2,000 | _____ | _____ |
| 2. T. Callaway | 35,000 | 3,600 | _____ | _____ |
| 3. J. Jones | 50,000 | 7,000 | _____ | _____ |
| 4. R. Welby | 43,000 | 6,200 | _____ | _____ |
| 5. O. Klein | 42,500 | 6,000 | _____ | _____ |

## FEDERAL INCOME TAX

The amount of federal income tax withheld from an employee's earnings is determined by the employee's gross pay, marital status, and number of withholding allowances claimed. An employee may claim a **withholding allowance** for himself or herself, a spouse, each dependent child, and other persons who rely upon the employee for financial support. Each withholding allowance, called an **exemption,** reduces the amount of income tax paid.

ANSWERS FOR PRACTICE PROBLEMS

| | Taxable Income | FICA Tax |
|---|---|---|
| (1) | $2,000 | $150.20 |
| (2) | 3,600 | 270.36 |
| (3) | 0 | 0 |
| (4) | 2,000 | 150.20 |
| (5) | 2,500 | 187.75 |

Before the first payday, each employee must complete an *Employee's Withholding Allowance Certificate,* Form W-4, indicating the number of withholding allowances he or she claims.

| Form **W-4** Department of the Treasury Internal Revenue Service | **Employee's Withholding Allowance Certificate** ▶ **For Privacy Act and Paperwork Reduction Act Notice, see instructions.** | OMB No. 1545-0010 **19--** |
|---|---|---|

**1** Type or print your full name  
Les D. Knox

Home address (number and street or rural route)  
6687 Sandy Lane

City or town, state, and ZIP code  
Anaheim, CA 92804

**2** Your social security number  
582-12-0916

**3** Marital Status  
☐ Single  ☒ Married  
☐ Married, but withhold at higher Single rate  
**Note:** *If married, but legally separated, or spouse is a nonresident alien, check the Single box.*

**4** Total number of allowances you are claiming (from the Worksheet on page 3) . . . . . . . . . . 2

**5** Additional amount, if any, you want deducted from each pay (see Step 4 on page 2) . . . . . $

**6** I claim exemption from withholding because (see Step 2 above and check boxes below that apply):
  **a** ☐ Last year I did not owe any Federal income tax and had a right to a full refund of **ALL** income tax withheld, **AND**
  **b** ☐ This year I do not expect to owe any Federal income tax and expect to have a right to a full refund of **ALL** income tax withheld. If both a and b apply, enter the year effective and "EXEMPT" here . . . ▶  Year 19
  **c** If you entered "EXEMPT" on line 6b, are you a full-time student? . . . . . . . . . . . . . ☐Yes  ☐No

Under penalties of perjury, I certify that I am entitled to the number of withholding allowances claimed on this certificate or, if claiming exemption from withholding, that I am entitled to claim the exempt status.
**Employee's signature ▶** Les D. Knox   **Date ▶** January 4  ,19 —

**7** Employer's name and address (Employer: Complete 7, 8, and 9 only if sending to IRS) | **8** Office code | **9** Employer identification number

The amount of federal income tax to be withheld is determined by referring to withholding tables like the partial ones shown on pages 124 and 125. Withholding tax tables, provided by the IRS for single and married taxpayers, cover weekly, biweekly, semimonthly, monthly, daily, or miscellaneous payroll periods. The procedure for any of these tables is the same.

In January the employer is required to furnish each employee a *Wage and Tax Statement,* Form W-2, showing the employee's gross earnings and total withholdings for federal income tax and FICA tax for the previous calendar year.

The employee files the Wage and Tax Statement with his or her federal income tax return.

| **1** Control number | | OMB No. 1545-0008 | **Copy D For employer** | |
|---|---|---|---|---|

**2** Employer's name, address, and ZIP code

Three Lakes Amusement Park
5115 Hobart Avenue
Los Angeles, CA 90027

**3** Employer's identification number  64-0462074
**4** Employer's state I.D. number  64-0462074

**5** Statutory employee ☐  Deceased ☐  Legal rep. ☐  942 emp. ☐  Subtotal ☐  Void ☐

**6** Allocated tips

**7** Advance EIC payment

**8** Employee's social security number  582-12-0916
**9** Federal income tax withheld  2,497.00

**10** Wages, tips, other compensation  22,754.25
**11** Social security tax withheld  1,708.84

**12** Employee's name, address, and ZIP code

Les D. Knox
6687 Sandy Lane
Anaheim, CA 92804

**13** Social security wages  22,754.25
**14** Social security tips

**16**
**16a** Fringe benefits incl. in Box 10

**17** State income tax  375.55
**18** State wages, tips, etc.
**19** Name of state  CA

**15** Employee's address and ZIP code

**20** Local income tax
**21** Local wages, tips, etc.
**22** Name of locality

Form **W-2** Wage and Tax Statement  **19--**
13-2581759

Department of the Treasury Internal Revenue Service

**TABLE 1**

## WEEKLY PAYROLL PERIOD — SINGLE PERSONS

| And the wages are— | | And the number of withholding allowances claimed is— | | | | | | | | | | |
|---|---|---|---|---|---|---|---|---|---|---|---|---|
| At least | But less than | 0 | 1 | 2 | 3 | 4 | 5 | 6 | 7 | 8 | 9 | 10 |
| | | The amount of income tax to be withheld shall be— | | | | | | | | | | |
| $170 | $180 | $23 | $18 | $12 | $7 | $2 | | | | | | |
| 180 | 190 | 25 | 19 | 14 | 8 | 3 | | | | | | |
| 190 | 200 | 26 | 21 | 15 | 10 | 4 | | | | | | |
| 200 | 210 | 28 | 22 | 17 | 11 | 6 | 1 | | | | | |
| 210 | 220 | 29 | 24 | 18 | 13 | 7 | 2 | | | | | |
| 220 | 230 | 31 | 25 | 20 | 14 | 9 | 3 | | | | | |
| 230 | 240 | 32 | 27 | 21 | 16 | 10 | 5 | | | | | |
| 240 | 250 | 34 | 28 | 23 | 17 | 12 | 6 | 1 | | | | |
| 250 | 260 | 35 | 30 | 24 | 19 | 13 | 8 | 3 | | | | |
| 260 | 270 | 37 | 31 | 26 | 20 | 15 | 9 | 4 | | | | |
| 270 | 280 | 38 | 33 | 27 | 22 | 16 | 11 | 5 | 1 | | | |
| 280 | 290 | 40 | 34 | 29 | 23 | 18 | 12 | 7 | 2 | | | |
| 290 | 300 | 41 | 36 | 30 | 25 | 19 | 14 | 8 | 3 | | | |
| 300 | 310 | 43 | 37 | 32 | 26 | 21 | 15 | 10 | 4 | | | |
| 310 | 320 | 44 | 39 | 33 | 28 | 22 | 17 | 11 | 6 | 1 | | |
| 320 | 330 | 46 | 40 | 35 | 29 | 24 | 18 | 13 | 7 | 2 | | |
| 330 | 340 | 47 | 42 | 36 | 31 | 25 | 20 | 14 | 9 | 3 | | |
| 340 | 350 | 50 | 43 | 38 | 32 | 27 | 21 | 16 | 10 | 5 | | |
| 350 | 360 | 53 | 45 | 39 | 34 | 28 | 23 | 17 | 12 | 6 | 2 | |
| 360 | 370 | 55 | 46 | 41 | 35 | 30 | 24 | 19 | 13 | 8 | 3 | |
| 370 | 380 | 58 | 48 | 42 | 37 | 31 | 26 | 20 | 15 | 9 | 4 | |
| 380 | 390 | 61 | 51 | 44 | 38 | 33 | 27 | 22 | 16 | 11 | 5 | 1 |
| 390 | 400 | 64 | 54 | 45 | 40 | 34 | 29 | 23 | 18 | 12 | 7 | 2 |
| 400 | 410 | 67 | 56 | 47 | 41 | 36 | 30 | 25 | 19 | 14 | 8 | 3 |
| 410 | 420 | 69 | 59 | 49 | 43 | 37 | 32 | 26 | 21 | 15 | 10 | 4 |
| 420 | 430 | 72 | 62 | 52 | 44 | 39 | 33 | 28 | 22 | 17 | 11 | 6 |
| 430 | 440 | 75 | 65 | 55 | 46 | 40 | 35 | 29 | 24 | 18 | 13 | 7 |
| 440 | 450 | 78 | 68 | 57 | 47 | 42 | 36 | 31 | 25 | 20 | 14 | 9 |
| 450 | 460 | 81 | 70 | 60 | 50 | 43 | 38 | 32 | 27 | 21 | 16 | 10 |
| 460 | 470 | 83 | 73 | 63 | 53 | 45 | 39 | 34 | 28 | 23 | 17 | 12 |
| 470 | 480 | 86 | 76 | 66 | 55 | 46 | 41 | 35 | 30 | 24 | 19 | 13 |
| 480 | 490 | 89 | 79 | 69 | 58 | 48 | 42 | 37 | 31 | 26 | 20 | 15 |
| 490 | 500 | 92 | 82 | 71 | 61 | 51 | 44 | 38 | 33 | 27 | 22 | 16 |
| 500 | 510 | 95 | 84 | 74 | 64 | 54 | 45 | 40 | 34 | 29 | 23 | 18 |
| 510 | 520 | 97 | 87 | 77 | 67 | 56 | 47 | 41 | 36 | 30 | 25 | 19 |
| 520 | 530 | 100 | 90 | 80 | 69 | 59 | 49 | 43 | 37 | 32 | 26 | 21 |
| 530 | 540 | 103 | 93 | 83 | 72 | 62 | 52 | 44 | 39 | 33 | 28 | 22 |
| 540 | 550 | 107 | 96 | 85 | 75 | 65 | 55 | 46 | 40 | 35 | 29 | 24 |
| 550 | 560 | 110 | 98 | 88 | 78 | 68 | 57 | 47 | 42 | 36 | 31 | 25 |
| 560 | 570 | 114 | 101 | 91 | 81 | 70 | 60 | 50 | 43 | 38 | 32 | 27 |
| 570 | 580 | 117 | 104 | 94 | 83 | 73 | 63 | 53 | 45 | 39 | 34 | 28 |
| 580 | 590 | 121 | 108 | 97 | 86 | 76 | 66 | 56 | 46 | 41 | 35 | 30 |
| 590 | 600 | 124 | 111 | 99 | 89 | 79 | 69 | 58 | 48 | 42 | 37 | 31 |
| 600 | 610 | 128 | 115 | 102 | 92 | 82 | 71 | 61 | 51 | 44 | 38 | 33 |
| 610 | 620 | 131 | 118 | 106 | 95 | 84 | 74 | 64 | 54 | 45 | 40 | 34 |
| 620 | 630 | 135 | 122 | 109 | 97 | 87 | 77 | 67 | 57 | 47 | 41 | 36 |
| 630 | 640 | 138 | 125 | 113 | 100 | 90 | 80 | 70 | 59 | 49 | 43 | 37 |
| 640 | 650 | 142 | 129 | 116 | 103 | 93 | 83 | 72 | 62 | 52 | 44 | 39 |
| 650 | 660 | 145 | 132 | 120 | 107 | 96 | 85 | 75 | 65 | 55 | 46 | 40 |
| 660 | 670 | 149 | 136 | 123 | 110 | 98 | 88 | 78 | 68 | 58 | 47 | 42 |
| 670 | 680 | 152 | 139 | 127 | 114 | 101 | 91 | 81 | 71 | 60 | 50 | 43 |
| 680 | 690 | 156 | 143 | 130 | 117 | 105 | 94 | 84 | 73 | 63 | 53 | 45 |
| 690 | 700 | 159 | 146 | 134 | 121 | 108 | 97 | 86 | 76 | 66 | 56 | 46 |
| 700 | 710 | 163 | 150 | 137 | 124 | 112 | 99 | 89 | 79 | 69 | 58 | 48 |
| 710 | 720 | 166 | 153 | 141 | 128 | 115 | 102 | 92 | 82 | 72 | 61 | 51 |
| 720 | 730 | 170 | 157 | 144 | 131 | 119 | 106 | 95 | 85 | 74 | 64 | 54 |
| 730 | 740 | 173 | 160 | 148 | 135 | 122 | 109 | 98 | 87 | 77 | 67 | 57 |
| 740 | 750 | 177 | 164 | 151 | 138 | 126 | 113 | 100 | 90 | 80 | 70 | 59 |
| 750 | 760 | 180 | 167 | 155 | 142 | 129 | 116 | 103 | 93 | 83 | 72 | 62 |
| 760 | 770 | 184 | 171 | 158 | 145 | 133 | 120 | 107 | 96 | 86 | 75 | 65 |
| 770 | 780 | 187 | 174 | 162 | 149 | 136 | 123 | 110 | 99 | 88 | 78 | 68 |
| 780 | 790 | 191 | 178 | 165 | 152 | 140 | 127 | 114 | 101 | 91 | 81 | 71 |
| 790 | 800 | 194 | 181 | 169 | 156 | 143 | 130 | 117 | 105 | 94 | 84 | 73 |
| 800 | 810 | 198 | 185 | 172 | 159 | 147 | 134 | 121 | 108 | 97 | 86 | 76 |
| 810 | 820 | 201 | 188 | 176 | 163 | 150 | 137 | 124 | 112 | 100 | 89 | 79 |

# TABLE 2

## WEEKLY PAYROLL PERIOD — MARRIED PERSONS

| And the wages are— | | And the number of withholding allowances claimed is— | | | | | | | | | | |
|---|---|---|---|---|---|---|---|---|---|---|---|---|
| At least | But less than | 0 | 1 | 2 | 3 | 4 | 5 | 6 | 7 | 8 | 9 | 10 |
| | | The amount of income tax to be withheld shall be— | | | | | | | | | | |
| $190 | $200 | $22 | $16 | $11 | $5 | $1 | | | | | | |
| 200 | 210 | 23 | 18 | 12 | 7 | 3 | | | | | | |
| 210 | 220 | 25 | 19 | 14 | 8 | 4 | | | | | | |
| 220 | 230 | 26 | 21 | 15 | 10 | 5 | 1 | | | | | |
| 230 | 240 | 28 | 22 | 17 | 11 | 6 | 2 | | | | | |
| 240 | 250 | 29 | 24 | 18 | 13 | 7 | 3 | | | | | |
| 250 | 260 | 31 | 25 | 20 | 14 | 9 | 4 | | | | | |
| 260 | 270 | 32 | 27 | 21 | 16 | 10 | 5 | 1 | | | | |
| 270 | 280 | 34 | 28 | 23 | 17 | 12 | 6 | 2 | | | | |
| 280 | 290 | 35 | 30 | 24 | 19 | 13 | 8 | 3 | | | | |
| 290 | 300 | 37 | 31 | 26 | 20 | 15 | 9 | 4 | | | | |
| 300 | 310 | 38 | 33 | 27 | 22 | 16 | 11 | 6 | 1 | | | |
| 310 | 320 | 40 | 34 | 29 | 23 | 18 | 12 | 7 | 3 | | | |
| 320 | 330 | 41 | 36 | 30 | 25 | 19 | 14 | 8 | 4 | | | |
| 330 | 340 | 43 | 37 | 32 | 26 | 21 | 15 | 10 | 5 | 1 | | |
| 340 | 350 | 44 | 39 | 33 | 28 | 22 | 17 | 11 | 6 | 2 | | |
| 350 | 360 | 46 | 40 | 35 | 29 | 24 | 18 | 13 | 7 | 3 | | |
| 360 | 370 | 47 | 42 | 36 | 31 | 25 | 20 | 14 | 9 | 4 | | |
| 370 | 380 | 49 | 43 | 38 | 32 | 27 | 21 | 16 | 10 | 5 | 1 | |
| 380 | 390 | 50 | 45 | 39 | 34 | 28 | 23 | 17 | 12 | 6 | 2 | |
| 390 | 400 | 52 | 46 | 41 | 35 | 30 | 24 | 19 | 13 | 8 | 3 | |
| 400 | 410 | 53 | 48 | 42 | 37 | 31 | 26 | 20 | 15 | 9 | 4 | |
| 410 | 420 | 55 | 49 | 44 | 38 | 33 | 27 | 22 | 16 | 11 | 6 | 2 |
| 420 | 430 | 56 | 51 | 45 | 40 | 34 | 29 | 23 | 18 | 12 | 7 | 3 |
| 430 | 440 | 58 | 52 | 47 | 41 | 36 | 30 | 25 | 19 | 14 | 8 | 4 |
| 440 | 450 | 59 | 54 | 48 | 43 | 37 | 32 | 26 | 21 | 15 | 10 | 5 |
| 450 | 460 | 61 | 55 | 50 | 44 | 39 | 33 | 28 | 22 | 17 | 11 | 6 |
| 460 | 470 | 62 | 57 | 51 | 46 | 40 | 35 | 29 | 24 | 18 | 13 | 7 |
| 470 | 480 | 64 | 58 | 53 | 47 | 42 | 36 | 31 | 25 | 20 | 14 | 9 |
| 480 | 490 | 65 | 60 | 54 | 49 | 43 | 38 | 32 | 27 | 21 | 16 | 10 |
| 490 | 500 | 67 | 61 | 56 | 50 | 45 | 39 | 34 | 28 | 23 | 17 | 12 |
| 500 | 510 | 68 | 63 | 57 | 52 | 46 | 41 | 35 | 30 | 24 | 19 | 13 |
| 510 | 520 | 70 | 64 | 59 | 53 | 48 | 42 | 37 | 31 | 26 | 20 | 15 |
| 520 | 530 | 71 | 66 | 60 | 55 | 49 | 44 | 38 | 33 | 27 | 22 | 16 |
| 530 | 540 | 73 | 67 | 62 | 56 | 51 | 45 | 40 | 34 | 29 | 23 | 18 |
| 540 | 550 | 74 | 69 | 63 | 58 | 52 | 47 | 41 | 36 | 30 | 25 | 19 |
| 550 | 560 | 76 | 70 | 65 | 59 | 54 | 48 | 43 | 37 | 32 | 26 | 21 |
| 560 | 570 | 77 | 72 | 66 | 61 | 55 | 50 | 44 | 39 | 33 | 28 | 22 |
| 570 | 580 | 79 | 73 | 68 | 62 | 57 | 51 | 46 | 40 | 35 | 29 | 24 |
| 580 | 590 | 81 | 75 | 69 | 64 | 58 | 53 | 47 | 42 | 36 | 31 | 25 |
| 590 | 600 | 84 | 76 | 71 | 65 | 60 | 54 | 49 | 43 | 38 | 32 | 27 |
| 600 | 610 | 87 | 78 | 72 | 67 | 61 | 56 | 50 | 45 | 39 | 34 | 28 |
| 610 | 620 | 90 | 80 | 74 | 68 | 63 | 57 | 52 | 46 | 41 | 35 | 30 |
| 620 | 630 | 93 | 82 | 75 | 70 | 64 | 59 | 53 | 48 | 42 | 37 | 31 |
| 630 | 640 | 95 | 85 | 77 | 71 | 66 | 60 | 55 | 49 | 44 | 38 | 33 |
| 640 | 650 | 98 | 88 | 78 | 73 | 67 | 62 | 56 | 51 | 45 | 40 | 34 |
| 650 | 660 | 101 | 91 | 81 | 74 | 69 | 63 | 58 | 52 | 47 | 41 | 36 |
| 660 | 670 | 104 | 94 | 83 | 76 | 70 | 65 | 59 | 54 | 48 | 43 | 37 |
| 670 | 680 | 107 | 96 | 86 | 77 | 72 | 66 | 61 | 55 | 50 | 44 | 39 |
| 680 | 690 | 109 | 99 | 89 | 79 | 73 | 68 | 62 | 57 | 51 | 46 | 40 |
| 690 | 700 | 112 | 102 | 92 | 82 | 75 | 69 | 64 | 58 | 53 | 47 | 42 |
| 700 | 710 | 115 | 105 | 95 | 84 | 76 | 71 | 65 | 60 | 54 | 49 | 43 |
| 710 | 720 | 118 | 108 | 97 | 87 | 78 | 72 | 67 | 61 | 56 | 50 | 45 |
| 720 | 730 | 121 | 110 | 100 | 90 | 80 | 74 | 68 | 63 | 57 | 52 | 46 |
| 730 | 740 | 123 | 113 | 103 | 93 | 83 | 75 | 70 | 64 | 59 | 53 | 48 |
| 740 | 750 | 126 | 116 | 106 | 96 | 85 | 77 | 71 | 66 | 60 | 55 | 49 |
| 750 | 760 | 129 | 119 | 109 | 98 | 88 | 78 | 73 | 67 | 62 | 56 | 51 |
| 760 | 770 | 132 | 122 | 111 | 101 | 91 | 81 | 74 | 69 | 63 | 58 | 52 |
| 770 | 780 | 135 | 124 | 114 | 104 | 94 | 84 | 76 | 70 | 65 | 59 | 54 |
| 780 | 790 | 137 | 127 | 117 | 107 | 97 | 86 | 77 | 72 | 66 | 61 | 55 |
| 790 | 800 | 140 | 130 | 120 | 110 | 99 | 89 | 79 | 73 | 68 | 62 | 57 |
| 800 | 810 | 143 | 133 | 123 | 112 | 102 | 92 | 82 | 75 | 69 | 64 | 58 |
| 810 | 820 | 146 | 136 | 125 | 115 | 105 | 95 | 84 | 76 | 71 | 65 | 60 |
| 820 | 830 | 149 | 138 | 128 | 118 | 108 | 98 | 87 | 78 | 72 | 67 | 61 |
| 830 | 840 | 151 | 141 | 131 | 121 | 111 | 100 | 90 | 80 | 74 | 68 | 63 |

**EXAMPLE**

Last week Don Clayton, a married employee claiming four withholding allowances, had gross pay of $455. How much should be withheld for federal income taxes?

**SOLUTION**

Use Table 2, "Weekly Payroll Period—Married Persons."

**Step 1** Read down the columns headed by the words "At Least" and "But Less Than," until you come to $450 and $460. Don's earnings of $455 fall between these two amounts.

**Step 2** Read across until you reach the column headed by the number 4. The amount, $39, is the federal income tax to be withheld.

**PRACTICE PROBLEMS**

Determine the amount of federal income tax to be withheld from each of these employee's weekly gross pay. Use the withholding tables on pages 124 and 125.

| Employee | Weekly Gross Pay | Marital Status | Exemptions | Federal Income Tax Withholding |
|---|---|---|---|---|
| **1.** S. Gordon | $415 | M | 3 | _____ |
| **2.** T. Brown | 425 | S | 1 | _____ |
| **3.** B. Harmon | 602 | M | 5 | _____ |
| **4.** J. Ling | 540 | M | 2 | _____ |
| **5.** C. Oakes | 258 | S | 0 | _____ |

## LESSON 4-3    EXERCISES

Calculate the taxable income and the amount of FICA tax to be withheld from each of the following employees this pay period. The FICA tax rate is 7.51% of the first $45,000 earned. Round to the nearest cent.

| Employee | Previous Gross Pay This Year | Gross Pay This Pay Period | Taxable Earnings | FICA Tax |
|----------|------------------------------|--------------------------|------------------|----------|
| 1. P. Razil | $15,620 | $1,824 | _____ | _____ |
| 2. J. Lucero | 25,620 | 2,425 | _____ | _____ |
| 3. K. Parsons | 42,765 | 3,815 | _____ | _____ |
| 4. T. Fuji | 43,050 | 3,960 | _____ | _____ |
| 5. D. Willis | 28,245 | 2,150 | _____ | _____ |

Use the federal income tax withholding tables on pages 124 and 125 to find the amount of federal income tax to be withheld from each of the following employee's gross pay.

| Employee | Weekly Gross Pay | Marital Status | Exemptions | Federal Income Tax Withheld |
|----------|------------------|----------------|------------|------------------------------|
| 6. B. Eddy | $704 | S | 1 | _____ |
| 7. A. Marino | 545 | M | 3 | _____ |
| 8. K. Frahm | 618 | S | 1 | _____ |
| 9. T. West | 618 | M | 1 | _____ |
| 10. W. Jewett | 378 | S | 0 | _____ |

### BUSINESS APPLICATIONS

11. Five years ago the maximum employee contribution for FICA was 6.65% of $29,700. Now the maximum FICA contribution is 7.51% of $45,000. How many dollars increase is this in the maximum contribution for FICA?

12. Ted Andrews earns $635 per week. He is married and claims five federal income tax withholding allowances. How much more or less will be withheld from Ted's gross pay for federal income taxes if he claims three exemptions instead of five?

13. Shelly O'Connor is single. Her sister, Marie, is married. Each of them earns $555 per week and claims one exemption. How much more or less is deducted from Shelly's weekly gross pay than from Marie's for federal income taxes?

14. Gerald Norman is married and claims five federal income tax exemptions. Currently, Gerald earns $450 per week. How much more will Gerald receive per week after federal income taxes are deducted if he takes a new job paying $550 per week?

15. David Fisher, a married employee claiming six exemptions, earns $625 per week. His wife just had a baby. How much more or less will he receive per week if he now claims seven exemptions?

16. Millie Nolan is single and earns $755 per week. She claims one federal income tax exemption. What percent of Millie's gross pay is deducted each week for federal income taxes? Round to the hundredths place.

17. Michelle Dayton earns $465 per week. She is single and claims one federal income tax withholding allowance. All of her earnings are subject to FICA tax. How much does Michelle receive per week after federal income taxes and FICA taxes are deducted?

18. Mary Gaetti is an employee who earned $41,000 last year. June Evans is self-employed and also earned $41,000 last year. How much more FICA tax did June pay than Mary?

Employers must keep accurate records of employees' earnings so that employees are paid the correct amount and the proper amounts can be deducted and submitted for FICA tax, federal income tax, and other purposes.

**PAYROLL REGISTER**

Businesses use a *payroll register,* a record of earnings and deductions for all company employees for each payroll period, to record various deductions and to determine an employee's net pay. The payroll register is the source from which paychecks are written.

### EXAMPLE

Payroll information for Les Knox is shown in the following payroll register. Each week he has $7 deducted for union dues and $15 deducted for savings. Complete the payroll register.

**Three Lakes Amusement Park**
**PAYROLL REGISTER**

**For Period Ending:** October 2, 19--

| Employee Data | | | | | Gross Earnings | | | Deductions | | | | | | |
|---|---|---|---|---|---|---|---|---|---|---|---|---|---|---|
| Name | Marital Status | No. of With. Allow. | Hours | Hourly Rate | Reg. | Over-time | Gross Pay | FICA Tax | Fed. Inc. Tax | State Inc. Tax | Union Dues | Other | Total Ded. | Net Pay |
| L. Knox | M | 2 | 42 | 10.00 | 400.00 | 30.00 | 430.00 | 32.29 | 47.00 | 7.05 | 7.00 | 15.00 | 108.34 | 321.66 |
| | | | | | a | b | c | d | e | f | g | h | i | j |

### SOLUTION

**Step 1**    Calculate gross pay.

**a.** Regular pay:      40 × $10                    $400.00
**b.** Overtime pay:    2 × $10 × 1.5              +    30.00
**c.** Gross pay:                                          $430.00

**Step 2**    Determine the individual deductions and total deductions.

**d.** FICA tax ($430 × 0.0751):              $ 32.29
**e.** Federal income tax (from table):          47.00
**f.** State income tax (provided for you):       7.05
**g.** Union dues:                                     7.00
**h.** Other (savings):                          +    15.00
**i.** Total deductions:                        $108.34

**Step 3**    Calculate net pay.

**j.** Gross Pay − Total Deductions = Net Pay
         $430.00   −      $108.34      = $321.66

Complete the remainder of the payroll register shown below. All employees are paid time and a half for hours worked in excess of 40 hours in a week. Use 7.51% as the FICA rate. No employee has reached the FICA maximum earnings for the year. Use the tables on pages 124 and 125 to determine federal income taxes. Round to the nearest cent.

Three Lakes Amusement Park
### PAYROLL REGISTER
**For Period Ending:** October 2, 19--

| | | Employee Data | | | | Gross Earnings | | | Deductions | | | | | | |
|---|---|---|---|---|---|---|---|---|---|---|---|---|---|---|---|
| | Name | Marital Status | No. of With. Allow. | Hours | Hourly Rate | Reg. | Over-time | Gross Pay | FICA Tax | Fed. Inc. Tax | State Inc. Tax | Union Dues | Other | Total Ded. | Net Pay |
| | L. Knox | M | 2 | 42 | 10.00 | 400.00 | 30.00 | 430.00 | 32.29 | 47.00 | 7.05 | 7.00 | 15.00 | 108.34 | 321.66 |
| 1. | M. Lee | S | 1 | 40 | 12.00 | | | | | | 11.85 | 7.00 | 20.00 | | |
| 2. | C. Mix | M | 4 | 38 | 9.00 | | | | | | 3.30 | 7.00 | 10.00 | | |
| 3. | J. Niles | S | 0 | 44 | 10.00 | | | | | | 12.45 | 7.00 | -0- | | |
| 4. | O. Orr | M | 6 | 42 | 12.00 | | | | | | 5.55 | -0- | 15.00 | | |
| 5. | J. Reed | M | 3 | 41 | 10.00 | | | | | | 5.70 | 7.00 | 5.00 | | |

## INTERNAL REVENUE SERVICE REPORTING

An employer must match all employees' contributions to social security and send the total amount to the Internal Revenue Service (IRS). The employer must also remit the total personal income tax collected from employees at the end of each payroll period to the IRS on a quarterly basis. Both payments usually are sent together.

### EXAMPLE

For the quarter ending March 30, the Earnest Upholstry Co. collected $1,832.17 in FICA tax and $2,323.86 in federal withholding tax from all its employees. Calculate total tax due to the government for the quarter.

### SOLUTION

| | |
|---|---|
| FICA collections from employees: | $1,832.17 |
| Matching FICA paid by Earnest: | 1,832.17 |
| Federal withholding tax: | + 2,323.86 |
| Total remittance to IRS: | $5,988.20 |

ANSWERS FOR PRACTICE PROBLEMS

| Employee | Regular Earnings | Over-time | Gross Pay | FICA Tax | Fed. Inc. Tax | Total Ded. | Net Pay |
|---|---|---|---|---|---|---|---|
| (1) M. Lee | $480.00 | $ 0 | $480.00 | $36.05 | $79.00 | $153.90 | $326.10 |
| (2) C. Mix | 342.00 | 0 | 342.00 | 25.68 | 22.00 | 67.98 | 274.02 |
| (3) J. Niles | 400.00 | 60.00 | 460.00 | 34.55 | 83.00 | 137.00 | 323.00 |
| (4) O. Orr | 400.00 | 116.00 | 516.00 | 38.75 | 37.00 | 96.30 | 419.70 |
| (5) J. Reed | 400.00 | 15.00 | 415.00 | 31.17 | 38.00 | 86.87 | 328.13 |

## LESSON 4-4    EXERCISES

Complete the payroll register shown below. All employees are paid time
and a half for hours worked over 40 hours in a week. Use 7.51% as the
FICA rate. No employees have reached the FICA maximum earnings. Use
the tables on pages 124 and 125 to calculate federal income taxes.

Wharton Industries
**PAYROLL REGISTER**

| | Name | Marital Status | No. of With. Allow. | Hours | Hourly Rate | Reg. | Over-time | Gross Pay | FICA Tax | Fed. Inc. Tax | State Inc. Tax | Other Ded. | Total Ded. | Net Pay |
|---|---|---|---|---|---|---|---|---|---|---|---|---|---|---|
| | | | | | | Gross Earnings | | | Deductions | | | | | |
| 1. | M. Barker | M | 4 | 40 | 12.00 | | | | | | 6.45 | 10.00 | | |
| 2. | D. Casey | M | 3 | 42 | 10.00 | | | | | | 6.15 | -0- | | |
| 3. | T. Dodd | S | 1 | 38 | 10.00 | | | | | | 7.65 | 5.00 | | |
| 4. | B. Frank | S | 1 | 44 | 9.00 | | | | | | 8.85 | 5.00 | | |
| 5. | M. Grant | M | 6 | 46 | 12.00 | | | | | | 7.05 | 20.00 | | |
| 6. | R. Idso | M | 5 | 40 | 9.00 | | | | | | 3.00 | 5.00 | | |
| 7. | D. Kizer | M | 2 | 40 | 11.50 | | | | | | 7.65 | 10.00 | | |
| 8. | B. Lyons | S | 0 | 40 | 8.50 | | | | | | 7.50 | -0- | | |
| 9. | P. Nichol | M | 4 | 45 | 10.00 | | | | | | 6.30 | -0- | | |
| 10. | T. Sully | S | 1 | 39 | 13.00 | | | | | | 12.60 | 10.00 | | |

The following employees at Garth Manufacturing are paid on the straight
piecework basis. Complete the payroll register. No employee has reached
the FICA maximum earnings.

Garth Manufacturing, Inc.
**PAYROLL REGISTER**

| | Name | Martial Status/ Exemptions | M | T | W | Th | F | Total Units | Rate per Unit | Gross Pay | FICA Tax | Fed. Inc. Tax | Other Ded. | Total Ded. | Net Pay |
|---|---|---|---|---|---|---|---|---|---|---|---|---|---|---|---|
| | | | Number of Units Produced | | | | | | | | | | | | |
| 11. | J. Abel | M–3 | 84 | 72 | 80 | 85 | 79 | | 1.05 | | | | 5.00 | | |
| 12. | S. Boman | M–2 | 95 | 86 | 97 | 87 | 90 | | .92 | | | | 10.00 | | |
| 13. | A. Gray | S–1 | 36 | 35 | 36 | 38 | 35 | | 3.20 | | | | 20.00 | | |
| 14. | J. Hayes | M–5 | 56 | 49 | 55 | 53 | 47 | | 2.25 | | | | 25.00 | | |
| 15. | R. Lynn | S–0 | 81 | 70 | 82 | 75 | 77 | | 1.10 | | | | 10.00 | | |
| 16. | C. Mier | S–1 | 60 | 58 | 64 | 60 | 58 | | 1.55 | | | | -0- | | |
| 17. | B. Norris | M–6 | 75 | 78 | 74 | 75 | 78 | | 1.05 | | | | -0- | | |
| 18. | S. Percy | M–2 | 20 | 22 | 21 | 20 | 22 | | 4.20 | | | | 5.00 | | |
| 19. | F. Vivone | M–3 | 55 | 56 | 55 | 56 | 53 | | 2.00 | | | | 10.00 | | |
| 20. | E. Wong | S–1 | 80 | 82 | 80 | 82 | 81 | | 1.10 | | | | 10.00 | | |

## BUSINESS APPLICATIONS

**21.** Sue Jacobson's gross pay was $775 last week. Her employer made the following deductions from her paycheck: FICA tax, $58.20; federal income tax, $124; city income tax, $11.63; and insurance premium payment, $25.00. What was Sue's net pay?

_____

**22.** Prior to this pay period, Andrew Petrie had gross earnings of $42,500. His gross pay this pay period was $4,500. Andrew's employer deducted $337.95 from his gross pay for FICA tax. Was this amount correct? If not, what amount should be deducted for FICA taxes?

_____

**23.** Edgar VanZant's total earnings to date shown on the employee earnings record for the pay period ending July 31 was $16,280. For the next pay period, ending August 31, Edgar's gross pay was $2,450 and his net pay was $1,845. What is Edgar's total earnings to date shown for the pay period ending August 31?

_____

**24.** Robert Cee had gross pay of $2,850 last month. His employer deducted the following amounts from his earnings: FICA taxes, $214.04; federal income taxes, $432; and state income taxes, $52. What percent was Robert's net pay of his gross pay? Round percent to two decimal places.

_____

**25.** Prior to this pay period, Shelly Marquart had gross earnings for the year of $43,100. This pay period Shelly had gross pay of $5,400. How much of Shelly's current gross pay is subject to FICA tax?

_____

## CHAPTER 4 REVIEW

### WAGES

Calculate the gross pay for each of the following employees. Overtime is $1\frac{1}{2}$ times the regular rate for hours worked over 40 hours in a week.

| Employee | Hours | Regular Rate | Gross Pay |
|---|---|---|---|
| 1. Helen Jerdes | 40 | $13.00 | _____ |
| 2. Jason Mallory | 44 | 12.45 | _____ |
| 3. Gloria Munson | 45 | 10.25 | _____ |

### STRAIGHT PIECEWORK

Calculate the gross pay of the following employees who are paid on the straight piecework basis.

| Employee | Number Produced | Rate per Item | Gross Pay |
|---|---|---|---|
| 4. Cheryl Pitney | 387 | $0.22 | _____ |
| 5. Dale Quinn | 374 | 0.20 | _____ |
| 6. Mike Rost | 415 | 0.25 | _____ |

### DIFFERENTIAL PIECEWORK

Use the following differential piecework scale to calculate the gross pay for the employees listed below.

### DIFFERENTIAL PIECEWORK SCALE

| Daily Items Produced | Rate |
|---|---|
| 1–50 | $0.40 |
| 51–100 | 0.50 |
| 101–150 | 0.60 |
| Over 150 | 0.70 |

| Employee | Number Produced | Gross Pay |
|---|---|---|
| 7. Joan Hudson | 180 | _____ |
| 8. Ron Hughes | 240 | _____ |
| 9. Becky Roe | 138 | _____ |

### SALARY

Convert the following employees' annual salaries to the amount earned for other pay periods. Round amounts to the nearest dollar.

| Employee | Annual Salary | Monthly Salary | Semimonthly Salary | Biweekly Salary | Weekly Salary |
|---|---|---|---|---|---|
| 10. S. Graham | $28,800 | _____ | _____ | _____ | _____ |
| 11. E. Jensen | 19,500 | _____ | _____ | _____ | _____ |
| 12. R. Jimenez | 38,580 | _____ | _____ | _____ | _____ |

## SALARY PLUS BONUS

Calculate the bonus and gross pay for each of the following sales managers, who are paid a salary and a bonus on sales above their quota.

| Manager | Annual Salary | Sales Quota | Actual Sales | Bonus Rate | Bonus | Gross Pay |
|---|---|---|---|---|---|---|
| 13. G. Giles | $32,000 | $800,000 | $923,615 | 4% | _____ | _____ |
| 14. M. Lopez | 35,000 | 900,000 | 982,105 | 6% | _____ | _____ |
| 15. T. Niflo | 30,000 | 600,000 | 959,374 | 3% | _____ | _____ |

## STRAIGHT COMMISSION

Calculate the gross pay for each of the following salespeople, who are paid a straight commission. Round to the nearest cent.

| Salesperson | Sales | Commission Rate | Gross Pay |
|---|---|---|---|
| 16. K. Jerret | $375,496 | 12.5% | _____ |
| 17. M. Kinkaid | 460,311 | 8.5% | _____ |
| 18. D. Pasa | 590,982 | 9.0% | _____ |

## SALARY PLUS COMMISSION

Find the commission and gross pay for each salesperson, who receives a guaranteed monthly salary plus a commission on all sales made. Round to the nearest cent.

| Employee | Base Salary | Sales | Commission Rate | Commission | Gross Pay |
|---|---|---|---|---|---|
| 19. C. Finley | $1,400 | $22,255 | 2.25% | _____ | _____ |
| 20. D. Parr | 1,600 | 25,408 | 1.55% | _____ | _____ |
| 21. T. Reisner | 2,000 | 31,060 | 3.75% | _____ | _____ |

## GRADUATED COMMISSION

Use the following graduated commission scale to calculate the gross pay for the salespeople shown below.

### GRADUATED COMMISSION SCALE

| Sales | Commission Rate |
|---|---|
| Up to $50,000 | 15% |
| 50,001–100,000 | 20% |
| Over 100,000 | 25% |

| Salesperson | Sales | Gross Pay |
|---|---|---|
| 22. P. Clemens | $91,321 | _____ |
| 23. W. Dalen | 220,016 | _____ |
| 24. E. Gratias | 205,497 | _____ |

## FICA TAXES

Determine the taxable earnings and the amount of FICA tax to be withheld from each of the following employee's gross pay this pay period. The FICA tax rate is 7.51% on maximum earnings of $45,000.

| | Employee | Previous Gross Pay This Year | Gross Pay This Pay Period | Taxable Earnings | FICA Tax |
|---|---|---|---|---|---|
| 25. | V. Packard | $21,000 | $2,000 | _____ | _____ |
| 26. | K. Schwab | 45,000 | 4,200 | _____ | _____ |
| 27. | S. Winslow | 42,000 | 3,500 | _____ | _____ |

## FEDERAL INCOME TAXES

Use the tax withholding tables on pages 124 and 125 to find the amount of federal income tax to be withheld from the following employees.

| | Employee | Weekly Gross Pay | Marital Status | Exemptions | Federal Income Tax Withheld |
|---|---|---|---|---|---|
| 28. | L. Corey | $628 | M | 4 | _____ |
| 29. | R. Evans | 415 | S | 1 | _____ |
| 30. | A. Russo | 710 | M | 3 | _____ |

## PAYROLL REGISTER

Complete the payroll register shown below. All employees are paid $1\frac{1}{2}$ times their regular rate for hours worked over 40 hours in a week. All employee earnings are subject to the FICA tax rate of 7.51%.

**PAYROLL REGISTER**

| | Employee Data | | | | | Gross Earnings | | | Deductions | | | | |
|---|---|---|---|---|---|---|---|---|---|---|---|---|---|
| | Name | Marital Status | No. of With. Allow. | Hours | Hourly Rate | Reg. | Over-time | Gross Pay | FICA Tax | Fed. Inc. Tax | Other Ded. | Total Ded. | Net Pay |
| 31. | J. Baxter | M | 2 | 40 | 8.00 | | | | | | –0– | | |
| 32. | T. Heki | S | 1 | 42 | 9.00 | | | | | | 5.00 | | |
| 33. | G. Mills | M | 3 | 43 | 8.25 | | | | | | 5.00 | | |
| 34. | O. Sims | S | 0 | 44 | 9.60 | | | | | | 15.00 | | |
| 35. | J. Hall | M | 2 | 45 | 10.25 | | | | | | 20.00 | | |
| | | | | | | | | | | | | | |
| | | | | | | | | | | | | | |
| | | | | | | | | | | | | | |
| | | | | | | | | | | | | | |
| | | | | | | | | | | | | | |

## BUSINESS APPLICATIONS

**36.** Mario Levine earns $13.75 per hour and is paid $1\frac{1}{2}$ times the regular rate for hours worked over 40 in a week. Last week Mario worked 47 hours. What percent of Mario's weekly gross pay came from overtime? Round gross pay to the nearest cent. Round percent to two decimal places.

_____

**37.** Mae Parker receives a $35,000 annual salary plus a bonus of 3.5% of sales made above a quota of $2,000,000. Last year sales were $2,642,350. What was Mae's gross pay for the year?

_____

**38.** Dave Paschall earns $2,850 per month. How much FICA tax will be withheld from Dave's gross pay in a year?

_____

**39.** Mario Gaetti earns $745 per week. Mario is married and claims three exemptions. Mario's employer withholds federal income tax and FICA tax and also withholds 5% of his weekly pay for investment in a savings account and $5.50 per week for union dues. What is Mario's weekly net pay? Round to the nearest cent.

_____

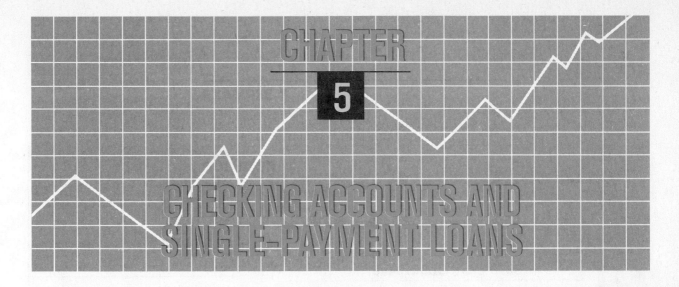

# CHAPTER 5

## CHECKING ACCOUNTS AND SINGLE-PAYMENT LOANS

### LEARNING OBJECTIVES

**1.** Prepare deposit slips, checks, and check records.

**2.** Reconcile checking accounts.

**3.** Calculate simple interest on loans.

**4.** Find the length of the loan periods and due dates of loans.

**5.** Find the interest and principal paid on partial loan payments.

Checking accounts, which allow money to be transferred easily and safely, are an important service provided by banks and other financial institutions.

In this chapter, you will learn how checking accounts operate and how to keep accurate checking account records. You also will learn how to calculate the cost of borrowing money and how to calculate interest saved when making a partial payment before a due date.

## LESSON 5-1    CHECKS AND CHECK RECORDS

Businesses like the Galaxy Theatre use a **checking account** as a safe and convenient way to make payments to others. Money is placed into the checking account, and amounts are then transferred to others by use of a written order called a **check.**

**DEPOSIT SLIPS**

Checks and cash to be deposited into a checking account are recorded on a **deposit slip.** Checks are identified on the deposit slip by their **transit number,** which is the number of the bank from which the check is drawn and the number of the federal reserve bank serving that area. The slip contains blanks for entering any currency or coin, as well as checks to be deposited. The sum of cash and checks deposited is called the **total deposit.**

Galaxy Theatre has the following amounts to deposit in its checking account on October 7: cash, $1,162.31; a check for $182.16 (transit no. 38-461); and a check for $48.50 (transit no. 16-229). Complete the deposit slip shown below.

**SOLUTION**

| Checking Account Deposits | | |
|---|---|---|
| | Dollars | Cents |
| Cash | 1,162 | 31 |
| Checks 38-461 | 182 | 16 |
| 16-229 | 48 | 50 |
| | | |
| | | |
| | | |
| Total | $ 1,392 | 97 |

**Summit Trust**

Deposited and accepted subject to the provisions of the Uniform Commercial Code and the rules and regulations governing this account.

October 7, 19 —

For Account of **Galaxy Theatre**

**9076 East Chain Ave.**

**Seattle, WA 98132**

Account Number

2 0 3  0 2 6 1  3 0 5

Checking Account Deposits

Add cash and check amounts to find the total.

## PRACTICE PROBLEM

On October 8 Galaxy Theatre has the following amounts to deposit in its checking account: cash, $875.61; checks for $126.14 (transit no. 15-340), $65.00 (transit no. 32-118), $12.38 (transit no. 19-569), $45.31 (transit no. 47-902), $78.21 (transit no. 52-423), and $64.10 (transit no. 10-641). Complete the following deposit slip.

**Summit Trust**

Deposited and accepted subject to the provisions of the Uniform Commercial Code and the rules and regulations governing this account.

_____ 19 _____

For Account of _____

Account Number

2 0 3  0 2 6 1  3 0 5

| Checking Account Deposits | | |
|---|---|---|
| | Dollars | Cents |
| Cash | | |
| Checks | | |
| | | |
| | | |
| | | |
| | | |
| Total | $ | |

Checking Account Deposits

ANSWER FOR
PRACTICE PROBLEM

Total deposit: $1,266.75

## CHECKING ACCOUNT SERVICES

Similar checking account services are offered by banks, savings and loan associations, credit unions, and other financial institutions. There are two major types of checking accounts: *personal* and *business.*

Banks often charge a fee, called a *service charge,* for processing checks and handling the account. *Business checking accounts* may receive more free services and may be charged lower fees because balances are often higher and transactions more complex. Some banks charge a flat fee per month, others charge for each check processed, and still others use a combination of these two systems. Some banks charge a fee based on the average balance of the period covered by the statement. Banks often do not charge a service charge if a minimum balance or average balance is maintained in the account.

Many banks and their counterparts have *automatic teller machines* (ATMs) at which any customer using a *personal banking card* and coding in a special identification code can withdraw cash, deposit cash and checks, and transfer money from one account to another at any time. Differences exist, however, in how any of these institutions charge for their services and handle other aspects of the account.

### EXAMPLE

Last month National Bank charged Deenee's Upholstery $7.50 plus $0.15 for each of the 37 checks it processed. What was the total service charge for the month?

### SOLUTION

$37 \times \$0.15 = \$5.55$     (individual check charge)
$\$7.50 + 5.55 = \$13.05$     (total service charge)

### PRACTICE PROBLEM

In October the People's Bank processed 38 checks through the Kramer Limousine Service checking account. The People's Bank charges $0.20 per check and a monthly service charge of $3.45. What is the total service charge for Kramer during October?

## CHECK RECORDS

A check directs the bank from which it is drawn, called the *drawee,* to make payment to the designated party, called the *payee.* The owner of the checking account is called the *drawer* or *depositor.*

The magnetic ink characters located at the bottom of the check can be read by a computer and can identify the federal reserve bank number, the drawee bank number, and the drawer account number.

Before a deposit is made or a check written, the drawer makes a record of the transaction and determines the *balance.* The balance is the amount in the checking account at any given time.

ANSWER FOR PRACTICE PROBLEM: Total service charge: $11.05

Copyright © 1989 by McGraw-Hill, Inc. All rights reserved.

LESSON 5-1 ■ 139

Some checkbooks have a *stub* attached to each check. The check stub should be fully completed before the check is detached from the stub. The check stub provides space for recording the check number, date, to whom the check is written, and the purpose. The balance from the previous stub, any deposits, the amount of the check, and the new balance are also recorded on the check stub.

## EXAMPLE

Complete the stub and check for Galaxy Theatre as shown below. Check no. 1337 was written on November 9 to Barclay Equipment Company for $800.00. Enter a forward balance of $3,910.20 and a $900.00 deposit on the stub.

## SOLUTION

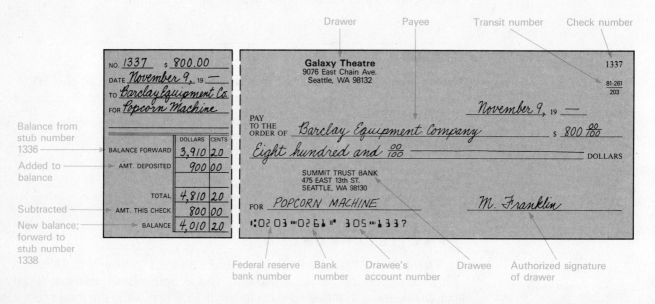

## PRACTICE PROBLEMS

Complete the following check stubs using the information shown below. The ending balance on check stub no. 1337 is $4,010.20.

| | Date | Check Number | Amount | To | For |
|---|---|---|---|---|---|
| **1.** | Nov. 9 | 1338 | $425.25 | Friesner Cartage Co. | Transportation |
| **2.** | Nov. 9 | 1339 | 37.90 | Al's Gift Shop | Miscellaneous |
| **3.** | Nov. 10 | Deposit | 642.50 | | |
| | Nov. 10 | 1340 | 931.08 | LeVoe Mfg. Co. | Merchandise |
| **4.** | Nov. 10 | 1341 | 89.07 | Popov Office Supply | Office Equip. |

| NO. _1338_ $ _____ | NO. _1339_ $ _____ | NO. _1340_ $ _____ | NO. _1341_ $ _____ |
|---|---|---|---|
| DATE _____ 19__ | DATE _____ 19__ | DATE _____ 19__ | DATE _____ 19__ |
| TO _____ | TO _____ | TO _____ | TO _____ |
| FOR _____ | FOR _____ | FOR _____ | FOR _____ |

| | DOLLARS | CENTS |
|---|---|---|
| BALANCE FORWARD | | |
| AMT. DEPOSITED | | |
| TOTAL | | |
| AMT. THIS CHECK | | |
| BALANCE | | |

## CHECK REGISTER

A *check register,* is also used to keep a record of all checking account transactions. The check register shows a list of the checks written and deposits made and a running balance after each transaction.

### EXAMPLE

On July 6 Kalo Company paid Arlo Realty rent of $975.00 by check no. 326. On July 7 Kalo deposited $1,200.00 into the checking account. On July 9 Kalo paid Alexander Publishing Company $325.25 for advertising by check no. 327. Record these transactions in Kalo's check register.

### SOLUTION

| CHECK NO. | DATE | CHECKS ISSUED TO | AMOUNT OF CHECK | | ✓ | AMOUNT OF DEPOSIT | | BALANCE | |
|---|---|---|---|---|---|---|---|---|---|
| | | BALANCE BROUGHT FORWARD → | | | | | | 1,670 | 50 |
| 326 | 7/6 | Arlo Realty/Rent | 975 | 00 | | | | 695 | 50 |
| | 7/7 | Deposit | | | | 1,200 | 00 | 1,895 | 50 |
| 327 | 7/9 | Alexander Publishing Co./Advertising | 325 | 25 | | | | 1,570 | 25 |
| | | | | | | | | | |
| | | | | | | | | | |
| | | | | | | | | | |

## PRACTICE PROBLEMS

A check register is presented below for Riverside Country Club. Record the following transactions and complete the check register.

| | Date | Check Number | Amount | To | For |
|---|---|---|---|---|---|
| **1.** | Aug. 6 | 416 | $485.50 | Milner Supply Co. | Supplies |
| **2.** | Aug. 7 | 417 | 146.90 | Nelson Bros. | Landscaping |
| **3.** | Aug. 9 | Deposit | 855.38 | | |
| **4.** | Aug. 10 | 418 | 625.80 | Trent Golf Company | Equipment |

| CHECK NO. | DATE | CHECKS ISSUED TO | AMOUNT OF CHECK | ✔ | AMOUNT OF DEPOSIT | BALANCE | |
|---|---|---|---|---|---|---|---|
| | | BALANCE BROUGHT FORWARD → | | | | 3,206 | 80 |
| | | | | | | | |
| | | | | | | | |
| | | | | | | | |
| | | | | | | | |
| | | | | | | | |
| | | | | | | | |

ANSWERS FOR PRACTICE PROBLEMS

| Date | Balance |
|---|---|
| **(1)** August 6 | $2,721.30 |
| **(2)** August 7 | 2,574.40 |
| **(3)** August 9 | 3,429.78 |
| **(4)** August 10 | 2,803.98 |

## LESSON 5-1     EXERCISES

1. On October 9 Global Imports has the following amounts to deposit in their checking account: cash, $543.18; a check for $132.05 (transit no. 12-326); and a check for $8.60 (transit no. 13-165). Complete the following deposit slip for Global Imports.

| Checking Account Deposits | | | |
|---|---|---|---|
| **Summit Trust** | Deposited and accepted subject to the provisions of the Uniform Commercial Code and the rules and regulations governing this account. | **Checking Account Deposits** | |
| | | Dollars | Cents |
| | _____19 _____ | Cash | |
| For | | Checks | |
| Account of _____ | | | |
| _____ | | | |
| _____ | | | |
| Account Number | | | |
| 2 0 3 0 2 6 1 3 0 5 | | Total   $ | |

2. Garr Enterprises listed checks for the following amounts on its deposit slip: $307.50, $215.80, $182.75, $155.15, and $93.60. Garr withheld $150.00 in cash and deposited the remainder. What was Garr's total deposit?

3. Complete the check stubs on page 144 using the information below. The ending balance on check stub no. 512 is $2,778.40.

| Date | Check Number | Amount | To | For |
|---|---|---|---|---|
| May 18 | 513 | $346.50 | Ace Mfg. | Equipment |
| May 19 | Deposit | 500.00 | | |
| May 19 | 514 | 860.25 | Zenor Advertising | Advertising |
| May 20 | 515 | 705.00 | Bagley Co. | Merchandise |
| May 21 | 516 | 310.45 | Ernest Energy Co. | Utilities |

NO. *513*  $ _____     NO. *514*  $ _____     NO. *515*  $ _____     NO. *516*  $ _____

| | | DOLLARS | CENTS |
|---|---|---|---|
| BALANCE FORWARD | | | |
| AMT. DEPOSITED | | | |
| TOTAL | | | |
| AMT. THIS CHECK | | | |
| BALANCE | | | |

**4.** Complete the check register shown below using the following information. The opening balance is $3,050.10.

| Date | Check Number | Amount | To | For |
|---|---|---|---|---|
| January 23 | 1821 | $ 146.20 | Western Lines | Telephone Expense |
| January 24 | 1822 | 452.80 | Comstock Transport | Transportation |
| January 25 | Deposit | 900.00 | | |
| January 27 | 1823 | 1,250.00 | Kathy's Decorating | Remodeling |
| January 28 | 1824 | 18.05 | Speedi-Delivery | Delivery |

| CHECK NO. | DATE | CHECKS ISSUED TO | AMOUNT OF CHECK | ✔ | AMOUNT OF DEPOSIT | BALANCE | |
|---|---|---|---|---|---|---|---|
| | | BALANCE BROUGHT FORWARD → | | | | | |
| | | | | | | | |
| | | | | | | | |
| | | | | | | | |
| | | | | | | | |
| | | | | | | | |
| | | | | | | | |

## BUSINESS APPLICATIONS

**5.** The ending balance on Soper Company's check stub no. 805 was $345.60. Check no. 806 was written for $187.25. What was Soper's ending balance on check stub no. 806?

**6.** Granite County State Bank charges its checking account customers $5.00 for a box of 100 blank checks. It also charges a flat monthly service charge of $10.00 and charges $0.10 for each check processed. Last month the bank processed 80 checks through Henley Company's checking account. What was Henley's total checking account cost?

**7.** Freeman Gifts averages writing 50 checks per month. Albany Bank charges $5.00 for a box of 100 blank checks and charges a flat monthly service charge fee of $5.00, regardless of how many checks are written. Cascade Bank has no charge for blank checks but charges $0.20 for each check processed. Which bank will result in the lowest monthly service charge for Freeman and how much lower is it?

**8.** On August 17 Norton Industries' checking account balance was $468.20. On the following days, these transactions occurred: August 18, wrote check no. 3971 for $126.70; August 19, deposited $750.00; August 20, wrote check no. 3972 for $810.50; August 21, wrote check no. 3973 for $15.00; August 22, deposited $500.00; August 23, wrote check no. 3974 for $310.55. What is Norton's checking account balance after check no. 3974 was written?

_____

**9.** On March 1 Crookston Auto Body Restoration had a checking account balance of $875.20. During March, Crookston made deposits of $1,685.80 and wrote checks totaling $1,940.70. Interest income of $3.75 was added to the account by the bank and $5.20 was deducted for bank service charges. What was Crookston's checking account balance on March 31?

_____

*Bank reconciliation* is a comparison of the depositor's checkbook record with the bank's record of the checking account to detect any discrepancies and to update the checkbook record.

Periodically, usually monthly, the bank sends the depositor a *bank statement.* All transactions, such as deposits made, checks written, interest earned, service charges, and collections of promissory notes made by the bank on the depositor's behalf, are shown on the bank statement.

Use a *bank reconciliation form,* usually found on the reverse side of the bank statement, to prepare the bank reconciliation. The starting point is the ending balances shown on the bank statement and on the depositor's checkbook record. A depositor with an active checking account will find that these two balances are rarely the same. Adjustments must be made to update these two balances. When the bank reconciliation has been properly completed, the *adjusted balances* will be equal. The discrepancies found on the bank statement and the checkbook records are due to one of the following reasons:

*Deposit in transit.* A deposit in transit is a deposit that is added to the checkbook balance but which arrived at the bank after the bank statement was prepared. The amount of the deposit has been added to the checkbook record but not to the bank statement.

---

### SUMMIT TRUST COMPANY
475 East 13th Street        Seattle, WA 98130

### BANK STATEMENT

Galaxy Theatre
9076 East Chain Avenue
Seattle, WA 98132

| | Date of This Statement |
| --- | --- |
| | Nov. 30, 19-- |
| | Date of Last Statement |
| Account No.  305 44 | Oct. 31, 19-- |

| Previous Ending Balance | Number of Checks/ Debits | Amount of Checks/ Debits | Number of Deposits/ Credits | Amount of Deposits/ Credits | Ending Balance |
| --- | --- | --- | --- | --- | --- |
| $4,182.10 | 126 | $3,190.15 | 8 | $2,155.20 | $3,147.15 |

| Date | Checks and Other Debits | Deposits and Other Credits | Balance |
| --- | --- | --- | --- |
| 11-1 | 150.00 ATM | | 4,032.10 |
| 11-2 | 75.47 | | 3,956.63 |
| 11-2 | 350.08 EFT | | 3,606.55 |
| 11-3 | | 750.00 | 4,356.55 |
| 11-4 | 5.00 CSF | 500.00 CM | 4,851.55 |
| 11-5 | 125.15 | | 4,726.40 |
| 11-29 | 5.16 | | 3,143.24 |
| 11-29 | 16.94 | | 3,126.30 |
| 11-30 | | 20.85 I | 3,147.15 |

Automatic teller machine withdrawal

Electronic fund transfer

Collection made by bank

Bank's fee for making collection

Interest income

| | |
| --- | --- |
| SC – Service Charge | CSF – Collection Service Fee |
| I – Interest Income | NSF – Not Sufficient Funds |
| CM – Collection Made | OD – Overdraft |
| EC – Error Correction | MS – Miscellaneous |
| EFT – Electronic Fund Transfer | |
| ATM – Automated Teller Machine Withdrawal | |

---

***Outstanding checks.*** Outstanding checks are checks written but not yet presented to the bank for payment. They have been deducted from the checkbook record but not from the bank statement.

***Interest and other deposits.*** Interest earned on checking accounts or automatic payroll deposits will appear on the bank statement but not in the checkbook records. Also banks sometimes collect payments on debts due the business and add the amounts collected to the depositor's account.

***Service charges and other deductions.*** Other transactions, such as ATM withdrawals and automatic withdrawals for insurance premium payments and mortgage payments, may appear on the bank statement but not on the checkbook record. All bank charges, such as fees for printing checks, use of automated teller machines, insufficient funds, service charges, use of electronic funds transfer (EFT), automatic loan payments, or fees for making collections appear only on the bank statement.

***Errors.*** The depositor may make an error in recording the amount of a deposit or check or in calculating the checkbook record balance. If this occurs, an amount is added to or subtracted from the checkbook record to yield the proper amount. If a bank error occurs, an adjustment is made to the bank statement side of the bank reconciliation.

## EXAMPLE

Galaxy Theatre's November 30 checkbook balance was $2,685.80. The bank statement balance on November 30 was $3,147.15. The following information was gathered for reconciling the bank statement. Prepare the bank reconciliation.

1. A $300.00 deposit mailed on November 29 was recorded in the checkbook but not on the bank statement.
2. The following checks were outstanding on November 30: check no. 1421, $210.00; check no. 1429, $50.00; and check no. 1430, $12.50.
3. Interest of $20.85 was earned on the account.
4. The bank collected a $500.00 promissory note for Galaxy.
5. The bank charged a $5.00 fee for collecting the promissory note.
6. Check no. 1426 was written for $85.00 but recorded as $58.00 in the checkbook record.

## SOLUTION

**Step 1**   Complete the section that starts with Balance Shown on Bank Statement.

a.   List the ending bank statement balance, $3,147.15.
b.   List the $300.00 deposit in transit. Add it to the Balance Shown on Bank Statement. The subtotal is $3,447.15.
c.   List the outstanding checks. Deduct the total, $272.50, from the subtotal, $3,447.15. The Adjusted Bank Balance, $3,174.65, is shown.

**Step 2**   Complete the section that starts with Balance Shown in Checkbook.

d.   List the ending checkbook balance, $2,685.80.
e.   Add the promissory note collected and the interest earned. Add the total, $520.85, to the Balance Shown in Checkbook. The subtotal is $3,206.65.
f.   List the collection fee, $5.00.

**g.** Calculate the error correction amount and list it.

| Correct Amount (Should Have Been Deducted) | − | Incorrect Amount (Was Deducted) | = | Additional Amount That Needs to Be Deducted |
|:---:|:---:|:---:|:---:|:---:|
| $85.00 | − | $58.00 | = | $27.00 |

**h.** Subtract the total deductions, $32.00, from the subtotal, $3,206.65. The Adjusted Checkbook balance, $3,174.65, is shown.

**Step 3** Compare the Adjusted Bank Statement balance and the Adjusted Checkbook balance. If they are now equal, the checkbook record is correct.

**Step 4** Update the checkbook record by adding $520.85 (promissory note collected and interest income earned) and deducting $32.00 (collection fee and correction of error in recording check no. 1426). The completed reconciliation form is shown below.

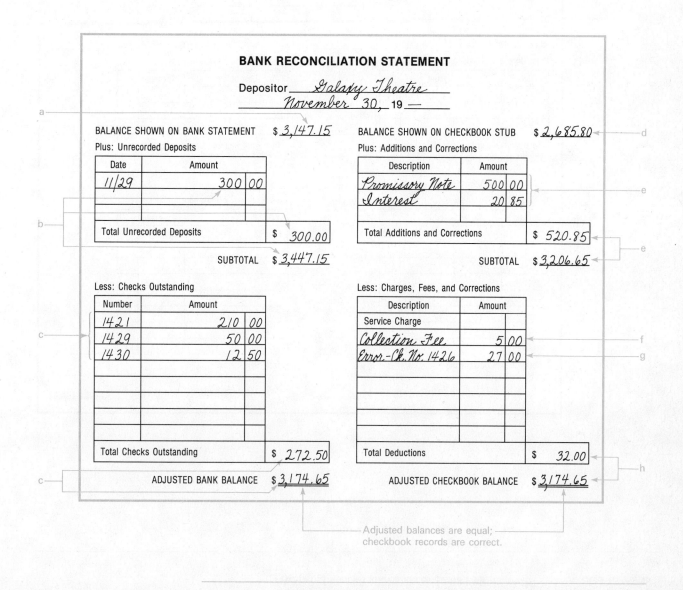

**BANK RECONCILIATION STATEMENT**

Depositor _Galaxy Theatre_
_November 30_, 19 ___

| BALANCE SHOWN ON BANK STATEMENT | $3,147.15 |
|---|---|

Plus: Unrecorded Deposits

| Date | Amount |
|---|---|
| 11/29 | 300 00 |
| | |
| | |
| Total Unrecorded Deposits | $ 300.00 |

SUBTOTAL $3,447.15

Less: Checks Outstanding

| Number | Amount |
|---|---|
| 1421 | 210 00 |
| 1429 | 50 00 |
| 1430 | 12 50 |
| | |
| | |
| | |
| | |
| Total Checks Outstanding | $ 272.50 |

ADJUSTED BANK BALANCE $3,174.65

| BALANCE SHOWN ON CHECKBOOK STUB | $2,685.80 |
|---|---|

Plus: Additions and Corrections

| Description | Amount |
|---|---|
| Promissory Note | 500 00 |
| Interest | 20 85 |
| Total Additions and Corrections | $ 520.85 |

SUBTOTAL $3,206.65

Less: Charges, Fees, and Corrections

| Description | Amount |
|---|---|
| Service Charge | |
| Collection Fee | 5 00 |
| Error – Ck. No. 1426 | 27 00 |
| Total Deductions | $ 32.00 |

ADJUSTED CHECKBOOK BALANCE $3,174.65

Adjusted balances are equal; checkbook records are correct.

# PRACTICE PROBLEM

On December 31 Knott Real Estate Co. had a bank statement balance of $2,815.20 and a checkbook balance of $2,023.20. Other information gathered for the bank reconciliation is as follows. Complete the bank reconciliation.

**a.** Deposit in transit: $350.00
**b.** Service charge: $8.00
**c.** Interest earned: $23.75
**d.** Error: $27.00 check entered on stub as $72.00
**e.** Outstanding checks: check no. 1246, $115.00; check no. 1251, $35.00; check no. 1267, $820.18; check no. 1268, $70.20; and check no. 1290, $40.87.

---

**BANK RECONCILIATION STATEMENT**

Depositor _____

_____ 19 _____

BALANCE SHOWN ON BANK STATEMENT   $ _____
Plus: Unrecorded Deposits

| Date | Amount | |
|------|--------|--|
|      |        |  |
|      |        |  |
|      |        |  |
| Total Unrecorded Deposits | $ | |

SUBTOTAL   $ _____

Less: Checks Outstanding

| Number | Amount | |
|--------|--------|--|
|        |        |  |
|        |        |  |
|        |        |  |
|        |        |  |
|        |        |  |
|        |        |  |
|        |        |  |
| Total Checks Outstanding | $ | |

ADJUSTED BANK BALANCE   $ _____

BALANCE SHOWN ON CHECKBOOK STUB   $ _____
Plus: Additions and Corrections

| Description | Amount | |
|-------------|--------|--|
|             |        |  |
|             |        |  |
| Total Additions and Corrections | $ | |

SUBTOTAL   $ _____

Less: Charges, Fees, and Corrections

| Description | Amount | |
|-------------|--------|--|
| Service Charge |     |  |
|             |        |  |
|             |        |  |
|             |        |  |
|             |        |  |
| Total Deductions | $ | |

ADJUSTED CHECKBOOK BALANCE   $ _____

---

## LESSON 5-2    EXERCISES

Find the adjusted balances for each of the following checking accounts.

| | Ending Bank Statement Balance | Ending Checkbook Balance | Deposit in Transit | Oustanding Checks | Interest Earned | Service Charge | Adjusted Balances |
|---|---|---|---|---|---|---|---|
| 1. | $ 629.50 | $ 652.52 | $250.00 | $146.50 84.90 | $ 2.08 | $ 6.50 | _____ |
| 2. | 3,020.00 | 2,767.15 | 500.00 | 620.00 132.60 | 12.50 | 12.25 | _____ |
| 3. | 1,656.10 | 1,309.55 | 0 | 348.50 | 5.85 | 7.80 | _____ |

4. On June 30 Kosack Industries had a bank statement balance of $1,796.50 and a checkbook balance of $3,307.54. Other information gathered for the bank reconciliation is on page 152. Complete the bank reconciliation.

**BANK RECONCILIATION STATEMENT**

Depositor _____

_____ 19 _____

BALANCE SHOWN ON BANK STATEMENT    $_____    BALANCE SHOWN ON CHECKBOOK STUB    $_____

Plus: Unrecorded Deposits

| Date | Amount | |
|---|---|---|
| | | |
| | | |
| | | |
| Total Unrecorded Deposits | $ | |

Plus: Additions and Corrections

| Description | Amount | |
|---|---|---|
| | | |
| | | |
| | | |
| Total Additions and Corrections | $ | |

SUBTOTAL    $_____        SUBTOTAL    $_____

Less: Checks Outstanding

| Number | Amount | |
|---|---|---|
| | | |
| | | |
| | | |
| | | |
| | | |
| Total Checks Outstanding | $ | |

Less: Charges, Fees, and Corrections

| Description | Amount | |
|---|---|---|
| Service Charge | | |
| | | |
| | | |
| | | |
| | | |
| | | |
| Total Deductions | $ | |

ADJUSTED BANK BALANCE    $_____        ADJUSTED CHECKBOOK BALANCE    $_____

a. Deposit in transit: $2,148.00.
b. Automatic account withdrawal for utilities: $916.25.
c. Outstanding checks: check no. 824, $1,421.16; check no. 832, $39.80.
d. Interest earned: $8.45.
e. Service charge: $16.20.
f. Error: $46.00 check entered on stub as $146.00.

## BUSINESS APPLICATIONS

5. In reconciling the checking account for Nishnabotna, Inc., Travis Williams found that the bank had collected a $2,000.00 promissory note on the company's behalf and had charged a $10.00 collection fee. Also, interest of $14.75 was earned on the account and a $8.50 service charge was assessed. What net amount should be added to or subtracted from the checkbook record to bring it up to date?

6. In reconciling the checking account for Ford Engineering, Kate McGovern found that a check for $23.45 had been recorded in the check register as $32.45. Also, interest income of $5.80 was earned and a service charge of $13.75 was assessed. What net amount should be added to or subtracted from the checkbook record to bring it up to date?

7. On the day Sandy Morris completed the bank reconciliation for Duncan and Company, the check register balance was $1,307.23. Sandy found that a $45.00 check had been recorded in the check register as $54.00, that $15.90 interest was earned on the account, and that a $10.00 service charge was assessed. What should the check register balance be after Sandy updates it with the bank reconciliation information?

Many businesses find it necessary or advisable to borrow money to take advantage of a sale on merchandise, to meet a payroll, to obtain a lower interest rate than on an existing loan, or to pay off a creditor. The agreement between the lender and borrower, which outlines the terms of the loan, is called a **promissory note.** In the illustration below, the borrower, called the **maker,** is Woodson Wholesale Company. The lender, called the **payee,** is First Atlanta State Bank.

The **principal** is the amount borrowed. **Interest** is the fee paid for borrowing money. The **Truth-in-Lending** law requires that lenders state the rate of interest as an **annual percentage rate (APR).** The annual percentage rate is the percent a year's interest is of the amount borrowed.

A short-term loan, a loan of up to one year, is traditionally a **single-payment** loan. A single-payment loan is one in which the borrower pays the lender the principal plus interest on the maturity date in one lump sum.

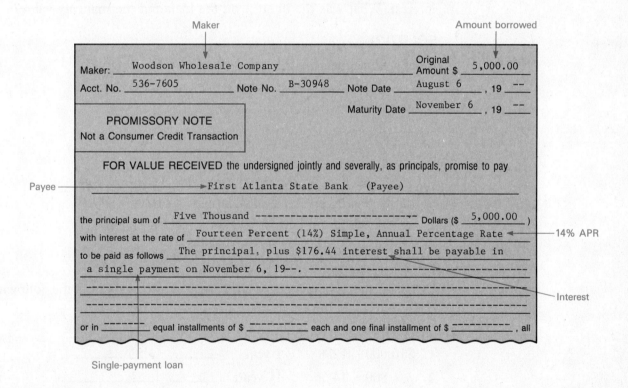

## CALCULATING SIMPLE INTEREST

**Simple interest** is interest calculated on the amount borrowed only and not on any accumulated interest. The amount of interest charged by a bank depends on three factors.

1. **Principal.** The amount borrowed.
2. **Rate.** The percent of the principal charged for use of the money.
3. **Time.** The length of time of the loan, which can be expressed in years, months, or days.

To calculate interest, multiply the principal by the rate by the time as shown in the illustration below.

Principal × Rate × Time = Interest

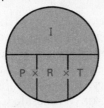

On the loan's due date, the principal plus the interest, called the **maturity value,** must be repaid.

Principal + Interest = Maturity Value

---

## EXAMPLE

Empire Bottling Company borrowed $15,000 for one year at 13% simple interest. What was the interest on the loan and the maturity value?

## SOLUTION

**Step 1**  Calculate the interest.

Principal × Rate × Time = Interest
$15,000 × 0.13 ×   1   = $1,950

        ↑
    13% = 0.13     One year

**Step 2**  Calculate the maturity value.

Principal + Interest = Maturity Value
$15,000 + $1,950 =     $16,950

---

## PRACTICE PROBLEMS

Calculate the simple interest and the maturity value on the following loans.

| | Principal | Rate | Time | Interest | Maturity Value |
|---|---|---|---|---|---|
| **1.** | $10,000 | 12% | 1 year | _____ | _____ |
| **2.** | 800 | 14% | 1½ years | _____ | _____ |
| **3.** | 5,000 | 12.75% | 2 years | _____ | _____ |

## TIME STATED AS MONTHS

Sometimes, the period of time of a loan is stated in months. When this occurs, express the time element of the interest formula as a fraction of a year. The numerator will be the number of months of the loan. The denominator will always be 12, the number of months in a year.

### EXAMPLE

Thomas Adams borrowed $1,400 on May 7 from the Portland Bank. The promissory note carried a 14% interest rate for 6 months. Calculate simple interest and maturity value using the simple-interest formula.

### SOLUTION

**Step 1**   Determine simple interest.

Principal × Rate × Time = Interest
$1,400  × 0.14 ×   $\frac{1}{2}$   =   $98    $\frac{6}{12} = \frac{1}{2}$

**Step 2**   Find maturity value.

Principal + Interest = Maturity Value
$1,400  +  $98  =  $1,498

### PRACTICE PROBLEMS

Calculate the amount of simple interest and the maturity value on the following single-payment loans. Round your answers to the nearest cent.

| | Principal | Rate | Time | Interest | Maturity Value |
|---|---|---|---|---|---|
| 1. | $ 2,500 | 12% | 6 months | _____ | _____ |
| 2. | 900 | 11% | 4 months | _____ | _____ |
| 3. | 20,000 | 10.75% | 14 months | _____ | _____ |

4. The Reade National Bank prepared a promissory note at $12\frac{1}{4}$% for 9 months in the amount of $1,000. What is the interest the borrower must pay to Reade? Round to the nearest cent.

5. A promissory note for $2,000 was executed by Rice Sterling Loan Association at $12\frac{3}{4}$% for a period of 18 months. Calculate the interest on this loan.

## EXACT INTEREST

In business it is common for loans to be due in a given number of days, such as a "90-day note." When time is given in days, it is expressed as a fraction of a year with the number of days of the loan over the number of days in a year.

The ***exact-interest method*** expresses time as a fraction, with the exact number of days in the loan period as the numerator. The denominator is the exact number of days in a year, 365, or in leap year, 366. Most loans with a time period of less than a year are calculated using the exact-interest method.

### EXAMPLE

J. K. Radosovich & Sons borrowed $10,000 at 12% interest for 180 days from Clay County State Bank. Using the exact-interest method, what is the interest on the loan? Round your answer to the nearest cent.

### SOLUTION

Principal × Rate × Time = Interest

$10,000 × 0.12 × $\frac{180}{365}$ = $591.78

Can also be expressed as $10,000 × 0.12 × 180 ÷ 365.

### PRACTICE PROBLEMS

Use the exact-interest method to calculate the amount of interest on the following loans. Round answers to the nearest cent.

| | Principal | Rate | Time, days | Interest |
|---|---|---|---|---|
| **1.** | $6,000 | 12.25% | 220 | _____ |
| **2.** | 5,000 | 11.75% | 180 | _____ |
| **3.** | 527 | 11.25% | 120 | _____ |
| **4.** | 5,787 | 12% | 79 | _____ |
| **5.** | 4,235 | 14% | 227 | _____ |

## ORDINARY INTEREST

The ***ordinary-interest method*** assumes a year has 360 days. Since this method does not produce an accurate amount of interest for a loan *because the stated interest rate is not a true annual percentage rate, APR,* it is rarely used in practice. However, since frequently used loan time periods, like $\frac{30}{360}$, $\frac{90}{360}$, and $\frac{180}{360}$, readily reduce to lower terms, they are easy to use in manual computations. The ordinary-interest is an excellent way to quickly estimate the amount of interest on a loan.

### EXAMPLE

Pioneer Beach Resort borrows $15,000 at 12% interest for 90 days. Use the ordinary-interest method to estimate the amount of interest on the loan.

    

## SOLUTION

Principal × Rate × Time = Ordinary Interest

$15,000 × 0.12 × $\frac{1}{4}$ = $450.00

Reduce $\frac{90}{360}$ to $\frac{1}{4}$

## PRACTICE PROBLEMS

Calculate the ordinary interest on the following loans. Round answers to the nearest cent.

| | Principal | Rate | Time | Ordinary Interest |
|---|---|---|---|---|
| 1. | $ 8,000 | 12.0% | 180 days | _____ |
| 2. | 20,000 | 11.5% | 90 days | _____ |
| 3. | 600 | 12.5% | 30 days | _____ |
| 4. | 3,600 | 14.0% | 240 days | _____ |
| 5. | 2,000 | 13.5% | 60 days | _____ |

## SIMPLE-INTEREST TABLES

The amount of exact interest on a loan can be calculated easily by using a simple-interest table like the one shown below. This table shows the amount of exact interest on $1 at various rates and number of days. To find the exact interest, multiply the principal by the appropriate interest table value.

### SIMPLE-INTEREST TABLE
### AMOUNT OF INTEREST ON $1

| Time in Days | Interest Rate | | | |
|---|---|---|---|---|
| | 10% | 12% | 14% | 16% |
| 30 | 0.0082192 | 0.0098630 | 0.0115068 | 0.0131507 |
| 31 | 0.0084932 | 0.0101918 | 0.0118904 | 0.0135890 |
| 60 | 0.0164384 | 0.0197260 | 0.0230137 | 0.0263014 |
| 100 | 0.0273973 | 0.0328767 | 0.0383562 | 0.0438356 |
| 120 | 0.0328767 | 0.0394521 | 0.0460274 | 0.0526027 |
| 150 | 0.0410959 | 0.0493151 | 0.0575342 | 0.0657534 |
| 180 | 0.0493151 | 0.0591781 | 0.0690411 | 0.0789041 |
| 220 | 0.0602740 | 0.0723288 | 0.0843836 | 0.0964384 |
| 250 | 0.0684932 | 0.0821918 | 0.0958904 | 0.1095890 |
| 300 | 0.0821918 | 0.0986301 | 0.1150685 | 0.1315068 |
| 365 | 0.1000000 | 0.1200000 | 0.1400000 | 0.1600000 |

ANSWERS FOR
PRACTICE PROBLEMS

(1) $480.00 (2) $575.00 (3) $6.25 (4) $336.00 (5) $45.00

## EXAMPLE

Mid-Town Entertainment Agency borrowed $15,000 at 14% interest for 300 days. Calculate the amount of exact interest on the loan using the simple-interest table. Round to the nearest cent.

### SOLUTION

**Step 1** Locate the interest table value. Read down the "Time in Days" column until you come to 300. Read across to the right until you come to the 14% column. The number, 0.1150685, is the amount of interest on $1 at 14% for 300 days.

**Step 2** Calculate the interest.

Principal × Table Value = Interest
$15,000 × 0.1150685 = $1,726.027 ≈ $1,726.03

To find interest for a time period longer than 365 days, combine the values on the interest table.

## EXAMPLE

Mid-Town borrowed $8,000 at 12% interest for 420 days. Calculate the amount of interest on the loan using the simple interest table.

### SOLUTION

**Step 1** Locate the interest on $1 at 12% for 420 days.

Interest on $1 at 12% for 300 days = 0.0986301
Interest on $1 at 12% for 120 days = 0.0394521

Interest on $1 at 12% for 420 days = 0.1380822

**Step 2** Calculate the interest.

Principal × Table Value = Interest
$8,000 × 0.1380822 = $1,104.657 ≈ $1,104.66

## PRACTICE PROBLEMS

Calculate the amount of interest on the following loans using the simple-interest table shown on page 157. Round to the nearest cent.

| | Principal | Rate | Time | Interest |
|---|---|---|---|---|
| **1.** | $ 515 | 10% | 180 days | _____ |
| **2.** | 12,300 | 14% | 250 days | _____ |
| **3.** | 14,392 | 16% | 425 days | _____ |
| **4.** | 810 | 12% | 495 days | _____ |
| **5.** | 7,540 | 12% | 365 days | _____ |

ANSWERS FOR
PRACTICE PROBLEMS

(1) $25.40  (2) $1,179.45  (3) $2,681.25  (4) $131.82  (5) $904.80

## FINDING THE PRINCIPAL

Occasionally, the principal, the rate, or the length of a loan may need to be calculated. When any three of the simple-interest formula elements are known, the fourth can be found by using the following interest formula illustration as a guide.

If the rate, time, and amount of interest are known, the principal can be found by using the formula:

$$\text{Principal} = \frac{\text{Interest}}{\text{Rate} \times \text{Time}}$$

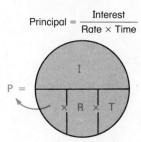

Remember, time in the interest formula is expressed in years or as a fraction of a year.

---

### EXAMPLE 1

The amount of interest on a 6-month loan at 10% is $1,000. Find the principal.

#### SOLUTION

State the formula and perform the calculation.

$$P = \frac{I}{R \times T} = \frac{\$1,000}{0.10 \times \dfrac{6}{12}} = \frac{1,000}{0.1 \times 0.5} = \frac{1,000}{0.05} = \$20,000 \text{ Principal}$$

$$\underset{12\overline{)6.0}}{\overset{.5}{\uparrow}}$$

### EXAMPLE 2

Find the principal of a note for 6 months if the exact interest is $1,500.00 and the interest rate is 12%.

#### SOLUTION

$$P = \frac{I}{R \times T} = \frac{\$1,500.00}{0.12 \times \dfrac{1}{2}} = \$25,000.00 \text{ Principal}$$

---

### PRACTICE PROBLEMS

Calculate the principal. Round to the nearest cent.

| | Exact Interest | Rate | Time | Principal |
|---|---|---|---|---|
| 1. | $ 800.00 | 14% | 2 years | _____ |
| 2. | 3,000.00 | 12% | 3 months | _____ |
| 3. | 8,000.00 | 14% | 3 months | _____ |

---

**(1)** $2,857.14  **(2)** $100,000.00  **(3)** $228,571.43

## FINDING THE RATE

If the principal, time, and interest are known, the rate can be found using the formula:

$$\text{Rate} = \frac{\text{Interest}}{\text{Principal} \times \text{Time}}$$

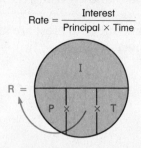

---

### EXAMPLE 1

Saunders Sporting Goods Company borrowed $15,000 for 7 months. The exact interest was $1,050. Calculate the rate of interest.

### SOLUTION

$$R = \frac{I}{P \times T} = \frac{\$1,050}{\$15,000 \times \dfrac{7}{12}} = \frac{\$1,050}{\$8,750} = 0.12 \text{ or } 12\% \text{ rate}$$

### EXAMPLE 2

Find the interest rate on a note with a face value of $12,000, due in 210 days, with ordinary interest of $910.

### SOLUTION

$$R = \frac{I}{P \times T} = \frac{\$910}{\$12,000 \times \dfrac{210}{360}} = \frac{\$910}{\$7,000} = 0.13 \text{ or } 13\%$$

---

### PRACTICE PROBLEMS

Calculate the rate of interest. Use a 360-day year or a 12-month year. Round to the nearest hundredth percent.

|    | Interest | Principal | Time | Rate |
|----|----------|-----------|------|------|
| 1. | $4,000   | $15,000   | 24 months | _____ |
| 2. | 2,000    | 28,000    | 180 days | _____ |
| 3. | 3,000    | 8,000     | 3 years | _____ |

---

If the principal, rate, and amount of interest are known, the time of the loan can be found using the formula:

$$\text{Time} = \frac{\text{Interest}}{\text{Principal} \times \text{Rate}}$$

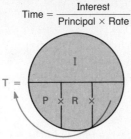

When calculating the time using the above formula, the result will be a decimal part of a year. It may be necessary to multiply the result by 365 to find the number of days in the loan period. Multiply the result by 12 to find the number of months in the loan period.

### EXAMPLE 1

Find the number of days in the loan period for a note with a face value of $25,000 at 12% interest if the exact interest is $2,400.

**SOLUTION**

**Step 1** $T = \dfrac{I}{P \times R} = \dfrac{\$2,400}{\$25,000 \times 0.12} = \dfrac{2,400}{3,000} = 0.8$ (decimal part of a year)

**Step 2** $0.8 \times 365 = 292$ days

### EXAMPLE 2

A promissory note for $45,000 with a 14% interest rate was signed by the Make-Over Designs Corporation. The interest charge is $1,575. Find the length of the loan in months.

**SOLUTION**

**Step 1** $T = \dfrac{I}{P \times R} = \dfrac{\$1,575}{\$45,000 \times 0.14} = \dfrac{1,575}{6,300} = 0.25$ (decimal part of a year)

**Step 2** $0.25 \times 12 = 3$ months

### PRACTICE PROBLEMS

The principal, rate, and amount of interest for several loans are shown below. Calculate the decimal part of a year and find the time of each loan in both days and months. Round answers to the nearest day or month.

| | Principal | Rate | Exact Interest | Decimal Part of a Year | Time in Days | Time in Months |
|---|---|---|---|---|---|---|
| **1.** | $20,000 | 14% | $1,400.00 | _____ | _____ | _____ |
| **2.** | 35,000 | 12% | 6,300.00 | _____ | _____ | _____ |
| **3.** | 5,000 | 11% | 137.50 | _____ | _____ | _____ |

ANSWERS FOR
PRACTICE PROBLEMS

| | Decimal Part of a Year | Time in Days | Time in Months |
|---|---|---|---|
| (1) | 0.50 | 183 | 6 |
| (2) | 1.50 | 548 | 18 |
| (3) | 0.25 | 91 | 3 |

## LESSON 5-3     EXERCISES

Calculate the simple interest and the maturity value due on the following loans. Round to the nearest cent.

| | Principal | Rate | Time | Interest | Maturity Value |
|---|---|---|---|---|---|
| **1.** | $ 4,320 | 10% | 8 months | _____ | _____ |
| **2.** | 7,500 | $12\frac{1}{2}$% | 10 months | _____ | _____ |
| **3.** | 16,780 | 13% | $1\frac{1}{2}$ years | _____ | _____ |
| **4.** | 18,000 | $12\frac{1}{4}$% | 7 months | _____ | _____ |
| **5.** | 2,020 | $14\frac{1}{2}$% | $2\frac{1}{4}$ years | _____ | _____ |

Calculate the interest on the following loans using the ordinary-interest method and the exact-interest method. Round to the nearest cent.

| | Principal | Rate | Time | Ordinary Interest | Exact Interest |
|---|---|---|---|---|---|
| **6.** | $ 5,000 | 14% | 90 days | _____ | _____ |
| **7.** | 8,500 | $12\frac{1}{2}$% | 180 days | _____ | _____ |
| **8.** | 9,000 | $11\frac{3}{4}$% | 60 days | _____ | _____ |
| **9.** | 23,400 | $13\frac{1}{2}$% | 280 days | _____ | _____ |
| **10.** | 18,000 | $14\frac{1}{4}$% | 120 days | _____ | _____ |

Use the simple-interest table on page 157 to calculate the amount of exact interest on the following loans. Round to the nearest cent.

| | Principal | Rate | Time | Interest |
|---|---|---|---|---|
| **11.** | $15,000 | 14% | 180 days | _____ |
| **12.** | 600 | 12% | 281 days | _____ |
| **13.** | 1,800 | 16% | 300 days | _____ |
| **14.** | 2,400 | 14% | 280 days | _____ |
| **15.** | 12,500 | 12% | 396 days | _____ |

Find the missing element for each of the following loans. Round answers 22 to 25 to the nearest day. Round amounts to the nearest cent.

| | Principal | Rate | Time | Interest |
|---|---|---|---|---|
| **16.** | _____ | 11.50% | 6 months | $ 92 |
| **17.** | _____ | 12.00% | 3 months | 225 |
| **18.** | _____ | 12.00% | 5 months | 400 |
| **19.** | $10,000 | _____ | 2 years | 3,200 |
| **20.** | 6,000 | _____ | 1.5 years | 1,125 |

| | Principal | Rate | Time | Interest |
|---|---|---|---|---|
| **21.** | 8,000 | _____ | 3 months | 220 |
| **22.** | 12,000 | 16.00% | _____ days | 384 |
| **23.** | 7,000 | 12.50% | _____ days | 350 |
| **24.** | 10,000 | 14.50% | _____ days | 870 |
| **25.** | 17,500 | 11.80% | _____ days | 2,065 |

## BUSINESS APPLICATIONS

**26.** Tri-State Express needs to borrow $30,000 for 200 days. The interest rate is 12.5%. Tri-State's accountant estimated the interest using the ordinary-interest method. How much more or less will the actual interest be, calculated with the exact-interest method? Round to the nearest cent.

**27.** Smathers's Costume Rental Store borrowed $10,000 at 12% for 1 year. At the end of the year, Smathers was unable to pay off the loan, and the bank gave Smathers a new loan for the total amount due. The new loan is at 14% for 1 year. What is the total amount Smathers will owe the bank at the end of the second year?

**28.** It will cost Stoddard Concert Promotions $150,000 to rent a stadium and hire a nationally known rock band for a concert. How much interest must Stoddard pay if it borrows the $150,000 at 12.5% interest for 180 days? Use the exact-interest method.

**29.** Zelco, Inc., borrowed $30,000 for $1\frac{1}{2}$ years. The amount of interest was $5,175. What was the rate of interest?

When borrowers obtain a loan, they often have a future date in mind when they will have the cash available to pay off the loan. The number of days between the loan date and the due date must be determined before interest can be calculated on the loan.

To calculate the number of days in the loan period, add the number of days in each month from the loan date to the due date.

The following chart shows the number of days in each month of the year.

### Days in Each Month

| | |
|---|---|
| January, 31 | July, 31 |
| February, 28* | August, 31 |
| March, 31 | September, 30 |
| April, 30 | October, 31 |
| May, 31 | November, 30 |
| June, 30 | December, 31 |

*In a leap year, 29 days.

## EXAMPLE

On May 10 Poholsky Insurance Agency borrowed $5,000 at 13% interest, to be repaid on August 15. Determine the number of days in the loan period and calculate the interest on the loan using the exact-interest method.

## SOLUTION

**Step 1** Add the number of days in each month from May 10 to August 15.

```
  21   days in May (May 31 − 10 = 21)
  30   days in June
  31   days in July
+ 15   days in August
  97   days in loan period
```

**Step 2** Calculate the interest on the loan.

Principal × Rate × Time = Interest
$5,000 × 0.13 × $\frac{97}{365}$ = $172.739 ≈ $172.74

## PRACTICE PROBLEMS

Determine the number of days in each of the following loan periods and calculate the interest on each loan. Use the exact-interest method. Round your answers to the nearest cent.

| | Principal | Rate | Loan Date | Due Date | Days in Loan Period | Interest |
|---|---|---|---|---|---|---|
| 1. | $ 6,000 | 12% | March 15 | July 20 | _____ | _____ |
| 2. | 8,000 | 12.25% | August 1 | December 15 | _____ | _____ |
| 3. | 10,000 | 11.50% | March 7 | June 14 | _____ | _____ |
| 4. | 7,500 | 12% | July 18 | November 2 | _____ | _____ |
| 5. | 4,000 | 13% | April 1 | September 1 | _____ | _____ |

## USING A TIME TABLE

The time table below shows the chronological number of each day of the year. Each day of the year is assigned a number. For example, May 12 is day 132 of the year (133 of a leap year). Find the number of days from one date to another by subtracting.

### TIME TABLE

| Day of Month | Jan. | Feb. | Mar. | Apr. | May | June | July | Aug. | Sept. | Oct. | Nov. | Dec. |
|---|---|---|---|---|---|---|---|---|---|---|---|---|
| 1 | 1 | 32 | 60 | 91 | 121 | 152 | 182 | 213 | 244 | 274 | 305 | 335 |
| 2 | 2 | 33 | 61 | 92 | 122 | 153 | 183 | 214 | 245 | 275 | 306 | 336 |
| 3 | 3 | 34 | 62 | 93 | 123 | 154 | 184 | 215 | 246 | 276 | 307 | 337 |
| 4 | 4 | 35 | 63 | 94 | 124 | 155 | 185 | 216 | 247 | 277 | 308 | 338 |
| 5 | 5 | 36 | 64 | 95 | 125 | 156 | 186 | 217 | 248 | 278 | 309 | 339 |
| 6 | 6 | 37 | 65 | 96 | 126 | 157 | 187 | 218 | 249 | 279 | 310 | 340 |
| 7 | 7 | 38 | 66 | 97 | 127 | 158 | 188 | 219 | 250 | 280 | 311 | 341 |
| 8 | 8 | 39 | 67 | 98 | 128 | 159 | 189 | 220 | 251 | 281 | 312 | 342 |
| 9 | 9 | 40 | 68 | 99 | 129 | 160 | 190 | 221 | 252 | 282 | 313 | 343 |
| 10 | 10 | 41 | 69 | 100 | 130 | 161 | 191 | 222 | 253 | 283 | 314 | 344 |
| 11 | 11 | 42 | 70 | 101 | 131 | 162 | 192 | 223 | 254 | 284 | 315 | 345 |
| 12 | 12 | 43 | 71 | 102 | 132 | 163 | 193 | 224 | 255 | 285 | 316 | 346 |
| 13 | 13 | 44 | 72 | 103 | 133 | 164 | 194 | 225 | 256 | 286 | 317 | 347 |
| 14 | 14 | 45 | 73 | 104 | 134 | 165 | 195 | 226 | 257 | 287 | 318 | 348 |
| 15 | 15 | 46 | 74 | 105 | 135 | 166 | 196 | 227 | 258 | 288 | 319 | 349 |
| 16 | 16 | 47 | 75 | 106 | 136 | 167 | 197 | 228 | 259 | 289 | 320 | 350 |
| 17 | 17 | 48 | 76 | 107 | 137 | 168 | 198 | 229 | 260 | 290 | 321 | 351 |
| 18 | 18 | 49 | 77 | 108 | 138 | 169 | 199 | 230 | 261 | 291 | 322 | 352 |
| 19 | 19 | 50 | 78 | 109 | 139 | 170 | 200 | 231 | 262 | 292 | 323 | 353 |
| 20 | 20 | 51 | 79 | 110 | 140 | 171 | 201 | 232 | 263 | 293 | 324 | 354 |
| 21 | 21 | 52 | 80 | 111 | 141 | 172 | 202 | 233 | 264 | 294 | 325 | 355 |
| 22 | 22 | 53 | 81 | 112 | 142 | 173 | 203 | 234 | 265 | 295 | 326 | 356 |
| 23 | 23 | 54 | 82 | 113 | 143 | 174 | 204 | 235 | 266 | 296 | 327 | 357 |
| 24 | 24 | 55 | 83 | 114 | 144 | 175 | 205 | 236 | 267 | 297 | 328 | 358 |
| 25 | 25 | 56 | 84 | 115 | 145 | 176 | 206 | 237 | 268 | 298 | 329 | 359 |
| 26 | 26 | 57 | 85 | 116 | 146 | 177 | 207 | 238 | 269 | 299 | 330 | 360 |
| 27 | 27 | 58 | 86 | 117 | 147 | 178 | 208 | 239 | 270 | 300 | 331 | 361 |
| 28 | 28 | 59 | 87 | 118 | 148 | 179 | 209 | 240 | 271 | 301 | 332 | 362 |
| 29 | 29 | | 88 | 119 | 149 | 180 | 210 | 241 | 272 | 302 | 333 | 363 |
| 30 | 30 | | 89 | 120 | 150 | 181 | 211 | 242 | 273 | 303 | 334 | 364 |
| 31 | 31 | | 90 | | 151 | | 212 | 243 | | 304 | | 365 |

*Note:* For a leap year, add 1 day to each number of days after February 28.

ANSWERS FOR PRACTICE PROBLEMS

| | Days in Loan Period | Interest |
|---|---|---|
| (1) | 127 | $250.52 |
| (2) | 136 | 365.15 |
| (3) | 99 | 311.92 |
| (4) | 107 | 263.84 |
| (5) | 153 | 217.97 |

## EXAMPLE 1

On March 6 Woodman Industries borrowed $8,000 at 13% interest to be repaid on August 15. Use the time table on page 166 to find the number of days in the loan period.

### SOLUTION

**Step 1** Read down the "March" column to the number 6 in the "Day of Month" column. The number 65 indicates that March 6 is the 65th day of the year.

**Step 2** Read down the "August" column to the number 15 in the "Day of Month" column. The number 227 indicates that August 15 is the 227th day of the year.

**Step 3** Find the number of days in the loan period.

Due Date's    Loan Date's     Days in
Day of Year − Day of Year = Loan Period
    227     −     65     =     162

August 15 ⎯⎯⎯⎯↑      ↑⎯⎯⎯⎯⎯⎯⎯⎯ March 6

## EXAMPLE 2

Find the number of days in the loan period from October 10 to February 20 of the following year using the time table on page 166.

### SOLUTION

**Step 1** Find the number of days in the first year.

                 Loan Date's     Days in
Days in a Year − Day of Year = First Year
    365       −      283     =     82

↑⎯October 10

**Step 2** Find the number of days in the second year. The due date, February 20, is day 51.

**Step 3** Find the number of days in the loan period.

Days in      Days in       Days in
First Year + Second Year = Loan Period
   82      +      51     =     133

## PRACTICE PROBLEMS

Using the time table shown on page 166, determine the number of days in the loan period for the following loans.

| | Loan Date | Due Date | Days in Loan Period |
|---|---|---|---|
| **1.** | July 8 | February 14 (following year) | _____ |
| **2.** | March 4 | August 4 | _____ |
| **3.** | February 6 (leap year) | June 8 | _____ |

## FINDING THE DUE DATE

When a loan is obtained for a certain number of months or days, the due date of the loan must be calculated so the borrower knows when to plan on making repayment.

When the loan period is stated in months, count that number of months forward from the loan date. The due date is the same day of the month as the loan date.

When the loan period is stated in days, the due date is found by adding the number of days in each month until the total equals the number of days in the loan period.

### EXAMPLE 1

On May 16 Maskee Sports Center obtained a 6-month loan. What is the loan's due date?

### SOLUTION

Count forward 6 months from May 16. The due date is November 16.

### EXAMPLE 2

On March 8 Dassel Popcorn Manufacturing Company obtained a 180-day loan. What is the due date of the loan.

### SOLUTION

Add the number of days in each month until the total equals 180 days.

|  |  |  |
|---|---|---|
|  | 23 | days left in March (March 31 − 8 = 23) |
|  | 30 | days in April |
|  | 31 | days in May |
|  | 30 | days in June |
|  | 31 | days in July |
| Due date, | 31 | days in August |
| September 4 | + 4 | days in September |
|  | 180 | days in loan period |

### PRACTICE PROBLEMS

Determine the due date for the following loans.

| | Loan Date | Loan Period | Due Date |
|---|---|---|---|
| 1. | March 16 | 3 months | _____ |
| 2. | January 20 | 8 months | _____ |
| 3. | July 25 | 120 days | _____ |
| 4. | August 5 | 90 days | _____ |
| 5. | December 24 | 60 days | _____ |

The time table shown on page 166 can be used to find the due date of a loan.

---

**EXAMPLE 1**

On March 10 Shelby Fiberglass Company obtained a loan for 210 days. Use the time table to find the loan's due date.

**SOLUTION**

March 10 is the 69th day of the year. Add the number of days in the loan period, 210, to the day which represents the loan date, 69.

$$
\begin{array}{c c c}
\text{Loan Date's} & \text{Days in} & \text{Due Date's} \\
& & \text{Day of} \\
\text{Day of Year} + \text{Loan Period} = & \text{Year} \\
69 & + \quad 210 \quad = & 279
\end{array}
$$

Locate 279 in the table. The 279th day of the year, October 6, is the due date of the loan.

**EXAMPLE 2**

Find the due date of a 60-day note dated December 20.

**SOLUTION**

**Step 1** Find the number of days in the first year.

$$
\begin{array}{c c c}
\text{Days} & \text{Loan Date's} & \text{Days in} \\
\text{in a Year} - \text{Day of Year} = & \text{First Year} \\
365 & - \quad 354 \quad = & 11
\end{array}
$$

⎢_____ December 20

**Step 2** Find the number of days in the second year.

$$
\begin{array}{c c c}
\text{Days} & \text{Days in} & \text{Days in} \\
\text{in Loan Period} - \text{First Year} = & \text{Second Year} \\
60 & - \quad 11 \quad = & 49
\end{array}
$$

**Step 3** Find the due date. Locate number 49 on the time table.

The 49th day of the year, February 18, is the due date.

---

**PRACTICE PROBLEMS**

Use the time table on page 166 to find the due date of the following loans.

| Loan Date | Days in Loan Period | Due Date |
|---|---|---|
| **1.** October 6 | 145 | _____ |
| **2.** June 20 | 180 | _____ |
| **3.** July 16 | 95 | _____ |
| **4.** December 10 | 60 | _____ |

Copyright © 1989 by McGraw-Hill, Inc. All rights reserved.

## EARLY PAYMENT OF SINGLE-PAYMENT LOANS

Sometimes a business is able to pay back all or part of a loan earlier than the due date. In most cases, it makes sense for the business to pay early, thereby reducing the total interest charged. The rules that apply to such early payments are governed by the **United States Rule.** Under this rule, the payment applies initially to any interest owed. The balance of the payment is then applied to reduce the principal of the loan.

### EXAMPLE 1

Jelco, Inc., signed a 60-day promissory note for $25,000 at 13% interest. Jelco paid the note in full after 40 days. Use the United States Rule to find the amount of interest saved.

#### SOLUTION

**Step 1**  Calculate the interest for 40 days using a 360-day year. Round to the nearest cent.

Principal × Rate × Time = Interest

$$\$25,000 \times 0.13 \times \frac{40}{360} = \$361.11$$

**Step 2**  Calculate the interest on the 60-day note.

Principal × Rate × Time =          Interest

$$\$25,000 \times 0.13 \times \frac{60}{360} = \$541.666 \approx \$541.67$$

**Step 3**  Calculate the interest charge saved by early payment.

$541.67 − $361.11 = $180.56 Interest saved

### EXAMPLE 2

The Graham Construction Company was hired by the Popham Paper Company to build an addition to the factory. On January 9, the Graham Construction Company borrowed $20,000 at 12% for 60 days from the Dulong Bank to purchase material for construction.

On February 10 Graham gave the Dulong Bank $9,000 as a partial payment on the promissory note. Using the principles of the United States Rule, find the amount of interest saved by early partial payment.

#### SOLUTION

**Step 1**  Calculate interest for 32 days (January 9 to February 10) using a 360-day year. Round to the nearest cent.

Principal × Rate × Time = Interest

$$\$20,000 \times 0.12 \times \tfrac{32}{360} = \$213.33$$

**Step 2**  Find the amount of principal paid for this period (32 days).

Partial          Principal
Payment − Interest =   Paid
$9,000 − $213.33 = $8,786.67

**Step 3**  Find the balance of the principal due.

$20,000 − $8,786.67 = $11,213.33 Balance of principal

**Step 4** Calculate interest on the remaining principal for 28 days (February 11 to March 10) using a 360-day year. Round to the nearest cent.

Principal $\times$ Rate $\times$ Time = Interest

$11,213.33 \times 0.12 \times \frac{28}{360} = \$104.657 \approx \$104.66$

**Step 5** Calculate the interest on the 60-day note.

Principal $\times$ Rate $\times$ Time = Interest

$20,000 \times 0.12 \times \frac{60}{360} = \$400$

**Step 6** Calculate the interest saved by making a partial payment.

$400 - (\$213.33 + \$104.66) = \$82.01$ Interest saved by early partial payment

## PRACTICE PROBLEM

On October 6 Cooper Company borrowed $50,000 from the Nuckolls Loan Corporation. The 75-day promissory note had a 15% annual interest rate. November 15, Stewart Stores paid Cooper Company $30,000 against an outstanding bill. Cooper applied this amount to the promissory note as partial payment. Calculate the adjusted principal using the United States Rule and the amount of interest saved by making a partial payment.

## LESSON 5-4　　EXERCISES

Calculate the number of days in the loan period and the exact interest.
Round the interest amount to the nearest cent.

| | Principal | Rate | Loan Date | Due Date | Days in Loan Period | Exact Interest |
|---|---|---|---|---|---|---|
| 1. | $12,000 | 12.00% | June 20 | August 20 | _____ | _____ |
| 2. | 8,500 | 11.50% | May 1 | September 30 | _____ | _____ |
| 3. | 3,000 | 12.50% | April 10 | November 10 | _____ | _____ |
| 4. | 10,000 | 14.00% | July 16 | February 16 | _____ | _____ |

Find the due date of the following loans.

| | Loan Date | Loan Period | Due Date |
|---|---|---|---|
| 5. | January 16 | 4 months | _____ |
| 6. | July 9 | 5 months | _____ |
| 7. | May 15 | 460 days | _____ |
| 8. | April 1 | 180 days | _____ |

### BUSINESS APPLICATIONS

Use the exact-interest method to solve the following problems unless
otherwise directed. Round answers to the nearest cent.

9. On June 10 Pace Construction Company borrowed $35,000 at 12.5%
interest to be repaid with interest on September 15. (a) How many
days are in the loan period, and (b) what is interest on the loan?

(a) _____

(b) _____

10. On February 7 Olaf Footwear Company borrowed money to be repaid
in 90 days. When that loan came due, Olaf was unable to pay and the
loan was extended another 60 days. Now when is the loan due?

_____

11. Feinson Furniture, Inc., signed a 120-day promissory note for $50,000 at 14% interest. After 90 days, Feinson paid the note in full. Using the United States Rule and a 360-day year, find the amount of interest saved.

_____

12. On April 23 Richard Records signed a 90-day promissory note for $15,000 at 12% interest. Richard Records paid the note in full after 45 days. Use the United States Rule and a 360-day year to find the amount of interest saved.

_____

13. Graham Granary signed a 140-day note on February 5 for $45,600 at 14% interest. On March 19 Graham pays $16,000 toward the note. Using the principles of the United States Rule and a 360-day year, find the amount of interest saved by early partial payment.

_____

14. Matheny Manufacturing took out a 120-day promissory note for $5,000 at 14% annual interest on May 1. On May 31 a partial payment of $1,000 is made. Use the United States Rule to find the interest saved by a partial payment. Use 360-day year.

_____

## CHAPTER 5 REVIEW

### CHECKBOOK RECORDS

1. Complete the following check stubs using the information shown below. The ending balance on check stub no. 486 is $705.16.

| Date | Check No. | Amount | To | For |
|------|-----------|--------|-----|-----|
| 11/3 | 487 | $146.15 | Allen Office Co. | Office Supplies |
| 11/5 | Deposit | 300.00 | | |
| 11/5 | 488 | 250.00 | Jane Treynor | Wages |

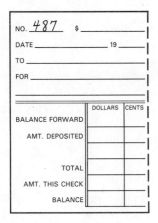

### BANK RECONCILIATION

Find the adjusted balances for each of the following checking accounts.

| | Ending Bank Statement Balance | Ending Checkbook Balance | Deposit in Transit | Outstanding Checks | Interest Income | Service Charge | Adjusted Balance |
|---|---|---|---|---|---|---|---|
| 2. | $ 805.60 | $830.40 | $200.00 | $105.40 73.10 | $5.20 | $ 8.50 | _____ |
| 3. | 1,205.90 | 632.35 | 0 | 426.50 138.15 | 8.90 | 0 | _____ |
| 4. | 520.15 | 285.27 | 500.00 | 321.50 316.80 105.00 | 1.78 | 10.20 | _____ |

## SIMPLE INTEREST

Calculate the simple interest on the following loans. Round answers to the nearest cent.

| | Principal | Rate | Time | Interest |
|---|---|---|---|---|
| 5. | $4,000 | 10% | 1 year | _____ |
| 6. | 7,500 | 12% | 2 years | _____ |
| 7. | 6,000 | 10% | 8 months | _____ |

## EXACT INTEREST

Use the exact-interest method to calculate the interest on the following loans. Round answers to the nearest cent.

| | Principal | Rate | Time | Interest |
|---|---|---|---|---|
| 8. | $10,000 | 12% | 180 days | _____ |
| 9. | 22,000 | 10% | 90 days | _____ |
| 10. | 900 | 11% | 300 days | _____ |

## ORDINARY INTEREST

Use the ordinary-interest method to calculate the approximate interest on each of the following loans. Round answers to the nearest cent.

| | Principal | Rate | Time | Interest |
|---|---|---|---|---|
| 11. | $ 2,000 | 12% | 90 days | _____ |
| 12. | 14,000 | 11% | 180 days | _____ |
| 13. | 25,000 | 10% | 240 days | _____ |

## SIMPLE-INTEREST TABLES

Use the simple-interest table on page 157 to calculate the amount of interest on each of the following loans.

| | Principal | Rate | Time | Interest |
|---|---|---|---|---|
| 14. | $12,000 | 14% | 180 days | _____ |
| 15. | 16,500 | 12% | 220 days | _____ |
| 16. | 9,800 | 16% | 120 days | _____ |

## FINDING PRINCIPAL, RATE, AND TIME

Find the missing element for each of the following loans. Round problem 19 to the nearest day.

| | Principal | Rate | Time | Interest |
|---|---|---|---|---|
| 17. | _____ | 12% | 1.5 years | $3,600.00 |
| 18. | 750 | _____ | 3 months | 26.25 |
| 19. | 14,000 | 14% | _____ | 2,352.00 |

## FINDING THE TIME PERIOD

Determine the number of days in each of the following loan periods and calculate the exact interest on each loan. Round answers to the nearest cent.

| | Principal | Rate | Loan Date | Due Date | Days in Loan Period | Interest |
|---|---|---|---|---|---|---|
| 20. | $ 7,800 | 12% | June 10 | October 15 | _____ | _____ |
| 21. | 12,500 | 13% | May 6 | November 10 | _____ | _____ |
| 22. | 1,350 | 11% | September 25 | December 5 | _____ | _____ |

## FINDING THE DUE DATE

Find the due date of the following loans.

| | Loan Date | Loan Period | Due Date |
|---|---|---|---|
| 23. | January 12 | 180 days | _____ |
| 24. | August 21 | 100 days | _____ |
| 25. | May 16 | 200 days | _____ |

## BUSINESS APPLICATIONS

Round amounts to the nearest cent.

26. Rose Travel Agency listed checks for the following amounts on its bank deposit slip: $485.62, $378.50, $792.46, $512.60, and $85.40. The travel agency withdrew $100.00 cash and deposited the rest in its checking account. What was the deposit in the checking account?

27. The bookkeeper at Flagstaff Trophy Company erroneously listed a $135.68 check as $168.35 on the check stub. In preparing the bank reconciliation, what adjustment should be made to correct this error?

28. An accounting trainee at Laughlin Landscaping Service calculated interest on a $12,000 loan at 10% interest for 180 days using the ordinary-interest method. How much more or less is the actual interest on the loan calculated with the exact-interest method?

**29.** Milford Mining Company needs to borrow $10,000 on June 15 to be repaid on October 15. Mt. Hood State Bank will make the loan at 14% interest. Ward County Credit Union will make the loan at 13% interest. How much interest will Milford Mining save by obtaining the loan from Ward County Credit Union? Use the exact-interest method. Round to the nearest cent.

_____

**30.** The Briggs Company signs a 150-day note on February 10. The note for $45,000 has an interest rate of 13%. On March 24, $18,000 is given as partial payment on the note. Find (a) the total interest on the note and (b) the amount paid on the due date. Use a 360-day year.

(a) _____

(b) _____

**31.** On May 10 Long Associates signed a 180-day promissory note for $50,000 at 15% interest. After 80 days, Long Associates paid the note in full. Use the United States Rule to find the amount of interest saved. Use a 360-day year.

_____

## CUMULATIVE REVIEW I

### NUMERATION

**1.** Write 5,128.63 in words. _____

_____

**2.** Round 24,479 to the nearest hundred. _____

### ADDITION

**3.** $23.9 + 1,030 + 1.928 + 417.23 =$

**4.**
$$13\frac{4}{9}$$
$$5\frac{1}{6}$$
$$+107\frac{2}{3}$$

### SUBTRACTION

**5.** $\$28,061.15 - \$9,507.39 =$

**6.**
$$24\frac{1}{3}$$
$$-\ 8\frac{4}{7}$$

### MULTIPLICATION

**7.**
$$\$16.09$$
$$\times\ \ \ 4.3$$

**8.** $\$62.58 \times 100 =$

**9.** $2\frac{4}{9} \times 2\frac{5}{11} =$

### DIVISION

**10.** Round to the nearest hundredth.

$0.82\overline{)52.20}$

**11.** $3\frac{4}{12} \div 3\frac{1}{8} =$

NAME _____ DATE _____

## BUSINESS APPLICATIONS

Round amounts to the nearest cent. Round percents to the nearest tenth.

12. The number of new homes being built in the county increased by 59% from last year to this year. If 179,500 homes were built last year, how many homes were built this year?

_____

13. Total expenses for the month are $145,600. Calculate the percent the salary expense of $57,350 is of the total expenses.

_____

14. Last year Bill Drull devoted 65% of his time to wholesale accounts and made sales of $227,500. What would Bill's total sales have been for the year if 100% of his time had been devoted to wholesale accounts?

_____

15. The Grover Company had a total advertising expense of $23,760 in March and $22,530 in April. Determine the percent of decrease from March to April in advertising expense.

_____

16. Martin Vagley worked 47 hours last week. His hourly rate is $22.50 an hour plus time and a half for hours worked over 40 in a week. Calculate Martin's net pay if his weekly payroll deductions total $312.13.

_____

17. Karen Hoffman receives an annual salary of $29,000 plus a 12% bonus on annual sales above $350,000. Calculate Karen's yearly gross earnings if her annual sales are $422,000.

_____

# CHAPTER 6

# PURCHASING GOODS

## LEARNING OBJECTIVES

1. Determine trade discount and net price using trade discount rate, series discount, and single equivalent discount rate.
2. Complete purchase orders and invoices.
3. Determine cash discount and cash price using various dating methods.
4. Determine credit given on partial payments.

Retailers are merchants who operate businesses such as clothing stores, department stores, appliance stores, and record shops that sell goods to consumers. **Retailers** buy their goods from manufacturers or wholesalers. **Wholesalers** buy goods or **merchandise** from **manufacturers** and sell to retailers.

Most manufacturers and wholesalers furnish retailers with a catalog which lists the products they have available for sale. Retailers select the items they want to buy and complete an order form called a **purchase order.** The manufacturer or wholesaler ships the goods to the retailer with a **sales invoice,** which lists the items being shipped and the amount due.

In this chapter, you will learn the calculations involved with the purchase of goods.

## LESSON 6-1    TRADE DISCOUNTS

### SINGLE TRADE DISCOUNTS

Most manufacturers' and wholesalers' catalogs show a suggested retail price for each product, called the **list price.** Retailers ordinarily receive a **trade discount,** which is a reduction from the list price. Trade discounts are often shown as a percent of the list price and appear either in the catalog or on a separate **discount sheet.** The latter is often used, if the manufacturer or wholesaler must raise or lower prices, because a new discount sheet can be printed and distributed without the necessity of printing new catalogs with revised list prices.

When only one trade discount is offered, it is called a **single trade discount.**

To find the trade discount, multiply the list price by the trade discount rate.

List Price × Trade Discount Rate = Trade Discount

The **net price** is what the retailer pays for goods and is calculated by deducting the trade discount from the list price.

List Price − Trade Discount = Net Price

## EXAMPLE

Empire Furniture Plaza ordered a grandfather clock with a $3,000 list price. The manufacturer's discount sheet shows a 45% trade discount. Find the net price.

## SOLUTION

**Step 1**  Calculate the trade discount.

List Price × Trade Discount Rate = Trade Discount
  $3,000  ×  0.45  =  $1,350

**Step 2**  Find the net price.

List Price − Trade Discount = Net Price
  $3,000  −  $1,350  = $1,650

## PRACTICE PROBLEMS

Calculate the trade discount and the net price for each of the following purchases.

|  | List Price | Trade Discount Rate | Trade Discount | Net Price |
|---|---|---|---|---|
| 1. | $1,259 | 40% | _____ | _____ |
| 2. | 3,547 | 45% | _____ | _____ |
| 3. | 513 | 36% | _____ | _____ |
| 4. | 859 | 55% | _____ | _____ |
| 5. | 5,236 | 42% | _____ | _____ |

**USING THE NET PRICE RATE**

Another method for finding the net price is to use the **net price rate.** The list price is considered to be 100%. The difference between 100% and the trade discount rate is the complement of the trade discount or the net price rate. The list price multiplied by the net price rate gives the net price. Using this method saves time, particularly if the net price rate can be determined mentally.

List Price × Net Price Rate = Net Price

ANSWERS FOR
PRACTICE PROBLEMS

Trade Discount  Net Price
(1)  $ 503.60  $ 755.40
(2)  1,596.15  1,950.85
(3)  184.68  328.32
(4)  472.45  386.55
(5)  2,199.12  3,036.88

## EXAMPLE

Downtown Electronics ordered an acoustic guitar with a $500 list price. The manufacturer offers a 48% trade discount. Calculate the net price using the net price rate.

## SOLUTION

**Step 1** Find the net price rate.

$$\begin{array}{ccc} & \text{Trade} & \\ 100\% - \text{Discount Rate} & = & \text{Net Price Rate} \\ 100\% - \quad 48\% & = & 52\% \end{array}$$

**Step 2** Find the net price.

$$\begin{array}{ccc} \text{List Price} \times \text{Net Price Rate} & = & \text{Net Price} \\ \$500 \quad \times \quad 0.52 & = & \$260 \end{array}$$

## PRACTICE PROBLEMS

Determine the net price rate and the net price for each of the following purchases.

| | List Price | Trade Discount Rate | Net Price Rate | Net Price |
|---|---|---|---|---|
| 1. | $ 1,471 | 38% | _____ | _____ |
| 2. | 719 | 48% | _____ | _____ |
| 3. | 44,123 | 50% | _____ | _____ |
| 4. | 37,564 | 55% | _____ | _____ |

## SERIES TRADE DISCOUNTS

Often, retailers are offered more than one trade discount on the same item. Two or more discounts are called a **series discount** or **chain discount.** A series discount, such as 40%, 10%, 5% (written 40, 10, 5) might be offered to encourage quantity buying, as a special promotion, or to adjust for price changes. The manufacturer may also offer a series discount if the customer is a wholesaler or retailer.

Series discounts may be deducted in any order, but each discount is subtracted from the balance *after* the preceding discount is deducted. For example, a series discount of 30, 20 means that a 30% discount is deducted from the list price, then a 20% discount is deducted from the remaining balance to find the net price.

ANSWERS FOR
PRACTICE PROBLEMS

| | Net Price Rate | Net Price |
|---|---|---|
| (1) | 62% | $ 912.02 |
| (2) | 52% | 373.88 |
| (3) | 50% | 22,061.50 |
| (4) | 45% | 16,903.80 |

Rhodes Furniture World is offered a series discount of 40, 10, 5 on a bedroom set with a $2,000 list price. What is the net price?

**SOLUTION**

**Step 1** Find the first trade discount and the remaining balance.

$$
\begin{array}{ccc}
 & \text{Trade} & \text{First} \\
\text{List Price} \times \text{Discount Rate} & = & \text{Trade Discount} \\
\$2,000 \quad \times \quad 0.40 & = & \$800
\end{array}
$$

$$
\begin{array}{ccc}
 & \text{First} & \\
\text{List Price} - \text{Trade Discount} & = & \text{Balance} \\
\$2,000 \quad - \quad \$800 & = & \$1,200
\end{array}
$$

**Step 2** Find the second trade discount and the remaining balance.

$$
\begin{array}{ccc}
\text{Balance After} & \text{Trade} & \text{Second} \\
\text{First Discount} \times \text{Discount Rate} & = & \text{Trade Discount} \\
\$1,200 \quad \times \quad 0.10 & = & \$120
\end{array}
$$

$$
\begin{array}{ccc}
\text{Balance After} & \text{Second} & \text{Balance After} \\
\text{First Discount} - \text{Trade Discount} & = & \text{Second Discount} \\
\$1,200 \quad - \quad \$120 & = & \$1,080
\end{array}
$$

**Step 3** Find the third trade discount and the net price.

$$
\begin{array}{ccc}
\text{Balance After} & \text{Trade} & \text{Third} \\
\text{Second Discount} \times \text{Discount Rate} & = & \text{Trade Discount} \\
\$1,080 \quad \times \quad 0.05 & = & \$54
\end{array}
$$

$$
\begin{array}{ccc}
\text{Balance After} & \text{Third} & \\
\text{Second Discount} - \text{Trade Discount} & = & \text{Net Price} \\
\$1,080 \quad - \quad \$54 & = & \$1,026
\end{array}
$$

Series discounts are never added together. For example, a series discount of 20, 15 is not the same as a discount of 35%.

**EXAMPLE**

Harry's Hardware can purchase lawn mowers either from Garden Suppliers, Inc., which offers a 35% discount, or from Home Improvements, Inc., who gives a series discount of 20, 15. Both offer the mowers at $125 each. Which is the better buy?

**SOLUTION**

**Step 1** Find Garden Suppliers' net price.

List Price × Trade Discount Rate = Trade Discount
$125 × 0.35 = $43.75

List Price − Trade Discount = Net Price
$125.00 − $43.75 = $81.25

**Step 2** Find Home Improvements' net price.

List Price − First Trade Discount = Balance After First Discount
$125 − $25 = $100

$125 × 0.20

|  | Balance After First Discount | − | Second Trade Discount | = | Net Price After Second Trade Discount |
|---|---|---|---|---|---|
| | $100 | − | $15 | = | $85 |

$100 × 0.15

**Step 3**  Compare the two net prices.

Garden Suppliers, Inc., offers the better discount.

$85.00 > $81.25

The symbol > means the first term is greater than the second term

## PRACTICE PROBLEMS

Find the net price for each of the following purchases and round to the nearest cent.

| | List Price | Series Discounts | Net Price |
|---|---|---|---|
| **1.** | $6,139 | 40, 10, 5 | _____ |
| **2.** | 9,528 | 50, 10, 10 | _____ |
| **3.** | 864 | 30, 10, 5 | _____ |
| **4.** | 1,432 | 45, 20, 10 | _____ |

## NET PRICE RATES

Like the single trade discount, when a series discount is offered, the net price can be easily calculated using the net price rates. First subtract each discount from 100% and change each difference to decimals. The product of the decimals is the decimal equivalent of the net price rates. Then multiply the decimal equivalent of the net price rate by the list price to find the net price. Round answers to four decimal places.

## EXAMPLE

Empire Furniture Plaza offers a series discount of 40, 20, and 5 on a dining room set with a $2,500 list price. What is the net price?

### SOLUTION

**Step 1**  Find the complement of the trade discounts and change to decimals.

| List Price Percent | − | Trade Discount Rate | = | Complement of the Trade Discount |
|---|---|---|---|---|
| 100% | − | 40% | = | 60% or 0.60 |
| 100% | − | 20% | = | 80% or 0.80 |
| 100% | − | 5% | = | 95% or 0.95 |

**Step 2**  Multiply to find the decimal equivalent of the net price rates.

| Complement of the Trade Discount = | Decimal Equivalent of Net Price Rate |
|---|---|
| 0.60 × 0.80 × 0.95 = | 0.456 |

**Step 3**  Find the net price.

| List Price × | Decimal Equivalent of Net Price Rate | = Net Price |
|---|---|---|
| $2,500 × | 0.456 | = $1,140 |

Calculate the decimal equivalent of the net price rate and the net price for each of the following purchases.

| | List Price | Series Discount | Decimal Equivalent of Net Price Rate | Net Price |
|---|---|---|---|---|
| 1. | $ 3,000 | 40, 30, 20 | _____ | _____ |
| 2. | 5,828 | 30, 10, 5 | _____ | _____ |

**SINGLE EQUIVALENT DISCOUNT**

A series or chain discount may also be expressed as a *single equivalent discount* (SED). To calculate the single equivalent discount rate, first find the net price rate, then subtract the net price rate from 100%. Buyers and sellers may compare a series discount with a single discount by converting a series discount to an SED. Multiply the SED rate by the list price to find the discount.

**EXAMPLE**

In the previous example, a dining room set lists for $2,500 and carries a series discount of 40, 20, 5; the decimal equivalent of the net price rate was found to be 0.456. Find the SED rate and discount.

**SOLUTION**

100% − Net Price Rate =    SED Rate
100% −     45.6%    = 54.4% or 0.544

List Price × SED Rate = Discount
  $2,500  ×   0.544   = $1,360

**PRACTICE PROBLEMS**

Calculate the net price rate, SED, discount amount, and net price. Round to nearest cent.

| | List Price | Series Discounts | Net Price Rate | SED | Discount Amount | Net Price |
|---|---|---|---|---|---|---|
| 1. | $21,234 | 15, 10, 10 | _____ | ____ | _____ | _____ |
| 2. | 1,671 | 40, 5, 5 | _____ | ____ | _____ | _____ |

## PURCHASE ORDERS

After a decision has been made to buy certain goods, the next step is to order them by completing a purchase order.

All items being ordered are listed on the purchase order along with the quantity of each item, a description, the item's catalog number, and the price per item, called the **unit price.** The total amount for each item, called the **extension,** is found by multiplying the quantity ordered by the unit price. The **net price** of the goods ordered is found by adding the extensions in the total column.

In some industries, purchase order amounts are shown as the net price. In others, the list price is shown, and trade discounts are deducted to reach the net price for goods ordered.

### EXAMPLE

Empire Furniture Plaza purchased items from Nu-Style Furniture Manufacturing Company with the following list prices: 5 desks at $432; 3 recliners at $296; 6 couches at $436; and 1 bedroom suite at $2,052. A 45% trade discount is granted. Complete the purchase order for Empire Furniture Plaza.

### PURCHASE ORDER

Empire Furniture Plaza
10641 Graham Ave.
Emmett, Idaho 83617

PO-5787

Nu-Style Furniture Manufacturing Co.
4050 East 10th St.
Portland, Oregon 97285

Purchase order number must appear on all letters and packages.

| Date Issued | Date Needed | Req. No. | Terms |
|---|---|---|---|
| 9/18/-- | 10/18/-- | 92412 | 2/10, net 30 |

| Via | Truck | FOB Destination |
|---|---|---|

| QUAN. RECD. | QUAN. ORDERED | STOCK NO. | DESCRIPTION | UNIT PRICE | TOTAL |
|---|---|---|---|---|---|
| | 5 | 10914 | Desks | $432.00 | **$2,160.00** |
| | 3 | 92770 | Recliners | 296.00 | **888.00** |
| | 6 | 12263 | Couches | 436.00 | **2,616.00** |
| | 1 | 31045 | Bedroom suite | 2,052.00 | **2,052.00** |
| | | | Total | | **$7,716.00** |
| | | | Trade Discount | 45% | **3,472.20** |
| | | | Net Price | | **4,243.80** |

By _Morgan Rice_
Purchasing Agent

Copy 1—Supplier

**SOLUTION**

**Step 1**  Extend the purchase order.

Quantity × Price = Extension
| | | |
5 × $ 432 = $2,160.00
3 × 296 = 888.00
6 × 436 = 2,616.00
1 × 2,052 = 2,052.00

**Step 2**  Total the extensions.

$2,160.00 + $888.00 + $2,616.00 + $2,052.00 = $7,716.00

**Step 3**  Calculate the trade discount.

List Price × Trade Discount Rate = Trade Discount
$7,716.00 ×            0.45            =       $3,472.20

**Step 4**  Calculate the net price.

List Price − Trade Discount = Net Price
$7,716.00 −       $3,472.20       = $4,243.80

PRACTICE PROBLEM

The following purchase order is partially completed. Extend the items and find the net price.

**PURCHASE ORDER**

DeLong Designs
226 West 13th St.
Westfield, MA 30249

PO–42352

French Furniture Co.
31146 Xavier Avenue
New York, NY 10009

Purchase order number must appear on all letters and packages.

| Date Issued | Date Needed | Req. No. | Terms |
|---|---|---|---|
| 4/12 | 5/15/–– | 13427 | 2/10, net 30 |

| Via | | Truck | | FOB | Shipping point |
|---|---|---|---|---|---|

| QUAN. RECD. | QUAN. ORDERED | STOCK NO. | DESCRIPTION | UNIT PRICE | TOTAL |
|---|---|---|---|---|---|
| | 30 | 5920 | Armchairs | $139.50 | |
| | 20 | 11963 | Rocking chairs | 198.50 | |
| | 50 | 4214 | Chairs (captain's) | 96.75 | |
| | | | | | |
| | | | Total Trade Discount | 48% | |
| | | | Net Price | | |

By _Arthur Read_
Purchasing Agent

Copy 1—Supplier

Extensions: $4,185.00, $3,970.00, $4,837.50; Total: $12,992.50; Trade discount: $6,236.40; Amount ordered: $6,756.10

**188** ■ LESSON 6-1

## LESSON 6-1          EXERCISES

Calculate the trade discount and the net price for each of the following purchases. Round to the nearest cent.

| Item | List Price | Trade Discount Rate | Trade Discount | Net Price |
|------|-----------|---------------------|----------------|-----------|
| 1. Machinery | $11,298 | 30% | _____ | _____ |
| 2. Tools | 437 | 40% | _____ | _____ |
| 3. Computer | 9,521 | 20% | _____ | _____ |
| 4. VCR | 750 | 25% | _____ | _____ |
| 5. Dune buggy | 3,600 | 45% | _____ | _____ |
| 6. Mink coat | 18,000 | 44% | _____ | _____ |
| 7. CD | 650 | 50% | _____ | _____ |

Determine the net price rate and the net price for each of the following purchases. Round to the nearest cent.

| Item | List Price | Trade Discount Rate | Net Price Rate | Net Price |
|------|-----------|---------------------|----------------|-----------|
| 8. Tractor | $ 6,234 | 30% | _____ | _____ |
| 9. Refrigerator | 1,519 | 40% | _____ | _____ |
| 10. Dishwasher | 423 | 44% | _____ | _____ |
| 11. Digital watch | 580 | 35% | _____ | _____ |
| 12. Designer suit | 700 | 25% | _____ | _____ |
| 13. Silver tea set | 3,800 | 45% | _____ | _____ |
| 14. Automobile | 16,000 | 38% | _____ | _____ |

Calculate the decimal equivalent of the net price rate and the net price for each of the following purchases. Round to the nearest cent.

| | List Price | Series Discounts | Decimal Equivalent of Net Price Rate | Net Price |
|---|-----------|------------------|--------------------------------------|-----------|
| 15. | $ 7,123 | 40, 20, 10 | _____ | _____ |
| 16. | 16,456 | 30, 10, 10 | _____ | _____ |
| 17. | 3,789 | 50, 10 | _____ | _____ |
| 18. | 850 | 30, 20, 5 | _____ | _____ |
| 19. | 1,800 | 20, 20, 10 | _____ | _____ |
| 20. | 500 | 20, 10, 10 | _____ | _____ |
| 21. | 2,200 | 40, 5 | _____ | _____ |

## BUSINESS APPLICATIONS

**22.** Boone Office Supply Company purchased five calculators with a list price of $125 each. The seller offers a 42% trade discount. What is Boone's total amount due for the calculators?

_____

**23.** Adams Manufacturing lists office desks at $310 with series trade discounts of 30 and 10%. Janerette Furniture Manufacturers lists comparable desks at $350 with a trade discount of 46%. How much more or less will each cost from Adams?

_____

**24.** The Bookworm, a retail bookstore, estimates it will sell 100 Maxwell Contemporary Dictionaries this year. The list price is $20 each. If the dictionaries are purchased from the publisher in small quantities throughout the year, Bookworm will receive a 44% trade discount. If Bookworm orders all 100 books at one time, the trade discount will be 48%. How much can Bookworm save on the cost of the 100 dictionaries by ordering them all at one time?

_____

**25.** The following amounts are listed on Andecker Jewelers' purchase order: 10 watches at $120, 15 bracelets at $100, and 20 rings at $350. A 50% trade discount is granted. What is the total amount due for the order?

_____

**26.** Boyd's Music City plans to buy some Foremost compact disc players for sale in its store. The list price is $400. If 1 to 20 are purchased, the trade discount is 40%. If 21 to 40 are purchased, the trade discount is 40 and 10%, and if 41 to 60 are purchased, the trade discount is 40, 10, and 10%. What is the cost per player at each of these quantities?

_____

Upon receiving a retailer's purchase order, the manufacturer or whole-saler fills the order and ships the goods. A printed document called an *invoice,* which identifies quantities shipped, the amount due, and other information, is sent along with the goods or is mailed to the buyer separately. A *sales invoice* from the Nu-Style Furniture Manufacturing Company to Empire Furniture Plaza is shown below.

Nu-Style Furniture
Manufacturing Co.
4050 East 10th St.
Portland, Oregon 97285

**INVOICE NO.** 3776

SOLD TO: Empire Furniture Plaza
10641 Graham Ave.
Emmett, Idaho 83617

Invoice Date: 10/16/--

SHIP TO: Same

Terms:    2/10, n/30

| Purchase Order No. | Date Shipped | Shipped Via | FOB | No. of Packages |
|---|---|---|---|---|
| 5787 | 10/17/-- | UPS | Destination | 15 |

| QUANTITY | STOCK NUMBER | DESCRIPTION | UNIT PRICE | AMOUNT | |
|---|---|---|---|---|---|
| 5 | 10914 | Desks | $  432.00 | $2,160.00 | 5 × $  432.00 |
| 3 | 92770 | Recliners | 296.00 | 888.00 | 3 ×    296.00 |
| 6 | 12263 | Couches | 436.00 | 2,616.00 | 6 ×    436.00 |
| 1 | 31045 | Bedroom suite | 2,052.00 | 2,052.00 | 1 ×  2,052.00 |
| | | Trade Discount 45% | | 3,472.20 | .45 ×  7,716.00 |
| | | Net Price | | $4,243.80 | $7,716.00 — 3,472.20 |

An important item on the sales invoice is the *terms of payment.* This identifies the date by which the buyer must pay for the goods. A *cash discount* may be offered for prompt or early payments. The amount paid is called the *cash price,* which is calculated by deducting the cash discount from the net price. In this lesson, you will learn about three types of cash discounts.

**ORDINARY DATING**

The *ordinary dating* discount method identifies a specific time span from the invoice date within which a cash discount is allowed. In addition, the last date full payment is allowed is also shown. The *terms* on the above invoice, 2/10, n/30, mean the following: The buyer may deduct a 2% cash discount if payment is made within 10 days of the invoice date, on or before October 26. The 10-day period is called the *discount period.* The net price is due 30 days from the invoice date, November 15. This is called the *credit period.*

Sometimes, terms are expressed as 3/10, 1/20, n/30. These terms mean that if payment is made within 10 days of the invoice date, a 3% cash discount is allowed, and if payment is made after 10 days but before 20 days of the invoice date, a 1% discount is allowed. Otherwise, the net price is due within 30 days of the invoice date.

## EXAMPLE

On January 4 Bookworm, Inc., purchased dictionaries with a $1,040 net price from Maxwell Publishing Company. Terms of sale were 2/10, n/30. Bookworm paid for the books on January 14. Find (a) the last day of the discount period, (b) the cash discount, and (c) the cash price.

## SOLUTION

**Step 1**  Find the last day of the discount period.

|  | Days in Cash | Last Day, |
| Invoice Date + | Discount Period = | Discount Period |
| January 4 + | 10 = | January 14 |

January 14 is within the discount period.

**Step 2**  Find the cash discount.

Net Price × Cash Discount Rate = Cash Discount
$1,040 ×          0.02          =     $20.80

**Step 3**  Find the cash price.

Net Price − Cash Discount = Cash Price
$1,040.00 −     $20.80     = $1,019.20

## PRACTICE PROBLEMS

For each of the following invoices, find the cash discount (if allowed) and the cash price.

|  | Net Price | Terms of Payment | Invoice Date | Date Paid | Cash Discount | Cash Price |
|---|---|---|---|---|---|---|
| 1. | $3,000 | 3/10, n/30 | September 25 | October 5 | _____ | _____ |
| 2. | 2,500 | 1/10, n/30 | October 26 | November 6 | _____ | _____ |
| 3. | 300 | 2/20, n/45 | April 17 | May 6 | _____ | _____ |
| 4. | 600 | 2/10, 1/20, n/30 | February 2 | February 12 | _____ | _____ |
| 5. | 1,200 | 3/10, 1/20, n/30 | March 4 | March 24 | _____ | _____ |

**END-OF-MONTH DATING**

In **end-of-month (EOM) dating,** the counting of days in the discount period starts from the end of the month in which the invoice is dated. Thus, if an invoice is dated June 15 with terms 2/10, n/30 EOM, the cash discount period ends on July 10 and the credit period ends on July 30.

ANSWERS FOR PRACTICE PROBLEMS

| | Cash Discount | Cash Price |
|---|---|---|
| (1) | $90.00 | $2,910.00 |
| (2) | 0 | 2,500.00 |
| (3) | 6.00 | 294.00 |
| (4) | 12.00 | 588.00 |
| (5) | 12.00 | 1,188.00 |

Ordinarily, if the invoice is dated on or after the twenty-sixth of the month, the counting of days in the cash discount period starts from the end of the following month. For instance, if an invoice is dated June 26 with terms 3/10, n/30 EOM, the cash discount period ends *August 10* and the credit period ends on *August 30*.

---

## EXAMPLE

On May 6, Bassano Electronics purchased 5 video cameras with a total net price of $6,375. Terms of payment were 2/10, n/30 EOM. Payment was made on June 7. Find the cash price.

## SOLUTION

**Step 1**  Find the last day of the cash discount period.

| Month Following Invoice Date | + | Days in Cash Discount Period | = | Last Day of Cash Discount Period |
|---|---|---|---|---|
| June | + | 10 | = | June 10 |

Payment was made within the discount period.

**Step 2**  Find the cash discount.

| Net Price × | Cash Discount Rate | = Cash Discount |
|---|---|---|
| $6,375 × | 0.02 | = $127.50 |

**Step 3**  Find the cash price.

Net Price − Cash Discount = Cash Price
$6,375  −  $127.50  = $6,247.50

---

## PRACTICE PROBLEMS

For the following invoices, find the cash discount (if allowable) and the cash price.

| | Net Price | Terms of Payment | Invoice Date | Date Paid | Cash Discount | Cash Price |
|---|---|---|---|---|---|---|
| 1. | $2,000 | 3/10, n/30 EOM | September 1 | October 30 | _____ | _____ |
| 2. | 750 | 1/10, n/30 EOM | March 16 | April 9 | _____ | _____ |
| 3. | 400 | 2/10, n/60 EOM | April 24 | May 10 | _____ | _____ |

## RECEIPT-OF-GOODS DATING

In *receipt-of-goods* (ROG) *dating,* the counting of days in the cash discount period starts from the date the goods are *received,* not the invoice date. This method is used when shipping time is long. Invoices may arrive overnight, but the merchandise may not arrive for a week or more.

### EXAMPLE

Douglas Sporting Goods Company received an invoice dated May 2 with terms 3/10, n/30 ROG for tennis rackets at the cost of $1,087. The goods were received on May 7. Find the last day of the cash discount period; the cash discount, if any, when the invoice is paid on May 17; and the cash price.

### SOLUTION

**Step 1**  Find the last day of the cash discount period.

| Date Goods Were Received | + | Days in Cash Discount Period | = | Last Day of Cash Discount Period |
|---|---|---|---|---|
| May 7 | + | 10 | = | May 17 |

Since the invoice was paid May 17, a cash discount may be taken.

**Step 2**  Calculate cash discount.

| Invoice Price | × | Cash Discount Rate | = | Cash Discount |
|---|---|---|---|---|
| $1,087 | × | 0.03 | = | $32.61 |

**Step 3**  Determine the cash price.

| Invoice Price | − | Cash Discount | = | Cash Price |
|---|---|---|---|---|
| $1,087 | − | $32.61 | = | $1,054.39 |

### PRACTICE PROBLEMS

| | Net Price | Terms of Payment | Invoice Date | Date Goods Received | Date Paid | Cash Discount | Cash Price |
|---|---|---|---|---|---|---|---|
| 1. | $3,500 | 2/10, n/30 ROG | 6/6 | 6/10 | 6/19 | _____ | _____ |
| 2. | 1,600 | 2/10, n/30 ROG | 1/2 | 1/8 | 1/18 | _____ | _____ |
| 3. | 2,100 | 3/10, n/30 ROG | 4/20 | 4/24 | 5/15 | _____ | _____ |

## LESSON 6-2    EXERCISES

For the following invoices, find the cash discount (if allowable) and the cash price.

| | Net Price | Terms of Payment | Invoice Date | Date Goods Received | Date Paid | Cash Discount | Cash Price |
|---|---|---|---|---|---|---|---|
| **1.** | $4,000 | 2/10, n/30 | 5/4 | 5/6 | 5/14 | _____ | _____ |
| **2.** | 1,500 | 3/10, 1/20, n/30 | 4/10 | 4/14 | 5/3 | _____ | _____ |
| **3.** | 2,000 | 2/10, n/30 EOM | 7/16 | 7/20 | 8/10 | _____ | _____ |
| **4.** | 3,500 | 3/10, n/30 ROG | 9/20 | 9/24 | 10/4 | _____ | _____ |
| **5.** | 600 | 2/10, 1/20, n/30 | 1/3 | 1/5 | 1/25 | _____ | _____ |

### BUSINESS APPLICATIONS

**6.** Ellsworth Jewelry purchased merchandise with a net price of $24,000. Terms of sale are 2/10, n/60. The invoice is dated July 8 and Ellsworth paid on July 17. How much should Ellsworth pay?

_____

**7.** Carmichael Office Supply Store purchased a filing cabinet with a $450 list price. The manufacturer offers a 40% trade discount and terms of 2/10, n/30. How much will Carmichael owe if they pay within the cash discount period?

_____

**8.** Thayer Sporting Goods Company purchased merchandise with a $1,500 net price. Thayer was granted a cash discount for prompt payment and submitted $1,455 in full payment within the discount period. What percent was Thayer's cash discount?

_____

9. Julie Saunders Decorating received an invoice dated August 10 for 12 wall decorations with a $150 list price each and a 50% trade discount. Terms of sale are 2/10, n/30. How much does Saunders Decorating owe if payment is made (a) on August 20 or (b) on September 9?

(a) _____

(b) _____

10. Mayville Auto Accessories received an invoice dated November 21 for $16,500. Terms were 3/10, 1/20, n/30. How much does Mayville owe if payment is made (a) on November 30 or (b) on December 12?

(a) _____

(b) _____

11. Venetta Boat Works received an invoice dated March 27 for boat accessories for $4,750. The goods were received April 3. Terms were 2/10, n/30 EOM. (a) What is the last day Venetta can take advantage of the cash discount? (b) How much should be paid on that date?

(a) _____

(b) _____

12. Sandy's Book Store purchased merchandise with a list price of $1,600 and a 44% trade discount. The invoice was dated December 1 and terms were 2/10, n/30. How much does Sandy's owe in full payment if payment is made on December 11?

_____

Goods are shipped from the seller to the buyer via truck, bus, air, rail, ship, or a combination of these methods. Smaller packages are often sent by the U.S. Postal Service or by private delivery companies such as United Parcel Service (UPS). Larger orders are shipped by freight companies. The shipping cost is determined by the method used, the package's weight, the distance to be shipped, and any special services selected, such as faster service. Of course the faster the service, the higher the cost.

**PARCEL POST**

*Parcel post,* also called *fourth-class mail,* is a service of the U.S. Postal Service that is commonly used for shipping packages weighing a maximum of 70 pounds and measuring no more than 108 inches in combined length and girth.

**PARCEL POST ZONE RATES**
**(Fourth-Class Mail)**

| Weight 1 Pound and not exceeding (pounds) | Local | Zones 1–2 | Zone 3 | Zone 4 | Zone 5 | Zone 6 | Zone 7 | Zone 8 |
|---|---|---|---|---|---|---|---|---|
| 2 | 1.35 | 1.41 | 1.51 | 1.66 | 1.89 | 2.13 | 2.25 | 2.30 |
| 3 | 1.41 | 1.49 | 1.65 | 1.87 | 2.21 | 2.58 | 2.99 | 3.87 |
| 4 | 1.47 | 1.57 | 1.78 | 2.08 | 2.54 | 3.03 | 3.57 | 4.74 |
| 5 | 1.52 | 1.65 | 1.92 | 2.29 | 2.86 | 3.47 | 4.16 | 5.62 |
| 6 | 1.58 | 1.74 | 2.05 | 2.50 | 3.18 | 3.92 | 4.74 | 6.49 |
| 7 | 1.63 | 1.82 | 2.19 | 2.71 | 3.51 | 4.37 | 5.32 | 7.36 |
| 8 | 1.69 | 1.90 | 2.32 | 2.92 | 3.83 | 4.82 | 5.91 | 8.25 |
| 9 | 1.75 | 1.99 | 2.46 | 3.13 | 4.15 | 5.26 | 6.49 | 9.12 |
| 10 | 1.80 | 2.07 | 2.59 | 3.34 | 4.48 | 5.71 | 7.07 | 10.00 |
| 11 | 1.85 | 2.13 | 2.70 | 3.49 | 4.71 | 6.03 | 7.49 | 10.62 |
| 12 | 1.90 | 2.20 | 2.79 | 3.65 | 4.94 | 6.34 | 7.89 | 11.23 |
| 13 | 1.94 | 2.26 | 2.89 | 3.79 | 5.15 | 6.63 | 8.27 | 11.78 |
| 14 | 1.98 | 2.32 | 2.98 | 3.92 | 5.35 | 6.90 | 8.62 | 12.30 |
| 15 | 2.02 | 2.37 | 3.06 | 4.04 | 5.54 | 7.15 | 8.94 | 12.79 |
| 16 | 2.06 | 2.42 | 3.14 | 4.16 | 5.71 | 7.39 | 9.25 | 13.24 |
| 17 | 2.10 | 2.48 | 3.22 | 4.27 | 5.88 | 7.61 | 9.54 | 13.67 |
| 18 | 2.14 | 2.52 | 3.29 | 4.38 | 6.03 | 7.83 | 9.81 | 14.08 |
| 19 | 2.18 | 2.57 | 3.36 | 4.48 | 6.18 | 8.03 | 10.07 | 14.46 |
| 20 | 2.21 | 2.62 | 3.43 | 4.58 | 6.33 | 8.22 | 10.32 | 14.83 |
| 21 | 2.25 | 2.67 | 3.49 | 4.67 | 6.47 | 8.41 | 10.56 | 15.18 |
| 22 | 2.28 | 2.71 | 3.56 | 4.76 | 6.60 | 8.59 | 10.79 | 15.51 |
| 23 | 2.32 | 2.75 | 3.62 | 4.85 | 6.73 | 8.76 | 11.01 | 15.83 |
| 24 | 2.35 | 2.80 | 3.68 | 4.94 | 6.85 | 8.92 | 11.22 | 16.14 |
| 25 | 2.39 | 2.84 | 3.74 | 5.02 | 6.97 | 9.08 | 11.42 | 16.43 |
| 61 | 3.44 | 4.08 | 5.35 | 7.15 | 9.90 | 12.87 | 16.16 | 23.22 |
| 62 | 3.47 | 4.11 | 5.38 | 7.20 | 9.96 | 12.94 | 16.25 | 23.35 |
| 63 | 3.50 | 4.14 | 5.42 | 7.24 | 10.02 | 13.02 | 16.35 | 23.48 |
| 64 | 3.52 | 4.17 | 5.46 | 7.29 | 10.08 | 13.10 | 16.44 | 23.61 |
| 65 | 3.55 | 4.20 | 5.49 | 7.33 | 10.14 | 13.17 | 16.53 | 23.74 |
| 66 | 3.58 | 4.23 | 5.53 | 7.38 | 10.19 | 13.24 | 16.62 | 23.86 |
| 67 | 3.60 | 4.26 | 5.57 | 7.42 | 10.25 | 13.32 | 16.71 | 23.99 |
| 68 | 3.63 | 4.29 | 5.60 | 7.47 | 10.31 | 13.39 | 16.80 | 24.11 |
| 69 | 3.66 | 4.32 | 5.64 | 7.51 | 10.37 | 13.46 | 16.88 | 24.23 |
| 70 | 3.68 | 4.35 | 5.67 | 7.56 | 10.43 | 13.53 | 16.97 | 24.35 |

A rate schedule like the sample shown on page 197 is available from the post office. *Zones* refer to shipping destinations, which are determined by the ZIP Codes of the receiver of the goods, called the **addressee.** Any fraction of a pound is charged for at the next higher weight classification.

## EXAMPLE

Mattson Book Specialists shipped a book to a customer via parcel post. The package weighed $3\frac{1}{4}$ pounds. The customer is located in Zone 7. What is the cost of shipping the package?

## SOLUTION

Round the weight up to the next weight classification, 4 pounds. On the parcel post rate schedule, read down the "weight" column until you come to 4. Read across to the right until you come to Zone 7. The charge is $3.57.

## PRACTICE PROBLEMS

Determine the shipping charges for the following packages sent via parcel post.

| | Package Weight, pounds | Addressee's Zone | Shipping Charge |
|---|---|---|---|
| **1.** | 5 | 3 | _____ |
| **2.** | $20\frac{1}{2}$ | 8 | _____ |
| **3.** | 65 | 4 | _____ |
| **4.** | 11 | Local | _____ |
| **5.** | $68\frac{1}{3}$ | 5 | _____ |

**OTHER DELIVERY SERVICES**

United Parcel Service (UPS) is one example of the many delivery services available. UPS offers fast delivery of packages weighing a maximum of 70 pounds and measuring no more than 108 inches in combined length and girth.

UPS offers three basic services. *Ground service* is delivery via truck. Destinations are divided into zones, similar to the system used by the U.S. Postal Service for parcel post shipments. *Next-day air* service provides for delivery the day following shipment. *Second-day air* service provides for delivery within 2 days following shipment.

Additional charges for services such as *COD (collect on delivery)* are identified at the bottom of the rate schedule.

As with parcel post shipments, any fraction of a pound is charged at the next higher weight classification.

ANSWERS FOR PRACTICE PROBLEMS

(1) $1.92  (2) $15.18  (3) $7.33  (4) $1.85  (5) $10.37

## GROUND SERVICE

### ZONES

| WEIGHT NOT TO EXCEED | 2 | 3 | 4 | 5 | 6 | 7 | 8 |
|---|---|---|---|---|---|---|---|
| 1 lb. | $1.23 | $1.32 | $1.46 | $1.52 | $1.59 | $1.67 | $1.74 |
| 2 " | 1.24 | 1.34 | 1.63 | 1.73 | 1.87 | 2.01 | 2.16 |
| 3 " | 1.32 | 1.48 | 1.80 | 1.95 | 2.15 | 2.36 | 2.57 |
| 4 " | 1.40 | 1.61 | 1.97 | 2.16 | 2.43 | 2.70 | 2.99 |
| 5 " | 1.49 | 1.76 | 2.13 | 2.37 | 2.70 | 3.05 | 3.40 |
| 6 " | 1.57 | 1.89 | 2.30 | 2.59 | 2.98 | 3.39 | 3.82 |
| 7 " | 1.65 | 2.02 | 2.47 | 2.80 | 3.26 | 3.74 | 4.24 |
| 8 " | 1.73 | 2.14 | 2.64 | 3.02 | 3.54 | 4.08 | 4.65 |
| 9 " | 1.82 | 2.27 | 2.81 | 3.23 | 3.82 | 4.43 | 5.07 |
| 10 " | 1.90 | 2.39 | 2.97 | 3.44 | 4.09 | 4.77 | 5.48 |
| 11 " | 1.98 | 2.52 | 3.14 | 3.66 | 4.37 | 5.12 | 5.90 |
| 12 " | 2.06 | 2.65 | 3.31 | 3.87 | 4.65 | 5.46 | 6.32 |
| 13 " | 2.15 | 2.77 | 3.48 | 4.09 | 4.93 | 5.81 | 6.73 |
| 14 " | 2.23 | 2.90 | 3.65 | 4.30 | 5.21 | 6.15 | 7.15 |
| 15 " | 2.31 | 3.02 | 3.81 | 4.51 | 5.48 | 6.50 | 7.56 |
| 16 " | 2.39 | 3.15 | 3.98 | 4.73 | 5.76 | 6.84 | 7.98 |
| 17 " | 2.48 | 3.28 | 4.15 | 4.94 | 6.04 | 7.19 | 8.40 |
| 18 " | 2.56 | 3.40 | 4.32 | 5.16 | 6.32 | 7.53 | 8.81 |
| 19 " | 2.64 | 3.53 | 4.49 | 5.37 | 6.60 | 7.88 | 9.23 |
| 20 " | 2.72 | 3.65 | 4.65 | 5.58 | 6.87 | 8.22 | 9.64 |
| 61 " | 5.43 | 7.78 | 10.17 | 12.60 | 15.97 | 19.56 | 23.26 |
| 62 " | 5.45 | 7.81 | 10.22 | 12.66 | 16.04 | 19.65 | 23.36 |
| 63 " | 5.47 | 7.84 | 10.26 | 12.71 | 16.11 | 19.74 | 23.47 |
| 64 " | 5.49 | 7.88 | 10.30 | 12.76 | 16.18 | 19.83 | 23.57 |
| 65 " | 5.51 | 7.91 | 10.35 | 12.82 | 16.25 | 19.92 | 23.67 |
| 66 " | 5.53 | 7.94 | 10.39 | 12.87 | 16.32 | 20.01 | 23.78 |
| 67 " | 5.55 | 7.97 | 10.43 | 12.93 | 16.39 | 20.10 | 23.88 |
| 68 " | 5.57 | 8.00 | 10.48 | 12.98 | 16.46 | 20.19 | 23.98 |
| 69 " | 5.59 | 8.03 | 10.52 | 13.04 | 16.53 | 20.28 | 24.09 |
| 70 " | 5.61 | 8.06 | 10.56 | 13.09 | 16.59 | 20.37 | 24.19 |

## NEXT DAY AIR

### ZONES

| PAKS & PACKAGES — WEIGHT NOT TO EXCEED | 48 & HAWAII — K | PUERTO RICO — M |
|---|---|---|
| 1 lb. | $11.50 | $14.00 |
| 2 " | 12.50 | 15.00 |
| 3 " | 13.50 | 16.00 |
| 4 " | 14.50 | 17.00 |
| 5 " | 15.50 | 18.00 |
| 6 " | 16.50 | 19.00 |
| 7 " | 18.00 | 20.50 |
| 8 " | 19.00 | 21.50 |
| 9 " | 20.00 | 22.50 |
| 10 " | 21.00 | 23.50 |
| 11 " | 22.00 | 24.50 |
| 12 " | 23.00 | 25.50 |
| 13 " | 24.00 | 26.50 |
| 14 " | 25.50 | 28.00 |
| 15 " | 26.50 | 29.00 |
| 16 " | 27.50 | 30.00 |
| 17 " | 28.50 | 31.00 |
| 18 " | 29.50 | 32.00 |
| 19 " | 30.50 | 33.00 |
| 20 " | 32.00 | 34.50 |
| 61 " | 67.00 | 69.50 |
| 62 " | 67.50 | 70.00 |
| 63 " | 67.50 | 70.00 |
| 64 " | 68.00 | 70.50 |
| 65 " | 68.50 | 71.00 |
| 66 " | 68.50 | 71.00 |
| 57 " | 69.00 | 71.50 |
| 68 " | 69.00 | 71.50 |
| 69 " | 69.50 | 72.00 |
| 70 " | 69.50 | 72.00 |

## 2ND DAY AIR

### ZONES

| WEIGHT NOT TO EXCEED | 48 STATES — A | HAWAII — D | PUERTO RICO — E |
|---|---|---|---|
| 1 lb. | $3.00 | $4.53 | $5.50 |
| 2 " | 4.00 | 5.71 | 6.50 |
| 3 " | 5.00 | 6.89 | 7.50 |
| 4 " | 6.00 | 8.07 | 8.50 |
| 5 " | 6.50 | 9.25 | 9.00 |
| 6 " | 7.50 | 10.43 | 10.00 |
| 7 " | 8.50 | 11.61 | 11.00 |
| 8 " | 9.50 | 12.79 | 12.00 |
| 9 " | 10.50 | 13.97 | 13.00 |
| 10 " | 11.50 | 15.15 | 14.00 |
| 11 " | 12.50 | 16.33 | 15.00 |
| 12 " | 13.50 | 17.51 | 16.00 |
| 13 " | 14.50 | 18.69 | 17.00 |
| 14 " | 15.50 | 19.87 | 18.00 |
| 15 " | 16.50 | 21.05 | 19.00 |
| 16 " | 17.50 | 22.23 | 20.00 |
| 17 " | 18.50 | 23.41 | 21.00 |
| 18 " | 19.50 | 24.59 | 22.00 |
| 19 " | 20.00 | 25.77 | 22.50 |
| 20 " | 21.00 | 26.95 | 23.50 |
| 61 " | 53.00 | 65.60 | 55.50 |
| 62 " | 53.50 | 65.89 | 56.00 |
| 63 " | 53.50 | 66.19 | 56.00 |
| 64 " | 53.50 | 66.48 | 56.00 |
| 65 " | 54.00 | 66.78 | 56.50 |
| 66 " | 54.00 | 67.07 | 56.50 |
| 67 " | 54.50 | 67.37 | 57.00 |
| 68 " | 54.50 | 67.66 | 57.00 |
| 69 " | 55.00 | 67.96 | 57.50 |
| 70 " | 55.00 | 68.25 | 57.50 |

ANY FRACTION OF A POUND OVER THE WEIGHT SHOWN TAKES THE NEXT HIGHER RATE

## NEXT DAY AIR LETTER $8.50 ZONE L

**ADDITIONAL CHARGES:**
For each COD received for collection — $1.90
For each Address Correction — $1.90
For each Acknowledgment of Delivery (AOD) — 30 cents
For each package with a declared value over $100 — 25 cents for each additional $100 or fraction thereof
Weekly service charge — $3.25

**WEIGHT AND SIZE LIMITS:**
Maximum weight per package — 70 POUNDS
Maximum size per package — 108 inches in length and girth combined
Minimum charge for a package measuring over 84 inches in length and girth combined will be equal to the charge for a package weighing 25 pounds.

**AIR RESTRICTIONS:**
The maximum value for an air service package is $25,000 and the maximum carrier liability is $25,000.
No Call Tag service provided in Air Service.
Hazardous materials are prohibited in Air Service.

Some Air Shipments may be shipped by Surface Transportation. Rates for Next Day Air Letter and Puerto Rico effective June 3, 1985.

## EXAMPLE 1

Harold's Photography of Shreveport, Louisiana, shipped several sample photo albums to a customer in Rock Island, Illinois, via UPS next-day air service. The package weighed $12\frac{1}{2}$ pounds. What was the charge for this shipment?

### SOLUTION

Round the weight up to the next weight classification, 13 pounds. On the next-day air schedule, read down the "Weight Not to Exceed" column until you come to 13. Read across to the right to the "Zones—48 States and Hawaii" column. The charge is $24.00.

## EXAMPLE 2

Canterberry Enterprises shipped a package valued at $489.50 via UPS. What was the additional charge because this package's value is above the $100 base value?

### SOLUTION

**Step 1** Refer to the "Additional Charges" section at the bottom of the UPS rate schedule. Determine the additional value.

Package's Value
(Rounded)    − $100 Base Value = Additional Value
    $500     −    $100     =     $400

**Step 2** Determine the additional charge.

    $100s of
Additional Value × Charge per $100 = Additional Charge
      4     ×     $0.25     =     $1.00

### PRACTICE PROBLEMS

Determine the shipping charges for the following packages sent via UPS.

| | Package Weight, pounds | Service | Destination | Shipping Charge |
|---|---|---|---|---|
| **1.** | 18 | Ground | Zone 5 | _____ |
| **2.** | $65\frac{1}{2}$ | Ground | Zone 8 | _____ |
| **3.** | 6 | Next-day air | Honolulu, Hawaii | _____ |
| **4.** | $17\frac{1}{4}$ | Second-day air | Hollywood, California | _____ |
| **5.** | 3 | Next-day air | Flagstaff, Arizona | _____ |

Determine the additional charges for each of the following UPS services.

| Service | Additional Charge |
|---|---|
| **6.** COD | _____ |
| **7.** Package value, $195 | _____ |
| **8.** Package value, $560 | _____ |
| **9.** AOD | _____ |
| **10.** Package value, $350 | _____ |

## LESSON 6-3    EXERCISES

Determine the shipping charges for the following packages sent via parcel post.

| | Package Weight, pounds | Addressee's Zone | Shipping Charges |
|---|---|---|---|
| **1.** | 7 | 3 | _____ |
| **2.** | $8\frac{1}{2}$ | 5 | _____ |
| **3.** | 24 | 8 | _____ |
| **4.** | 67 | 2 | _____ |
| **5.** | 13 | Local | _____ |

Determine the shipping charges for the following packages sent via UPS.

| | Package Weight, pounds | Service | Destination | Shipping Charges |
|---|---|---|---|---|
| **6.** | 3 | Next-day air | Billings, Montana | _____ |
| **7.** | 65 | Ground | Zone 5 | _____ |
| **8.** | $8\frac{1}{2}$ | Next-day air | Cheyenne, Wyoming | _____ |
| **9.** | 70 | Ground | Zone 8 | _____ |
| **10.** | 11 | Second-day air | Kailua, Hawaii | _____ |

Determine the total shipping charges for the following packages sent via UPS, which have additional charges for extra services.

| | Package Weight, pounds | Service | Destination | Additional Services | Shipping Charges |
|---|---|---|---|---|---|
| **11.** | $65\frac{1}{2}$ | Ground | Zone 5 | Package value, $850 | _____ |
| **12.** | 10 | Next-day air | Spokane, Washington | COD | _____ |
| **13.** | $18\frac{1}{4}$ | Second-day air | Massillon, Ohio | AOD | _____ |
| **14.** | 7 | Next-day air | Enid, Oklahoma | Package value, $285 | _____ |
| **15.** | $62\frac{1}{2}$ | Ground | Zone 3 | Package value, $470 | _____ |

## BUSINESS APPLICATIONS

**16.** Layfayette, Inc., plans to ship a $19\frac{1}{2}$-pound package to a customer in Tulsa, Oklahoma, which is in UPS ground service Zone 6. What will the shipping charges be if the merchandise is shipped via UPS (a) ground service, (b) second-day air, (c) next-day air?

(a) _____

(b) _____

(c) _____

**17.** Insurance Education Specialists plans to ship a 16-pound box of materials to Oscar Oslow Insurance Agency of St. Cloud, Minnesota. How much less will the shipping charges be if UPS second-day air is used instead of next-day air?

_____

**18.** Torgy Manufacturing Company plans to ship a 65-pound box of merchandise to a customer in St. Louis, Missouri. The customer is in Zone 5 of both the parcel post and UPS ground service schedules. Which service will cost less and by how much?

_____

**19.** David Carey's company is transferring him to Los Angeles. A moving company will move his possessions for $1,050. David's company will pay him that amount, and he can move by any means he chooses, pocketing any savings in moving cost. David, an accountant, has determined that he can pack all of his belongings in 82 boxes weighing 70 pounds each. How much more or less will it cost to ship David's belongings via parcel post instead of by the moving company? Los Angeles is in parcel post Zone 5.

_____

The buyer is responsible for paying for the goods according to the terms of payment shown on the purchase invoice. Payment varies according to such factors as discounts and shipping charges. In addition, if the buyer does not have enough cash on hand to make payment within the cash discount period, the buyer might have the option to make a partial payment or might borrow the cash to take advantage of the cash discount.

**FREIGHT CHARGES**

Freight charges for shipping the goods are paid by either the seller or the buyer, according to the *shipping terms* shown on the sales invoice. If the shipping terms are stated as *FOB destination,* the seller pays the freight charges. If the terms are *FOB shipping point,* the buyer pays. FOB stands for *free on board.*

If the buyer is to pay the freight charges, the seller will often make arrangements with the shipper and pay the shipper on the buyer's behalf. The seller then bills the buyer for the freight charges on the invoice.

Freight charges prepaid by the seller are not subject to a cash discount. If freight charges are included in the invoice total, they must be deducted before applying any cash discount. The freight charges are then added to the cash price to find the total amount due.

**EXAMPLE**

Halonen Giftwares received a $5,300 invoice dated August 3 with terms 2/10, n/30. The invoice includes freight charges of $300 and shipping terms of FOB shipping point. What is the total amount due if Halonen makes payment within the cash discount period?

**SOLUTION**

**Step 1**  Find the net price for the goods.

Invoice Total − Freight Charges = Net Price
   $5,300    −    $300    =  $5,000

**Step 2**  Calculate the cash price.

Net Price × Cash Discount Rate = Cash Discount
   $5,000   ×      0.02       =    $100

Net Price − Cash Discount = Cash Price
   $5,000  −    $100    =  $4,900

**Step 3**  Find the total amount due.

Cash Price + Freight Charges = Total Amount Due
   $4,900  +    $300    =      $5,200

Find the cash discount and the total amount due on each of the following invoices. All shipping terms are FOB shipping point. The freight charges are included in the invoice total.

| | Invoice Total | Freight Charges | Terms of Payment | Invoice Date | Date Paid | Cash Discount | Total Amount Due |
|---|---|---|---|---|---|---|---|
| 1. | $4,800 | $200 | 2/10, n/30 | July 2 | July 12 | _____ | _____ |
| 2. | 1,950 | 50 | 3/10, n/30 EOM | March 15 | May 9 | _____ | _____ |
| 3. | 185 | 15 | 2/10, n/30 | October 3 | December 2 | _____ | _____ |
| 4. | 3,780 | 150 | 2/10, 1/20, n/30 | May 4 | May 24 | _____ | _____ |
| 5. | 2,475 | 125 | 2/10, n/30 EOM | June 28 | August 10 | _____ | _____ |

**PARTIAL PAYMENT**   A customer may not have enough cash to pay the invoice amount within the discount period. If the customer makes a *partial payment* within the discount period, the seller may allow a cash discount on the partial payment. The seller must then deduct the amount of credit granted for the payment from the net price to determine the amount that the buyer still owes.

To find the amount of credit given, first calculate the cash price rate, then divide the partial payment by the cash price rate.

---

### EXAMPLE

Florine's Gifts received a shipment of porcelain gift items. The invoice showed a net price of $25,000 and terms of 3/10, n/30. On the last day of the cash discount period, Florine's made a cash payment of $15,000. Calculate (a) the amount of credit received for partial payment and (b) the remaining balance owed. Round your answers to the nearest cent.

### SOLUTION

**Step 1**   Calculate the cash price percent.

Net Price Rate − Cash Discount Percent = Cash Price Rate

100% − 3% = 97%

**Step 2**   Calculate the credit earned for the partial payment.

Amount Paid ÷ Cash Price Rate = Credit Earned

$15,000 ÷ 0.97 = $15,463.92

**Step 3**   Calculate the remaining balance.

Net Price − Credit Earned = Remaining Balance

$25,000.00 − $15,463.92 = $9,536.08

---

ANSWERS FOR
PRACTICE PROBLEMS

| | Cash Discount | Total Amount Due |
|---|---|---|
| (1) | $92.00 | $4,708.00 |
| (2) | 0 | 1,950.00 |
| (3) | 0 | 185.00 |
| (4) | 36.30 | 3,743.70 |
| (5) | 47.00 | 2,428.00 |

Each of the following partial payments was made within the cash discount period. Calculate the cash discount rate, the credit earned, and the remaining balance owed on the invoice. Round your answers to the nearest cent.

| | Net Price | Terms | Partial Payment | Cash Price Rate | Credit Earned | Remaining Balance |
|---|---|---|---|---|---|---|
| 1. | $10,000 | 2/10, n/30 | $ 6,000 | ____ | _____ | _____ |
| 2. | 800 | 1/10, n/30 | 500 | ____ | _____ | _____ |
| 3. | 6,000 | 3/10, n/45 | 2,000 | ____ | _____ | _____ |
| 4. | 15,000 | 2/10, n/30 | 10,000 | ____ | _____ | _____ |
| 5. | 2,000 | 3/10, n/60 | 1,200 | ____ | _____ | _____ |

## BORROWING MONEY TO OBTAIN CASH DISCOUNTS

Is a 1%, 2%, or 3% cash discount worth taking? Is it worth borrowing money at 10%, 12%, or 14% interest to take advantage of the cash discount? These are decisions that business managers must make, and the answer to these questions is often yes. If the buyer does not have sufficient cash on hand to take advantage of a cash discount, it may be worth borrowing the money if the amount of interest on the loan is lower than the amount of the cash discount. Before deciding to borrow the money in order to take advantage of the cash discount, the cash discount should be compared to the interest on the loan.

It is assumed in this discussion that the business would follow the prudent practice of paying for the merchandise on the last day of the discount period and that a loan would be repaid at the end of the credit period. For example, with terms 2/10, n/30 the loan period would be 20 days.

### EXAMPLE

Caledonia Souvenir Center received an invoice dated July 15 for goods with a $10,000 net price. The terms were 2/10, n/30. Caledonia can borrow the cash price for 20 days at 12% interest, calculated with the exact-interest method. Should Caledonia borrow money in order to take advantage of the discount? If so, what would be the savings? Round to the nearest cent.

### SOLUTION

**Step 1**  Calculate the cash discount and cash price.

Net Price × Cash Discount Rate = Cash Discount
$10,000 ×          0.02          =     $200

Net Price − Cash Discount = Cash Price

$10,000 −      $200     =   $9,800

**Step 2**   Calculate interest on the loan.

Principal (Cash Price) × Rate × Time = Interest

$$\$9,800 \qquad \times\ 0.12 \times \frac{20}{365} = \$64.44$$

**Step 3**   Find the savings.

Cash Discount − Interest = Savings

  $200.00    − $64.44 = $135.56

By borrowing $9,800 and paying the invoice on the last day of the discount period, the Caledonia Souvenir Center will save $135.56. In this example, it is more advantageous to obtain a loan to pay the invoice within the discount period.

## PRACTICE PROBLEMS

Calculate the amount of cash discount, the interest for borrowing the cash price from the end of the cash discount period to the end of the credit period, and the savings (+) or loss (−) from borrowing to obtain the cash discount. Use a 12% interest rate and the exact-interest method to calculate interest. Round to the nearest cent.

|     | Net Price | Terms | Cash Discount | Interest | Savings (+) or Loss (−) |
|-----|-----------|-------|---------------|----------|-------------------------|
| 1.  | $ 8,000   | 2/10, n/30 | _____ | _____ | _____ |
| 2.  | 15,000    | 1/10, n/60 | _____ | _____ | _____ |
| 3.  | 600       | 3/10, n/30 | _____ | _____ | _____ |
| 4.  | 1,800     | 2/20, n/45 | _____ | _____ | _____ |
| 5.  | 2,000     | 1/30, n/60 | _____ | _____ | _____ |

## LESSON 6-4    EXERCISES

Find the cash discount and the total amount due on each of the following invoices. All shipping terms are FOB shipping point. The freight charges are included in the invoice total.

| | Invoice Total | Freight Charges | Terms of Payment | Invoice Date | Date Paid | Cash Discount | Total Amount Due |
|---|---|---|---|---|---|---|---|
| 1. | $ 8,350 | $350 | 1/10, n/30 | 3/16 | 4/15 | _____ | _____ |
| 2. | 7,250 | 375 | 3/10, n/30 | 1/12 | 1/22 | _____ | _____ |
| 3. | 460 | 75 | 2/10, n/30 | 7/18 | 7/28 | _____ | _____ |
| 4. | 15,080 | 480 | 2/10, 1/20, n/30 | 8/4 | 8/24 | _____ | _____ |
| 5. | 1,360 | 200 | 2/10, n/30 EOM | 3/20 | 5/10 | _____ | _____ |

Each of the following partial payments was made within the cash discount period. Calculate (a) the credit earned and (b) the remaining balance owed on the invoice. Round your answers off to the nearest cent.

| | Net Price | Terms | Partial Payment | Credit Earned | Remaining Balance |
|---|---|---|---|---|---|
| 6. | $ 1,600 | 1/10, n/45 | $ 1,000 | _____ | _____ |
| 7. | 3,500 | 2/20, n/60 | 2,000 | _____ | _____ |
| 8. | 14,500 | 2/10, n/30 | 5,000 | _____ | _____ |
| 9. | 17,600 | 1/10, n/45 | 10,000 | _____ | _____ |
| 10. | 8,200 | 3/10, n/60 | 6,000 | _____ | _____ |

Shown below are the net price and terms for several invoices. Calculate the amount of cash discount, the interest for borrowing the cash price from the end of the cash discount period to the end of the credit period, and the savings or loss from borrowing to take advantage of the cash discount. Use a 12% interest rate and the exact-interest method to calculate interest. Round to the nearest cent.

| | Net Price | Terms | Cash Discount | Interest | Savings (+) or Loss (−) |
|---|---|---|---|---|---|
| 11. | $ 3,500 | 2/10, n/30 | _____ | _____ | _____ |
| 12. | 1,800 | 1/10, n/60 | _____ | _____ | _____ |
| 13. | 12,000 | 2/20, n/45 | _____ | _____ | _____ |
| 14. | 15,000 | 1/10, n/30 | _____ | _____ | _____ |
| 15. | 1,400 | 3/10, n/30 | _____ | _____ | _____ |

## BUSINESS APPLICATIONS

**16.** Fairchild Awards and Trophies purchased goods with an invoice total of $2,850, which included $90 in freight charges. The invoice was dated March 14 and terms were 2/10, n/30. Fairchild made full payment on March 24. How much did Fairchild owe?

_____

**17.** J.J.'s Fishing Equipment Shop received an $1,800 invoice with terms of 3/10, n/30. On the last day of the cash discount period, J.J.'s only has $700 it can use to make a partial payment. How much will J.J.'s still owe on the invoice after making this payment? Round to the nearest cent.

_____

**18.** Burlington Publishing Company received an invoice dated July 16 for $45,000 from Clinton Printing Company. Terms were 2/10, n/60. On July 26, Burlington paid $20,000. How much credit did Burlington receive for this payment? Round to the nearest cent.

_____

**19.** Burns Auto Supplies received goods with a $40,000 net price. Terms were 3/10, n/60. On the last day of the cash discount period, Burns does not have cash available to take advantage of the cash discount. Burns can borrow the cash price at 13% interest for 50 days, calculated using the exact-interest method. How much will Burns save or lose by borrowing to obtain the cash discount? Round to the nearest cent.

_____

## CHAPTER 6 REVIEW

### SINGLE TRADE DISCOUNTS

Calculate the trade discount and the net price for each of the following purchases.

|  | List Price | Trade Discount Rate | Trade Discount | Net Price |
|---|---|---|---|---|
| 1. | $3,000 | 40% | _____ | _____ |
| 2. | 7,000 | 45% | _____ | _____ |
| 3. | 900 | 48% | _____ | _____ |
| 4. | 1,250 | 44% | _____ | _____ |

### TRADE DISCOUNTS—USING THE NET PRICE RATE

Determine the net price rate and the net price for each of the following purchases.

|  | List Price | Trade Discount Rate | Net Price Rate | Net Price |
|---|---|---|---|---|
| 5. | $ 300 | 30% | _____ | _____ |
| 6. | 1,500 | 40% | _____ | _____ |
| 7. | 3,200 | 44% | _____ | _____ |
| 8. | 2,800 | 46% | _____ | _____ |

### SERIES TRADE DISCOUNTS

Find the SED, discount amount, and net price for each of the following purchases.

|  | List Price | Series Discounts | SED | Discount Amount | Net Price |
|---|---|---|---|---|---|
| 9. | $5,000 | 40%, 20%, 10% | _____ | _____ | _____ |
| 10. | 8,000 | 30%, 20%, 5% | _____ | _____ | _____ |
| 11. | 1,500 | 40%, 10%, 10% | _____ | _____ | _____ |
| 12. | 1,000 | 30%, 20%, 10% | _____ | _____ | _____ |

### CASH DISCOUNTS

For the invoices listed on page 210, find the cash discount, if allowed, and the cash price.

| | Net Price | Terms of Payment | Invoice Date | Date Goods Received | Date Paid | Cash Discount | Cash Price |
|---|---|---|---|---|---|---|---|
| 13. | $ 2,000 | 2/10, n/30 | 6/6 | 6/8 | 6/10 | _____ | _____ |
| 14. | 15,000 | 3/10, n/30 EOM | 5/14 | 5/18 | 6/9 | _____ | _____ |
| 15. | 7,500 | 1/10, n/30 ROG | 8/2 | 8/4 | 8/14 | _____ | _____ |
| 16. | 400 | 2/10, n/60 | 9/21 | 9/24 | 11/1 | _____ | _____ |

## FREIGHT CHARGES

Determine the shipping charges for each of the following sent via either parcel post or UPS. Use the appropriate rate schedules in this chapter.

| | Package Weight, pounds | Type of Service | Addressee's Location | Shipping Charge |
|---|---|---|---|---|
| 17. | 6 | UPS next-day air | Iowa | _____ |
| 18. | 11 | UPS second-day air | Hawaii | _____ |
| 19. | 16 | Parcel post | Zone 8 | _____ |
| 20. | 68 | UPS ground service | Zone 5 | _____ |

## PAYING FREIGHT CHARGES

Find the cash discount and the total amount due on each of the following invoices. All shipping terms are FOB shipping point. The freight charges are included in the invoice total.

| | Invoice Total | Freight Charges | Terms of Payment | Invoice Date | Date Paid | Cash Discount | Total Amount Due |
|---|---|---|---|---|---|---|---|
| 21. | $5,200 | $400 | 2/10, n/30 | 5/7 | 5/17 | _____ | _____ |
| 22. | 3,750 | 150 | 3/10, n/60 | 6/20 | 6/30 | _____ | _____ |
| 23. | 875 | 75 | 1/10, n/30 | ·7/10 | 7/20 | _____ | _____ |
| 24. | 1,680 | 180 | 2/10, n/45 | 11/8 | 11/18 | _____ | _____ |

## PARTIAL PAYMENT

Each of the following partial payments was made within the cash discount period. Calculate the amount of credit earned and the remaining balance. Round your answers to the nearest cent.

| | Net Price | Terms | Partial Payment | Credit Earned | Remaining Balance |
|---|---|---|---|---|---|
| 25. | $ 8,000 | 2/10, n/45 | $ 2,000 | _____ | _____ |
| 26. | 1,300 | 3/10, n/30 | 600 | _____ | _____ |
| 27. | 22,000 | 1/20, n/60 | 12,000 | _____ | _____ |
| 28. | 2,400 | 2/10, n/30 | 1,000 | _____ | _____ |

## BORROWING MONEY TO OBTAIN CASH DISCOUNTS

Calculate the amount of cash discount, the interest for borrowing the cash price from the end of the cash discount period to the end of the credit period, and the savings or loss from borrowing to obtain the cash discount. Use a 12% interest rate and the exact-interest method to calculate interest. Round to the nearest cent.

| | Net Price | Terms | Cash Discount | Interest | Savings (+) or Loss (−) |
|---|---|---|---|---|---|
| 29. | $ 7,500 | 2/10, n/30 | _____ | _____ | _____ |
| 30. | 10,000 | 3/10, n/45 | _____ | _____ | _____ |
| 31. | 8,000 | 1/10, n/60 | _____ | _____ | _____ |
| 32. | 5,000 | 2/10, n/90 | _____ | _____ | _____ |

## BUSINESS APPLICATIONS

33. The following quantities and list prices were shown on Senn Company's purchase order. The entire order is subject to a 40% trade discount. What is the net price for the order?

| Quantity | Unit Price | Extension |
|---|---|---|
| 8 | $12.50 | _____ |
| 10 | 15.50 | _____ |
| 6 | 18.00 | _____ |
| 9 | 15.00 | _____ |
| | | _____ |

34. Lendl's Computerland purchased goods with a $14,000 list price. The manufacturer grants a 45% trade discount. Terms are 2/10, n/60. If Lendl pays within the cash discount period, what is the amount owed?

**35.** The following items were shown at their list price on an invoice received by Macmillan and Company: 8 at $100 each, 15 at $200 each, and 20 at $180 each. Series trade discounts of 40% and 10% are allowed. In addition, freight charges of $350 are listed. Terms of payment are 3/10, n/30. Shipping terms are FOB shipping point. How much does Macmillan owe if it pays within the cash discount period?

**36.** In the past year, Haroldson Department Stores purchased goods costing $20,600,000. Cash discounts of 2% were available on all goods purchased, but Haroldson did not obtain any of these. How much could Haroldson have saved by taking advantage of the cash discounts?

**37.** Donovan Manufacturing offers to sell Furniture City a dining room set at a $2,800 list price and 48% trade discount. Terms are 1/10, n/30. Pinewood Furniture Manufacturing Company offers to sell Furniture City a similar set at a $3,000 list price and 50% trade discount. Terms are 3/10, n/30. Which manufacturer offers the lowest cost to Furniture City if payment is made within the cash discount period and how much lower is it?

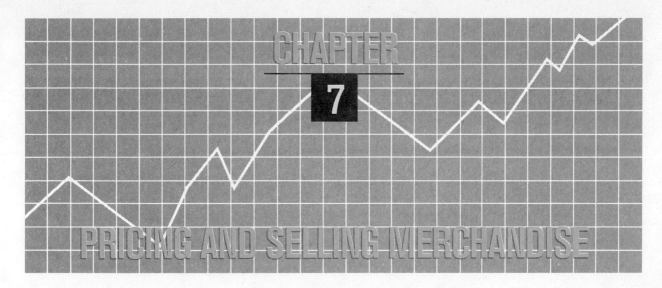

# CHAPTER 7

## PRICING AND SELLING MERCHANDISE

### LEARNING OBJECTIVES

1. Calculate the markup, markup rate, cost, and selling price when markup is based on cost.
2. Calculate the markup, markup rate, cost, and selling price when markup is based on selling price.
3. Calculate the markdown and markdown rate.
4. Calculate sales tax and complete sales records.

A retailer buys merchandise from manufacturers and wholesalers and earns income by reselling the merchandise to customers. A retailer must sell the goods at a price high enough to cover their cost and operating expenses to earn a *profit.* This chapter presents methods for calculating selling price, reducing selling price, calculating sales tax and excise taxes, and completing sales records.

## LESSON 7-1    MARKUP BASED ON COST

**THE MARKUP FORMULA**

The amount a business pays for merchandise, including freight charges, is called the *cost.* An amount called *markup* is then added to the cost to determine the price at which goods will be sold to customers, called *selling price.* If any two elements of the markup formula—cost plus markup equals selling price—are known, the third can be calculated.

Cost + Markup = Selling Price

**EXAMPLE 1**

The Wedding Chalet purchased a wedding dress from its supplier for $350 and marked it up $300. What is the selling price?

**SOLUTION**

Cost Price + Markup = Selling Price
$350 + $300 = $650

## EXAMPLE 2

Waterworld's selling price for an aquarium is $150. The markup is $60. What was the cost?

### SOLUTION

Selling Price − Markup = Cost
   $150    −  $60   = $90

## EXAMPLE 3

The selling price of a gold necklace is $250. The jeweler's cost was $130. What is the markup?

### SOLUTION

Selling Price − Cost  = Markup
   $250   − $130 =  $120

---

When determining the amount of markup on a product, a business manager takes into consideration the amount of operating expenses connected with the sale of that product, such as wages, rent, and advertising, and the desired net profit from the sale. **Net profit** is the income generated for the business owners after all expenses have been deducted. The selling price is the sum of the cost, operating expenses, and net profit.

          Operating    Net    Selling
Cost + Expenses + Profit = Price

Operating Expenses + Net Profit = Markup

---

## EXAMPLE

Elwee Plastics Manufacturers produced $75,000 worth of plastic sheeting last month. Operating expenses incurred to produce the item totaled $15,000. Calculate the selling price and the markup if the company plans to make a $10,000 profit.

### SOLUTION

**Step 1** Find the selling price.

  Cost   + Operating Expenses + Net Profit = Selling Price
$75,000 +       $15,000      + $10,000 =  $100,000

**Step 2** Find the markup.

Operating     Net
 Expenses +  Profit  = Markup
 $15,000  + $10,000 = $25,000

| Selling Price |
| :---: |
| $100,000 |

| Cost  + Operating Expenses + Net Profit |
| :---: |
| $75,000 +     $15,000     + $10,000 |
| Markup |
| $25,000 |

Find the missing amount.

| | Cost + Markup = Selling Price |
|---|---|
| 1. | $574 + $_____ = $839 |
| 2. | $1,017 + $952 = $_____ |
| 3. | $_____ + $1,213 = $2,537 |
| 4. | $111 + $88 = $_____ |

5. Owners of the Smith & Clark Video Store calculated that their projected expenses for the month total $20,400. What will be the total selling price for merchandise costing $59,200 if they expect a net profit of $18,080?

## MARKUP BASED ON COST

Most manufacturers and wholesalers as well as some retailers base markup on cost. Since cost price can be taken directly from the invoice, and their inventory records usually are based on cost, this is a practical method for manufacturers to use.

In all markup problems, the cost, markup, and selling price have percent values as well as amounts. To solve markup problems, first determine if the percent of markup is based on cost or on selling price. Then, solve for either base, rate, or part using the percentage formula.

When markup is based on cost, the cost has a value of 100%. To find the markup based on cost, multiply the cost by the markup rate. Use the formula $B \times R = P$ where the markup is the part, the cost is the base, and the markup rate is the rate.

Cost Rate $(B) \times$ Markup Rate $(R) =$ Markup $(P)$

### EXAMPLE

Buckeye Auto Parts Wholesale Company purchased a shipment of auto radios for $82 each. Buckeye's markup rate is 40% based on cost. What is the amount of markup and the selling price?

### SOLUTION

**Step 1** Calculate the markup.

Cost $(B) \times$ Markup Rate $(R) =$ Markup $(P)$
$82 $\times$ 0.40 = $32.80

**Step 2** Find the selling price.

Cost + Markup = Selling Price
$82 + $32.80 = $114.80

The following illustration shows the percent and amount relationship of cost and markup to selling price in the above example. Cost plus markup always equals selling price, both with dollar amounts and percent values.

The selling price rate (140%) is found by adding the cost rate (100%) and the markup rate (40%). This illustration will be used in the examples throughout the chapter.

| Cost + Markup = Selling Price |
|---|
| 100% + 40% = 140% |
| $82.00 + $32.80 = $114.80 |

## PRACTICE PROBLEMS

Calculate the amount of markup and the selling price. Markup is based on cost. Round amounts to the nearest cent.

| Item | Cost | Markup Based on Cost | Markup | Selling Price |
|---|---|---|---|---|
| 1. Bookcase | $238.00 | 82½% | $_____ | $_____ |
| 2. Man's suit | 156.90 | 92% | _____ | _____ |
| 3. Basketball | 13.80 | 64% | _____ | _____ |

Calculate the missing quantities. Round amounts to the nearest cent.

| | Cost + Markup = Selling Price |
|---|---|
| 4. | 100% + 75% = _____% |
| | $26 + $_____ = $_____ |
| 5. | 100% + 33⅓% = _____% |
| | $3.15 + $_____ = $_____ |
| 6. | 100% + 120% = _____% |
| | $89.00 + $_____ = $_____ |

## MARKUP RATE BASED ON COST

To find the markup rate based on cost, divide the markup by the cost. Use the formula $R = P/B$.

$$\text{Markup Rate } (R) = \frac{\text{Markup } (P)}{\text{Cost } (B)}$$

### EXAMPLE 1

A sleeper sofa costing $472 has a markup based on cost of $205. Calculate the markup rate, rounded to the nearest tenth of a percent.

ANSWERS FOR
PRACTICE PROBLEMS

|  | Markup | Selling Price |
|---|---|---|
| (1) | $196.35 | $434.35 |
| (2) | 144.35 | 301.25 |
| (3) | 8.83 | 22.63 |

(4) 175%, $19.50, $45.50 (5) 133⅓%, $1.05, $4.20 (6) 220%, $106.80, $195.80

**SOLUTION**

**Step 1**   List the known elements of the markup formula.

```
 Cost  +   Markup   = Selling Price
 100% + _____ % = _____%
 $472 +     $205    = $_____
```

**Step 2**   Find the markup rate.

$$\text{Markup Rate } (R) = \frac{\text{Markup } (P)}{\text{Cost } (B)} = \frac{\$205}{\$472} = 0.4343 \text{ or } 43.4\%$$

```
 Cost  + Markup = Selling Price
 100%  + 43.4%  =   143.4%
 $472  + $205   =    $677
```

**EXAMPLE 2**

A set of tools costing $229.50 sold for $383.95. Find the markup rate based on cost rounded to the nearest tenth of a percent.

**SOLUTION**

**Step 1**   List the known elements of the markup formula.

```
 Cost    +   Markup   = Selling Price
 100%   + _____ % = _____%
 $229.50 + $_____ =   $383.95
```

**Step 2**   Find the amount of markup.

```
Selling Price −   Cost   = Markup
  $383.95    − $229.50 = $154.45
```

**Step 3**   Find the markup rate based on cost.

$$\text{Markup Rate } (R) = \frac{\text{Markup } (P)}{\text{Cost } (B)} = \frac{\$154.45}{\$229.50} = 0.6729 \text{ or } 67.3\%$$

```
 Cost    + Markup   = Selling Price
 100%   + 67.3%    =   167.3%
 $229.50 + $154.45 =   $383.95
```

**PRACTICE PROBLEMS**

Find the missing quantities. Round amounts to the nearest cent and percents to the nearest tenth.

| Item | Cost | Markup | Markup Rate on Cost | Selling Price |
|------|------|--------|---------------------|---------------|
| 1. Telephone | $ 53.50 | $_____ | _____% | $ 73.83 |
| 2. Necklace | 383.80 | 407.36 | _____% | _____ |
| 3. Clock | 27.35 | _____ | _____% | 47.15 |
| 4. Book | 42.70 | 25.10 | _____% | _____ |

|   | Cost + | Markup | = Selling Price |
|---|--------|--------|-----------------|
| **5.** | 100% + | _____% = | _____% |
|   | $36 + | $41.58 | = $_____ |
| **6.** | 100% + | _____% = | _____% |
|   | $1,057 + | $_____ | = $1,363.53 |
| **7.** | 100% + | _____% = | _____% |
|   | $2.05 + | $1.67 | = $_____ |

## COST WHEN MARKUP IS BASED ON COST

When markup is based on cost, the cost can be found by dividing the markup by the markup rate. Use the formula $B = P/R$.

$$\text{Cost } (B) = \frac{\text{Markup } (P)}{\text{Markup Rate } (R)}$$

### EXAMPLE

The markup on a typewriter is $121. This amount is a 55% markup based on cost. Calculate the cost.

### SOLUTION

**Step 1**  List the known elements of the markup formula.

| Cost | + Markup | = Selling Price |
|------|----------|-----------------|
| 100% | + 55% | = _____% |
| $_____ | + $121 | = $_____ |

**Step 2**  Calculate the cost.

$$\text{Cost } (B) = \frac{\text{Markup } (P)}{\text{Markup Rate } (R)} = \frac{\$121}{0.55} = \$220$$

| Cost | + Markup | = Selling Price |
|------|----------|-----------------|
| 100% + | 55% = | 155% |
| $220 + | $121 = | $341 |

ANSWERS FOR PRACTICE PROBLEMS

**Starting on page 217:**

| | Markup Rate on Cost | Markup on Cost | Selling Price |
|---|---|---|---|
| **(1)** | 38% | | $20.33 |
| **(2)** | 106.1% | | $791.16 |
| **(3)** | 72.4% | 19.80 | |
| **(4)** | 58.8% | | 67.80 |
| **(5)** | 115.5%, 215.5%, $77.58 | | |
| **(6)** | 29%, 129%, $306.53 | | |
| **(7)** | 81.5%, 181.5%, $3.72 | | |

**218** ■ LESSON 7-1

Copyright © 1989 by McGraw-Hill, Inc. All rights reserved.

Sometimes both the price at which goods can be sold and the markup rate on cost are already determined as a matter of company or industry-wide policy. The task, then, is to determine the cost at which it is feasible to buy the item. First determine the percent that selling price is of cost. This is called the **selling price rate.** Then divide selling price by selling price rate.

$$\text{Cost } (B) = \frac{\text{Selling Price } (P)}{\text{Selling Price Rate } (R)}$$

### EXAMPLE

Sound City, Inc., has determined through comparative shopping that it can sell an acoustic guitar for $480. Sound City requires a markup of 60% on cost. What must be Sound City's cost to meet its goals?

### SOLUTION

**Step 1** List the known elements of the markup formula.

| Cost | + | Markup | = Selling Price |
|------|---|--------|-----------------|
| 100% | + | 60% | = _____ % |
| $_____ | + $_____ | = | $480 |

**Step 2** Find the selling price rate.

Cost Rate + Markup Rate = Selling Price Rate
  100%   +   60%   =   160%

**Step 3** Calculate the cost.

$$\text{Cost } (B) = \frac{\text{Selling Price } (P)}{\text{Selling Price Rate } (R)} = \frac{\$480}{1.60} = \$300$$

| Cost | + Markup | = Selling Price |
|------|----------|-----------------|
| 100% + | 60% | = 160% |
| $300 + | $180 | = $480 |

### PRACTICE PROBLEMS

Calculate the cost and selling price. Round to the nearest cent.

| | Markup Rate Based on Cost | Markup | Cost | Selling Price |
|---|---|---|---|---|
| 1. | 22.25% | $ 39.56 | $_____ | $_____ |
| 2. | 129% | 255.50 | _____ | _____ |
| 3. | 43.5% | 702.25 | _____ | _____ |

Calculate the cost and the markup. Round amounts to the nearest cent.

| | Markup Rate Based on Cost | Selling Price | Cost | Markup |
|---|---|---|---|---|
| 4. | 37% | $1,215.29 | $_____ | $_____ |
| 5. | 109.5% | 391.79 | _____ | _____ |
| 6. | $66\frac{2}{3}\%$ | 915.43 | _____ | _____ |

Find the missing quantities. Round amounts to the nearest cent and percents to the nearest hundredth.

|   | Cost | + | Markup | = | Selling Price |
|---|------|---|--------|---|---------------|
| **7.** | 100% | + | 41.5% | = | _____% |
|   | $_____ | + | $93.50 | = | $_____ |
| **8.** | 100% | + | 221% | = | _____% |
|   | $_____ | + | $_____ | = | $6,759.95 |
| **9.** | 100% | + | 81% | = | _____% |
|   | $_____ | + | $119.49 | = | $_____ |

**220** ■ LESSON 7-1

## LESSON 7-1    EXERCISES

Round money amounts to the nearest cent. Round percents to the nearest tenth.

Find the missing amounts.

| | | Markup | | |
| | Cost | Operating Expenses | Net Profit | Selling Price |
|---|---|---|---|---|
| 1. | $64,800 | $_____ | $20,200 | $ 120,000 |
| 2. | 26,000 | 35,000 | 9,500 | _____ |
| 3. | 59,200 | 20,400 | _____ | 97,680 |
| 4. | _____ | 62,500 | 32,500 | 250,000 |
| 5. | 764,500 | 375,300 | _____ | 1,390,000 |

Find the missing quantities. Markup is based on cost.

Cost  +  Markup  = Selling Price

**6.** 100%  +  35%  = _____

_____ + _____ =  $51.04

**7.** 100%  + _____ =  172%

$803.33 + _____ = _____

**8.** 100%  + _____ = _____

$49.90  +  $16.62  = _____

**9.** 100%  +  112%  = _____

_____ + $2,399.04 = _____

**10.** 100%  + _____ =  116%

_____ + _____ =  $4.74

Find the missing quantities.

| | Item | Cost | Markup | Markup Rate on Cost | Selling Price |
|---|---|---|---|---|---|
| 11. | Compact disk player | $349.95 | $150.48 | _____ | $_____ |
| 12. | Cassette tape case | 19.39 | _____ | 55% | _____ |
| 13. | Bicycle | _____ | 67.44 | 48% | _____ |
| 14. | Clock/radio | 45.35 | _____ | _____ | 74.37 |
| 15. | Jewelry | _____ | _____ | 175% | 1,976.98 |

| Item | Cost | Markup | Markup Rate on Cost | Selling Price |
|---|---|---|---|---|
| **16.** Personal computer | 986.99 | 394.80 | _____ | _____ |
| **17.** Jacket | _____ | 24.97 | _____ | 81.72 |
| **18.** Briefcase | 34.99 | _____ | 71.7% | _____ |
| **19.** Shoes | 19.50 | _____ | _____ | 39.00 |
| **20.** Table lamp | _____ | 17.90 | _____ | 80.69 |

## BUSINESS APPLICATIONS

Round amounts to the nearest cent and percents to the nearest tenth.

**21.** Williams Brothers Appliance Store purchased a refrigerator for $450 from a supplier and marked it up $300. Find (a) the selling price and (b) the markup rate based on cost.

(a) _____

(b) _____

**22.** Travel Time, Inc., marks up leather luggage 120% on its cost. If Travel Time paid $167 for a leather suitcase, what is Travel Time's (a) amount of markup and (b) selling price for the suitcase?

(a) _____

(b) _____

**23.** Belling Hardware purchases paneling at $6.50 a sheet from its supplier. The company requires a $4.25 markup on paneling. Calculate the markup rate based on cost.

_____

**24.** Tyson's Bicycle Shop purchased a shipment of bicycles for $98.60 each. Tyson's will mark up the bicycles 75.5% on its cost. Calculate (a) the amount of markup and (b) the selling price of the bicycle.

(a) _____

(b) _____

**25.** Saunders' Art Gallery marked Marcell Bedell's painting "Gone Again" with a $1,800 selling price. If Saunders purchased the painting for $700, what is the percent of markup based on cost?

_____

**26.** The Popham Camera Shop's markup on a camera is $126.95. If the markup is 50.2% based on cost, what is (a) the cost and (b) the selling price of the camera?

(a) _____

(b) _____

**27.** Graham Greeting Cards purchased a box of 25 greeting cards for $15. Graham sold all of the cards at $1.25 each. Find (a) the amount of markup on the box of cards and (b) the percent of markup based on cost.

(a) _____

(b) _____

**28.** Culver Communications' marketing department has determined that it can sell 100 mobile auto phones at $980 each. Culver requires a 75% markup on its cost and anticipates total selling expenses of $6,000. How much profit will Culver earn after deducting cost and selling expenses if it sells all 100 units?

_____

**29.** Margo's Health Center would like to sell the Model RD2 exercise bike with a $175 markup. The markup is 54.7% of the cost. Calculate (a) the cost and (b) the selling price.

(a) _____

(b) _____

**30.** To stay competitive within the industry, Kids' Stuff must sell their Model 200-56 swingset for $315.00. If the company requires a 75% markup on cost, what must the company pay for the swingset?

_____

**31.** The Baxter Wholesale Distributors has purchased $1,397,200 worth of merchandise. The company with operating expenses of $400,200 hopes to make a profit of $70,000 on the merchandise. Calculate (a) the amount of markup, (b) the selling price, and (c) the markup rate based on cost.

(a) _____

(b) _____

(c) _____

**32.** From January 1 to June 30, the Newell Manufacturers sold merchandise for $519,485. The merchandise cost $289,300, and operating expenses of $109,000 were incurred. Calculate (a) the net profit on the sale of the merchandise and (b) the percent of markup based on cost.

(a) _____

(b) _____

**MARKUP AND COST WHEN MARKUP IS BASED ON SELLING PRICE**

Many retail businesses express markup as a percent of the selling price. When markup is based on selling price, selling price is the base and has a value of 100%.

To find the markup based on selling price, multiply the selling price by the markup rate. Use the formula $B \times R = P$, where the markup is the part, the selling price is the base, and the markup rate is the rate:

Selling Price $(B) \times$ Markup Rate $(R) =$ Markup $(P)$

---

**EXAMPLE**

What is the markup and the cost of chocolate fudge that sells for $4.29 a pound if the markup rate is 33.5% based on selling price? Round to the nearest cent.

**SOLUTION**

**Step 1**   List the known elements of the markup formula.

| Cost | + | Markup | = | Selling Price | |
|---|---|---|---|---|---|
| _____ % + | | 33.5% | = | 100% ← | Selling price is now 100% |
| $_____ + | $ | _____ | = | $4.29 | |

**Step 2**   Find the markup.

Selling Price $(B) \times$ Markup Rate $(R) =$ Markup $(P)$
$4.29 $\times$ 0.335 $\approx$ $1.44

**Step 3**   Find the cost.

Selling Price $-$ Markup $=$ Cost
$4.29 $-$ $1.44 $=$ $2.85

The cost also may be found by multiplying the selling price by the cost rate. The **cost rate** is the difference between the selling price rate and the markup rate.

Selling Price $(B) \times$ Cost Rate $(R) =$ Cost $(P)$
$4.29 $\times$ 0.665 $\approx$ $2.85

| Cost | + Markup | = Selling Price |
|---|---|---|
| 66.5% | + 33.5% = | 100% |
| $2.85 | + $1.44 = | $4.29 |

---

**PRACTICE PROBLEMS**

Calculate the amount of markup and cost. Round to the nearest cent.

| Item | Selling Price | Markup Rate Based on Selling Price | Markup | Cost |
|---|---|---|---|---|
| 1. Luggage | $114.89 | 53% | $_____ | $_____ |
| 2. Lamp | 31.47 | $28\frac{1}{2}$% | _____ | _____ |
| 3. Clock | 443.55 | 69% | _____ | _____ |

Find the missing amounts and percents. Round amounts to the nearest cent and percents to the nearest tenth.

| | Cost | + | Markup | = Selling Price |
|---|---|---|---|---|
| **4.** | _____% | + | 22% | = 100% |
| | $_____ | + | $_____ | = $25.64 |
| **5.** | _____% | + | 27% | = 100% |
| | $_____ | + | $_____ | = $375.28 |
| **6.** | _____% | + | 42% | = 100% |
| | $_____ | + | $_____ | = $390.48 |

## MARKUP RATE BASED ON SELLING PRICE

To find the markup rate based on selling price when the markup and selling price are known, divide the markup by the selling price, using the formula $R = P/B$.

$$\text{Markup Rate } (R) = \frac{\text{Markup } (P)}{\text{Selling Price } (B)}$$

### EXAMPLE 1

What is the markup rate based on selling price of a remote control model airplane selling for $118 if the markup is $31.86?

### SOLUTION

**Step 1**  List the known elements of the markup formula.

| Cost | + | Markup | = Selling Price |
|---|---|---|---|
| _____% | + | _____% | = 100% |
| $_____ | + | $31.86 | = $118.00 |

**Step 2**  Find the markup rate.

$$\text{Markup Rate } (R) = \frac{\text{Markup } (P)}{\text{Selling Price } (B)} = \frac{\$31.86}{\$118.00} = 0.27 \text{ or } 27\%$$

| Cost | + Markup | = Selling Price |
|---|---|---|
| 73% | + 27% | = 100% |
| $86.14 | + $31.86 | = $118.00 |

## EXAMPLE 2

An electric blanket costs $29.90 and sells for $42.59. Find the markup rate based on selling price.

### SOLUTION

**Step 1**   List the known elements of the markup formula.

| Cost | + | Markup | = Selling Price |
|------|---|--------|-----------------|
| _____% + | | _____% = | 100% |
| $29.90 | + $_____ | | = $42.59 |

**Step 2**   Find the markup.

Selling Price − Cost = Markup
$42.59 − $29.90 = $12.69

**Step 3**   Find the markup rate.

$$\text{Markup Rate } (R) = \frac{\text{Markup } (P)}{\text{Selling Price } (B)} = \frac{\$12.69}{\$42.59} = 0.2979 \text{ or } 29.8\%$$

| Cost + Markup = Selling Price |
|-------------------------------|
| 70.2% + 29.8% = 100% |
| $29.90 + $12.69 = $42.59 |

## PRACTICE PROBLEMS

Calculate the markup rate based on selling price. Round to the nearest tenth.

1. Item selling for $2,850.90 with a $850 markup.
   Markup rate: _____%
2. Item selling for $8.15 with a $3.94 markup.
   Markup rate: _____%

Find the missing quantities. Round amounts to the nearest cent and percents to the nearest tenth.

| | Cost | + | Markup | = Selling Price |
|----|------|---|--------|-----------------|
| **3.** | _____% + | | _____% = | 100% |
| | $68.30 | + $_____ | | = $115.49 |
| **4.** | _____% + | | _____% = | 100% |
| | $_____ | + | $19.66 | = $95.29 |

## SELLING PRICE WHEN MARKUP IS BASED ON SELLING PRICE

To find the selling price when the markup and markup rate based on selling price are known, divide the markup by the markup rate using the formula $B = P/R$.

$$\text{Selling Price } (B) = \frac{\text{Markup } (P)}{\text{Markup Rate } (R)}$$

### EXAMPLE

The markup on a pair of socks is $2.62, which is a markup of 44.2% based on selling price. Calculate the selling price.

### SOLUTION

**Step 1** List the known elements of the markup formula.

| Cost | + Markup | = Selling Price |
|------|----------|-----------------|
| _____% | + 44.2% | = 100% |
| $_____ | + $2.62 | = $_____ |

**Step 2** Calculate the selling price.

$$\text{Selling Price } (B) = \frac{\text{Markup } (P)}{\text{Markup Rate } (R)} = \frac{\$2.62}{0.442} = \$5.927 \approx \$5.93$$

| Cost | + Markup | = Selling Price |
|------|----------|-----------------|
| 55.8% | + 44.2% | = 100% |
| $3.31 | + $2.62 | = $5.93 |

To find the selling price if the cost and markup rate based on selling price are known, divide the cost by the cost rate. The cost rate is the difference between the selling price rate (100%) and the markup rate.

$$\text{Selling Price } (B) = \frac{\text{Cost } (P)}{\text{Cost Rate } (R)}$$

### EXAMPLE

Crawley Department Store's markup rate on chairs is 43% of selling price. Recently Crawley purchased a rocking chair at a cost of $75. Calculate the selling price.

### SOLUTION

**Step 1** List the known elements of the markup formula.

| Cost | + Markup | = Selling Price |
|------|----------|-----------------|
| _____% | + 43% | = 100% |
| $75 | + $_____ | = $_____ |

**Step 2** Find the cost rate.

Selling Price − Markup Rate = Cost Rate

100% − 43% = 57%

**Step 3** Calculate the selling price.

$$\text{Selling Price } (B) = \frac{\text{Cost } (P)}{\text{Cost Rate } (R)} = \frac{\$75}{0.57} = \$131.578 \approx \$131.58$$

| Cost + Markup = Selling Price |
|---|
| 57% + 43% = 100% |
| $75 + $56.58 = $131.58 |

## PRACTICE PROBLEMS

Calculate the selling price. Round amounts to the nearest cent.

1. Item with a markup of $72.50, which is 19.5% of selling price.
   Selling price: $_____
2. Item with a markup of $1.03, which is 41% of selling price.
   Selling price: $_____
3. Item costing $82.90 with a 30% markup based on selling price.
   Selling price: $_____
4. Item costing $413.20 with a 23% markup based on selling price.
   Selling price: $_____

Find the missing quantities. Round amounts to the nearest cent and percents to the nearest tenth.

| | Cost | + | Markup | = Selling Price |
|---|---|---|---|---|
| **5.** | _____% + | | 39% | = 100% |
| | $46.10 | + $_____ | | = $_____ |
| **6.** | _____% + | | 11.5% | = 100% |
| | $_____ | + | $0.79 | = $_____ |

## COMPARING MARKUP RATES

When business operators refer to markup rate, some mean markup based on cost, and others mean markup based on selling price. It is important, therefore, to clarify which procedure is being used so that everyone can communicate on the same basis. The following example shows the difference in markup rates for the same product when markup is based on cost and when it is based on selling price.

### EXAMPLE

What is the markup rate based on cost and the markup rate based on selling price of equipment that costs $1,400 and sells for $2,026?

### SOLUTION

**Step 1** Find the markup.

Selling Price − Cost = Markup
$2,026 − $1,400 = $626

**Step 2** Calculate the markup rate based on cost.

$$\text{Markup Rate Based on Cost } (R) = \frac{\text{Markup } (P)}{\text{Cost } (B)} = \frac{\$626}{\$1,400} = 0.4471 \text{ or } 44.7\%$$

**Step 3** Calculate the markup rate based on selling price.

$$\text{Markup Rate Based on Selling Price } (R) = \frac{\text{Markup } (P)}{\text{Selling Price } (B)} = \frac{\$626}{\$2,026} = 0.3089 \text{ or } 30.9\%$$

Markup rates based on cost are always higher than equivalent markup rates based on selling price.

## PRACTICE PROBLEMS

Calculate the amount of markup and the equivalent markup rates. Round percents to the nearest tenth.

|   | Cost | Selling Price | Markup | Markup Rate on Cost | Markup Rate on Selling Price |
|---|------|---------------|--------|---------------------|------------------------------|
| 1. | $ 50.90 | $ 72.59 | $_____ | _____ % | _____ % |
| 2. | 193.40 | 220.95 | _____ | _____ % | _____ % |
| 3. | 519.50 | 702.89 | _____ | _____ % | _____ % |
| 4. | 9.20 | 12.49 | _____ | _____ % | _____ % |

## LESSON 7-2    EXERCISES

Find the missing quantities. Round amounts to the nearest cent. Round percents to the nearest tenth.

|   | Cost | + | Markup | = | Selling Price |
|---|------|---|--------|---|---------------|
| **1.** | _____ | + | 36.3% | = | 100% |
|   | _____ | + | $44.96 | = | _____ |
| **2.** | 78.5% | + | _____ | = | 100% |
|   | _____ | + | _____ | = | $5,195.22 |
| **3.** | _____ | + | 43% | = | 100% |
|   | $219.98 | + | _____ | = | _____ |
| **4.** | _____ | + | _____ | = | 100% |
|   | _____ | + | $4.14 | = | $22.00 |
| **5.** | 44% | + | _____ | = | 100% |
|   | $672.30 | + | _____ | = | _____ |

Find the missing quantities. Markup is based on selling price.

| | Item | Cost | Markup | Markup Rate on Selling Price | Selling Price |
|---|------|------|--------|------------------------------|---------------|
| **6.** | Typewriter | $ _____ | $ _____ | 31.9% | $  308.50 |
| **7.** | Video camera | 784.35 | _____ | _____ | 1,120.50 |
| **8.** | Electric shaver | _____ | 5.00 | _____ | 24.99 |
| **9.** | VHS recorder | 129.50 | _____ | 55.3% | _____ |
| **10.** | Light bulb | 0.92 | 0.75 | _____ | _____ |
| **11.** | Steel-belted tire | _____ | 11.52 | 38.4% | _____ |
| **12.** | Computer printer | _____ | _____ | 25.1% | 500.70 |
| **13.** | Crystal vase | _____ | 25.92 | _____ | 57.60 |
| **14.** | Desk | 185.51 | 108.95 | 37% | _____ |
| **15.** | Air conditioner | 922.08 | _____ | 25% | _____ |

Calculate the amount of markup, the markup rate based on cost, and the markup rate based on selling price. Round to the nearest tenth of a percent.

| | Cost | Selling Price | Markup | Markup Rate on Cost | Markup Rate on Selling Price |
|---|---|---|---|---|---|
| **16.** | $ 530.30 | $ 646.90 | _____ | _____ | _____ |
| **17.** | 110.04 | 318.88 | _____ | _____ | _____ |
| **18.** | 862.32 | 1,032.10 | _____ | _____ | _____ |
| **19.** | 22.56 | 32.70 | _____ | _____ | _____ |
| **20.** | 2,248.00 | 2,698.60 | _____ | _____ | _____ |

## BUSINESS APPLICATIONS

Round amounts to the nearest cent and percents to the nearest tenth.

**21.** Recently, Marcell Clothing purchased a shipment of suits for $182 each and sold them for $350 each. Calculate the markup rate based on selling price.

_____

**22.** The Clinton Creative Arts store requires a 44% markup on selling price on artists' easels. The store estimates it can sell easels at $35.95 each. How much should the store budget to purchase 15 easels?

_____

**23.** Popham's Sporting Goods has purchased for resale a home fitness center for $346.57. Popham's plans to sell the fitness center with a $229.50 markup. Calculate the markup rate based on selling price.

_____

**24.** Dulong Enterprises maintains a markup rate of 45% of selling price on office equipment it sells. Expenses incurred from selling 217 desks for $225.70 each totaled $9,125. Calculate the net profit per desk.

_____

**25.** Doran Marina requires a 36% markup on selling price on all sailboats sold. Calculate (a) the selling price and (b) the cost if the markup on a sailboat is $3,094.47.

(a) _____

(b) _____

**26.** Rojas Fireplace Company sells a metal fireplace for $477. Calculate the markup rate based on the selling price if the amount of markup is $318.

_____

**27.** Larue's Jewelry Store's selling price of a 14 karat gold ring is $498.99. Larue requires a markup of 75% of selling price on its gold rings. Calculate (a) the unit cost and (b) the amount of profit earned on 118 rings after deducting total operating expenses of $19,785.

(a) _____

(b) _____

**28.** Hazel Hardware's markup rate on snow shovels is 34% of selling price. If the amount of markup per shovel is $7.70, what is the selling price?

_____

**29.** Last year High Tower Appliances purchased 25-inch color TVs from the manufacturer for $380.45 and sold them for $689.99 each. This year the cost from the manufacturer has increased 10%. If High Tower keeps the same percent of markup on selling price as last year, what should be the retail selling price of the TVs?

_____

**30.** Whitaker Electronics can purchase a shipment of Model TY100 electronic typewriters with a $782 list price each and a 45% trade discount. Whitaker requires a 42.6% markup on selling price. At what selling price should Whitaker mark each typewriter?

_____

**MARKDOWN AND MARKDOWN RATE**

*Markdown* is a reduction in price from the original selling price to a sale price. The amount of markdown is calculated by multiplying the original selling price by the percent of markdown, called the *markdown rate.* The selling price minus the markdown is called the *reduced price.*

Selling Price − Markdown = Reduced Price

To find the amount of markdown, multiply the original selling price by the markdown rate.

Original Selling Price $(B)$ × Markdown Rate $(R)$ = Markdown $(P)$

---

**EXAMPLE**

Huston Sporting Goods sells hammocks for $32.59. The hammock is marked down 40 percent. What is the markdown and the reduced price?

**SOLUTION**

**Step 1**   Find the markdown.

Original Selling Price $(B)$ × Markdown Rate $(R)$ = Markdown $(P)$

$32.59 × 0.4 = $13.036 ≈ $13.04

**Step 2**   Find the reduced price.

Original Selling Price − Markdown = Reduced Price

$32.59 − $13.04 = $19.55

---

If the original selling price of an item is reduced by a dollar amount, the markdown rate may be found by dividing the amount of markdown by the original selling price using the formula below.

$$\text{Markdown Rate } (R) = \frac{\text{Markdown } (P)}{\text{Original Selling Price } (B)}$$

---

**EXAMPLE**

Terry Togs sold their remaining inventory of last year's swimsuits at a total reduced price of $890. The original selling price totaled $1,625. What was the percent of markdown on the original selling price?

**SOLUTION**

**Step 1**   Find the markdown.

Original Selling Price − Reduced Price = Markdown

$1,625 − $890 = $735

**Step 2**   Find the markdown rate.

$$\frac{\text{Markdown}}{\text{Rate } (R)} = \frac{\text{Markdown } (P)}{\text{Original Selling Price } (B)} = \frac{\$735}{\$1,625} = 0.4523 \text{ or } 45.2\%$$

---

Find the missing quantities.

| | Selling Price | Markdown Rate | Markdown | Reduced Price |
|---|---|---|---|---|
| **1.** | $ 450 | 22% | $ _____ | $ _____ |
| **2.** | 1,300 | _____ % | _____ | 910 |
| **3.** | 925 | 34% | _____ | _____ |

## ORIGINAL SELLING PRICE

The original selling price of an item can be found by dividing the reduced price by the reduced price rate. The **reduced price rate** is the difference between the original selling price rate (100%) and the markdown rate.

$$\text{Original Selling Price } (B) = \frac{\text{Reduced Price } (P)}{\text{Reduced Price Rate } (R)}$$

### EXAMPLE

Gerrard's is selling all wool women's suits for $155, which is a 35% markdown from the original price. Calculate the original sales price.

### SOLUTION

**Step 1** Find the reduced price rate.

| Original Selling Price Rate | − | Markdown Rate | = | Reduced Price Rate |
|---|---|---|---|---|
| 100% | − | 35% | = | 65% |

**Step 2** Calculate the original selling price.

$$\text{Original Selling Price } (B) = \frac{\text{Reduced Price } (P)}{\text{Reduced Price Rate } (R)} = \frac{\$155}{0.65} \approx \$238.46$$

### PRACTICE PROBLEMS

Find original selling price and markdown. Round to the nearest cent.

| | Item | Original Selling Price | Markdown Rate | Markdown | Reduced Price |
|---|---|---|---|---|---|
| **1.** | Calculator | $ _____ | 22% | $ _____ | $136.00 |
| **2.** | Shoes | _____ | 34% | _____ | 44.00 |
| **3.** | Lamp | _____ | 35% | _____ | 52.50 |

**Bottom: (1)** $174.36   $ 38.36
**(2)** 66.67   22.67
**(3)** 80.77   28.27

Original Selling Price   Markdown

**Top: (1)** $99, $351   **(2)** 30%, $390   **(3)** $314.50, $610.50

## SALES TAX

Most states and some local governments require businesses to collect **sales tax** on retail sales. The tax rate and the items covered by the sales tax vary from state to state. For example, some states have no sales tax on consumable groceries and pharmaceuticals, while others have no sales tax on clothing.

Sales that are exempt from sales tax are deducted from the total sales before sales tax is calculated. Sales tax is calculated by multiplying taxable sales by the sales tax rate using the formula below.

Taxable Sales ($B$) × Sales Tax Rate ($R$) = Sales Tax ($P$)

Sales tax is paid only by the customer. Retail businesses are exempt from paying sales tax on merchandise purchased for resale.

### EXAMPLE

Emery Food and Drug sold goods totaling $125.75. Of this sale, $36.50 was exempt from state sales tax. The state sales tax rate is 5%. Calculate the taxable sales, the sales tax, and the total amount due. Round amounts to the nearest cent.

### SOLUTION

**Step 1**  Find the taxable sales.

Total Sales − Exempt Sales = Taxable Sales
  $125.75  −   $36.50   =   $89.25

**Step 2**  Calculate the sales tax.

 Taxable      Sales Tax
Sales ($B$) × Rate ($R$) = Sales Tax ($P$)
 $89.25  ×   0.05   =     $4.46

**Step 3**  Find the total amount due.

Total Sales + Sales Tax = Total Amount Due
  $125.75  +   $4.46   =      $130.21

### PRACTICE PROBLEMS

Calculate the taxable sales, sales tax, and total amount due. Round amounts to the nearest cent.

| | Total Sales | Sales Tax Rate | Tax-Exempt Sales | Taxable Sales | Sales Tax | Total Amount Due |
|---|---|---|---|---|---|---|
| **1.** | $  570.20 | 4.5% | $130.50 | $_____ | $_____ | $_____ |
| **2.** | 1,150.12 | 5.0% | 0 | _____ | _____ | _____ |
| **3.** | 261.45 | 6.0% | 14.56 | _____ | _____ | _____ |

## SALES TAX RATE

When the amount of sales tax and the purchase price is known, the tax rate can be found by dividing the sales tax by the purchase price using the following formula.

$$\text{Sales Tax Rate } (R) = \frac{\text{Sales Tax } (P)}{\text{Purchase Price } (B)}$$

### EXAMPLE

The purchase price of a microwave oven is $329.95. The sales tax on the oven is $21.45. What is the sales tax rate? Round to the nearest tenth of a percent.

### SOLUTION

Find the sales tax rate.

$$\text{Sales Tax Rate } (R) = \frac{\text{Sales Tax } (P)}{\text{Purchase Price } (B)} = \frac{\$21.45}{\$329.95} = 0.065 = 6.5\%$$

### PRACTICE PROBLEMS

Calculate the sales tax rate. Round percents to the nearest tenth.

| | Selling Price | Sales Tax | Sales Tax Rate |
|---|---|---|---|
| **1.** | $ 29.80 | $ 1.54 | _____ % |
| **2.** | 507.35 | 30.44 | _____ % |

## SALES RECORDS

All businesses prepare some form of written record when a sale occurs. In many stores, an optical character reader scans a product's bar code, as shown below, to electronically record the sale and to automatically deduct the item from the company's record of goods on hand.

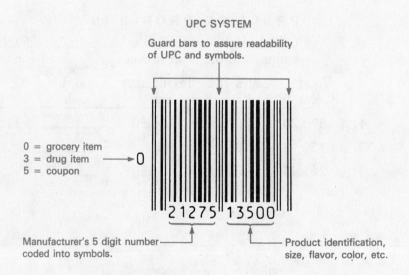

UPC SYSTEM

Guard bars to assure readability of UPC and symbols.

0 = grocery item
3 = drug item
5 = coupon

0  21275 13500

Manufacturer's 5 digit number coded into symbols.

Product identification, size, flavor, color, etc.

A customer usually receives a cash register receipt or a sales slip for cash purchases like the one shown below. The cash sales slip contains the customer's name and address as well as the following information: The **price,** or **unit price,** is the cost of one item. The amount, also called an **extension,** is the product of the quantity and the unit price. The description defines the purchases. The subtotal is the sum of the extensions. The **total** is the sum of the subtotal and sales tax.

### REDLIN DEPARTMENT STORE
2410 E. WOODCREST
TOPEKA, KS 66618

Customer's Order No. _192837_          DATE _May 7,_ 19 _—_

SOLD TO _Shena Ellis_

ADDRESS _652 W. Thoreson St._

Salesperson _J. Tempio_          TERMS _—_

| CASH ✓ | CHARGE | C.O.D. | PAID OUT | RETD. MDSE. | RECD. ON ACCT. |
|---|---|---|---|---|---|

| QUAN. | DESCRIPTION | PRICE | AMOUNT | |
|---|---|---|---|---|
| 2 | Pairs of slacks | $35.98 | $71 | 96 | ← 2 × $35.98 = $71.96 |
| 1 | Dress | 97.50 | 97 | 50 | |
| 3 | Pairs of socks | 3.50 | 10 | 50 | ← 3 × $3.50 = $10.50 |
| | Subtotal | | 179 | 96 | ← $71.96 + $97.50 + $10.50 |
| | Sales tax | | 9 | 00 | ← $179.96 × 0.5 |
| | Total | | 188 | 96 | ← $179.96 + $9.00 |

ALL Claims and Returned Goods MUST Be Accompanied By This Receipt

SIGNATURE _____

If payment is made with a credit card, the customer receives a credit sales slip like the one shown below. The calculations and information recorded on a credit sales slip is similar to that recorded on a cash sales slip.

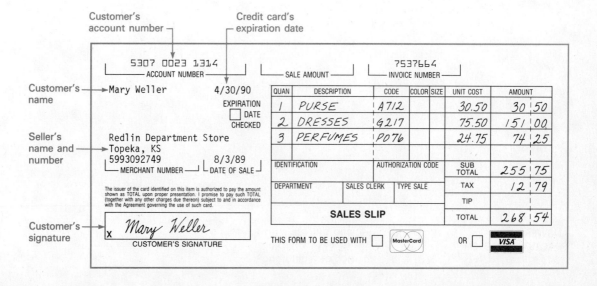

## PRACTICE PROBLEM

Round to the nearest cent.

Redlin Department Store sold the following items to Gloria Dallas for cash: 3 pairs of slacks at $42.50 each, 2 dresses at $94.50 each, and 2 pairs of shoes at $75.50 each. The sales tax of 5% applies to the total sale. Complete the following sales slip for this sale.

**REDLIN DEPARTMENT STORE**
2410 E. WOODCREST
TOPEKA, KS 66618

Customer's Order No. _192838_    DATE _May 8,_ 19 _—_

SOLD TO _Gloria Dallas_

ADDRESS _256 E. Charles St._

Salesperson _J. Tempio_    TERMS _—_

| CASH | CHARGE | C.O.D. | PAID OUT | RETD. MDSE. | RECD. ON ACCT. |
|------|--------|--------|----------|-------------|----------------|
|      |        |        |          |             |                |

| QUAN. | DESCRIPTION | PRICE | AMOUNT |
|-------|-------------|-------|--------|
|       |             |       |        |
|       |             |       |        |
|       |             |       |        |
|       |             |       |        |
|       |             |       |        |
|       |             |       |        |
|       |             |       |        |
|       |             |       |        |

ALL Claims and Returned Goods MUST Be Accompanied By This Receipt

SIGNATURE _____

ANSWERS FOR
PRACTICE PROBLEM

Subtotal: $467.50
Sales tax: $23.38
Total amount due: $490.88

## LESSON 7-3    EXERCISES

Find the missing quantities. Round amounts to the nearest cent and percents to the nearest tenth.

| Item | Original Selling Price | Markdown Rate | Markdown | Reduced Price |
|------|------------------------|---------------|----------|---------------|
| 1. Telescope | $ _____ | _____ | $  10.07 | $  54.88 |
| 2. Sailboat | 8,596.90 | 25% | _____ | _____ |
| 3. Golf clubs | _____ | 44% | _____ | 197.51 |

Calculate the taxable sales, sales tax, and total amount due. Round amounts to the nearest cent.

| | Total Sales | Sales Tax Rate | Tax-Exempt Sales | Taxable Sales | Sales Tax | Total Amount Due |
|---|-------------|----------------|------------------|---------------|-----------|------------------|
| 4. | $  237.42 | 4.5% | $39.36 | _____ | _____ | _____ |
| 5. | 37.11 | 6.0% | 29.84 | _____ | _____ | _____ |
| 6. | 3,195.49 | 5.5% | 0 | _____ | _____ | _____ |

Calculate the sales tax rate. Round percents to the nearest tenth.

| | Selling Price | Sales Tax | Sales Tax Rate |
|---|---------------|-----------|----------------|
| 7. | $712.32 | $49.86 | _____ |
| 8. | 91.12 | 5.01 | _____ |
| 9. | 5.90 | 0.35 | _____ |

10. Ellerbroek Furniture sold the following items to Carol Hemmingway: 1 desk at $1,150, 2 chairs at $475 each, 4 chairs at $139 each, and 3 wall hangings at $74.50 each. A 6% sales tax applies to the entire order. Carol charged the purchase on her credit card. Complete the following sales slip for this sale. You can leave the "Code," "Color," and "Size" columns blank.

11. Ellerbroek Furniture sold the following items to James Heppner: 4 chairs at $105 each, 1 table at $375, 2 wall hangings at $98.50 each, 3 pillows at $15.75 each, and 2 bookcases at $125 each. The entire order is subject to a 5% sales tax. James paid cash. Complete the following sales slip for the sale.

**ELLERBROEK FURNITURE**
816 HARRISON STREET
ALEXANDRIA, VA 22316

Customer's Order No. _192839_    DATE _May 9,_ 19 __

SOLD TO _James Hoeppner_

ADDRESS _829 E. Langdon St._

Salesperson _J. Tempio_    TERMS ___

| CASH | CHARGE | C.O.D. | PAID OUT | RETD. MDSE. | RECD. ON ACCT. |
|------|--------|--------|----------|-------------|----------------|
|      |        |        |          |             |                |

| QUAN. | DESCRIPTION | PRICE | AMOUNT |
|-------|-------------|-------|--------|
|       |             |       |        |
|       |             |       |        |
|       |             |       |        |
|       |             |       |        |
|       |             |       |        |
|       |             |       |        |
|       |             |       |        |
|       |             |       |        |

ALL Claims and Returned Goods MUST Be Accompanied By This Receipt

SIGNATURE _____

## BUSINESS APPLICATIONS

Round amounts to the nearest cent and percents to the nearest tenth.

12. Clovis Department Store sold Marilyn Weiss goods for $105.80. Of this, $24.60 was exempt from the 4% state sales tax. What was the total amount due?

13. Sikolski Gifts marked down a lawn fountain from $450 to $360. What percent markdown is this, based on the original selling price?

**14.** The Wicker Shoppe sold Leonard Martelli the following items: 2 baskets at $27.50 each, 2 chairs at $179.50 each, and 2 wall hangings at $14.98 each. The entire order was subject to a 6% sales tax. What is the total amount due?

**15.** Ellis Decorating sold Sean Ritter the following items, all listed at their original retail prices: 2 floor lamps at $115 each, 2 stained glass decorations at $175 each, and 4 framed photos at $90 each. Ellis granted a 20% discount on the entire order. The entire order is subject to a 5% state sales tax. What is the total amount Ellis should collect from Sean?

**16.** Gwenn McDonald purchased goods costing $235.00. The sales clerk added $14.10 sales tax to Gwenn's sales slip. What was the sales tax rate?

**17.** Last year Sportsland, Inc., purchased plastic skateboards for $23.95 each and sold them at a 45% markup based on cost. Since they did not sell well, Sportsland marked them down 47%. Calculate (a) the original selling price and (b) the reduced price.

**(a)** _____

**(b)** _____

**18.** Last month Music World sold their remaining inventory of compact disks at a total reduced price of $285,590. The total original selling price was $428,430. Calculate the percent of markdown on the original selling price.

_____

## CHAPTER 7 REVIEW

Round money amounts to the nearest cent and percents to the nearest tenth.

**1.** Cranston Office Supply sold the following items to Monica Galik: 4 boxes of envelopes at $1.58 each, and 6 pens at $4.50 each. A 6% sales tax applies to the entire order. Complete the following sales slip for this sale to Monica.

| CRANSTON OFFICE SUPPLY |
|---|
| 3400 LAKESHORE AVENUE |
| SHEBOYGAN, WI 53081 |

Customer's Order No. _5787_          DATE _October 6,_ 19 __

SOLD TO _Monica Galik_

ADDRESS _423 Dumont St._

Salesperson _C. Graham_          TERMS __

| CASH ✓ | CHARGE | C.O.D. | PAID OUT | RETD. MDSE. | RECD. ON ACCT. |
|---|---|---|---|---|---|

| QUAN. | DESCRIPTION | PRICE | AMOUNT |
|---|---|---|---|
| | | | |
| | | | |
| | | | |
| | | | |
| | | | |
| | | | |
| | | | |
| | | | |

ALL Claims and Returned Goods MUST Be Accompanied By This Receipt

SIGNATURE _____

**2.** Gomez Carpet Store purchased a roll of carpet at a cost of $8.00 per square yard. Gomez's markup rate on the carpet is 60% on cost. What is the selling price per square yard?

3. Burlington Office Supply has determined that it can sell an executive desk for $1,250. Burlington requires a 60% markup on its cost. At what price must Burlington purchase the desk to meet its goals?

_____

4. The Bike Shop's markup rate on Italian 10-speed bicycles is 40% of selling price. Recently, the Bike Shop purchased 30 Italian bicycles at $300 each. What was the Bike Shop's total selling price for the 30 bicycles?

_____

5. Toyland purchased a battery-operated doll buggy for $80 and marked it up 75% of cost. Since the buggy did not sell, it was marked down 40%. What is the reduced price of the buggy?

_____

6. Midland Emporium sold the following items to Perry Hirshall: 4 compact disks at $15.95 each, 3 framed photos at $29.75 each, 2 pairs of slacks at $34.50 each, and groceries for $35.60. The state sales tax is 5%. Only the groceries are exempt from sales tax. What is the total amount due from Perry?

_____

7. Peck's Music Store is offering $2.00 off the regular price of $13.99 on any compact disks in stock. Calculate the markdown rate.

_____

8. The International Motors Company has a $9,840 markup on cost on a tractor costing $21,959. Calculate the markup rate.

_____

9. The cost of a refrigerator is $215. The refrigerator has a retail price of $395.99. Calculate the markup rate (a) based on cost and (b) based on selling price.

(a) _____

(b) _____

10. The U-Finish Oak Shop sells their unfinished oak furniture with a 35% markup rate based on cost. The markup on a five-drawer dresser is $98. Calculate (a) the cost and (b) the selling price.

(a) _____

(b) _____

11. The selling price of a leather jacket at Leather Man is $245.50. The selling price includes a 36% markup based on selling price. Calculate the cost.

_____

12. The Miller Equipment Company requires a 42% markup on cost on all their metal products. Last month the cost of goods sold was $1,309,000 and operating expenses were $326,000. Calculate (a) net profit and (b) selling price.

(a)  _____

(b) _____

13. Health-Aid Drugs sells a 2.5 fluid ounce bottle of Everyoung hair coloring for $3.99. The selling price includes a $1.17 markup on selling price. Calculate the markup rate.

_____

14. The Beylow Equipment Company manufactured drilling equipment at a cost of $619,500 and sold the equipment for $951,800. Calculate the markup rate on cost.

_____

15. The Green Acre Farms marks up their produce by 26.7% on selling price. The markup on a head of cauliflower is $0.49. Calculate the selling price.

_____

16. The Kitchen Korner marked down their entire stock of dishes by 35%. Calculate the original selling price of a set of dishes selling at a reduced price of $39.95.

_____

17. Janell Building and Remodeling renovated a kitchen at a cost of $1,340. Janell charged the customer $2,195 for the work. What was the markup rate based on cost for the renovation job?

_____

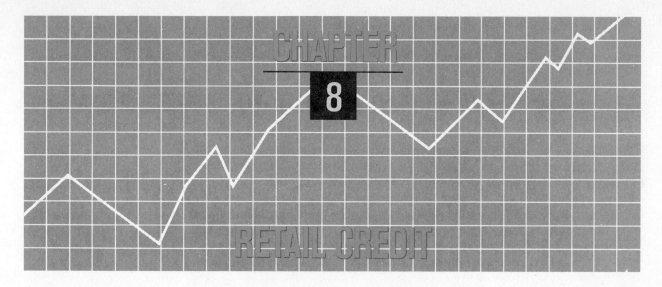

## LEARNING OBJECTIVES

1. Calculate the average daily balance on revolving charge accounts.
2. Calculate the finance charge on charge accounts and retail installment contracts.
3. Calculate the approximate APR.
4. Find the APR using an APR table.
5. Calculate interest on installment loans and prepare a loan amortization schedule.
6. Calculate early payment of installment loans.

**M**any individuals and businesses purchase products and services and agree to pay for them at a later date. This purchasing agreement is called *buying on credit.* Retail businesses extend credit to customers as a service and to increase sales. The finance charges collected from credit purchases and installment loans represent a large portion of the profits for many businesses, banks, and other lending institutions. This chapter will discuss terms associated with credit purchases and installment loans and present procedures for computing amounts such as finance charges and annual percentage rates (APR).

## LESSON 8-1    OPEN-END CREDIT

**REVOLVING CHARGE ACCOUNTS**

Consumers use an open-end credit account to purchase items on credit. With *open-end credit,* a customer may continue to make credit purchases up to a specified credit limit even though the previous balance is not paid off. Since the entire balance is not required to be paid at the end of the billing cycle, such an account is called a *revolving charge account.*

Bank credit cards, such as VISA and MasterCard, and retail credit cards, such as Sears and Exxon, are revolving charge accounts. Credit card holders usually pay a *finance charge* for the use of the credit card. Finance charges usually include interest on the unpaid balance and/or an annual membership fee.

Monthly finance charges are added to any previous unpaid balance and monthly payments or credits are deducted. The credit card holder is usually required to pay a minimum amount each month. If the full amount is paid the same month it is charged, there is no finance charge on that amount.

## CALCULATING THE AVERAGE DAILY BALANCE

Most companies that offer revolving charge accounts calculate the finance charge as a percent of the *average daily balance*. To find the average daily balance, divide the sum of the daily balances by the number of days in the monthly billing cycle. The daily balance is found by adding the previous balance, cash advances, and purchases and then subtracting payments.

$$\frac{\text{Average}}{\text{Daily Balance}} = \frac{\text{Sum of Daily Balances}}{\text{Number of Days in Billing Cycle}}$$

### EXAMPLE

Find the average daily balance for the following credit card transactions during a 31-day billing cycle. The billing date is May 1. Round to the nearest cent.

### SOLUTION

**Step 1** Find the daily balance after each transaction.

| Date | Transaction | Amount | Daily Balance | |
|------|-------------|--------|---------------|--|
| 5/1 | Previous balance | | $259.20 | |
| 5/5 | Purchase | $ 82.45 | 341.65 | ($259.20 + 82.45) |
| 5/11 | Cash advance | 100.00 | 441.65 | ($341.65 + 100) |
| 5/17 | Payment | −200.00 | 241.65 | ($441.65 − 200) |
| 5/28 | Purchase | 45.65 | 287.30 | ($241.65 + 45.65) |

**Step 2** Find the number of days for each balance.

| Date | Daily Balance | Number of Days for Each Balance | |
|------|---------------|---------|-------------------------------|
| 5/1 | $259.20 | 4 days | (May 1 through May 4) |
| 5/5 | 341.65 | 6 days | (May 5 through May 10) |
| 5/11 | 441.65 | 6 days | (May 11 through May 16) |
| 5/17 | 241.65 | 11 days | (May 17 through May 27) |
| 5/28 | 287.30 | 4 days | (May 28 through May 31) |
| | | 31 days in billing cycle | |

**Step 3** Find the sum of the daily balances.

| Daily Balance × | Number of Days | = Extensions |
|-----------------|------|------------|
| $259.20 × | 4 | = $1,036.80 |
| 341.65 × | 6 | = 2,049.90 |
| 441.65 × | 6 | = 2,649.90 |
| 241.65 × | 11 | = 2,658.15 |
| 287.30 × | 4 | = 1,149.20 |
| | | $9,543.95  Sum of daily balances |

**Step 4**  Find the average daily balance.

$$\frac{\text{Average}}{\text{Daily Balance}} = \frac{\text{Sum of Daily Balances}}{\text{Number of Days in Billing Cycle}} = \frac{\$9,543.95}{31} = \$307.87$$

## PRACTICE PROBLEM

Find the average daily balance for the following credit card account. The billing cycle is 31 days from the billing date of January 4.

| Date | Transactions | Amount |
|------|-------------|--------|
| 1/4 | Previous balance | $894.53 |
| 1/10 | Purchase | 45.62 |
| 1/25 | Purchase | 208.12 |
| 1/31 | Payment | 425.50 |

## CALCULATING THE FINANCE CHARGE

Every month, charge card holders receive a statement similar to the one shown below. The statement shows the previous balance, advances, purchases, credits, payments, and new balance. According to the Truth-in-Lending Act, the statement must show the amount of the finance charge, the balance subject to finance charge, the monthly periodic rate, and the annual percentage rate (APR) of the finance charge.

### VISA STATEMENT

ALL TRANSACTIONS SHOWN ARE AS OF THE BILLING DATE INDICATED BELOW;
TRANSACTIONS POSTED AFTER THE BILLING DATE WILL APPEAR ON YOUR NEXT STATEMENT.

| ACCOUNT RECAP | PREVIOUS BALANCE (+) | ADVANCES (+) PURCHASES (+) | DEBIT ADJUSTMENTS (+) | CREDITS (−) | PAYMENTS (−) | FINANCE CHARGE | NEW BALANCE |
|---|---|---|---|---|---|---|---|
| | 717.25 | 73.40 | .00 | 5.31 | 300.00 | 10.60 | 495.94 |

| BALANCE SUBJECT TO FINANCE CHARGE AND THE APPLICABLE RATES. | AVERAGE DAILY BALANCE | MONTHLY PERIODIC RATE | ANNUAL PERCENTAGE RATE | | |
|---|---|---|---|---|---|
| | 706.98 | 1.500 % | 18 % | ACCOUNT NUMBER | 1811 0318 0017 |

TO AVOID FINANCE CHARGES ON YOUR NEXT STATEMENT, ON PURCHASES ONLY, THE AGGREGATE OF YOUR RETURNS, CREDITS AND PAYMENTS MUST EQUAL OR EXCEED THE NEW BALANCE AND BE RECEIVED ON OR BEFORE THE DUE DATE.

| BILLING DATE | # DAYS IN CYCLE | PAST DUE AMOUNT | CURRENT PAYMENT | MINIMUM AMOUNT DUE | PAYMENT DUE DATE | CREDIT LIMIT | AVAILABLE CREDIT |
|---|---|---|---|---|---|---|---|
| 02/11/-- | 29 | .00 | 25.00 | 25.00 | 03/08/-- | 1,500 | 824 |

| TRANS. DATE | REFERENCE NUMBER | POSTING DATE | DESCRIPTION | AMOUNT |
|---|---|---|---|---|
| 01 15 | 4425000127WT909JBA | 01 28 | LA FIESTA RESTAURANT  6    DAYTONA BCH  FL | 11 46 |
| 02 01 | 439900020739F6APVY | 02 10 | HALS MARKET                      PORT ORANGE  FL | 49 70 |
| 02 03 | 4861000020603705089 | 02 06 | SPEEDWAY MOTOR INN       DAYTONA BEACH FL | 12 24 |
| 02 11 | 4861000021141808862 | 02 11 | PAYMENT RECEIVED - THANK YOU | -300 00 |
| 12 27 | 4861000114W5603191 | 01 14 | RADIO STORE NO. 018881  PORT ORANGE    FL | -5 31 |

TAX DEDUCTIBLE FINANCE CHARGES BILLED
FOR CALENDAR YEAR 19-- -$106.76

The **annual percentage rate (APR),** stated on retail installment contracts, represents the true annual rate of interest charged by lenders or retail creditors. The Truth-in-Lending Act requires that the APR be stated on credit agreements.

On most revolving charge plans, the finance charge is calculated by multiplying the average daily balance by the monthly periodic rate.

$$\text{Average Daily Balance} \times \text{Monthly Periodic Rate} = \text{Finance Charge}$$

To find the monthly periodic rate, divide the APR by 12.

$$\text{Monthly Periodic Rate} = \frac{\text{Annual Percentage Rate (APR)}}{12}$$

The new balance shown is the total of any previous balance, advances, purchases, and finance charges minus credits and payments made.

$$\text{Previous Balance} + \text{Advances \& Purchases} + \text{Finance Charge} - \text{Credits \& Payments} = \text{New Balance}$$

---

### EXAMPLE

Calculate the monthly periodic rate, monthly finance charge, and the new balance on the above VISA statement.

### SOLUTION

**Step 1**  Find the monthly periodic rate.

$$\text{Monthly Periodic Rate} = \frac{\text{Annual Percentage Rate}}{12} = \frac{0.18}{12} = 0.015 = 1.5\%$$

**Step 2**  Find the monthly finance charge and the new balance.

$$\text{Average Daily Balance} \times \text{Monthly Periodic Rate} = \text{Finance Charge}$$
$$\$706.98 \times 0.015 = \$10.60$$

$$\text{Previous Balance} + \text{Advances \& Purchases} + \text{Finance Charge} - \text{Credits} - \text{Payments} = \text{New Balance}$$
$$\$717.25 + \$73.40 + \$10.60 - \$5.31 - \$300.00 = \$495.94$$

---

### PRACTICE PROBLEMS

Find the finance charge on the daily balance and the new balance. Round to the nearest cent.

| | Average Daily Balance | Monthly Interest Rate | Previous Balance | Advances & Purchases | Finance Charge | Credits & Payments | New Balance |
|---|---|---|---|---|---|---|---|
| **1.** | $952.29 | 1.75% | $555.56 | $598.52 | _____ | $300.00 | _____ |
| **2.** | 509.17 | 1.5% | 556.53 | 75.78 | _____ | 250.00 | _____ |

---

## LESSON 8-1    EXERCISES

Find the average daily balance for the following credit card accounts. The billing cycle is 31 days from the billing date. Calculate the finance charge on the average daily balance using a monthly periodic rate of 1.6%.

**1.** Billing date is August 10.

| Date | Transactions | Amount |
|------|-------------|--------|
| 8/10 | Previous balance | $ 98.67 |
| 8/17 | Purchase | 11.24 |
| 8/23 | Cash advance | 100.00 |
| 8/28 | Payment | −50.00 |

Average daily balance $_____        Finance charge $_____

**2.** Billing date is July 15.

| Date | Transactions | Amount |
|------|-------------|--------|
| 7/15 | Previous balance | $ 208.58 |
| 7/19 | Credit | −21.15 |
| 7/22 | Purchase | 143.50 |
| 7/31 | Payment | −250.00 |
| 8/10 | Purchase | 62.31 |

Average daily balance $_____        Finance charge $_____

**3.** Billing date is January 1.

| Date | Transactions | Amount |
|------|-------------|--------|
| 1/1 | Previous balance | $290.60 |
| 1/7 | Cash advance | 175.00 |
| 1/18 | Payment | −85.70 |
| 1/20 | Purchase | 238.65 |
| 1/26 | Credit | −87.50 |

Average daily balance $_____        Finance charge $_____

Find the monthly periodic rate, finance charge on the daily balance, and new balance for the following credit card accounts. Round amounts to the nearest cent and percents to the nearest hundredth.

| | Average Daily Balance | APR | Monthly Periodic Rate | Finance Charge | Previous Balance | Advances & Purchases | Credits & Payments | New Balance |
|---|------|-----|------|------|------|------|------|------|
| **4.** | $ 345.00 | 20% | _____ | _____ | $ 159.32 | $ 82.49 | $247.57 | $___0___ |
| **5.** | 56.20 | 21% | _____ | _____ | 190.72 | 72.93 | 100.00 | _____ |
| **6.** | 211.90 | 15% | _____ | _____ | 52.94 | 284.59 | 48.00 | _____ |
| **7.** | 3,056.90 | 20% | _____ | _____ | 2,073.39 | 378.40 | 435.00 | _____ |
| **8.** | 211.68 | 24% | _____ | _____ | 67.20 | 15.98 | 87.41 | _____ |

## BUSINESS APPLICATIONS

Round amounts to the nearest cent and percents to the nearest hundredth. Assume a 31-day billing cycle.

9. For the month of June, Jay Kline's VISA statement showed an average daily balance of $394.50, a previous balance of $298.67, purchases of $189.29, and payments of $45.60. The credit card company charges an APR of 15% for the year. Find (a) the monthly periodic rate, (b) the finance charge on the average daily balance, and (c) the new balance.

(a) _____

(b) _____

(c) _____

10. On March 1 Janice Ramona's credit card account showed a balance of $145. On March 15 a payment of $145 was credited. Janice made no additional purchases or payments for the remaining 31-day billing cycle. The monthly periodic rate is 1.5% on the daily balance. Find (a) the average daily balance and (b) the finance charge.

(a) _____

(b) _____

11. For the month of July, Dr. Glen Murphy's credit account showed an average daily balance of $1,920, a previous balance of $567.80, purchases of $239.40, cash advance of $290, and payments of $765.50. The credit card company charges a monthly interest rate of 1.75% on the average daily balance. Find (a) the finance charge and (b) the new balance.

(a) _____

(b) _____

**12.** On August 1 Roy Elwood's credit card statement showed a beginning balance of $450.00. Purchases of $320.40 and $94.00 were charged on August 7 and 19, respectively, and a payment of $270.00 was credited on August 21. The annual percentage rate is 12%. Find (a) the average daily balance for the 31-day billing cycle beginning August 1, (b) the finance charge on the average daily balance, and (c) the new balance.

(a) _____

(b) _____

(c) _____

**13.** Helen Wheeler's MasterCard statement had a beginning balance of $145.65. Her purchases totaled $451.72 and her cash advance was $110. If the statement showed a new balance of $167.37, what was her payment?

**14.** On December 10 Maria's credit card statement showed a previous balance of $45.60. On December 12 her purchases totaled $645.63, and on December 20 her purchases totaled $895.67. She received a cash advance of $300 on December 28 and made a payment of $300 on January 6. Find (a) Maria's average daily balance for the 31-day billing cycle beginning December 10. Then find (b) the finance charge on the average daily balance if the monthly periodic rate is 1.67% and (c) the new balance. Calculate (d) by how much Maria's new balance is extended over her credit limit of $1,500.

(a) _____

(b) _____

(c) _____

(d) _____

## INSTALLMENT PURCHASES

*Retail installment contracts* are used to buy expensive items, such as an automobile or a major appliance. With this type of credit agreement, the buyer usually is required to make a down payment toward the purchase and make periodic payments (usually monthly) over a period of time. Commercial banks, savings and loan institutions, personal loan companies, and credit unions offer installment loans to individuals and businesses. An *installment loan,* repaid in a series of equal payments over a period of time, is similar to a retail installment contract.

The amount of credit extended and interest charged depends on such factors as the borrower's past credit record, present financial condition, current and future earning power, and valuables pledged as security.

The charge for obtaining credit is called a *finance charge* and is included in the payments. According to the Truth-in-Lending Act, the creditor must provide the buyer with the following information in writing on a retail credit contract.

1. *The annual percentage rate (APR).* The cost of credit as a yearly rate.
2. *The finance charge.* The finance charge is the interest computed on the amount financed, including credit report fees, banking fees, and credit insurance.
3. *The amount financed.* The amount financed is the purchase price minus the down payment.
4. *The total of payments.* The amount paid after all payments have been made as scheduled.
5. *The total sale price.* The total cost of the purchase on credit, including the down payment, the amount financed, and the finance charges.

### EXAMPLE

The Sun Country Auto dealership is advertising a truck with a price of $10,900. The truck can be purchased on the installment plan for $250 down and $349.08 a month for 36 months. Find the amount financed, the finance charge, and the total sale price.

### SOLUTION

**Step 1**  Find the amount financed.

Purchase Price − Down Payment = Amount Financed
$10,900 − $250 = $10,650

**Step 2**  Find the finance charge.

Total of  Amount  Finance
Payments − Financed = Charge
$12,566.88 − $10,650 = $1,916.88

$349.08 × 36 ⟶

**Step 3**  Find the total sale price.

Down  Amount  Finance  Total Sale
Payment + Financed +  Charge  =  Price
$250 + $10,650 + $1,916.88 = $12,816.88

Find the amount financed, the finance charge, and the total sale price on the following installment purchases. Round to the nearest cent.

| | Purchase Price | Down Payment | Monthly Payments | Number of Payments | Amount Financed | Finance Charge | Total Sale Price |
|---|---|---|---|---|---|---|---|
| **1.** | $ 895 | $130 | $ 38.25 | 24 | _____ | _____ | _____ |
| **2.** | 1,200 | 250 | 35.49 | 36 | _____ | _____ | _____ |
| **3.** | 365 | 0 | 34.07 | 12 | _____ | _____ | _____ |
| **4.** | 2,300 | 500 | 109.50 | 20 | _____ | _____ | _____ |

## CALCULATING THE APPROXIMATE APR

If a 12-month loan with a 10% finance charge were repaid with a single payment at the end of 12 months, the APR would also be 10%. If the loan, with the same amount of total interest, were repaid in 12 monthly installments, the APR would be greater than 10%. Since a payment is made each month reducing the balance on an installment loan, the borrower is actually paying more than a 10% finance charge on the outstanding balance.

Computation of the APR is fairly complicated and is usually determined by a computer program or by a table available at any bank or lending institution. The APR of an installment contract calling for monthly payments can be approximated by using the following formula.

2 × number of payments per year ⟶

$$\text{Approximate APR} = \frac{24 \times \text{Finance Charge}}{\text{Amount Financed} \times (\text{Number of Payments} + 1)}$$

### EXAMPLE

Use the formula to find the approximate APR for an installment purchase if the amount financed is $5,975, the 8% finance charge is $1,434, and the number of monthly installment payments is 36.

### SOLUTION

$$\begin{aligned}\text{Approximate APR} &= \frac{24 \times \text{Finance Charge}}{\text{Amount Financed} \times (\text{Number of Payments} + 1)} \\ &= \frac{24 \times \$1,434}{\$5,975 \times (36 + 1)} = 0.15567 \approx 15.57\%\end{aligned}$$

ANSWERS FOR PRACTICE PROBLEMS

| | Amount Financed | Finance Charge | Total Sale Price |
|---|---|---|---|
| **(1)** | $ 765.00 | $153.00 | $1,048.00 |
| **(2)** | 950.00 | 327.64 | 1,527.64 |
| **(3)** | 365.00 | 43.84 | 408.84 |
| **(4)** | 1,800.00 | 390.00 | 2,690.00 |

Find the approximate APR using the formula. Round to the nearest tenth of a percent.

| | Amount Financed | Finance Charge | Number of Monthly Payments | Approximate APR |
|---|---|---|---|---|
| **1.** | $ 675 | $131.60 | 24 | _____ |
| **2.** | 2,750 | 969.38 | 36 | _____ |

## FINDING THE APR USING A TABLE

The approximate APR calculated using the formula is not accurate enough to satisfy the requirements of the Truth-in-Lending Act. The APR found using the annual percentage rate table on page 260 is accurate to the nearest quarter of a percent, which meets federal law requirements.

Before using the table, find the finance charge per $100 of the amount financed. To find the finance charge per $100, divide the finance charge by the amount financed, then multiply the result by 100.

$$\text{Finance Charge per } \$100 = \frac{\text{Finance Charge}}{\text{Amount Financed}} \times 100$$

### EXAMPLE

An air-conditioning unit can be purchased on the installment plan by financing $2,150 over 18 monthly installments with a finance charge of $290.26. Use the APR table to find the APR for the installment purchase.

### SOLUTION

**Step 1**   Find the finance charge per $100.

$$\text{Finance Charge per } \$100 = \frac{\text{Finance Charge}}{\text{Amount Financed}} \times 100 = \frac{\$290.26}{\$2,150} \times 100 = \$13.5$$

**Step 2**   Find the APR using the table on page 260. Read down the left-hand column to 18 payments; then move across until you find the number nearest 13.5. The nearest number is 13.57. Read up the column to find the APR, 16.50%.

### PRACTICE PROBLEMS

Find the APR using the APR table.

| | Amount Financed | Finance Charge | Number of Payments | APR |
|---|---|---|---|---|
| **1.** | $3,400 | $765.00 | 30 | _____ |
| **2.** | 935 | 126.23 | 18 | _____ |

| Number of Payments | 16.00% | 16.25% | 16.50% | 16.75% | 17.00% | 17.25% | 17.50% | 17.75% | 18.00% | 18.25% | 18.50% | 18.75% | 19.00% | 19.25% | 19.50% | 19.75% |
|---|---|---|---|---|---|---|---|---|---|---|---|---|---|---|---|---|
| | | | | | | | (Finance Charge per $100 of Amount Financed) | | | | | | | | | |
| 1 | 1.33 | 1.35 | 1.37 | 1.40 | 1.42 | 1.44 | 1.46 | 1.48 | 1.50 | 1.52 | 1.54 | 1.56 | 1.58 | 1.60 | 1.62 | 1.65 |
| 2 | 2.00 | 2.04 | 2.07 | 2.10 | 2.13 | 2.16 | 2.19 | 2.22 | 2.26 | 2.29 | 2.32 | 2.35 | 2.38 | 2.41 | 2.44 | 2.48 |
| 3 | 2.68 | 2.72 | 2.76 | 2.80 | 2.85 | 2.89 | 2.93 | 2.97 | 3.01 | 3.06 | 3.10 | 3.14 | 3.18 | 3.23 | 3.27 | 3.31 |
| 4 | 3.36 | 3.41 | 3.46 | 3.51 | 3.57 | 3.62 | 3.67 | 3.73 | 3.78 | 3.83 | 3.88 | 3.94 | 3.99 | 4.04 | 4.10 | 4.15 |
| 5 | 4.04 | 4.10 | 4.16 | 4.23 | 4.29 | 4.35 | 4.42 | 4.48 | 4.54 | 4.61 | 4.67 | 4.74 | 4.80 | 4.86 | 4.93 | 4.99 |
| 6 | 4.72 | 4.79 | 4.87 | 4.94 | 5.02 | 5.09 | 5.17 | 5.24 | 5.32 | 5.39 | 5.46 | 5.54 | 5.61 | 5.69 | 5.76 | 5.84 |
| 7 | 5.40 | 5.49 | 5.58 | 5.66 | 5.75 | 5.83 | 5.92 | 6.00 | 6.09 | 6.18 | 6.26 | 6.35 | 6.43 | 6.52 | 6.60 | 6.69 |
| 8 | 6.09 | 6.19 | 6.29 | 6.38 | 6.48 | 6.58 | 6.67 | 6.77 | 6.87 | 6.96 | 7.06 | 7.16 | 7.26 | 7.35 | 7.45 | 7.55 |
| 9 | 6.78 | 6.89 | 7.00 | 7.11 | 7.22 | 7.32 | 7.43 | 7.54 | 7.65 | 7.76 | 7.87 | 7.97 | 8.08 | 8.19 | 8.30 | 8.41 |
| 10 | 7.48 | 7.60 | 7.72 | 7.84 | 7.96 | 8.08 | 8.19 | 8.31 | 8.43 | 8.55 | 8.67 | 8.79 | 8.91 | 9.03 | 9.15 | 9.27 |
| 11 | 8.18 | 8.31 | 8.44 | 8.57 | 8.70 | 8.83 | 8.96 | 9.09 | 9.22 | 9.35 | 9.49 | 9.62 | 9.75 | 9.88 | 10.01 | 10.14 |
| 12 | 8.88 | 9.02 | 9.16 | 9.30 | 9.45 | 9.59 | 9.73 | 9.87 | 10.02 | 10.16 | 10.30 | 10.44 | 10.59 | 10.73 | 10.87 | 11.02 |
| 13 | 9.58 | 9.73 | 9.89 | 10.04 | 10.20 | 10.35 | 10.50 | 10.66 | 10.81 | 10.97 | 11.12 | 11.28 | 11.43 | 11.59 | 11.74 | 11.90 |
| 14 | 10.79 | 10.45 | 10.67 | 10.78 | 10.95 | 11.11 | 11.28 | 11.45 | 11.61 | 11.78 | 11.95 | 12.11 | 12.28 | 12.45 | 12.61 | 12.78 |
| 15 | 11.00 | 11.17 | 11.35 | 11.53 | 11.71 | 11.88 | 12.06 | 12.24 | 12.42 | 12.59 | 12.77 | 12.95 | 13.13 | 13.31 | 13.49 | 13.67 |
| 16 | 11.71 | 11.90 | 12.09 | 12.28 | 12.46 | 12.65 | 12.84 | 13.03 | 13.22 | 13.41 | 13.60 | 13.80 | 13.99 | 14.18 | 14.37 | 14.56 |
| 17 | 12.42 | 12.67 | 12.83 | 13.03 | 13.23 | 13.43 | 13.63 | 13.83 | 14.04 | 14.24 | 14.44 | 14.64 | 14.85 | 15.05 | 15.25 | 15.46 |
| 18 | 13.14 | 13.35 | 13.57 | 13.78 | 13.99 | 14.21 | 14.42 | 14.64 | 14.85 | 15.07 | 15.28 | 15.49 | 15.71 | 15.93 | 16.14 | 16.36 |
| 19 | 13.86 | 14.09 | 14.31 | 14.54 | 14.76 | 14.99 | 15.22 | 15.44 | 15.67 | 15.90 | 16.12 | 16.35 | 16.58 | 16.81 | 17.03 | 17.26 |
| 20 | 14.59 | 14.82 | 15.06 | 15.30 | 15.54 | 15.77 | 16.01 | 16.25 | 16.49 | 16.73 | 16.97 | 17.21 | 17.45 | 17.69 | 17.93 | 18.17 |
| 21 | 15.31 | 15.56 | 15.81 | 16.06 | 16.31 | 16.56 | 16.81 | 17.07 | 17.32 | 17.57 | 17.82 | 18.07 | 18.33 | 18.58 | 18.83 | 19.09 |
| 22 | 16.04 | 16.30 | 16.57 | 16.83 | 17.09 | 17.36 | 17.62 | 17.88 | 18.15 | 18.41 | 18.68 | 18.94 | 19.21 | 19.47 | 19.74 | 20.01 |
| 23 | 16.78 | 17.05 | 17.32 | 17.60 | 17.86 | 18.15 | 18.43 | 18.70 | 18.98 | 19.26 | 19.54 | 19.81 | 20.09 | 20.37 | 20.65 | 20.93 |
| 24 | 17.51 | 17.80 | 18.09 | 18.37 | 18.66 | 18.95 | 19.24 | 19.53 | 19.82 | 20.11 | 20.40 | 20.69 | 20.98 | 21.27 | 21.56 | 21.86 |
| 25 | 18.25 | 18.55 | 18.85 | 19.15 | 19.45 | 19.75 | 20.05 | 20.36 | 20.66 | 20.96 | 21.27 | 21.57 | 21.87 | 22.18 | 22.48 | 22.79 |
| 26 | 18.99 | 19.30 | 19.62 | 19.93 | 20.24 | 20.56 | 20.87 | 21.19 | 21.50 | 21.82 | 22.14 | 22.45 | 22.77 | 23.09 | 23.41 | 23.73 |
| 27 | 19.74 | 20.06 | 20.39 | 20.71 | 21.04 | 21.37 | 21.69 | 22.02 | 22.35 | 22.68 | 23.01 | 23.34 | 23.67 | 24.00 | 24.33 | 24.67 |
| 28 | 20.48 | 20.82 | 21.16 | 21.50 | 21.84 | 22.18 | 22.52 | 22.86 | 23.20 | 23.55 | 23.89 | 24.23 | 24.58 | 24.92 | 25.27 | 25.61 |
| 29 | 21.23 | 21.58 | 21.94 | 22.29 | 22.64 | 22.99 | 23.35 | 23.70 | 24.06 | 24.41 | 24.77 | 25.13 | 25.49 | 25.84 | 26.20 | 26.56 |
| 30 | 21.99 | 22.35 | 22.72 | 23.08 | 23.45 | 23.81 | 24.18 | 24.55 | 24.92 | 25.29 | 25.66 | 26.03 | 26.40 | 26.77 | 27.14 | 27.52 |
| 31 | 22.74 | 23.12 | 23.50 | 23.88 | 24.26 | 24.64 | 25.02 | 25.40 | 25.78 | 26.16 | 26.55 | 26.93 | 27.32 | 27.70 | 28.09 | 28.47 |
| 32 | 23.50 | 23.89 | 24.28 | 24.68 | 25.07 | 25.46 | 25.86 | 26.25 | 26.65 | 27.04 | 27.44 | 27.84 | 28.24 | 28.64 | 29.04 | 29.44 |
| 33 | 24.26 | 24.67 | 25.07 | 25.48 | 25.88 | 26.29 | 26.70 | 27.11 | 27.52 | 27.93 | 28.34 | 28.75 | 29.16 | 29.57 | 29.99 | 30.40 |
| 34 | 25.03 | 25.44 | 25.86 | 26.28 | 26.70 | 27.12 | 27.54 | 27.97 | 28.39 | 28.81 | 29.24 | 29.66 | 30.09 | 30.52 | 30.95 | 31.37 |
| 35 | 25.79 | 26.23 | 26.66 | 27.09 | 27.52 | 27.96 | 28.39 | 28.83 | 29.27 | 29.71 | 30.14 | 30.58 | 31.02 | 31.47 | 31.91 | 32.35 |
| 36 | 26.57 | 27.01 | 27.46 | 27.90 | 28.35 | 28.80 | 29.25 | 29.70 | 30.15 | 30.60 | 31.05 | 31.51 | 31.96 | 32.42 | 32.87 | 33.33 |
| 37 | 37.34 | 27.80 | 28.26 | 28.72 | 29.18 | 29.64 | 30.10 | 30.57 | 31.03 | 31.50 | 31.97 | 32.43 | 32.90 | 33.37 | 33.84 | 34.32 |
| 38 | 28.11 | 28.59 | 29.06 | 29.53 | 30.01 | 30.49 | 30.96 | 31.44 | 31.92 | 32.40 | 32.88 | 33.37 | 33.85 | 34.33 | 34.82 | 35.30 |
| 39 | 28.89 | 29.38 | 29.87 | 30.36 | 30.85 | 31.34 | 31.83 | 32.32 | 32.81 | 33.31 | 33.80 | 34.30 | 34.80 | 35.30 | 35.80 | 36.30 |
| 40 | 29.68 | 30.18 | 30.68 | 31.18 | 31.68 | 32.19 | 32.69 | 33.20 | 33.71 | 34.22 | 34.73 | 35.24 | 35.75 | 36.26 | 36.78 | 37.29 |

## LESSON 8-2    EXERCISES

Find the amount financed, finance charge, and total sale price on the following installment purchases. Round to the nearest cent.

| | Item | Purchase Price | Down Payment | Monthly Payment | Number of Payments | Amount Financed | Finance Charge | Total Sale Price |
|---|---|---|---|---|---|---|---|---|
| 1. | VCR | $ 456 | $ 50 | $ 49.17 | 9 | _____ | _____ | _____ |
| 2. | Stereo | 782 | 85 | 76.67 | 10 | _____ | _____ | _____ |
| 3. | Computer | 1,800 | 400 | 130.67 | 12 | _____ | _____ | _____ |
| 4. | Lawn tractor | 2,890 | 325 | 132.53 | 24 | _____ | _____ | _____ |
| 5. | Furniture | 3,475 | 350 | 204.86 | 18 | _____ | _____ | _____ |

Find the amount financed and the finance charge. Round amounts to the nearest cent. Find the approximate APR using the formula. Round percents to the nearest hundredth.

| | Purchase Price | Down Payment | Monthly Payment | Number of Payments | Amount Financed | Finance Charge | Approximate APR |
|---|---|---|---|---|---|---|---|
| 6. | $7,200 | $700 | $229.31 | 36 | _____ | _____ | _____ |
| 7. | 1,320 | 220 | 101.67 | 12 | _____ | _____ | _____ |
| 8. | 6,100 | 640 | 273.00 | 24 | _____ | _____ | _____ |
| 9. | 815 | 0 | 52.41 | 18 | _____ | _____ | _____ |
| 10. | 1,575 | 75 | 65.00 | 30 | _____ | _____ | _____ |

Find the APR using the APR table. Round amounts to the nearest cent.

| | Amount Financed | Credit Period | Finance Charge | APR |
|---|---|---|---|---|
| 11. | $4,070 | 40 months | $1,322.75 | _____ |
| 12. | 935 | 3 years | 308.55 | _____ |
| 13. | 2,500 | 24 months | 437.50 | _____ |
| 14. | 4,100 | 9 months | 345.94 | _____ |
| 15. | 780 | 10 months | 58.50 | _____ |

## BUSINESS APPLICATIONS

Round amounts to the nearest cent. Round percents to the nearest hundredth.

16. Rosa Romera signed an installment loan for $950. The loan calls for 18 monthly payments with a $121.13 finance charge. Find (a) the total amount due, (b) the monthly payments, and (c) the approximate APR.

(a) _____

(b) _____

(c) _____

17. Valley New and Used Cars sold Henry Kemp a car costing $10,600 on the installment plan. According to the contract, the down payment is $400 and the 36 monthly payments are $332.21 each. Find (a) the amount financed, (b) the finance charge, and (c) the total sale price.

(a) _____

(b) _____

(c) _____

18. The total finance charge on a $4,500 installment loan is $1,275. If the loan is to be repaid in 34 monthly installments, find the APR using the APR table.

_____

19. During an inventory sale at Home Supply, Eric Finn purchased a television set for $416. He signed an installment contract calling for a $25 down payment and 10 monthly payments of $42.68 each. Find (a) the finance charge, and (b) the approximate APR using the formula.

(a) _____

(b) _____

20. The Economy Savings and Loan advertises a 12-month loan of $1,500 with only $135 interest. The Federal Savings and Loan advertise an 18-month loan of $1,000 with only $135 interest. Which loan has the lower APR? Use the APR table.

_____

21. The Molinni Construction Company purchased $8,900 worth of equipment on the installment plan. The company put $800 down and agreed to pay $364.17 per month for 30 months. Find (a) the finance charge and (b) the total sale price.

(a) _____

(b) _____

22. The Hanheizer Repair Company signed an installment loan contract for $6,500. The contract called for a finance charge of $2,340 to be paid over a 4-year period. Find the approximate APR using the formula if the loan is to be repaid in monthly payments.

_____

**23.** The Dodge Lumber Company purchased 5 radial arm saws for a total of $6,000 from the Wholesale Distributors. The Wholesale Distributors agreed to finance the purchase by accepting $1,000 down and 12 payments of $458.34 each. Find (a) the finance charge, (b) the total sale price, and (c) the APR using the APR table.

(a) _____

(b) _____

(c) _____

## USING THE LOAN AMORTIZATION TABLE

Installment loans usually are **amortized,** that is, paid off with a series of equal periodic payments that include both interest and principal. Since calculating installment payments is a lengthy process, a computer, a programmable calculator, or a table similar to the one shown below is used to find the monthly payment. The loan amortization table shows the monthly payments for various installment loan amounts at 10%, 11%, and 12% interest rates.

**LOAN AMORTIZATION TABLE**
**Monthly Payment Required to Amortize a Loan**

| Amount | 10% 5 Years | 10% 20 Years | 10% 25 Years | 11% 5 Years | 11% 20 Years | 11% 25 Years | 12% 5 Years | 12% 20 Years | 12% 25 Years |
|---|---|---|---|---|---|---|---|---|---|
| 500 | 10.63 | 4.83 | 4.55 | 10.88 | 5.17 | 4.91 | 11.13 | 5.51 | 5.27 |
| 1,000 | 21.25 | 9.66 | 9.09 | 21.75 | 10.33 | 9.81 | 22.25 | 11.02 | 10.54 |
| 2,000 | 42.50 | 19.31 | 18.18 | 43.49 | 20.65 | 19.61 | 44.49 | 22.03 | 21.07 |
| 3,000 | 63.75 | 28.96 | 27.27 | 65.23 | 30.97 | 29.41 | 66.74 | 33.04 | 31.60 |
| 4,000 | 84.99 | 38.61 | 36.35 | 86.97 | 41.29 | 39.21 | 88.98 | 44.05 | 42.13 |
| 5,000 | 106.24 | 48.26 | 45.44 | 108.72 | 51.61 | 49.01 | 111.23 | 55.06 | 52.67 |
| 6,000 | 127.49 | 57.91 | 54.53 | 130.46 | 61.94 | 58.81 | 133.47 | 66.07 | 63.20 |
| 7,000 | 148.73 | 67.56 | 63.61 | 152.20 | 72.26 | 68.61 | 155.72 | 77.08 | 73.73 |
| 8,000 | 169.98 | 77.21 | 72.70 | 173.94 | 82.58 | 78.41 | 177.96 | 88.09 | 84.26 |
| 9,000 | 191.23 | 86.86 | 81.79 | 195.69 | 92.90 | 88.22 | 200.21 | 99.10 | 94.80 |
| 10,000 | 212.48 | 96.51 | 90.88 | 217.43 | 103.22 | 98.02 | 222.45 | 110.11 | 105.33 |
| 11,000 | 233.72 | 106.16 | 99.96 | 239.17 | 113.55 | 107.82 | 244.69 | 121.12 | 115.86 |
| 12,000 | 254.97 | 115.81 | 109.05 | 260.91 | 123.87 | 117.62 | 266.94 | 132.14 | 126.39 |
| 13,000 | 276.22 | 125.46 | 118.14 | 282.66 | 134.19 | 127.42 | 289.18 | 143.15 | 136.92 |
| 14,000 | 297.46 | 135.11 | 127.22 | 304.40 | 144.51 | 137.22 | 311.43 | 154.16 | 147.46 |
| 15,000 | 318.71 | 144.76 | 136.31 | 326.14 | 154.83 | 147.02 | 333.67 | 165.17 | 157.99 |
| 20,000 | 424.95 | 193.01 | 181.75 | 434.85 | 206.44 | 196.03 | 444.89 | 220.22 | 210.65 |
| 25,000 | 531.18 | 241.26 | 227.18 | 543.57 | 258.05 | 245.03 | 556.12 | 275.28 | 263.31 |
| 30,000 | 637.42 | 289.51 | 272.62 | 652.28 | 309.66 | 294.04 | 667.34 | 330.33 | 315.97 |
| 35,000 | 743.65 | 337.76 | 318.05 | 760.99 | 361.27 | 343.04 | 778.56 | 385.39 | 368.63 |
| 40,000 | 849.89 | 386.01 | 363.49 | 869.70 | 412.88 | 392.05 | 889.78 | 440.44 | 421.29 |
| 45,000 | 956.12 | 434.26 | 408.92 | 978.41 | 464.49 | 441.06 | 1,001.01 | 495.49 | 473.96 |
| 50,000 | 1,062.36 | 482.52 | 454.36 | 1,087.13 | 516.10 | 490.06 | 1,112.23 | 550.55 | 526.62 |
| 55,000 | 1,168.59 | 530.77 | 499.79 | 1,195.84 | 567.71 | 539.07 | 1,223.45 | 605.60 | 579.28 |
| 60,000 | 1,274.83 | 579.02 | 545.23 | 1,304.55 | 619.32 | 588.07 | 1,334.67 | 660.66 | 631.94 |
| 65,000 | 1,381.06 | 627.27 | 590.66 | 1,413.26 | 670.93 | 637.08 | 1,445.89 | 715.71 | 684.60 |
| 70,000 | 1,487.30 | 675.52 | 636.10 | 1,521.97 | 722.54 | 686.08 | 1,557.12 | 770.77 | 737.26 |
| 75,000 | 1,593.53 | 723.77 | 681.53 | 1,630.69 | 774.15 | 735.09 | 1,668.34 | 825.82 | 789.92 |
| 80,000 | 1,699.77 | 772.02 | 726.97 | 1,739.40 | 825.76 | 784.10 | 1,779.56 | 880.87 | 842.58 |
| 85,000 | 1,806.00 | 820.27 | 772.40 | 1,848.11 | 877.37 | 833.10 | 1,890.78 | 935.93 | 895.25 |
| 90,000 | 1,912.24 | 868.52 | 817.84 | 1,956.82 | 928.97 | 882.11 | 2,002.01 | 990.98 | 947.91 |
| 95,000 | 2,018.47 | 916.78 | 863.27 | 2,065.54 | 980.58 | 931.11 | 2,113.23 | 1,046.04 | 1,000.57 |
| 100,000 | 2,124.71 | 965.03 | 908.71 | 2,174.25 | 1,032.19 | 980.12 | 2,224.45 | 1,101.09 | 1,053.23 |

Carmac Industries purchased an office complex with a purchase price of $82,945. The company paid $7,945 down and financed the remaining balance at 12% for 20 years. Calculate the monthly payment, the total amount due, and the finance charge.

**SOLUTION**

**Step 1**   Find the amount financed.

Cash Price − Down Payment = Amount Financed
  $82,945   −   $7,945   =   $75,000

**Step 2**   Find the monthly payment using the loan amortization table on page 265. The monthly payment on a loan of $75,000 for 20 years at 12% interest is $825.82.

**Step 3**   Find the total amount due.

Number of   Monthly     Total
Payments × Payments = Amount Due

20 months × 12 years ⟶ 240   × $825.82 = $198,196.80

**Step 4**   Find the finance charge.

  Total        Amount      Finance
Amount Due − Financed =   Charge
$198,196.80 − $75,000 = $123,196.80

Installment loans for the purchase of real estate, such as land and buildings, are called **mortgage loans.** The mortgage loan in the above example is called a **fixed-rate mortgage** since the monthly mortgage payment will not change over the life of the loan. During times of inflation, fixed-rate mortgage loans are more advantageous to the borrower since the value of the real estate will continue to rise while the mortgage payments remain steady.

Many lenders protect themselves against inflation by offering **variable-interest-rate** loans. These loans initially have a lower interest rate than fixed-rate loans, but the rate can vary up or down at stated periods over the life of the loan.

**PRACTICE PROBLEMS**

Calculate the monthly payment, the total amount due, and the finance charge for the following installment loans. Use the loan amortization table on page 265.

| | Amount Financed | Rate | Years | Monthly Payment | Total Amount Due | Finance Charge |
|---|---|---|---|---|---|---|
| 1. | $50,000 | 12% | 25 | _____ | _____ | _____ |
| 2. | 75,000 | 10% | 20 | _____ | _____ | _____ |
| 3. | 10,000 | 11% | 5 | _____ | _____ | _____ |

Traditionally, mortgage loans have been for 25 or 30 years but loan periods of 15 to 20 years have become increasingly popular. Loans with shorter payoff periods have higher monthly payments, but the total interest charge is substantially lower.

## EXAMPLE

Simmons Corporation financed $95,000 on a new retail outlet. The monthly payment at 11% is $904.40 for 30 years or $1,080.15 for 15 years. How much less interest will be charged if a 15-year loan is obtained instead of a 30-year loan?

## SOLUTION

**Step 1** Find the total amount due for the 30-year loan.

Number of Payments × Monthly Payment = Total Amount Due

12 months × 30 years ⟶ 360 × $904.40 = $325,584

**Step 2** Find the total amount due for the 15-year loan.

Number of Payments × Monthly Payment = Total Amount Due

12 months × 15 years ⟶ 180 × $1,080.15 = $194,427

**Step 3** Find the difference in the finance charge.

| Total Amount Due, 30-Year Loan | − | Total Amount Due, 15-Year Loan | = | Difference in Finance Charge |
|---|---|---|---|---|
| $325,584 | − | $194,427 | = | $131,157 |

Although the monthly payment for the 15-year loan is $175.75 more a month ($1,080.15 − $904.40) than for the 30-year loan, the total interest charged is $131,157 less.

## PRACTICE PROBLEMS

The following amounts were financed at 10% interest. Find the monthly installment payment for a 20-year loan and for a 25-year loan. Calculate the difference in the total finance charge. Use the loan amortization table on page 265.

|  | Amount Financed | 20-Year Payment | 25-Year Payment | Difference in Finance Charge |
|---|---|---|---|---|
| 1. | $ 35,000 | _____ | _____ | _____ |
| 2. | 100,000 | _____ | _____ | _____ |
| 3. | 75,000 | _____ | _____ | _____ |

ANSWERS FOR
PRACTICE PROBLEMS

| 20-Year Payment | 25-Year Payment | Difference in Finance Charge |
|---|---|---|
| (1) $337.76 | $318.05 | $14,352.60 |
| (2) 965.03 | 908.71 | 41,005.80 |
| (3) 723.77 | 681.53 | 30,754.20 |

A lending institution usually provides the borrower with a loan payment schedule called an **amortization schedule.** The amortization schedule shows the portion of each payment that is interest and the portion of each payment that is principal. The schedule also shows the balance of the principal after each payment. The amount financed is called the **principal.** To find the amount of interest that applies to each payment, use the simple interest formula on the unpaid balance.

### EXAMPLE

Rebow Industries borrowed $15,000 at 11% interest for 5 years to purchase manufacturing equipment. Monthly payments on the loan are $326.14. Calculate the amount of interest paid, the amount of principal paid, and the balance of the principal for the first two payments.

### SOLUTION

**Step 1**   Find the amount of interest paid for the first month.

Principal $\times$ Rate $\times$ Time = Interest Paid
$15,000 $\times$ 0.11 $\times$ $\frac{1}{12}$ =   $137.50

**Step 2**   Find the amount of principal paid for the first month.

Monthly    First Month's    Principal
Payment $-$    Interest    =    Paid
$326.14 $-$    $137.50    = $188.64

**Step 3**   Find the balance of the principal at the end of the first month.

                 Principal    Balance of
Principal $-$    Paid    =    Principal
$15,000 $-$ $188.64 = $14,811.36

**Step 4**   Find the amount of interest paid for the second month.

  Principal   $\times$ Rate $\times$ Time = Interest Paid
$14,811.36 $\times$ 0.11 $\times$   $\frac{1}{12}$   =   $135.77

**Step 5**   Find the amount of principal paid for the second month.

Payment $-$ Interest = Principal Paid
$326.14 $-$ $135.77 =    $190.37

**Step 6**   Find the balance of the principal at the end of the second month.

                 Principal    Balance of
  Principal   $-$    Paid    =    Principal
$14,811.36 $-$ $190.37 = $14,620.99

The preceding amounts are included in the following amortization schedule prepared using a computer program. The amortization schedule shows that in the early years of an installment loan, a large amount of the payment applies toward interest. The interest gradually decreases so that during the last years most of the payment is applied to the principal.

## LOAN AMORTIZATION SCHEDULE

PRINCIPAL: $15,000
INTEREST RATE: 11%
TERM: 5 Years

| Payment Number | Amount of Payment | Portion of Payment for: | | Balance of Principal |
|---|---|---|---|---|
| | | Interest | Principal | |
| 1 | $326.14 | $137.50 | $188.64 | $14,811.36 |
| 2 | $326.14 | $135.77 | $190.37 | $14,620.99 |
| 3 | $326.14 | $134.03 | $192.11 | $14,428.88 |
| 4 | $326.14 | $132.26 | $193.88 | $14,235.00 |
| 5 | $326.14 | $130.49 | $195.65 | $14,039.35 |
| 6 | $326.14 | $128.69 | $197.45 | $13,841.90 |
| 7 | $326.14 | $126.88 | $199.26 | $13,642.65 |
| 8 | $326.14 | $125.06 | $201.08 | $13,441.57 |
| 9 | $326.14 | $123.21 | $202.93 | $13,238.64 |
| 10 | $326.14 | $121.35 | $204.79 | $13,033.85 |
| 11 | $326.14 | $119.48 | $206.66 | $12,827.19 |
| 12 | $326.14 | $117.58 | $208.56 | $12,618.63 |
| 13 | $326.14 | $115.67 | $210.47 | $12,408.16 |
| 14 | $326.14 | $113.74 | $212.40 | $12,195.76 |
| 15 | $326.14 | $111.79 | $214.35 | $11,981.42 |
| 42 | $326.14 | $51.91 | $274.23 | $5,388.84 |
| 43 | $326.14 | $49.40 | $276.74 | $5,112.10 |
| 44 | $326.14 | $46.86 | $279.28 | $4,832.82 |
| 45 | $326.14 | $44.30 | $281.84 | $4,550.98 |
| 46 | $326.14 | $41.72 | $284.42 | $4,266.56 |
| 47 | $326.14 | $39.11 | $287.03 | $3,979.53 |
| 48 | $326.14 | $36.48 | $289.66 | $3,689.87 |
| 49 | $326.14 | $33.82 | $292.32 | $3,397.55 |
| 50 | $326.14 | $31.14 | $295.00 | $3,102.55 |
| 51 | $326.14 | $28.44 | $297.70 | $2,804.85 |
| 52 | $326.14 | $25.71 | $300.43 | $2,504.43 |
| 53 | $326.14 | $22.96 | $303.18 | $2,201.24 |
| 54 | $326.14 | $20.18 | $305.96 | $1,895.28 |
| 55 | $326.14 | $17.37 | $308.77 | $1,586.51 |
| 56 | $326.14 | $14.54 | $311.60 | $1,274.92 |
| 57 | $326.14 | $11.69 | $314.45 | $960.46 |
| 58 | $326.14 | $8.80 | $317.34 | $643.13 |
| 59 | $326.14 | $5.90 | $320.24 | $322.88 |
| 60 | $326.14 | $2.96 | $323.18 | $0.00 |

## PRACTICE PROBLEMS

Calculate the interest paid, the principal paid, and the remaining principal balance for each monthly payment. Round to the nearest cent.

| | Previous Principal Balance | Monthly Payment | Rate | Interest Paid | Principal Paid | Remaining Principal Balance |
|---|---|---|---|---|---|---|
| 1. | $59,833.20 | $631.94 | 12.5% | _____ | _____ | _____ |
| 2. | 21,792.10 | 424.95 | 10.75% | _____ | _____ | _____ |
| 3. | 67,725.82 | 835.09 | 13.25% | _____ | _____ | _____ |

4. Prepare an amortization schedule for the first four payments on a loan of $8,000 with monthly payments of $524.28 and an annual percentage rate of 21.6%.

| Payment Number | Monthly Payment | Interest Paid | Principal Paid | Remaining Principal Balance |
|---|---|---|---|---|
| 1 | $524.28 | _____ | _____ | _____ |
| 2 | $524.28 | _____ | _____ | _____ |
| 3 | $524.28 | _____ | _____ | _____ |
| 4 | $524.28 | _____ | _____ | _____ |

## USING A REMAINING BALANCE TABLE

A **remaining balance table**, similar to the one shown on page 271, can be used to determine the balance of the principal at the end of any year of a loan. The numbers in the table represent the percent of the remaining balance. To find the remaining principal balance, multiply the amount financed by the appropriate table value.

Amount Financed × Percent of Remaining Balance = Remaining Balance

ANSWERS FOR PRACTICE PROBLEMS

(4)
| Payment Number | Interest Paid | Principal Paid | Remaining Principal Balance |
|---|---|---|---|
| 1 | $144.00 | $380.28 | $7,619.72 |
| 2 | 137.15 | 387.13 | 7,232.59 |
| 3 | 130.19 | 394.09 | 6,838.50 |
| 4 | 123.09 | 401.19 | 6,437.31 |

| | Interest Paid | Principal Paid | Remaining Principal Balance |
|---|---|---|---|
| (1) | $623.26 | $ 8.68 | $59,824.52 |
| (2) | 195.22 | 229.73 | 21,562.37 |
| (3) | 747.81 | 87.28 | 67,638.54 |

**REMAINING BALANCE TABLE**
Original Term in Years

| Age of Loan | 1 | 2 | 3 | 4 | 5 | 6 | 7 | 8 | 9 | 10 | 15 | 20 | 25 | 30 | 35 | 40 |
|---|---|---|---|---|---|---|---|---|---|---|---|---|---|---|---|---|
| | | | | | | | | 12.00% | | | | | | | | |
| 1 | 0. | 52.98 | 70.56 | 79.28 | 84.47 | 87.89 | 90.29 | 92.07 | 93.43 | 94.49 | 97.46 | 98.72 | 99.32 | 99.64 | 99.80 | 99.89 |
| 2 | | 0. | 37.38 | 55.94 | 66.97 | 74.24 | 79.36 | 83.13 | 86.02 | 88.27 | 94.60 | 97.27 | 98.56 | 99.23 | 99.58 | 99.77 |
| 3 | | | 0. | 29.64 | 47.25 | 58.86 | 67.03 | 73.06 | 77.67 | 81.27 | 91.38 | 95.65 | 97.71 | 98.77 | 99.33 | 99.63 |
| 4 | | | | 0. | 25.04 | 41.53 | 53.15 | 61.72 | 68.26 | 73.39 | 87.75 | 93.81 | 96.74 | 98.25 | 99.05 | 99.48 |
| 5 | | | | | 0. | 22.00 | 37.50 | 48.93 | 57.66 | 64.50 | 83.65 | 91.74 | 95.65 | 97.66 | 98.73 | 99.31 |
| 6 | | | | | | 0. | 19.87 | 34.53 | 45.72 | 54.48 | 79.04 | 89.42 | 94.43 | 97.00 | 98.37 | 99.11 |
| 7 | | | | | | | 0. | 18.29 | 32.26 | 43.20 | 73.84 | 86.79 | 93.05 | 96.26 | 97.97 | 98.89 |
| 8 | | | | | | | | 0. | 17.09 | 30.48 | 67.99 | 83.83 | 91.49 | 95.42 | 97.51 | 98.64 |
| 9 | | | | | | | | | 0. | 16.15 | 61.39 | 80.50 | 89.73 | 94.48 | 97.00 | 98.36 |
| 10 | | | | | | | | | | 0. | 53.95 | 76.75 | 87.76 | 93.42 | 96.42 | 98.04 |
| 11 | | | | | | | | | | 0. | 45.58 | 72.52 | 85.53 | 92.22 | 95.77 | 97.69 |
| 12 | | | | | | | | | | 0. | 36.13 | 67.75 | 83.02 | 90.87 | 95.04 | 97.29 |
| 13 | | | | | | | | | | 0. | 25.50 | 62.37 | 80.19 | 89.35 | 94.21 | 96.84 |
| 14 | | | | | | | | | | 0. | 13.51 | 56.32 | 77.00 | 87.64 | 93.28 | 96.33 |
| 15 | | | | | | | | | | | 0. | 49.50 | 73.41 | 85.71 | 92.23 | 95.75 |
| 16 | | | | | | | | | | | 0. | 41.81 | 69.36 | 83.53 | 91.05 | 95.11 |
| 17 | | | | | | | | | | | 0. | 33.15 | 64.80 | 81.08 | 89.72 | 94.38 |
| 18 | | | | | | | | | | | 0. | 23.39 | 59.66 | 78.32 | 88.22 | 93.56 |
| 19 | | | | | | | | | | | 0. | 12.39 | 53.87 | 75.20 | 86.52 | 92.63 |
| 20 | | | | | | | | | | | | 0. | 47.35 | 71.69 | 84.62 | 91.59 |
| 21 | | | | | | | | | | | | 0. | 40.00 | 67.74 | 82.47 | 90.42 |
| 22 | | | | | | | | | | | | 0. | 31.71 | 63.29 | 80.05 | 89.09 |
| 23 | | | | | | | | | | | | 0. | 22.37 | 58.27 | 77.32 | 87.60 |
| 24 | | | | | | | | | | | | 0. | 11.85 | 52.61 | 74.25 | 85.92 |
| 25 | | | | | | | | | | | | | 0. | 46.24 | 70.78 | 84.03 |
| 26 | | | | | | | | | | | | | 0. | 39.06 | 66.88 | 81.90 |
| 27 | | | | | | | | | | | | | 0. | 30.97 | 62.48 | 79.49 |
| 28 | | | | | | | | | | | | | 0. | 21.85 | 57.53 | 76.78 |
| 29 | | | | | | | | | | | | | 0. | 11.58 | 51.95 | 73.73 |
| 30 | | | | | | | | | | | | | | 0. | 45.65 | 70.29 |
| 35 | | | | | | | | | | | | | | | 0. | 45.34 |
| 40 | | | | | | | | | | | | | | | | 0. |

## EXAMPLE

Strawberry Point Nursery obtained an $85,000 loan at 12% interest to be paid back over 25 years. To date, the nursery has made monthly payments for 5 years. Use the remaining balance table to determine how much of the original $85,000 principal remains unpaid.

## SOLUTION

**Step 1**  Find the percent of the remaining balance. Read down the Age of Loan column to 5 and across to the term of loan column heading 25 to find 95.65%.

**Step 2**  Calculate the remaining balance.

| Amount Financed | × | Percent of Remaining Balance | = | Remaining Balance |
|---|---|---|---|---|
| $85,000.00 | × | 0.9565 | = | $81,302.50 |

Calculate the remaining principal balance for each of the following loans. Use the remaining balance table on page 271.

| | Amount Financed | Term of Loan | Rate | Age of Loan | Remaining Principal Balance |
|---|---|---|---|---|---|
| 1. | $63,500 | 25 | 12% | 10 years | _____ |
| 2. | 25,600 | 20 | 12% | 5 years | _____ |
| 3. | 17,300 | 10 | 12% | 7 years | _____ |
| 4. | 54,900 | 30 | 12% | 23 years | _____ |

5. The purchase price of Kelly McNaught's new car was $15,200. Kelly McNaught paid $2,000 down and financed the remaining amount at 12% for 6 years. If Kelly has paid 24 monthly payments, what is the remaining principal balance on the loan?

## EARLY PAYMENT OF INSTALLMENT LOANS

If a borrower pays an installment loan before it is due, the finance charge on the unpaid payments is deducted from the balance due on the loan. This deduction, called a **finance charge refund**, usually is calculated using the **Rule of 78**. This method assumes that a large portion of the finance charge is assigned to the early payments and progressively smaller amounts to the later payments. Therefore, the interest refunded on the remaining payments of an installment loan is relatively small.

To find the finance charge refund using the Rule of 78, multiply the total finance charge by a fraction whose numerator is the sum of the number of payments remaining and whose denominator is the sum of the total number of payments in the loan.

$$\frac{\text{Sum of Number of Payments Remaining}}{\text{Sum of Total Number of Payments}} \times \frac{\text{Finance}}{\text{Charge}} = \frac{\text{Finance Charge}}{\text{Refund}}$$

To find the balance due on the loan, subtract the finance charge refund from the total amount of the remaining payments.

$$\text{Total of Payments Remaining} - \text{Finance Charge Refund} = \text{Balance Due}$$

## EXAMPLE

Norma Stevens signed an installment contract calling for 12 payments of $34.44 each. The finance charge is $160. She wants to pay the loan in full with 3 payments remaining. Find the finance charge refund and the balance due using the Rule of 78. Round to the nearest cent.

(1) $55,727.60 (2) $23,485.44 (3) $7,473.60 (4) $31,990.23 (5) $9,799.68

**SOLUTION**

**Step 1**  Find the sum of the number of payments remaining.

$1 + 2 + 3 = 6$

**Step 2**  Find the sum of the total number of payments in the loan.

$1 + 2 + 3 + 4 + 5 + 6 + 7 + 8 + 9 + 10 + 11 + 12 = 78$

**Step 3**  Find the finance charge refund.

$$\frac{\text{Sum of Number of Payments Remaining}}{\text{Sum of Total Number of Payments}} \times \frac{\text{Finance}}{\text{Charge}} = \frac{\text{Finance Charge}}{\text{Refund}}$$

$$\frac{6}{78} \quad\quad \times \quad \$160 \quad = \$12.307 \approx \$12.31$$

**Step 4**  Find the balance due.

|  | Total of Payments Remaining | − | Finance Charge Refund | = Balance Due |
|---|---|---|---|---|
| $\$34.44 \times 3 \longrightarrow$ | \$103.32 | − | \$12.31 | = \$91.01 |

The sums for the numerator and denominator in the finance charge refund formula can be found more easily by using the following formula:

$$\frac{n \times (n + 1)}{2}$$

When finding the numerator, let $n$ represent the number of payments remaining. When finding the denominator, let $n$ represent the total number of payments.

---

**EXAMPLE**

Terry Pike has an installment loan for \$2,600 to be repaid in 24 monthly payments. With 10 payments of \$133.25 remaining, Terry decides to pay the loan in full. Find the finance charge, finance charge refund, and balance due on the loan.

**SOLUTION**

**Step 1**  Find the finance charge.

| | Total Amount Due | − Amount Financed | = Finance Charge |
|---|---|---|---|
| $\$133.25 \times 24 \longrightarrow$ | \$3,198.00 | − \$2,600 | = \$598 |

**Step 2**  Find the sum of the number of remaining payments.

$$\frac{\text{Sum of Number of}}{\text{Payments Remaining}} = \frac{n \times (n + 1)}{2} = \frac{10 \times (10 + 1)}{2} = 55$$

**Step 3**  Find the sum of the total number of payments.

$$\frac{\text{Sum of Total}}{\text{Number of Payments}} = \frac{n \times (n + 1)}{2} = \frac{24 \times (24 + 1)}{2} = 300$$

**Step 4**  Find the finance charge refund.

$$\frac{\text{Sum of Number of Payments Remaining}}{\text{Sum of Total Number of Payments}} \times \frac{\text{Finance}}{\text{Charge}} = \frac{\text{Finance Charge}}{\text{Refund}}$$

$$\frac{55}{300} \times \$598 = \$109.633 \approx \$109.63$$

**Step 5**  Find the balance due.

| Total of Payments Remaining | − | Finance Charge Refund | = Balance Due |
|---|---|---|---|

$\$133.25 \times 10 \longrightarrow \$1,332.50 \quad - \quad \$109.63 \quad = \quad \$1,222.87$

## PRACTICE PROBLEMS

The following installment loans were paid in full before the end of the installment contract. Find the finance charge refund and balance due using the Rule of 78. Round to the nearest cent.

| | Monthly Payment | Finance Charge | Total Number of Payments | Remaining No. of Payments | Finance Charge Refund | Balance Due |
|---|---|---|---|---|---|---|
| **1.** | $59.63 | $ 98.70 | 12 | 5 | _____ | _____ |
| **2.** | 35.72 | 207.80 | 24 | 8 | _____ | _____ |
| **3.** | 64.67 | 187.50 | 18 | 11 | _____ | _____ |

4. Perry Snyder has 9 payments of $64 each remaining on a 24-month installment loan with a $336 finance charge. Perry has come into some extra money and wants to pay off the loan. Find (a) the finance charge refund, and (b) the balance due on the loan using the Rule of 78.
5. Angela Vito bought a piano with a purchase price of $4,200 on the installment plan. The agreement calls for 36 monthly payments of $151.11 each and a finance charge of $1,239.96. Angela decides to repay the loan in full after 25 payments. Find (a) the finance charge refund and (b) the balance due. Use the Rule of 78.

## LESSON 8-3     EXERCISES

Determine the monthly payment, the total amount due, and the finance charge. Use the loan amortization table on page 265.

| | Amount Financed | Rate | Years | Monthly Payment | Total Amount Due | Finance Charge |
|---|---|---|---|---|---|---|
| 1. | $ 45,000 | 11% | 20 | _____ | _____ | _____ |
| 2. | 50,000 | 12% | 25 | _____ | _____ | _____ |
| 3. | 30,000 | 11% | 20 | _____ | _____ | _____ |
| 4. | 80,000 | 10% | 25 | _____ | _____ | _____ |
| 5. | 5,000 | 12% | 5 | _____ | _____ | _____ |
| 6. | 90,000 | 11% | 20 | _____ | _____ | _____ |
| 7. | 60,000 | 12% | 25 | _____ | _____ | _____ |
| 8. | 100,000 | 11% | 5 | _____ | _____ | _____ |
| 9. | 12,000 | 11% | 20 | _____ | _____ | _____ |
| 10. | 8,000 | 10% | 5 | _____ | _____ | _____ |

The following amounts were financed at various interest rates. Determine the monthly installment payment for a 20-year loan and a 25-year loan. Calculate the difference in the total interest charged. Use the loan amortization table on page 265.

| | Amount Financed | Rate | 20-Year Payment | 25-Year Payment | Difference in Interest Charge |
|---|---|---|---|---|---|
| 11. | $20,000 | 10% | _____ | _____ | _____ |
| 12. | 6,000 | 12% | _____ | _____ | _____ |
| 13. | 40,000 | 11% | _____ | _____ | _____ |
| 14. | 11,000 | 12% | _____ | _____ | _____ |
| 15. | 55,000 | 11% | _____ | _____ | _____ |

Calculate the interest paid, principal paid, and remaining principal balance for each monthly payment. Round to the nearest cent.

| | Previous Principal Balance | Monthly Payment | Rate | Interest Paid | Principal Paid | Remaining Principal Balance |
|---|---|---|---|---|---|---|
| 16. | $39,182 | $ 670.93 | 10.75% | _____ | _____ | _____ |
| 17. | 19,670 | 781.75 | 16.5% | _____ | _____ | _____ |
| 18. | 88,704 | 1,800.57 | 21.0% | _____ | _____ | _____ |
| 19. | 13,072 | 525.76 | 18.75% | _____ | _____ | _____ |

**20.** Prepare an amortization schedule for the first 4 payments of $230.52 each for an installment loan of $27,000 at 14% interest.

| Payment Number | Monthly Payment | Interest Paid | Principal Paid | Remaining Principal Balance |
|---|---|---|---|---|
| 0 | 0 | 0 | 0 | $27,000.00 |
| 1 | _____ | _____ | _____ | _____ |
| 2 | _____ | _____ | _____ | _____ |
| 3 | _____ | _____ | _____ | _____ |
| 4 | _____ | _____ | _____ | _____ |

Use the remaining balance table on page 271 to determine the remaining balance on each of the following loans.

| | Amount Financed | Term of Loan | Rate | Age of Loan | Remaining Principal Balance |
|---|---|---|---|---|---|
| **21.** | $75,000 | 25 years | 12% | 1 year | _____ |
| **22.** | 95,000 | 20 years | 12% | 15 years | _____ |
| **23.** | 40,000 | 15 years | 12% | 10 years | _____ |
| **24.** | 10,000 | 20 years | 12% | 2 years | _____ |
| **25.** | 60,000 | 25 years | 12% | 20 years | _____ |
| **26.** | 55,000 | 25 years | 12% | 4 years | _____ |
| **27.** | 30,000 | 20 years | 12% | 5 years | _____ |
| **28.** | 65,000 | 15 years | 12% | 10 years | _____ |
| **29.** | 45,000 | 10 years | 12% | 5 years | _____ |
| **30.** | 90,000 | 25 years | 12% | 24 years | _____ |

The following installment loans were paid in full before the end of the installment contract. Find the finance charge refund and balance due using the Rule of 78. Round to the nearest cent.

| | Monthly Payment | Finance Charge | Total Number of Payments | Remaining Number of Payments | Finance Charge Refund | Balance Due |
|---|---|---|---|---|---|---|
| **31.** | $ 51.00 | $ 324 | 24 | 8 | _____ | _____ |
| **32.** | 616.39 | 7,490 | 36 | 15 | _____ | _____ |
| **33.** | 490.00 | 1,620 | 18 | 6 | _____ | _____ |
| **34.** | 573.33 | 3,010 | 24 | 10 | _____ | _____ |
| **35.** | 238.50 | 1,855 | 30 | 20 | _____ | _____ |
| **36.** | 481.83 | 882 | 12 | 3 | _____ | _____ |
| **37.** | 817.14 | 12,320 | 42 | 12 | _____ | _____ |
| **38.** | 478.17 | 5,814 | 36 | 18 | _____ | _____ |

## BUSINESS APPLICATIONS

**39.** Crown Shoe Manufacturing Company plans to borrow $50,000 for 20 years at 12% interest for the purchase of a warehouse. Calculate (a) the monthly payment, (b) the total amount due, and (c) the finance charge.

(a) _____

(b) _____

(c) _____

**40.** For some time, Dakken Industries has been planning to borrow $75,000 to build an addition onto its office. A year ago, a 25-year loan could have been obtained at 10% interest. Now, the interest rate is 12%. How much more total interest will Dakken pay over the 25 years by obtaining a 12% loan instead of a 10% loan?

_____

**41.** Merriland Amusement Park can obtain a $100,000 construction loan at 11% for either 20 or 25 years. How much less total interest will be charged if a 20-year loan is obtained instead of a 25-year loan?

_____

**42.** Tjaden Athletic Equipment Company borrowed $54,500 at 12% interest for 25 years. How much principal does Tjaden still owe on the loan after making 228 monthly payments?

_____

**43.** On September 11 Wellington Beverage Distributing Company obtained a $60,000 loan at 11% interest for 20 years. Determine (a) the monthly payment, (b) interest paid on the first payment, (c) principal paid on the first payment, and (d) remaining principal balance for the first loan payment.

(a) _____

(b) _____

(c) _____

(d) _____

**44.** Central Emporium plans to borrow $80,000 to be repaid over 20 years. Northern States Bank will make the loan at 12% interest. Midland National Bank will charge 11%. How much will Central Emporium save in the 20 years by obtaining the loan from Midland National?

_____

**45.** James Baldwin, president of the Magic Carpet Stores, purchased a summer home for $150,000. He made a $50,000 down payment and financed the remaining amount at 12% interest for 25 years. Calculate (a) the amount financed, (b) the total amount due, and (c) the finance charge.

(a) _____

(b) _____

(b) _____

**46.** Maxfield Aeronautics is planning to buy a building to use as a repair shop. A 10% interest, 20-year loan can be obtained. Maxfield is thinking of borrowing either $80,000 or $100,000. Find the difference in the total amount due between the $100,000 loan and the $80,000 loan.

_____

**47.** Goody's Bakery borrowed $40,000 at 12% interest for 20 years. They have made 120 monthly payments. How much of the loan's principal has been paid?

_____

**48.** Klondike Exploration Company obtained a $75,000 loan at 12% interest for 25 years. They have 120 payments remaining on the loan. Determine (a) the total amount paid in loan payments so far, and (b) the amount of principal that has been paid on the loan.

**(a)** _____

**(b)** _____

**49.** The Lloyd's Auto Dealership purchased a $85,000 computer system. The company financed the purchase price at 12% for 20 years. Prepare an amortization schedule showing the first 3 monthly payments. Round to the nearest cent.

| Payment Number | Monthly Payment | Interest Paid | Principal Paid | Remaining Principal Balance |
|---|---|---|---|---|
| 0 | $0 | $0 | $0 | $85,000 |
| 1 | _____ | _____ | _____ | _____ |
| 2 | _____ | _____ | _____ | _____ |
| 3 | _____ | _____ | _____ | _____ |

**50.** Jenkins Realty obtained a $14,000 installment loan for 5 years at 16% interest. The loan will be repaid in 10 semiannual payments of $2,086.42. Prepare an amortization schedule showing the first 5 payments. Round to the nearest cent.

| Payment Number | Semiannual Payment | Interest Paid | Principal Paid | Remaining Principal Balance |
|---|---|---|---|---|
| 0 | $0 | $0 | $0 | $14,000 |
| 1 | _____ | _____ | _____ | _____ |
| 2 | _____ | _____ | _____ | _____ |
| 3 | _____ | _____ | _____ | _____ |
| 4 | _____ | _____ | _____ | _____ |
| 5 | _____ | _____ | _____ | _____ |

**51.** The Lake View School system purchased 25 copy machines for $34,750. They made a down payment of $9,750 and financed the remaining amount at 10% interest for 25 years. Prepare an amortization schedule showing the first 6 monthly payments. Round to the nearest cent.

| Payment Number | Monthly Payment | Interest Paid | Principal Paid | Remaining Principal Balance |
|---|---|---|---|---|
| 0 | $0 | $0 | $0 | _____ |
| 1 | _____ | _____ | _____ | _____ |
| 2 | _____ | _____ | _____ | _____ |
| 3 | _____ | _____ | _____ | _____ |
| 4 | _____ | _____ | _____ | _____ |
| 5 | _____ | _____ | _____ | _____ |
| 6 | _____ | _____ | _____ | _____ |

**52.** Wee-Care Day Care purchased $9,800 worth of playground equipment on the installment plan with a $2,352 finance charge and monthly payments of $506.33 each. After paying 18 of the 24 payments due on the loan, they decide to pay the loan in full. Calculate (a) the finance charge refund, (b) the total of payments remaining, and (c) the balance due using the Rule of 78.

(a) _____

(b) _____

(c) _____

**53.** Harry Walters paid off a $4,230 installment loan after making 15 of the 36 payments of $259.80 each. Calculate (a) the finance charge, (b) the finance charge refund, and (c) the balance due using the Rule of 78. Round to the nearest cent.

(a) _____

(b) _____

(c) _____

## CHAPTER 8 REVIEW

Round amounts to the nearest cent. Round percents to the nearest hundredth.

1. Newton Finway's VISA statement shows an average daily balance of $456.72 and a monthly periodic rate of 2%. Find the finance charge based on the average daily balance.

2. Find the average daily balance for the following credit card transactions. Use a 31-day billing cycle beginning October 3.

   October 3, previous balance of $189.21
   October 9, purchase of $34.57
   October 15, payment of $200
   October 20, purchase of $109.67
   October 31, purchase of $94.71

3. Find the monthly periodic rate for a credit card account if the APR is 21%.

4. What is the new balance on a credit card statement if the previous balance is $405.21, advances and purchases are $717.23, finance charges are $22.56, and credits and payments are $367.50?

5. The Algers Travel Trailer Company is financing a motor home costing $20,000 for $2,000 down and 48 monthly payments of $510 each. Find (a) the amount financed, (b) the finance charge, and (c) the total sale price.

(a) _____

(b) _____

(c) _____

6. Find the APR on a 30-payment installment purchase agreement with a finance charge of $204.50 if the amount financed is $904.58. Use the APR table.

_____

7. Find the approximate APR using the formula on a 33-month install- ment agreement if the amount financed is $1,290 and the finance charge is $354.75.

_____

8. Rhonda Henning signed a 24-month installment loan agreement for $2,340 with a finance charge of $421.20. Find the APR using (a) the APR formula and (b) the APR table.

(a) _____

(b) _____

9. Premier Industries obtained a $55,000 loan to purchase a warehouse. The monthly payment is $530.77 for 20 years. Find (a) the total amount due and (b) the finance charge.

(a) _____

(b) _____

10. After making a down payment, Diamond Rentals wants to finance the remaining $13,510 on the purchase price of a new van. The monthly payment at 18% interest is $488.39 for 3 years or $396.79 for 4 years. How much less interest will be charged if a 3-year loan is obtained instead of a 4-year loan?

_____

11. Prepare an amortization schedule showing the first 5 monthly payments on an installment loan of $15,000 at 11% interest for 5 years.

| Payment Number | Monthly Payment | Interest Paid | Principal Paid | Remaining Principal Balance |
|---|---|---|---|---|
| 0 | $0 | $0 | $0 | $15,000 |
| 1 | _____ | _____ | _____ | _____ |
| 2 | _____ | _____ | _____ | _____ |
| 3 | _____ | _____ | _____ | _____ |
| 4 | _____ | _____ | _____ | _____ |
| 5 | _____ | _____ | _____ | _____ |

12. The Wichersham Law Firm purchased a new office complex for $118,490. They made a $23,490 down payment and financed the remaining balance at 12% for 25 years. Prepare an amortization schedule showing the first 4 monthly payments.

| Payment Number | Monthly Payment | Interest Paid | Principal Paid | Remaining Principal Balance |
|---|---|---|---|---|
| 0 | $0 | $0 | $0 | _____ |
| 1 | _____ | _____ | _____ | _____ |
| 2 | _____ | _____ | _____ | _____ |
| 3 | _____ | _____ | _____ | _____ |
| 4 | _____ | _____ | _____ | _____ |

**13.** The Unger Construction Company obtained an installment loan to finance the purchase of new equipment. The company financed $33,500 for 12 years at 12% interest. The quarterly payments are $1,580.53. Prepare an amortization schedule showing the first 7 payments.

| Payment Number | Quarterly Payment | Interest Paid | Principal Paid | Remaining Principal Balance |
|---|---|---|---|---|
| 0 | $0 | $0 | $0 | $33,500 |
| 1 | _____ | _____ | _____ | _____ |
| 2 | _____ | _____ | _____ | _____ |
| 3 | _____ | _____ | _____ | _____ |
| 4 | _____ | _____ | _____ | _____ |
| 5 | _____ | _____ | _____ | _____ |
| 6 | _____ | _____ | _____ | _____ |
| 7 | _____ | _____ | _____ | _____ |

**14.** Janowitz Builders took out a $10,000 installment loan for 4 years with monthly payments of $283.33 and a finance charge of $3,600. After making payments for 3 years, Janowitz paid the loan in full. Calculate (a) the finance charge refund and (b) the balance due on the loan using the Rule of 78.

(a) _____

(b) _____

**15.** Art Hinto financed $12,700 for 2½ years to purchase a sail boat. After paying the $592.67 monthly payment for 22 months, Art sold the boat and paid the loan in full. Calculate (a) the finance charge, (b) the finance charge refund, and (c) the balance due using the Rule of 78.

(a) _____

(b) _____

(c) _____

# CHAPTER

## 9

# BUSINESS INSURANCE AND PROPERTY TAXES

## LEARNING OBJECTIVES

**1.** Calculate life insurance premiums.

**2.** Calculate fire insurance policies premiums.

**3.** Calculate short-rate premiums.

**4.** Calculate the refund if a fire insurance policy is canceled.

**5.** Calculate the loss paid using the coinsurance formula.

**6.** Calculate the loss paid by each multiple carrier.

**7.** Calculate the annual motor vehicle insurance premium.

**8.** Calculate the assessed valuation and tax rate for real property.

Insurance policies are available to cover a variety of financial risks such as loss of income due to death or disability, financial losses from accidents or illness, and property loss or damage. Businesses and individuals pay property taxes, which are used as a source of revenue for public services provided by state, county, and other local governments. This chapter will present the basic types of business insurance protection and the various ways of determining insurance premiums and property taxes.

## LESSON 9-1    LIFE INSURANCE

The **insurance policy** is a written contract between the insurance company and the insured stating benefits and costs of the protection purchased. The **policyholder** is the person or business that pays the insurance company. The **face value** of a policy is the amount of insurance provided. The **premium** is the amount charged for the insurance.

Insurance is based on the concept of shared risk. Thousands of insurance premiums are pooled into a common fund. The money in the fund is used by the insurance company to cover the losses of individual policyholders. In this way, each policyholder pays a little, and no one individual or business is responsible for the entire financial loss.

## CALCULATING PREMIUMS ON TERM LIFE INSURANCE

In many businesses, the untimely death of key personnel vital to the company's operation would result in a financial loss to the company. Therefore, many businesses protect themselves by obtaining business life insurance. Business partners may obtain life insurance on one another to provide funds to buy out the other partner's share in case of death. Businesses also provide life insurance plans as fringe benefits for their employees.

There are two basic types of life insurance coverage: term and permanent. **Term insurance,** the least expensive insurance, provides pure life insurance protection for a specified term such as 5, 10, or 20 years. If the insured lives to the end of the term, coverage expires and the policyholder ordinarily receives no money back (cash value). If the policyholder dies during the term of the policy, the beneficiary is paid the face value of the policy. The **beneficiary** is the person named in the policy to receive the face value, also called **proceeds** or **death benefits.** Usually at the end of the term, the policy can be renewed at a higher rate or converted to another type of insurance.

Term insurance is popular with employers who provide group life insurance as a fringe benefit. Group life insurance provides employees with life insurance at a much lower cost than what each individual could purchase separately. If the premium is paid by the employee, a monthly deduction from earnings is made. Total premiums are then paid for all employees at the same time by the employer. The premium is usually stated as a dollar amount per $1,000 of insurance coverage (face value). To find the annual life insurance premium, divide the face value by 1,000; then multiply by the rate per $1,000 of coverage.

$$\text{Annual Life Insurance Premium} = \frac{\text{Face Value}}{1,000} \times \text{Rate}$$

### EXAMPLE

The Arlington Manufacturing Company offers its employees a group life insurance plan with benefits equal to $1\frac{1}{2}$ times his or her annual salary at a rate of $2.95 per $1,000 of insurance. Helen White decides to take advantage of the plan. If her annual salary is $37,350, what will be her total insurance coverage and her monthly salary payroll deduction?

### SOLUTION

**Step 1**   Find the total insurance coverage (face value).

Annual Salary × 1.5 = Face Value
$37,350   × 1.5 =   $56,025

**Step 2**   Find the annual premium.

$$\text{Annual Life Insurance Premium} = \frac{\text{Face Value}}{1,000} \times \text{Rate} = \frac{\$56,025}{1,000} \times \$2.95 = \$165.27$$

**Step 3**   Find the monthly payroll deduction.

$$\text{Monthly Payroll Deduction} = \frac{\text{Annual Premium}}{12} = \frac{\$165.27}{12} = \$13.77$$

Find the face value of the policy and the monthly premium deduction for the following employees. Each employee is insured for $1\frac{3}{4}$ times his or her annual salary. Round to the nearest cent.

| Employee | Annual Salary | Rate per $1,000 | Face Value | Monthly Premium Deduction |
|---|---|---|---|---|
| **1.** E. Callahan | $29,750 | $1.85 | $_____ | $_____ |
| **2.** C. Hanna | 39,450 | 2.59 | _____ | _____ |
| **3.** K. Bailey | 40,300 | 3.09 | _____ | _____ |
| **4.** L. Getty | 33,700 | 3.15 | _____ | _____ |

## CALCULATING PREMIUMS ON PERMANENT LIFE INSURANCE

Permanent insurance provides protection with a cash savings feature. *Whole life,* also called *straight life* and *ordinary life,* is the predominant type of permanent life insurance. Whole life insurance provides death benefits combined with a savings plan called *cash value.* Therefore, the premium for whole life is higher than the premium for the same amount of term life. With most whole life policies, the premium remains steady and the insurance protection remains in force for the insured's lifetime, unless the policy is canceled.

*Limited payment* life insurance provides permanent life insurance coverage and a cash value similar to ordinary life, but premiums are paid for a limited number of years, usually 10 or 20. This type of policy appeals to an individual, such as a professional athlete, who anticipates a high income for a number of years and then a decline. Since premiums are paid for a limited number of years, they are usually higher than those for the same amount of whole life insurance and cash value builds up faster.

*Universal life,* another popular type of permanent insurance, provides adjustable rates and increased cash value depending on economic conditions. A variable universal life policy, similar to a universal life policy, provides the policyholder with a choice of investments—stock funds, money market funds, bond funds—for accumulating cash value.

The amount of the life insurance premium is determined by several factors: (1) the type of policy, (2) the age and sex of the insured, (3) the amount of coverage, and (4) the frequency of premium payments. The life insurance premium may be found using a table similar to the following, which shows life insurance rates per $1,000 of coverage. To find the life insurance premium, multiply the insurance rate by the number of thousands of dollars of life insurance purchased.

$$\text{Annual Premium} = \frac{\text{Face Value}}{1,000} \times \text{Rate}$$

ANSWERS FOR
PRACTICE PROBLEMS

| | Face Value | Monthly Premium Deduction |
|---|---|---|
| **(1)** | $52,062.50 | $ 8.03 |
| **(2)** | 69,037.50 | 14.90 |
| **(3)** | 70,525.00 | 18.16 |
| **(4)** | 58,975.00 | 15.48 |

## ANNUAL LIFE INSURANCE PREMIUM RATES PER $1,000 OF FACE VALUE

| Age* | | Type of Policy | | |
| --- | --- | --- | --- | --- |
| Male | Female | Term, 10-year | Whole Life | Limited Payment, 20-year |
| 20 | 23 | $ 5.70 | $12.02 | $17.86 |
| 25 | 28 | 6.35 | 14.09 | 19.89 |
| 30 | 33 | 7.80 | 16.75 | 22.26 |
| 35 | 38 | 10.10 | 20.30 | 25.14 |
| 40 | 43 | 14.01 | 24.82 | 28.62 |
| 45 | 48 | 20.45 | 30.25 | 37.97 |
| 50 | 53 | 26.87 | 37.08 | 43.20 |

*Because of longer female life expectancy, rates for females are the same as for males 3 years younger.

## EXAMPLE

The management of Ariel Dantley Cosmetics wants to purchase a life insurance policy for $200,000 to insure the life of its chief chemist, Antonios Stavros, who is 35 years old. Find the amount of the annual premium on a 10-year term policy, a whole life policy, and a 20-year limited payment policy. Use the annual life insurance premium rates table.

## SOLUTION

**Step 1** Find the annual premium for a 10-year term policy. The rate for a 10-year term policy for a 35-year-old male is $10.10.

$$\frac{\text{Annual}}{\text{Premium}} = \frac{\text{Face Value}}{1,000} \times \text{Rate} = \frac{\$200,000}{1,000} \times \$10.10 = \$2,020$$

**Step 2** Find the annual premium for a whole life policy. The rate for a whole life policy for a 35-year-old male is $20.30.

$$\frac{\text{Annual}}{\text{Premium}} = \frac{\text{Face Value}}{1,000} \times \text{Rate} = \frac{\$200,000}{1,000} \times \$20.30 = \$4,060$$

**Step 3** Find the annual premium for a 20-year limited payment policy. The rate for a 20-year limited payment policy for a 35-year-old male is $25.14.

$$\frac{\text{Annual}}{\text{Premium}} = \frac{\text{Face Value}}{1,000} \times \text{Rate} = \frac{\$200,000}{1,000} \times \$25.14 = \$5,028$$

Insurance premiums may be paid annually, semiannually, quarterly, or monthly. The more frequent the payment, the higher the cost. To find the semiannual, quarterly, or monthly payment, multiply the annual payment by the following premium factors:

| Payment | Percent of Annual Payment |
| --- | --- |
| Semiannual | 51% |
| Quarterly | 26% |
| Monthly | 9% |

## EXAMPLE

Ariel Dantley Cosmetics decided to purchase the whole life policy for Antonios Stavros with an annual premium of $4,060. Calculate the monthly premium and the total amount due for the year. How much would they save if they paid an annual premium instead of monthly premiums?

**SOLUTION**

**Step 1** Find the monthly premium.

$$
\begin{array}{ccc}
\text{Annual} & \text{Monthly} & \text{Monthly} \\
\text{Premium} \times & \text{Premium Percent} = & \text{Premium} \\
\$4,060 \times & 0.09 = & \$365.40
\end{array}
$$

**Step 2** Find the total amount due.

$$
\begin{array}{ccc}
\text{Monthly} & 12 & \text{Total} \\
\text{Premium} \times & \text{Months} = & \text{Amount Due} \\
\$365.40 \times & 12 = & \$4,384.80
\end{array}
$$

**Step 3** Find the difference between the annual premium and the total amount due.

$$
\begin{array}{ccc}
\text{Total Amount} & \text{Annual} & \\
\text{Due} - & \text{Premium} = & \text{Difference} \\
\$4,384.80 - & \$4,060 = & \$324.80
\end{array}
$$

## PRACTICE PROBLEMS

Find the annual life insurance premium for 10-year term, whole life, and 20-year limited payment. Round to the nearest cent.

| | Insured | Sex | Age | Face Value | 10-year Term | Whole Life | 20-year Limited Payment |
|---|---|---|---|---|---|---|---|
| **1.** | E. Hauley | F | 28 | $250,000 | $_____ | $_____ | $_____ |
| **2.** | M. Fleming | M | 20 | $ 75,000 | _____ | _____ | _____ |
| **3.** | V. LaPanta | M | 35 | $200,000 | _____ | _____ | _____ |
| **4.** | E. Evalon | F | 53 | $150,000 | _____ | _____ | _____ |

Find the annual, semiannual, quarterly, and monthly premiums for the following life insurance policies. Use the tables on page 288. Round to the nearest cent.

| | Face Value | Age | Sex | Type of Insurance | Premium Annual | Semiannual | Quarterly | Monthly |
|---|---|---|---|---|---|---|---|---|
| **5.** | $50,000 | 20 | M | Whole life | $_____ | $_____ | $_____ | $_____ |
| **6.** | $87,000 | 48 | F | 10-year term | _____ | _____ | _____ | _____ |
| **7.** | $95,000 | 25 | M | 20-year limited pay | _____ | _____ | _____ | _____ |

ANSWERS FOR
PRACTICE PROBLEMS

| | Annual | Semiannual | Quarterly | Monthly |
|---|---|---|---|---|
| **(5)** | $ 601.00 | $306.51 | $156.26 | $ 54.09 |
| **(6)** | 1,779.15 | 907.37 | 462.58 | 160.12 |
| **(7)** | 1,889.55 | 963.67 | 491.28 | 170.06 |

| | 10-year Term | Whole Life | 20-year Limited Payment |
|---|---|---|---|
| **(1)** | $1,587.50 | $3,522.50 | $4,972.50 |
| **(2)** | 427.50 | 901.50 | 1,339.50 |
| **(3)** | 2,020.00 | 4,060.00 | 5,028.00 |
| **(4)** | 4,030.50 | 5,562.00 | 6,480.00 |

Permanent life insurance policies accumulate a **cash value,** which is the amount of money available if the insured voluntarily cancels the policy. This amount, also called the loan value or the cash surrender value, can be borrowed against the policy. To find the cash value of a permanent life insurance policy, divide the face value by 1,000 and multiply the result by the dollar value found on a cash value table similar to the one shown below.

$$\text{Cash Value} = \frac{\text{Face Value}}{1,000} \times \text{Cash Value per } \$1,000$$

### WHOLE LIFE INSURANCE CASH VALUE PER $1,000

| Age When Policy Acquired | | End of Policy Year | | | | | | |
|---|---|---|---|---|---|---|---|---|
| Male | Female | 1 | 2 | 3 | 5 | 10 | 15 | 20 |
| 20 | 23 | $0 | $ 0 | $ 2 | $ 18 | $ 61 | $111 | $168 |
| 25 | 28 | 0 | 0 | 6 | 25 | 77 | 135 | 200 |
| 30 | 33 | 0 | 0 | 11 | 34 | 95 | 162 | 236 |
| 35 | 38 | 0 | 4 | 17 | 44 | 116 | 194 | 277 |
| 40 | 43 | 0 | 9 | 24 | 55 | 139 | 229 | 321 |
| 45 | 48 | 0 | 14 | 32 | 69 | 166 | 267 | 368 |
| 50 | 53 | 0 | 19 | 41 | 84 | 194 | 303 | 411 |

### EXAMPLE

Ten years ago the Donnely Medical Research Company purchased a $250,000 whole life policy on its president, Johanson Drake, who was 30 years old at the time. What is the cash value of the policy today?

### SOLUTION

**Step 1**  Find the cash value per $1,000 of face value. Read down the male column to age 30 and across to column 10 in the End of Policy Year heading to find the cash value per $1,000, which is $95.

**Step 2**  Find the policy's cash value.

$$\text{Cash Value} = \frac{\text{Face Value}}{\$1,000} \times \frac{\text{Cash Value}}{\text{per } \$1,000} = \frac{\$250,000}{1,000} \times \$95 = \$23,750$$

### PRACTICE PROBLEMS

Calculate the cash value for the following whole life policies. Use the cash value table above. Round to the nearest cent.

| Insured | Sex | Age When Policy Acquired | Face Value | End of Policy Year | Cash Value |
|---|---|---|---|---|---|
| 1. M. Hernandez | M | 25 | $125,000 | 15 | $_____ |
| 2. J. Larsen | F | 28 | $115,000 | 20 | _____ |
| 3. S. Orlando | M | 40 | $250,000 | 10 | _____ |

**(1)** $16,875   **(2)** $23,000   **(3)** $34,750

## LESSON 9-1   EXERCISES

The following employees are enrolled in a payroll deduction life insurance program. The face value of the policy is $2\frac{1}{2}$ times each employee's annual salary. Find the face value and the monthly payroll deduction. Round to the nearest cent.

| | Employee | Annual Salary | Rate per $1,000 | Face Value | Monthly Payroll Deduction |
|---|---|---|---|---|---|
| 1. | W. Quigley | $19,850 | $1.98 | _____ | _____ |
| 2. | A. Steiner | 27,600 | 2.05 | _____ | _____ |
| 3. | L. Levinson | 32,300 | 2.22 | _____ | _____ |
| 4. | K. Kelly | 37,190 | 2.95 | _____ | _____ |
| 5. | L. Hoisey | 43,900 | 3.07 | _____ | _____ |
| 6. | J. Peachy | 47,250 | 3.32 | _____ | _____ |
| 7. | G. Walters | 51,900 | 3.47 | _____ | _____ |
| 8. | M. Harding | 57,300 | 3.73 | _____ | _____ |

Find the annual, semiannual, quarterly, and monthly premium using the life insurance rates table and premium factor table on page 288. Round to the nearest cent.

| | Face Value | Age | Sex | Type of Insurance | Annual Premium | Semiannual Premium | Quarterly Premium | Monthly Premium |
|---|---|---|---|---|---|---|---|---|
| 9. | $150,000 | 35 | M | 10-year term | _____ | _____ | _____ | _____ |
| 10. | 100,000 | 25 | M | Whole life | _____ | _____ | _____ | _____ |
| 11. | 75,000 | 53 | F | 10-year term | _____ | _____ | _____ | _____ |
| 12. | 130,000 | 48 | F | 20-year limited pay | _____ | _____ | _____ | _____ |
| 13. | 250,000 | 20 | M | Whole life | _____ | _____ | _____ | _____ |
| 14. | 80,000 | 33 | F | 20-year limited pay | _____ | _____ | _____ | _____ |
| 15. | 125,900 | 30 | M | 10-year term | _____ | _____ | _____ | _____ |
| 16. | 282,000 | 33 | F | 10-year term | _____ | _____ | _____ | _____ |

Calculate the cash value for the following whole life policies. Use the cash value table on page 290. Round to the nearest cent.

| Insured | Sex | Age When Policy Acquired | Face Value | End of Policy Year | Cash Value |
|---|---|---|---|---|---|
| 17. U. Unger | M | 35 | $159,000 | 20 | _____ |
| 18. R. Laars | F | 43 | 270,000 | 5 | _____ |
| 19. N. Ling | F | 28 | 115,000 | 3 | _____ |
| 20. H. Carmine | M | 40 | 50,000 | 10 | _____ |
| 21. Y. Stavros | M | 25 | 25,000 | 15 | _____ |
| 22. T. Airs | F | 33 | 85,000 | 2 | _____ |
| 23. C. Crouch | M | 50 | 200,000 | 15 | _____ |
| 24. I. Quigley | F | 23 | 130,000 | 10 | _____ |

## BUSINESS APPLICATIONS

Round amounts to the nearest cent. Use the tables found on pages 288 and 290.

25. Terry Whilton is planning to enroll in her company group life insurance plan. The plan offers life insurance coverage of 2.75 times Terry's annual salary at a rate of $3.11 per $1,000 of insurance. If her annual salary is $29,800, find (a) her total insurance coverage and (b) her monthly salary payroll deduction.

(a) _____

(b) _____

26. Frank and Marvin Gainer are joint owners of a manufacturing company. The company purchased 20-year limited payment life insurance policy with a face value of $200,000 for each of them. Calculate the annual premium for (a) Frank who is 25 and (b) Marvin who is 35.

(a) _____

(b) _____

27. The Raber Modeling Agency purchased a $1,000,000, 10-year term life insurance policy on their top model, Darleen Jenkins, who is 23 years old. Find (a) the quarterly premium payment and (b) the total amount due for the year, if premiums are paid quarterly.

(a) _____

(b) _____

28. Waverly Shipping purchased a $375,000 whole life insurance policy on Donald Howell, president of the company, who was 35 years old. Fifteen years later, Mr. Howell resigned from the company and the company canceled the policy. How much cash value did they receive?

_____

29. Arleen Evans purchased a whole life policy with a face value of $75,000. Arleen is 43 years old. Find (a) the monthly premium if the monthly premium is 9% of the annual premium and (b) the total amount due for the year, if premiums are paid monthly.

(a) _____

(b) _____

30. The management of Drexler Toy Manufacturing purchased a 20-year limited payment life insurance policy for $100,000 on their chief engineer, Helen Hines, who is 38 years old. Find (a) the semiannual payment if it is 51% of the annual premium and (b) the total amount due for the year, if premiums are paid semiannually.

(a) _____

(b) _____

**31.** Lake Distributors purchased a $350,000 whole life policy on their executive vice-president. He was 45 years old at the time the policy was purchased. What is the cash value of the policy 5 years later?

_____

**32.** Gina Pike purchased an $85,000, 10-year term life insurance policy. She is 28 years old. How much would she save per year if she paid an annual premium instead of monthly premiums, which are 9% of the annual premium?

_____

**33.** Harold Heuss's employer is offering a group life insurance plan with benefits equal to $2\frac{3}{4}$ times his annual salary at a rate of $2.80 per $1,000 of insurance. If Harold's annual salary is $36,750, what will be (a) his total insurance coverage and (b) his monthly payroll deduction?

(a) _____

(b) _____

**34.** How much whole life insurance can a 33 year-old woman purchase for an annual premium of $500?

_____

**CALCULATING PROPERTY INSURANCE PREMIUMS**

Property insurance provides coverage against damage due to fire, windstorm, flood, vandalism, or accident. Businesses usually insure buildings and their contents by purchasing a standard fire-insurance policy with extended coverage.

The premium for fire insurance is based on a building's rating and classification. The rating and classification assigned to a building depend on such factors as the building's construction (wood, brick, etc.), age, type and location of fire protection available, and contents. The annual fire insurance premium may be found using a table similar to the following.

**ANNUAL FIRE INSURANCE RATES PER $100 OF COVERAGE FOR BUILDINGS AND CONTENTS**

| Area Rating | Classification | | | |
|---|---|---|---|---|
| | Class A | | Class B | |
| | Building | Contents | Building | Contents |
| 1 | $0.27 | $0.34 | $0.38 | $0.51 |
| 2 | $0.32 | $0.46 | $0.49 | $0.59 |
| 3 | $0.40 | $0.49 | $0.60 | $0.64 |
| 4 | $0.51 | $0.53 | $0.76 | $0.78 |

The premium is usually stated as a dollar amount per $100 of insurance coverage (face value). To find the annual premium, multiply the insurance rate by the number of hundreds of dollars of fire insurance purchased (face value).

$$\text{Fire Insurance Premium} = \frac{\text{Face Value}}{100} \times \text{Rate}$$

**EXAMPLE**

The Breakstead Company owns a building in area rating 2, class A. The owners would like to insure the building for $130,000 and the contents for $31,000 against fire damage. Find the total annual premium for the building and the contents.

**SOLUTION**

**Step 1**  Find the premium for the building.
The rate for area 2, class A building is $0.32.

$$\text{Premium} = \frac{\text{Face Value}}{100} \times \text{Rate} = \frac{\$130,000}{100} \times \$0.32 = \$416$$

**Step 2**  Find the premium for the contents.
The rate for area 2, class A contents is $0.46.

$$\text{Premium} = \frac{\text{Face Value}}{100} \times \text{Rate} = \frac{\$31,000}{100} \times \$0.46 = \$142.60$$

**Step 3**  Find the total annual premium.

| Building Premium | + | Contents Premium | = | Total Annual Premium |
|---|---|---|---|---|
| $416 | + | $142.60 | = | $558.60 |

Property insurance policies are generally issued for 1- to 5-year terms. If the policy is to run for more than 1 year, most companies offer reduced rates. For example, if a policy ran for 3 years at 2.7 times the one-year premium of $558.60, the 3-year premium would be $1,508.22.

$$\begin{array}{ccc} & \text{3-Year} & \\ \text{Annual} & \text{Premium} & \text{3-Year} \\ \text{Premium} \times & \text{Factor} = & \text{Premium} \\ \$558.60 \times & 2.7 & = \$1,508.22 \end{array}$$

## PRACTICE PROBLEMS

Find the 1-year premium, 2-year premium, and the 3-year premium for the following fire insurance policies. The 2-year premium factor is 1.80 times and the 3-year premium factor is 2.75 times the 1-year premium. Round to the nearest cent.

| | Area Rating | Class | Building | Contents | 1-Year Premium | 2-Year Premium | 3-Year Premium |
|---|---|---|---|---|---|---|---|
| **1.** | 2 | B | $ 75,900 | $ 39,000 | $_____ | $_____ | $_____ |
| **2.** | 1 | A | 450,000 | 190,400 | _____ | _____ | _____ |
| **3.** | 3 | B | 284,300 | 300,000 | _____ | _____ | _____ |
| **4.** | 4 | A | 137,000 | 250,000 | _____ | _____ | _____ |
| **5.** | 3 | A | 435,000 | 125,000 | _____ | _____ | _____ |

## FINDING SHORT-RATE PREMIUMS

If a policyholder cancels insurance coverage or wants insurance coverage for less than 1 year, the standard short-rate table is used to compute the rates. Short-term policies have higher rates than annual policies because the insurance company's expenses for processing the paperwork is the same, although the number of payments received is less. To find the short-rate premium, multiply the annual premium by the appropriate short rate found on the following short-rate and cancellation table.

Annual Premium × Short Rate = Short-Rate Premium

### SHORT-RATE AND CANCELLATION TABLE

| Months in Force | Percent of Annual Rate to Be Charged | Months in Force | Percent of Annual Rate to Be Charged |
|---|---|---|---|
| 1 | 19% | 7 | 67% |
| 2 | 27% | 8 | 74% |
| 3 | 35% | 9 | 80% |
| 4 | 43% | 10 | 87% |
| 5 | 52% | 11 | 94% |
| 6 | 60% | 12 | 100% |

ANSWERS FOR PRACTICE PROBLEMS

| | 1-Year Premium | 2-Year Premium | 3-Year Premium |
|---|---|---|---|
| (1) | $ 602.01 | $1,083.62 | $1,655.53 |
| (2) | 1,862.36 | 3,352.25 | 5,121.49 |
| (3) | 3,625.80 | 6,526.44 | 9,970.95 |
| (4) | 2,023.70 | 3,642.66 | 5,565.18 |
| (5) | 2,352.50 | 4,234.50 | 6,469.38 |

## EXAMPLE

The North Star Company wants to purchase an 8-month fire insurance policy on merchandise worth $16,000. If the merchandise is located in area 3, class B, find the 8-month, short-rate premium.

## SOLUTION

**Step 1** Find the annual premium.
The rate for area 3, class B is $0.64.

$$\frac{\text{Annual}}{\text{Premium}} = \frac{\text{Face Value}}{100} \times \text{Rate} = \frac{\$16,000}{100} \times \$0.64 = \$102.40$$

**Step 2** Find the short-rate premium for 8 months.
The short rate for 8 months is 74%.

Annual Premium × Short Rate = Short-Rate Premium
$102.40      ×      0.74      ≈           $75.78

When a policyholder cancels the policy, the amount of refund is found by subtracting the short-rate premium from the annual premium that was paid.

Annual Premium − Short-Rate Premium = Refund

## EXAMPLE

The Bridge Port Warehouse canceled a fire insurance policy after 6 months. If they paid an annual premium of $501, what is the amount of their refund?

## SOLUTION

**Step 1** Find the short-rate premium for 6 months.

Annual    Short    Short-Rate
Premium × Rate  =  Premium
 $501    × 0.60 =   $300.60

**Step 2** Find the refund.

Annual    Short-Rate
Premium − Premium  =  Refund
 $501   − $300.60  = $200.40

## PRACTICE PROBLEMS

Find the short-rate premium for the following policies. Use the short-rate and cancellation table on page 296.

| | Annual Premium | Length of Policy | Short-Rate Premium |
|---|---|---|---|
| 1. | $2,560 | 4 months | $_____ |
| 2. | $7,030 | 6 months | _____ |
| 3. | $5,600 | 9 months | _____ |

Find the refund on the following policies.

| | Annual Premium | Canceled After | Refund |
|---|---|---|---|
| **4.** | $ 895 | 3 months | $_____ |
| **5.** | $3,980 | 11 months | _____ |
| **6.** | $1,300 | 7 months | _____ |

## CALCULATING COVERAGE USING THE COINSURANCE FORMULA

Since a fire often damages only a portion of a building and its contents, many business owners may be inclined to insure a building for less than its full value to save on insurance premium costs. Then, if a partial loss occurs, the business would expect the insurance company to pay the full amount of the loss.

To encourage property owners to insure their property for its full value, many insurance companies offer a policy containing a **coinsurance clause.** The **80% coinsurance clause,** which is popular, requires coverage for 80% of the property's replacement value to receive full recovery upon a loss. **Replacement value** is the amount it would cost to replace the building if it were destroyed.

If the property owner insures the property for less than 80% of its replacement value, the property owner becomes a coinsurer along with the insurance company and must bear a pro rata amount of the loss.

The following formula is used to calculate the amount of recovery a property owner will receive from a loss under a coinsurance policy.

$$\text{Recovery} = \frac{\text{Face Value of Policy}}{\text{Insurance Required}} \times \text{Loss}$$

80% coinsurance requirement

### EXAMPLE

The National Packaging Company owns a building that has a $500,000 replacement value. A $100,000 loss resulted from fire. The building was insured by a policy with an 80% coinsurance clause. Calculate the amount of insurance required (face value) to be fully insured, the recovery from the insurance company if fully insured at the 80% coinsurance requirement, the amount of recovery from the insurance company if the building were insured for $300,000, and the amount National Packaging Company will have to pay of the loss if the building were insured for $300.000.

### SOLUTION

**Step 1** Calculate insurance required to be fully insured.

| Replacement Value | × | 80% Coinsurance Rate | = Insurance Required |
|---|---|---|---|
| $500,000 | × | 0.80 | = $400,000 |

**Step 2** Calculate recovery if insured at the 80% coinsurance requirement.

$$\text{Recovery} = \frac{\text{Face Value of Policy}}{\text{Insurance Required}} \times \text{Loss}$$

$$\$100{,}000 = \frac{\$400{,}000}{\$400{,}000} \times \$100{,}000$$

**Step 3** Calculate recovery if insured for $300,000.

$$\text{Recovery} = \frac{\text{Face Value of Policy}}{\text{Insurance Required}} \times \text{Loss}$$

$$\$75{,}000 = \frac{\$300{,}000}{\$400{,}000} \times \$100{,}000$$

Insured for $\frac{3}{4}$, or 75% of amount required

**Step 4** Calculate insured's share of the loss if insured for $300,000.

$$\text{Loss} - \frac{\text{Recovery from}}{\text{Insurance Company}} = \frac{\text{Insured's Share}}{\text{of Loss}}$$

$$\$100{,}000 - \$75{,}000 = \$25{,}000$$

National Packaging pays $25,000

## PRACTICE PROBLEMS

Calculate 80% of the replacement value and the portion of the loss paid by the insurance company and the portion of the loss paid by the insured. Round to the nearest cent.

| Property | Replacement Value | Face Value | Amount of Loss | 80% of Replacement Value | Recovery from Insurance Company | Paid by Insured |
|---|---|---|---|---|---|---|
| 1. Office building | $800,000 | $640,000 | $ 42,000 | $_____ | $_____ | $_____ |
| 2. Warehouse | 229,300 | 183,440 | 53,000 | _____ | _____ | _____ |
| 3. Factory | 950,000 | 570,000 | 100,000 | _____ | _____ | _____ |
| 4. Apartment building | 475,000 | 228,000 | 98,300 | _____ | _____ | _____ |

**INSURING WITH MORE THAN ONE COMPANY**

It is not uncommon for a business to have property insurance policies on the same property with more than one company. This may occur when additional insurance is purchased over a period of time or when the value of the property insured may be so high that no one insurance company would want to assume all of the risk. Also, if a mortgage is held on the property, the mortgage company may require more than one insurer.

ANSWERS FOR PRACTICE PROBLEMS

| 80% of Replacement Value | Recovery from Insurance Company | Paid by Insured |
|---|---|---|
| (1) $640,000 | $42,000 | $ 0 |
| (2) 183,440 | 53,000 | 0 |
| (3) 760,000 | 75,000 | 25,000 |
| (4) 380,000 | 58,980 | 39,320 |

If a piece of property is insured with two or more companies, called *multiple carriers,* and a loss occurs, each of the insurance companies will pay a portion of the total loss. To find the amount each multiple carrier will pay, first divide the face value of the individual policy by the total amount of insurance coverage; then multiply by the amount of the total loss.

$$\text{Loss Paid by Insurer} = \frac{\text{Face Value of Policy}}{\text{Total Insurance Coverage}} \times \text{Loss}$$

### EXAMPLE

Ginneco Distributors has property insurance policies with the following companies: Mutual, $115,000; National, $100,000; and Fidelity, $72,500. A fire causing $150,600 worth of damage occurs. What portion of the loss will each company pay? Assume that the coinsurance requirement is met.

### SOLUTION

**Step 1**  Determine the total amount of insurance on the property.

$115,000 + $100,000 + $72,500 = $287,500 Total insurance coverage

**Step 2**  Find the amount of loss paid by each insurer. Round to the nearest hundred dollars.

$$\frac{\text{Loss Paid}}{\text{by Insurer}} = \frac{\text{Face Value of Policy}}{\text{Total Insurance Coverage}} \times \text{Loss}$$

Mutual:  $60,200 $\approx$  $\dfrac{\$115,000}{\$287,500}$  $\times$ $150,600

National:  $52,400 $\approx$  $\dfrac{\$100,000}{\$287,500}$  $\times$ $150,600

Fidelity:  $38,000 $\approx$  $\dfrac{\$72,500}{\$287,500}$  $\times$ $150,600

**Step 3**  The sum of the losses paid by each insurance company should equal the total loss.

$60,200 + $52,400 + $38,000 = $150,600 Total loss

### PRACTICE PROBLEM

The Tower Retirement Complex sustained $272,000 worth of fire damage. The building was insured by the following companies: (a) Western for $105,000, (b) Mutual for $275,000, and (c) Farmers for $230,000. Calculate the amount each multiple carrier will pay for the loss due to fire. Assume that the coinsurance requirement is met. Round to the nearest hundred dollars.

ANSWERS FOR
PRACTICE PROBLEM

(a) Western, $46,800  (b) Mutual, $122,600  (c) Farmers, $102,600

## LESSON 9-2     EXERCISES

Find the annual fire insurance premium for the building and its contents using the table on page 295.

| | Area Rating | Class | Coverage Building | Contents | Total Annual Premium |
|---|---|---|---|---|---|
| **1.** | 3 | A | $220,000 | $ 49,000 | _____ |
| **2.** | 4 | B | 743,000 | 190,000 | _____ |
| **3.** | 2 | A | 189,000 | 89,000 | _____ |

Find the total fire insurance premium for the building and its contents using the following schedule for policies of more than one year.
  The 2-year premium is 1.85 times the 1-year premium.
  The 3-year premium is 2.7 times the 1-year premium.
  The 4-year premium is 3.55 times the 1-year premium.
  The 5-year premium is 4.4 times the 1-year premium.

| | Area Rating | Class | Length of Policy | Coverage Building | Contents | Total Premium |
|---|---|---|---|---|---|---|
| **4.** | 1 | A | 2 years | $104,000 | $ 35,600 | _____ |
| **5.** | 4 | A | 4 years | 285,000 | 147,000 | _____ |
| **6.** | 2 | B | 2 years | 582,000 | 213,000 | _____ |

Find the annual fire insurance premium and the short-rate premium for the following building contents. Use the tables on pages 295 and 296.

| | Area Rating | Class | Contents Coverage | Length of Policy | Annual Premium | Short-Rate Premium |
|---|---|---|---|---|---|---|
| **7.** | 2 | B | $120,000 | 2 months | _____ | _____ |
| **8.** | 4 | A | 87,000 | 7 months | _____ | _____ |
| **9.** | 2 | A | 52,000 | 6 months | _____ | _____ |

The following policies were canceled by the policyholders. Find the short-rate premium and the refund. Use the short-rate and cancellation table on page 296.

| | Annual Premium | Canceled After | Short-Rate Premium | Refund |
|---|---|---|---|---|
| **10.** | $1,290 | 3 months | _____ | _____ |
| **11.** | 2,080 | 8 months | _____ | _____ |
| **12.** | 956 | 4 months | _____ | _____ |

Calculate the insurance coverage required to meet the 80% coinsurance clause and the portion of the loss paid by the insurance company. Round to the nearest cent.

| | Property | Replacement Value | Face Value | Amount of Loss | 80% of Market Value | Recovery from Insurance Company | Paid by Insured |
|---|---|---|---|---|---|---|---|
| 13. | Office complex | $562,500 | $450,000 | $116,000 | _____ | _____ | _____ |
| 14. | Apartment building | 895,000 | 716,000 | 82,000 | _____ | _____ | _____ |
| 15. | Factory | 310,000 | 186,000 | 15,800 | _____ | _____ | _____ |

Calculate the amount each multiple carrier will pay for the loss due to fire. Assume that the coinsurance requirement is met. Round amounts to the nearest hundred.

| | Insurance Loss | Companies | Coverage | Amount Paid |
|---|---|---|---|---|
| 16. | $ 20,000 | A | $ 75,000 | _____ |
| | | B | 15,000 | _____ |
| | | C | 25,500 | _____ |
| 17. | $ 95,000 | A | $105,000 | _____ |
| | | B | 75,000 | _____ |
| | | C | 35,000 | _____ |
| 18. | $126,000 | A | $112,000 | _____ |
| | | B | 100,000 | _____ |
| | | C | 62,000 | _____ |
| 19. | $275,000 | A | $250,000 | _____ |
| | | B | 110,000 | _____ |
| | | C | 80,000 | _____ |
| 20. | $ 45,000 | A | $ 97,000 | _____ |
| | | B | 50,000 | _____ |
| | | C | 15,000 | _____ |

## BUSINESS APPLICATIONS

Round to the nearest cent unless otherwise directed. Use the tables on pages 295 and 296.

21. The studio for station WKRT is in rating area 3, class B. The owners would like to insure the building for $380,000 and the contents for $915,000. Calculate the total annual premium for the building and the contents.

**22.** The Yardley Paint company purchased a 4-year fire insurance policy for their warehouse for $89,000 and the contents for $756,000. The warehouse is in rating area 2, class B. Find the amount of the 4-year premium if it is 3.65 times the annual premium.

_____

**23.** Manfred Textiles purchased a 5-month fire insurance policy for $623,000 on merchandise temporarily stored in a warehouse until it can be shipped. The warehouse is in area 4, class A. How much is the 5-month premium?

_____

**24.** Heller Well Drilling stores its equipment in a building in area 1, class B. The building is insured for $56,000 and the equipment is insured for $307,000. Find the total annual premium.

_____

**25.** The Western Boot and Saddle Company incurred $98,000 worth of damage from a fire. The owner has a $126,000 fire insurance policy with an 80% coinsurance clause. The building has a replacement value of $180,000. Find (a) the portion of the loss paid by the insurer and (b) the portion paid by the Western Boot and Saddle Company.

(a) _____

(b) _____

**26.** After 8 months the Peoples Savings and Loan canceled the fire insurance policy on their property. If the annual premium is $2,090, what is the amount of refund?

_____

27. Carmen's Italian Restaurant has a replacement value of $150,000 and is insured for $102,500 with an 80% coinsurance clause. The restaurant suffered $32,000 worth of fire damage. Find (a) what amount of the loss is paid by the insurer, and (b) what amount is paid by the owner of the restaurant.

(a) _____

(b) _____

28. Fire damage to Howell's Auto Parts totaled $37,000. The owner had fire insurance with the following companies: Tri-State for $52,000, Atlantic for $76,000, and Fidelity for $100,000. What portion of the loss is paid by (a) Tri-state, (b) Atlantic, and (c) Fidelity? Assume the 80% coinsurance requirement is met. Round to the nearest hundred dollar.

(a) _____

(b) _____

(c) _____

29. The owner of Apple Grove Apartments has fire insurance with the following companies: Jenkins for $45,000, Kenmore for $83,000, and Union for $112,000. The apartments suffered a fire loss of $27,000. Find the amount of loss paid by (a) Jenkins, (b) Kenmore, and (c) Union. Assume the coinsurance requirement has been met. Round to the nearest hundred dollar.

(a) _____

(b) _____

(c) _____

**TYPES OF
COVERAGE**

Motor vehicle insurance protects against financial loss due to damage or injury from accidents, vandalism, or natural causes. Most businesses own one or more automobiles, delivery vans, or trucks. Business vehicle insurance usually is provided under a master policy with each vehicle listed separately. The amount of the premium for motor vehicle insurance depends on factors such as the type and amount of insurance purchased, age and sex of the drivers, age and type of vehicle, amount the vehicle is driven and for what purpose, and area in which the vehicle is driven.

If the policyholder is at fault when an accident occurs, he or she must pay any claims that exceed the amount of insurance coverage. For instance, if a policyholder has coverage for $100,000 and sustains $125,000 of damages, the insurance company will pay $100,000 and the policyholder must pay the remaining $25,000. Therefore, it is important for businesses and individuals to carry adequate amounts of insurance protection.

Most states require that individuals and businesses obtain some kind of motor vehicle insurance. The various types of insurance available are described below.

*Bodily injury liability insurance* provides coverage for medical expenses, lost wages, court fees, and other costs resulting from injuries to others incurred in an accident for which the policyholder's vehicle is responsible. Bodily injury liability coverage is often stated as amounts such as 250/500. This means that the insurance company will pay up to $250,000 for injuries to one person and a maximum of $500,000 for two or more persons involved in a single accident.

*Property damage liability insurance* provides coverage for legal fees and repairs for damages to someone else's property caused by the policyholder's vehicle. Many businesses carry liability coverage that protects them to well over $1,000,000 for any one accident.

*Collision insurance* covers the insured's vehicle if it is damaged from a collision. *Comprehensive insurance* covers damage to the insured's vehicle from hail, fire, theft, glass breakage, and other such events. Businesses usually select deductible amounts of $250, $500, or $1,000 when purchasing collision and comprehensive insurance. This means that the business pays for repairs to their vehicle up to the amount of the deductible, and the insurance company pays the remaining amount. The higher the amount of the deductible, the lower the amount of the premium.

*Uninsured motorist insurance* provides payment for medical care to occupants of the insured's vehicle if there is a collision with a vehicle not insured or underinsured. This coverage is mandatory in most states.

**CALCULATING
MOTOR VEHICLE
INSURANCE
PREMIUMS**

The table at the top of the next page quotes premiums similar to those an insurance company might provide to a business. The business selects the type of coverage wanted and calculates the annual premium per vehicle by adding together the selected premiums.

## MOTOR VEHICLE INSURANCE RATES

| Type of Vehicle | Bodily Injury Liability, Amount of Coverage | | | Property Damage Liability, Amount of Coverage | | |
|---|---|---|---|---|---|---|
| | 50/100 | 100/300 | 300/500 | $300,000 | $500,000 | $1,000,000 |
| Auto | $186 | $228 | $ 372 | $192 | $240 | $384 |
| Delivery van | 222 | 276 | 444 | 228 | 288 | 456 |
| Truck-tractor | 576 | 708 | 1,152 | 258 | 324 | 516 |
| Truck-trailer | 180 | 222 | 360 | 108 | 138 | 216 |

| Type of Vehicle | Collision Deductible | | Comprehensive Deductible | | Uninsured Motorist |
|---|---|---|---|---|---|
| | $ 250 | $ 500 | $ 0 | $250 | |
| Auto | $ 360 | $ 342 | $264 | $228 | $25 |
| Delivery van | 432 | 408 | 318 | 276 | 34 |
| Truck-tractor | 1,098 | 1,320 | 780 | 708 | 39 |
| Truck-trailer | 216 | 258 | 120 | 108 | –0– |

### EXAMPLE

Rogette Building Supply received the quotations shown in the table above. Calculate the annual premium for a delivery van if the following coverages are selected: bodily injury, 300/500; property damage, $500,000; collision, $250 deductible; comprehensive, no deductible; uninsured motorist. Use the motor vehicle insurance rate table.

### SOLUTION

Locate "delivery truck" under "type of vehicle" and move to the right to select the correct amounts. Total the amounts to find the annual premium.

| Bodily Injury Liability | + | Property Damage Liability | + | Collision Deductible | + | Comprehensive Deductible | + | Uninsured Motorist | = | Annual Premium |
|---|---|---|---|---|---|---|---|---|---|---|
| $444 | + | $288 | + | $432 | + | $318 | + | $34 | = | $1,516 |

### PRACTICE PROBLEMS

Calculate the annual premium for the following. Use the insurance premium quotations from the motor vehicle insurance rates table.

| | Bodily Injury Liability | Property Damage Liability | Collision Deductible | Comprehensive Deductible | Uninsured Motorist | Annual Premium |
|---|---|---|---|---|---|---|
| **1.** Delivery van coverage | 50/100 | $ 300,000 | $500 | $250 | No | |
| Premium | ———— | ———— | ———— | ———— | ———— | ———— |
| **2.** Automobile coverage | 300/500 | $1,000,000 | $250 | $ 0 | Yes | |
| Premium | ———— | ———— | ———— | ———— | ———— | ———— |

## LESSON 9-3     EXERCISES

Calculate the amount of premium for the vehicle insurance coverage shown below and the total annual premium. Use the insurance quotations table on page 306.

| | Bodily Injury Liability | Property Damage Liability | Collision Deductible | Comprehensive Deductible | Uninsured Motorist | Annual Premium |
|---|---|---|---|---|---|---|
| **1.** Delivery van coverage | 100/300 | $ 300,000 | $500 | $250 | Yes | |
| Premium | _____ | _____ | _____ | _____ | _____ | _____ |
| **2.** Auto coverage | 50/100 | $ 500,000 | $250 | $ 0 | Yes | |
| Premium | _____ | _____ | _____ | _____ | _____ | _____ |
| **3.** Truck-tractor coverage | 300/500 | $1,000,000 | $500 | $250 | Yes | |
| Premium | _____ | _____ | _____ | _____ | _____ | _____ |
| **4.** Auto coverage | 100/300 | $ 300,000 | $500 | $250 | Yes | |
| Premium | _____ | _____ | _____ | _____ | _____ | _____ |
| **5.** Delivery van coverage | 50/100 | $1,000,000 | $250 | $ 0 | No | |
| Premium | _____ | _____ | _____ | _____ | _____ | _____ |
| **6.** Truck-tractor coverage | 100/300 | $ 500,000 | $250 | $250 | Yes | |
| Premium | _____ | _____ | _____ | _____ | _____ | _____ |
| **7.** Truck-trailer coverage | 100/300 | $ 300,000 | $500 | $ 0 | No | |
| Premium | _____ | _____ | _____ | _____ | _____ | _____ |
| **8.** Delivery van coverage | 300/500 | $ 500,000 | $250 | $250 | Yes | |
| Premium | _____ | _____ | _____ | _____ | _____ | _____ |

## BUSINESS APPLICATIONS

Solve the following problems. Use the table on page 306.

9. Calculate the annual premium for a truck-tractor with bodily injury of 50/100, property damage of $300,000, collision with a $250 deductible, comprehensive with no deductible, and uninsured motorist.

**10.** The Young Bottling Company has two delivery vans and one company car. Find the total annual premium on the three vehicles if the following coverages are selected: bodily injury for 100/300; property damage for $500,000; collision deductible for $500; comprehensive deductible for $250; and uninsured motorist.

_____

**11.** United Chemical insures their truck-tractors for 300/500 bodily injury liability, $1,000,000 property damage liability, $500 collision deductible, $250 comprehensive deductible, and uninsured motorist. What is the total annual insurance premium on seven truck-tractors?

_____

**12.** T & K Plumbing has two delivery vans insured for 100/300 bodily injury liability, $300,000 property damage liability, $250 collision deductible, and comprehensive with no deductible. Find (a) the monthly premium for the two vans and (b) the total amount paid for the year if the monthly premium is 9% of the annual premium.

(a) _____

(b) _____

**13.** The driver of a service truck belonging to Gardener's Lawn Service hit a car. The judge awarded the driver of the car $87,500 for injuries. The Gardener's Lawn Service has bodily injury liability insurance of 50/100. Find (a) the amount the insurance company will pay, and (b) the amount Garden Lawn Service will pay.

(a) _____

(b) _____

**14.** The driver of a car hit a truck causing $27,000 worth of injuries to the truck driver and $10,000 worth of damage to the truck. The driver of the car has 25/50 bodily injury liability coverage and $15,000 property damage liability coverage. Find (a) the amount the insurance company will pay, and (b) the amount the driver of the auto will pay.

(a) _____

(b) _____

**CALCULATING THE ASSESSED VALUE AND TAX RATE**

Owners of land and buildings (real property) must pay a **property tax** imposed by county, state, and local governments. The property tax is a major source of revenue for police and fire protection, public schools, street and highway maintenance, and other public services. In most states, nonprofit organizations, such as schools, churches, and charities, do not pay property tax.

The amount of tax paid is based on the **assessed valuation** of the property. In most jurisdictions, the assessed valuation is a certain percentage of the **market value.** The market value is the amount for which the property reasonably could be sold. The **assessment rate** usually ranges from 25% to 100% of the market value. To find the assessed valuation, multiply the market value by the assessment rate.

Market Value × Assessment Rate = Assessed Valuation

### EXAMPLE

Kline's Delicatessen has a market value of $275,000. Find the assessed valuation of the property if the assessment rate is 62.5% of the market value.

### SOLUTION

Find the assessed valuation.

| Market Value | × | Assessment Rate | = | Assessed Valuation |
|---|---|---|---|---|
| $275,000 | × | 0.625 | = | $171,875 |

Government officials determine the total amount of money needed to operate a city, county, or state by making a yearly budget. If the revenue generated for the budget comes from property taxes, the **tax rate** is determined by dividing the total amount needed in the budget by the total assessed valuation of the property in the area. In some states, however, part of the budget is funded through other sources.

$$\text{Tax Rate} = \frac{\text{Total Tax Amount Needed}}{\text{Total Assessed Valuation}}$$

### EXAMPLE

The assessed valuation of real property in Litchfield County is $680,000,000. The county budget is $44,200,000. Find the county tax rate.

### SOLUTION

Find the tax rate.

$$\frac{\text{Tax}}{\text{Rate}} = \frac{\text{Total Tax Amount Needed}}{\text{Total Assessed Valuation}} = \frac{\$44,200,000}{\$680,000,000} = 0.065 \text{ or } 6.5\%$$

Find the assessed valuation of the following property.

| | Market Value | Rate of Assessment | Assessed Valuation |
|---|---|---|---|
| 1. | $   800,000 | $53\frac{1}{2}\%$ | $_____ |
| 2. | $    67,800 | 67% | _____ |
| 3. | $1,270,000 | $78\frac{1}{4}\%$ | _____ |

Find the tax rate for the following counties. Round to the nearest tenth.

| County | Total Tax Amount Needed | Total Assessed Valuation | Tax Rate |
|---|---|---|---|
| 4. Cascade | $16,800,000 | $270,100,000 | $_____ |
| 5. Brunswick | $21,600,000 | $950,000,000 | _____ |
| 6. Cross | $37,900,000 | $715,094,000 | _____ |

**CALCULATING THE AMOUNT OF PROPERTY TAX**

Tax rates can be expressed in various forms. When the tax rate is *expressed as a percent,* multiply the assessed valuation of the property by the percent to find the annual property tax.

---

**EXAMPLE**

Mr. and Mrs. Chew own a business with a market value of $227,500. Property in the city is assessed at 75% of the market value with a tax rate of $3\frac{3}{4}\%$. What is the assessed value and the annual property tax on their business? Round to the nearest cent.

**SOLUTION**

**Step 1**   Find the assessed valuation.

| Market Value | × | Assessment Rate | = | Assessed Valuation |
|---|---|---|---|---|
| $227,500 | × | 0.75 | = | $170,625 |

**Step 2**   Find the annual property tax.

| Assessed Valuation | × | Tax Rate | = | Annual Property Tax |
|---|---|---|---|---|
| $170,625 | × | 0.0375 | ≈ | $6,398.44 |

When the tax rate is expressed as a certain number of *dollars per $100,* divide the assessed valuation by 100 and multiply the result by the tax rate to find the annual property tax.

---

**EXAMPLE**

Walter Howard owns a tract of land with an assessed valuation of $125,000. The tax rate in the county is $7.95 per $100 of assessed valuation. Find the annual county property tax on Mr. Howard's land.

---

**SOLUTION**

$$\begin{aligned}\text{Annual} \atop \text{Property Tax} &= \frac{\text{Assessed Valuation}}{100} \times \text{Tax Rate per \$100} \\[2mm] &= \frac{\$125,000}{100} \quad\times\quad \$7.95 \quad= \$9,937.50\end{aligned}$$

When the tax rate is expressed as a certain number of **dollars per $1,000,** divide the assessed valuation by 1,000 and multiply the result by the tax rate to find the annual property tax.

**EXAMPLE**

Child Craft Day Care owns a building with an assessed valuation of $173,300. The county property tax is $56 per $1,000 of assessed valuation. What is the annual county property tax on the building?

**SOLUTION**

$$\begin{aligned}\text{Annual} \atop \text{Property Tax} &= \frac{\text{Assessed Valuation}}{1,000} \times \text{Tax Rate per \$1,000} \\[2mm] &= \frac{\$173,300}{1,000} \quad\times\quad \$56 \quad= \$9,704.80\end{aligned}$$

It is also common to find tax rates expressed as a certain number of **mills per dollar.** A mill is one-tenth of a cent or one-thousandth of a dollar ($0.001). When the tax rate is expressed in mills per dollar, convert the mills to dollars and multiply the result by the assessed valuation to find the annual property tax.

**EXAMPLE**

The O'Reilly family owns a dairy farm with an assessed valuation of $1,825,000. If the county tax rate is 47 mills per dollar, find the annual county property tax.

**SOLUTION**

**Step 1**   Convert mills to dollars by multiplying by $0.001.

$47 \times \$0.001 = \$0.047$

**Step 2**   Find the property tax.

$$\begin{array}{ccc}\text{Assessed} & \text{Mills per} & \text{Property} \\ \text{Valuation} \times & \text{Dollar} = & \text{Tax} \\ \$1,825,000 \times & \$0.047 = & \$85,775\end{array}$$

Although each of the tax rates below is expressed in a different form, the amount of tax on property assessed at $87,000 is the same.

| Tax Rate | Property Tax |
|---|---|
| 5.3% | $87,000 × 0.053 = $4,611 |
| $5.30 per $100 | $870 × $5.30 = $4,611 |
| $53 per $1,000 | $87 × $53 = $4,611 |
| 53 mills per $1 | $87,000 × $0.053 = $4,611 |

Find the annual property tax on the following property. Round to the nearest cent.

| Assessed Valuation | Tax Rate | Annual Property Tax |
|---|---|---|
| 1. $210,000 | $6.50 per $100 | _____ |
| 2. $400,000 | 75 mills | _____ |
| 3. $ 52,000 | 4.2% | _____ |
| 4. $185,000 | $62.50 per $1,000 | _____ |

## LESSON 9-4     EXERCISES

Find the assessed valuation of the following property.

| Property | Market Value | Rate of Assessment | Assessed Valuation |
|---|---|---|---|
| 1. Home | $ 543,000 | 65% | _____ |
| 2. Building | 1,560,000 | 80% | _____ |
| 3. Warehouse | 42,700 | 45% | _____ |

Find the tax rate. Round to the nearest tenth.

| County | Total Tax Amount Needed | Total Assessed Valuation | Tax Rate |
|---|---|---|---|
| 4. Noble | $39,400,000 | $437,800,000 | _____ |
| 5. Drake | 22,700,000 | 412,700,000 | _____ |
| 6. Brower | 13,200,000 | 170,300,000 | _____ |

Find the annual property tax on the following property.

| Property | Assessed Valuation | Tax Rate | Annual Property Tax |
|---|---|---|---|
| 7. Tract of land | $ 90,000 | 62 mills | _____ |
| 8. Office complex | 2,500,000 | $8.06 per $100 | _____ |
| 9. Home | 76,000 | 9.2% | _____ |

## BUSINESS APPLICATIONS

Solve the following problems. Round amounts to the nearest cent and percents to the nearest tenth.

10. The Darwin Optical Center Research Building has a market value of $3,295,000. Find the assessed valuation of the building if the assessment rate is 73.5% of the market value.

_____

11. Rawlings County wants to raise $17.7 million through property taxes. If the total assessed value of the county property is $321,000,000, find the tax rate expressed (a) as a percent and (b) as mills.

(a) _____

(b) _____

12. The Sunset Trailer Park has a market value of $950,000. Property in the city is assessed at 45% of the market value with a tax rate of $91.65 per $1,000. Find (a) the assessed valuation and (b) the annual property tax on the park.

(a) _____

(b) _____

13. Heller Development Corporation owns land in Wadsworth County worth $3,750,000. Property in the county is assessed at 67% of market value with a tax rate of 65 mills. Find (a) the assessed value and (b) the annual property tax on the land.

(a) _____

(b) _____

14. Marion County has a 4.5% property tax rate. If Twin River Apartments has an assessed value of $785,600, what is the annual state property tax?

_____

15. The All-Night Food Mart has a market value of $145,000. The county assessment rate is $73\frac{3}{4}$% of the market value and the tax rate is $5.75 per $100 of assessed valuation. Find (a) the assessed valuation and (b) the annual property tax.

(a) _____

(b) _____

16. If the assessment rate on a tract of land with a market value of $92,000 increases from 50% to 55%, how much more will the taxes be if the tax rate remains at $9.27 per $100?

_____

## CHAPTER 9 REVIEW

Round to the nearest cent unless otherwise directed.

1. Employees at Northern Power & Light may apply for term life insurance through the company. Each employee is insured for $2\frac{1}{4}$ times his or her annual salary at $3.09 per $1,000 of insurance. What is Alex Rawling's monthly payroll deduction if his annual salary is $46,500?

_____

2. The Zodiac Book Company owns a class A building in rating area 2. The owners would like to purchase fire insurance on the building for $330,000 and the contents for $127,000. Find the amount of the 4-year premium if it is 3.55 times the annual premium.

_____

3. The Rochelle Manufacturing Company wants to purchase a 10-month fire insurance policy on merchandise worth $219,500. If the merchandise is located in area 1, classification A, find the amount of the short-rate premium.

_____

4. The Regency Apartment Complex canceled their fire insurance policy after 3 months. If they paid an annual premium of $3,250, what is the amount of their refund?

_____

5. American Investments owns a building with a replacement value of $1,250,000. The owners of the company purchased a fire insurance policy with a face value of $900,000 containing an 80% coinsurance clause. The building sustained $500,000 worth of fire damage. Find (a) the portion of the loss paid by the insurance company and (b) the portion paid by American Investments.

(a) _____

(b) _____

6. The Riverview Mall purchased fire insurance from the following companies: $250,000 from Independent, $315,000 from United, and $405,000 from Mutual. The mall sustained $232,000 in fire damage. Find the amount paid by (a) Independent, (b) United, and (c) Mutual. Assume the coinsurance clause is met. Round amounts to the nearest hundred dollar.

(a) _____

(b) _____

(c) _____

7. Mr. Rollin's service truck hit an automobile. Mr. Rollins has bodily injury liability of 30/50 and property damage liability of $15,000. The driver of the automobile is awarded $33,000 for injuries and $17,300 for property damage. Find (a) the amount the insurance company will pay and (b) the amount Mr. Rollins will pay.

(a) _____

(b) _____

8. Rebecca Lowell wants to purchase a 20-year limited payment life insurance policy for $80,000. If Rebecca is 33 years old, how much is her annual premium?

_____

9. Find (a) the annual premium and (b) the semiannual premium payment on a 10-year term life insurance policy with a face value of $120,000 for a 20-year old man. The semiannual premium factor is 51%.

(a) _____

(b) _____

10. Jackson County's budget calls for $42.5 million in property taxes. The total assessed value of property in the county is $1.7 billion. Find the tax rate expressed (a) as a percent and (b) as dollars per $1,000.

(a) _____

(b) _____

11. Longville Investment firm owns property valued at $873,200. Property in the city is assessed at 57% of the market value with a tax rate of $7.56 per $100. Find (a) the assessed valuation and (b) the annual property tax on the property.

(a) _____

(b) _____

**12.** Jonathon Hensley purchased land worth $195,000 in Henderson County with a tax rate of 37 mills based on assessed valuation of 77% of the market value. He also purchased land with a $195,000 market value in Fairfield County with a tax rate of $4.333 per $100 based on 65% of the market value. Find his annual property tax in (a) Henderson County and (b) in Fairfield County. Find (c) which county has the lower taxes.

(a) _____

(b) _____

(c) _____

**13.** Roma Pizza purchased the following insurance on their two delivery vans: bodily injury liability of 50/100, property damage liability of $500,000, collision deductible of $250, and uninsured motorist insurance. If the quarterly premium factor is 26%, find (a) the annual premium and (b) the quarterly premium for both vans.

(a) _____

(b) _____

## CUMULATIVE REVIEW II

Round amounts to the nearest hundredth. Round percents to the nearest tenth.

1. Jackie Roberts wants to purchase gloves priced at $35.99, a scarf priced at $12.49, leather boots priced at $129.99, and coat priced at $140.69. Estimate by rounding each item to the nearest dollar the total purchase price of the items.

2. At Smith Manufactures, the list price of a machine is $7,525 less a series discount of 40, 10, 5. At Bowers Industries the list price of the same machine is $6,900 less a trade discount of 45%. Which company offers the lower net price?

3. An invoice dated October 2 has terms of 2/10, 1/20, n/30, and freight charges of $450. The invoice total is $8,195. Calculate the amount due if the invoice was paid on October 12.

4. Little Wholesale Tires purchased steel belted radial tires for $42 each. The retail price includes a markup of $20. Calculate the rate of markup based on cost.

5. Rosa Vasquez purchased a ring with a purchase price of $49.50. Calculate the total cash payment if a 5% sales tax and a 7.5% excise tax are levied on the jewelry.

6. Hudson Company purchased equipment under the installment plan. The cash price was $12,695.99. Hudson paid $2,500 down and agreed to make 24 monthly installment payments of $560.78 each. Calculate the total sale price of the equipment.

_____

7. Roylie Company financed $65,000 for a new telephone system. The monthly payment at 14% interest is $865.64 for 15 years and $808.29 for 20 years. How much more interest is charged on the 20-year loan than on the 15-year loan?

_____

8. Wanda Carnes purchased a whole life insurance policy with a face value of $125,000. Calculate the annual premium if the rate is $14.15 per $1,000 of face value.

_____

9. Baxter Retail Shoes suffered a fire loss of $47,000. The owners carry a $120,000 coinsurance policy. Calculate the portion of the loss the insurance company will pay if the building has a replacement value of $160,000.

_____

10. The assessed value of the Carmen Candy Factory is $315,000. Calculate the annual property tax if the tax rate is 52 mils.

_____

# CHAPTER
## 10

# CALCULATING INTEREST ON INVESTMENTS

## LEARNING OBJECTIVES

1. Calculate interest on passbook savings accounts and savings certificates.
2. Calculate compound amount and compound interest manually and using a table.
3. Calculate the effective rate of interest.
4. Calculate present value using a table.
5. Calculate the amount of an ordinary annuity and an annuity due.
6. Calculate the present value of an ordinary annuity.
7. Calculate the amount of the annuity payment for a sinking fund.

A business owner or individual may invest a lump sum of money over a period of time that will earn simple or compound interest. Money also may be invested over a period of time as a series of equal deposits or payments at regular intervals that will earn compound interest at a specific rate.

In this chapter, you will learn to calculate interest on different types of investments and use tables to calculate present values, annuities, and sinking fund payments.

## LESSON 10-1    INTEREST-BEARING INVESTMENTS

Interest-bearing investments commonly available through banks and other financial institutions include passbook savings accounts and savings certificates. For years, federal law set various guidelines for these investments including the rate of interest that could be paid. These guidelines no longer apply. Financial institutions can now set their own interest rates and offer a variety of interest-bearing investment opportunities. Competition for investors' dollars, however, has led to very similar offerings available from most financial institutions.

When investments yield *simple interest,* interest is calculated only on the principal. When investments yield *compound interest,* interest is calculated on principal plus all accumulated interest. Compound interest is greater than simple interest because the interest for each compounding period is added to the principal before calculating interest for the next compounding period.

The *compounding period* is the length of time for which interest has been calculated. Compound interest can be calculated annually, semiannually, quarterly, or daily. The more frequently interest is compounded, the greater the amount of interest earned.

The *compound amount* represents the sum of the original principal plus the accumulated interest. To find the amount of compound interest, subtract the original principal from the compound amount.

## EXAMPLE

When Dr. Robert Flyer invested $10,000 at 6.5% simple interest for 1 year, he earned $650 interest ($10,000 × 0.065 × 1). How much interest would Dr. Flyer have earned if he had invested $10,000 at 6.5% interest compounded semiannually? Round to the nearest cent.

## SOLUTION

**Step 1**  Calculate interest for 6 months, and find the new principal.

Principal × Rate × Time = Interest
$10,000 × 0.065 × 0.5 = $325

Principal + Interest = Principal
$10,000 + $325 = $10,325

**Step 2**  Find the interest for the second 6 months.

Principal × Rate × Time = Interest
$10,325 × 0.065 × 0.5 = $335.56

**Step 3**  Find the compound amount and compound interest.

Principal + Interest = Compound Amount
$10,325 + $335.56 = $10,660.56

Compound Amount − Original Principal = Compound Interest
$10,660.56 − $10,000 = $660.56

If Dr. Flyer's investment had earned compound interest instead of simple interest, he would have earned $10.56 more ($660.56 − $650).

## PRACTICE PROBLEMS

Manually calculate the compound interest. Round to the nearest cent.

| | Principal | Investment Terms | Compound Interest |
|---|---|---|---|
| **1.** | $12,000 | 1 year at 5% compounded semiannually | |
| **2.** | 13,000 | 3 months at 6.25% compounded monthly | |

**322** ■ LESSON 10-1

## PASSBOOK SAVINGS ACCOUNTS

The most common form of savings account offered by banks, savings and loans associations, and credit unions is a *passbook savings account.* Money deposited in these accounts, also called *regular savings accounts,* is guaranteed by an agency of the federal government such as FDIC and FSLIC for amounts up to $100,000.

### INTEREST COMPOUNDED DAILY

Most financial institutions use a compound interest table, which lists the daily compound amount of $1.00 at various interest rates and time periods. To find the daily compound amount, multiply the principal by the appropriate table value.

**DAILY COMPOUND INTEREST (Amount of $ Compounded Daily)**

| Days | STATED RATE | | | | | |
|------|-------|-------|-------|-------|-------|-------|
| | 5.00% | 5.25% | 5.50% | 5.75% | 6.00% | 6.25% |
| 30 | 1.0041177 | 1.0043240 | 1.0045304 | 1.0047368 | 1.0049432 | 1.0051497 |
| 31 | 1.0042553 | 1.0044685 | 1.0046818 | 1.0048951 | 1.0051084 | 1.0053218 |
| 60 | 1.0082524 | 1.0086668 | 1.0090814 | 1.0094961 | 1.0099109 | 1.0103260 |
| 90 | 1.0124022 | 1.0130284 | 1.0136529 | 1.0142779 | 1.0149032 | 1.0155289 |
| 120 | 1.0165730 | 1.0174088 | 1.0182452 | 1.0190824 | 1.0199202 | 1.0207587 |
| 180 | 1.0249623 | 1.0262265 | 1.0274923 | 1.0287595 | 1.0300286 | 1.0312991 |
| 365 | 1.0512674 | 1.0538985 | 1.0565362 | 1.0591804 | 1.0618313 | 1.0644887 |

### EXAMPLE

On October 15 House of Imports invested $5,000 in a passbook savings account at $5\frac{1}{4}\%$ interest, compounded daily. Thirty-one days later they deposited $2,500. What is the balance in the savings account 60 days after the last deposit? Round to the nearest cent.

### SOLUTION

**Step 1** Use the daily compound interest table. Read down the days column until you come to 31. Read across to the right until you come to the column headed 5.25%.

**Step 2** Calculate the interest.

Principal Invested × Table Value = Interest Earned
$5,000 × 0.004468538 = $22.34

**Step 3** Calculate the balance after the following 60 days.

Principal × Table Value = Compound Amount

$5,022.34 + $2,500.00 ⟶ $7,522.34 × 1.086668 = $7,587.53

### PRACTICE PROBLEMS

Use the daily compound interest table to find the interest earned and the savings account balance. Round to the nearest cent.

| | Principal | Rate | Days Invested | Savings Account Balance | Interest |
|---|-----------|------|---------------|-------------------------|----------|
| **1.** | $666 | 6.00% | 30 | _____ | _____ |
| **2.** | 454 | 5.00% | 180 | _____ | _____ |

A *savings certificate* earns interest at a set rate which is usually higher than a passbook savings account, provided the depositor leaves a minimum amount of money on deposit for a specific period of time set by the financial institution. There are several forms of savings certificates, including *money-market certificates* and *certificates of deposit (CDs).*

Traditionally, financial institutions require that a minimum amount be on deposit for savings certificates and that the money be invested for a specified time period, such as 6 months, 1 year, or 2 years. The longer the investment period, the higher the interest rate. Savings certificates ordinarily pay simple interest or interest compounded quarterly.

If money is withdrawn from the savings certificate before its due date, there is a substantial penalty, such as the forfeiture of 3 months' interest on a savings certificate of 1 year or less, and 6 months' interest on a certificate of more than 1 year. The total amount received at the end of the period is called the *proceeds.*

### EXAMPLE

Saville, Inc., invested $2,500 in a 6-month savings certificate that earns 8% simple interest. Saville's bank assesses a penalty of 3 months' interest if withdrawal is made before the due date. What are Saville's proceeds if it withdraws the full amount after 2 months? Round to the nearest cent.

### SOLUTION

**Step 1** Calculate the investment value after 2 months.

Principal × Rate × Time ≈ Interest
$2,500 × 0.08 × $\frac{1}{6}$ ≈ $33.33

Principal + Interest = Investment Value
$2,500 + $33.33 = $2,533.33

**Step 2** Calculate the penalty of 3 months' interest.

Principal × Rate × Time = Penalty
$2,500 × 0.08 × $\frac{1}{4}$ = $50.00

**Step 3** Calculate the proceeds.

Investment Value − Penalty = Proceeds
$2,533.33 − $50.00 = $2,483.33

### PRACTICE PROBLEMS

Complete the following table. Round to the nearest cent. A penalty of 3 months' interest is assessed for early withdrawal.

| | Amount Invested | Interest Rate | Length of Certificate | Time Held | Interest | Investment Value | Penalty | Proceeds |
|---|---|---|---|---|---|---|---|---|
| 1. | $5,000 | 8.25% | 1 year | 8 months | _____ | _____ | _____ | _____ |
| 2. | 7,500 | 9.25% | 6 months | 2 months | _____ | _____ | _____ | _____ |

## LESSON 10-1  EXERCISES

Calculate the compound interest. Round to the nearest cent.

| | Principal | Investment Terms | Compound Interest |
|---|---|---|---|
| **1.** | $  220 | 2 years at 7%<br>compounded annually | _____ |
| **2.** | 1,350 | 1 year at 8.5%<br>compounded quarterly | _____ |
| **3.** | 4,600 | $1\frac{1}{2}$ years at 6.5%<br>compounded semiannually | _____ |
| **4.** | 9,100 | 2 years at 5%<br>compounded semiannually | _____ |
| **5.** | 11,200 | 4 months at 10%<br>compounded monthly | _____ |

Calculate the simple interest, the compound interest compounded annually, and the difference between the two. Round to nearest cent.

| | Principal | Rate | Number of Years | Simple Interest | Compound Interest | Difference |
|---|---|---|---|---|---|---|
| **6.** | $  234 | 7.25% | 2 | _____ | _____ | _____ |
| **7.** | 720 | 12.75% | 3 | _____ | _____ | _____ |
| **8.** | 1,050 | 6.5% | 4 | _____ | _____ | _____ |
| **9.** | 490 | 11% | 2 | _____ | _____ | _____ |
| **10.** | 890 | 9% | 2 | _____ | _____ | _____ |

Use the daily compound interest table to find the interest earned and the
ending balance on the following passbook savings account deposits.
Round to the nearest cent.

| | Principal | Rate | Days Invested | Ending Balance | Interest |
|---|---|---|---|---|---|
| **11.** | $3,000 | 5.50% | 30 | _____ | _____ |
| **12.** | 1,500 | 5.00% | 120 | _____ | _____ |
| **13.** | 750 | 6.00% | 180 | _____ | _____ |
| **14.** | 4,000 | 5.75% | 31 | _____ | _____ |
| **15.** | 5,000 | 5.25% | 60 | _____ | _____ |

Each savings certificate pays simple interest. A penalty of 3 months' interest is assessed for early withdrawal of a certificate of 1 year or less; 6
months' interest is assessed for early withdrawal of a certificate of more

than 1 year. Calculate the interest, the complement value, the penalty assessed, and the proceeds for each investment. Round to the nearest cent.

| | Amount Invested | Rate | Length of Certificate | Time Held | Interest | Investment Value | Penalty | Proceeds |
|---|---|---|---|---|---|---|---|---|
| 16. | $6,000 | 9.375% | 1 year | 1 year | _____ | _____ | _____ | _____ |
| 17. | 1,500 | 8.00% | 6 months | 3 months | _____ | _____ | _____ | _____ |
| 18. | 7,000 | 11.625% | 2 years | 2 years | _____ | _____ | _____ | _____ |
| 19. | 850 | 8.00% | 6 months | 1 month | _____ | _____ | _____ | _____ |
| 20. | 5,000 | 7.75% | 18 months | 1 month | _____ | _____ | _____ | _____ |
| 21. | 11,500 | 12.5% | 2 years | 18 months | _____ | _____ | _____ | _____ |

## BUSINESS APPLICATIONS

Round to the nearest cent.

22. Tatman Industries deposited $5,000 in a passbook savings account for 180 days. Interest is compounded daily at 5.50%. What is the balance in the account at the end of the 180 days?

23. On April 1 Kingsland Computer Center deposited $500 in a passbook savings account. No other deposits or withdrawals were made to the account. The bank pays 5.25% interest, compounded quarterly, and adds interest on January 2, April 1, July 1, and October 1. (a) How much interest was added to the account on July 1? (b) How much interest was added to the account on October 1? (c) What is the account's balance on October 1 after the interest is added?

(a) _____

(b) _____

(c) _____

**24.** Global Publishing Company has $15,000 to invest for 90 days (a quarter of a year). Century City Bank offers a savings account that pays 5.50% compounded daily. Terrace Park Credit Union pays 6.00% interest compounded quarterly. Which financial institution will pay the greater amount of interest and how much greater is it?

_____

**25.** Milburn Plastics has $8,000 to invest for 1 year (365 days). Cleghorn National Bank offers a passbook savings account that pays 5.25% interest compounded daily. Cleghorn also has available a 1-year savings certificate that pays 6.5% simple interest. Which investment medium will pay Milburn Plastics the greatest interest and how much greater is it?

_____

**26.** RussCo, Inc., invested $12,000 in a 2-year savings certificate that pays 10.5% simple interest. There is a 6 months' interest penalty for early withdrawal. (a) What will be RussCo's proceeds if the investment is left until the due date? (b) What will be RussCo's proceeds if the investment is withdrawn after 15 months?

(a) _____

(b) _____

27. Billings Company has $10,000 to invest for 18 months. Total interest of $778.04 will be earned if the money is invested in a passbook savings account paying 5% interest compounded daily. A 24-month savings certificate, which carries a penalty of 6 months' interest for early withdrawal, pays 8.5% simple interest. How much more or less interest will be earned by investing in the savings certificate for 18 months instead of in the passbook savings account?

28. Bayrum Corporation invested $15,000 in a 6-month savings certificate that paid 9.50% simple interest. At the end of the 6 months, Bayrum invested the proceeds for 90 days in a passbook savings account that paid 5.25% interest compounded daily. What was Bayrum's investment worth at the end of this time?

29. Janice Roomer deposited $3,030 in an account that pays 7.25% interest compounded annually. Calculate (a) the balance in the account after 4 years and (b) the interest earned.

(a) _____

(b) _____

30. Calculate the interest earned after 2 years on $15,000 if deposited in (a) a certificate of deposit paying 8% simple interest or (b) a passbook savings account paying 7.5% interest compounded semiannually. (c) Which is the better investment at the end of 2 years?

(a) _____

(b) _____

(c) _____

## FINDING COMPOUND INTEREST USING A TABLE

Compound interest for various time periods, compounding periods, and interest rates can be calculated quickly using a compound interest table like the partial one shown below.

**COMPOUND INTEREST**
**(Amount of $1 Compounded Annually)**

| Period | 1% | 1½% | 2% | 3% | 4% | 5% | 6% | 7% | 8% | 9% | 10% | 11% | 12% | 14% |
|--------|------|------|------|------|------|------|------|------|------|------|------|------|------|------|
| 1 | 1.0100 | 1.0150 | 1.0200 | 1.0300 | 1.0400 | 1.0500 | 1.0600 | 1.0700 | 1.0800 | 1.0900 | 1.1000 | 1.1100 | 1.1200 | 1.1400 |
| 2 | 1.0201 | 1.0302 | 1.0404 | 1.0609 | 1.0816 | 1.1025 | 1.1236 | 1.1449 | 1.1664 | 1.1881 | 1.2100 | 1.2321 | 1.2544 | 1.2996 |
| 3 | 1.0303 | 1.0457 | 1.0612 | 1.0927 | 1.1249 | 1.1576 | 1.1910 | 1.2250 | 1.2597 | 1.2950 | 1.3310 | 1.3676 | 1.4049 | 1.4815 |
| 4 | 1.0406 | 1.0614 | 1.0824 | 1.1255 | 1.1699 | 1.2155 | 1.2625 | 1.3108 | 1.3605 | 1.4116 | 1.4641 | 1.5181 | 1.5735 | 1.6890 |
| 5 | 1.0510 | 1.0773 | 1.1041 | 1.1593 | 1.2167 | 1.2763 | 1.3382 | 1.4026 | 1.4693 | 1.5386 | 1.6105 | 1.6851 | 1.7623 | 1.9254 |
| 6 | 1.0615 | 1.0934 | 1.1262 | 1.1941 | 1.2653 | 1.3401 | 1.4185 | 1.5007 | 1.5869 | 1.6771 | 1.7716 | 1.8704 | 1.9738 | 2.1950 |
| 7 | 1.0721 | 1.1098 | 1.1487 | 1.2299 | 1.3159 | 1.4071 | 1.5036 | 1.6058 | 1.7138 | 1.8280 | 1.9487 | 2.0762 | 2.2107 | 2.5023 |
| 8 | 1.0829 | 1.1265 | 1.1717 | 1.2668 | 1.3686 | 1.4775 | 1.5938 | 1.7182 | 1.8509 | 1.9926 | 2.1436 | 2.3045 | 2.4760 | 2.8526 |
| 9 | 1.0937 | 1.1434 | 1.1951 | 1.3048 | 1.4233 | 1.5513 | 1.6895 | 1.8385 | 1.9990 | 2.1719 | 2.3579 | 2.5580 | 2.7731 | 3.2519 |
| 10 | 1.1046 | 1.1605 | 1.2190 | 1.3439 | 1.4802 | 1.6289 | 1.7908 | 1.9672 | 2.1589 | 2.3674 | 2.5937 | 2.8394 | 3.1058 | 3.7072 |
| 11 | 1.1157 | 1.1780 | 1.2434 | 1.3842 | 1.5395 | 1.7103 | 1.8983 | 2.1049 | 2.3316 | 2.5804 | 2.8531 | 3.1518 | 3.4785 | 4.2262 |
| 12 | 1.1268 | 1.1960 | 1.2682 | 1.4258 | 1.6010 | 1.7959 | 2.0122 | 2.2522 | 2.5182 | 2.8127 | 3.1384 | 3.4985 | 3.8960 | 4.8179 |
| 13 | 1.1381 | 1.2135 | 1.2936 | 1.4685 | 1.6651 | 1.8856 | 2.1329 | 2.4098 | 2.7196 | 3.0658 | 3.4523 | 3.8833 | 4.3635 | 5.4924 |
| 14 | 1.1495 | 1.2318 | 1.3195 | 1.5126 | 1.7317 | 1.9799 | 2.2609 | 2.5785 | 2.9372 | 3.3417 | 3.7975 | 4.3104 | 4.8871 | 6.2613 |
| 15 | 1.1610 | 1.2502 | 1.3459 | 1.5580 | 1.8009 | 2.0789 | 2.3966 | 2.7590 | 3.1722 | 3.6425 | 4.1772 | 4.7846 | 5.4736 | 7.1379 |
| 16 | 1.1726 | 1.2690 | 1.3728 | 1.6047 | 1.8730 | 2.1829 | 2.5404 | 2.9522 | 3.4259 | 3.9703 | 4.5950 | 5.3109 | 6.1304 | 8.1372 |
| 17 | 1.1843 | 1.2880 | 1.4002 | 1.6528 | 1.9479 | 2.2920 | 2.6928 | 3.1588 | 3.7000 | 4.3276 | 5.0545 | 5.8951 | 6.8660 | 9.2765 |
| 18 | 1.1961 | 1.3073 | 1.4282 | 1.7024 | 2.0258 | 2.4066 | 2.8543 | 3.3799 | 3.9960 | 4.7171 | 5.5599 | 6.5436 | 7.6900 | 10.575 |
| 19 | 1.2081 | 1.3270 | 1.4568 | 1.7535 | 2.1068 | 2.5270 | 3.0256 | 3.6165 | 4.3157 | 5.1417 | 6.1159 | 7.2633 | 8.6128 | 12.055 |
| 20 | 1.2202 | 1.3469 | 1.4859 | 1.8061 | 2.1911 | 2.6533 | 3.2071 | 3.8697 | 4.6610 | 5.6044 | 6.7275 | 8.0623 | 9.6463 | 13.743 |
| 21 | 1.2324 | 1.3671 | 1.5157 | 1.8603 | 2.2788 | 2.7860 | 3.3996 | 4.1406 | 5.0338 | 6.1088 | 7.4002 | 8.9492 | 10.803 | 15.667 |
| 22 | 1.2447 | 1.3876 | 1.5460 | 1.9161 | 2.3699 | 2.9253 | 3.6035 | 4.4304 | 5.4365 | 6.6586 | 8.1403 | 9.9336 | 12.100 | 17.861 |
| 23 | 1.2572 | 1.4084 | 1.5769 | 1.9736 | 2.4647 | 3.0715 | 3.8197 | 4.7405 | 5.8715 | 7.2579 | 8.9543 | 11.026 | 13.552 | 20.361 |
| 24 | 1.2697 | 1.4295 | 1.6084 | 2.0328 | 2.5633 | 3.2251 | 4.0489 | 5.0724 | 6.3412 | 7.9111 | 9.8497 | 12.239 | 15.178 | 23.212 |
| 25 | 1.2824 | 1.4510 | 1.6406 | 2.0938 | 2.6658 | 3.3864 | 4.2919 | 5.4274 | 6.8485 | 8.6231 | 10.835 | 13.585 | 17.000 | 26.461 |
| 26 | 1.2953 | 1.4727 | 1.6734 | 2.1566 | 2.7725 | 3.5557 | 4.5494 | 5.8074 | 7.3964 | 9.3992 | 11.918 | 15.080 | 19.040 | 30.166 |
| 27 | 1.3082 | 1.4948 | 1.7069 | 2.2213 | 2.8834 | 3.7335 | 4.8223 | 6.2139 | 7.9881 | 10.245 | 13.110 | 16.739 | 21.324 | 34.389 |
| 28 | 1.3213 | 1.5172 | 1.7410 | 2.2879 | 2.9987 | 3.9201 | 5.1117 | 6.6488 | 8.6271 | 11.167 | 14.421 | 18.580 | 23.883 | 39.204 |
| 29 | 1.3345 | 1.5400 | 1.7758 | 2.3566 | 3.1187 | 4.1161 | 5.4184 | 7.1143 | 9.3173 | 12.172 | 15.863 | 20.624 | 26.749 | 44.693 |
| 30 | 1.3478 | 1.5631 | 1.8114 | 2.4273 | 3.2434 | 4.3219 | 5.7435 | 7.6123 | 10.062 | 13.267 | 17.449 | 22.892 | 29.959 | 50.950 |

Before using a compound interest table, find the **number of compounding periods** and the **rate per compounding period** for the investment or loan. To find the number of compounding periods, multiply the number of years by the number of times interest is compounded per year. To find the rate per compounding period, divide the annual rate by the number of times interest is compounded per year.

When interest is compounded semiannually, there are two compounding periods per year, and the interest rate per period is one-half the annual rate.

When interest is compounded quarterly, there are four compounding periods per year, and the interest rate per period is one-fourth the annual interest rate.

When interest is compounded monthly, there are twelve interest periods per year, and the interest rate per period is one-twelfth the annual interest rate.

Years × Periods per Year = Compounding Periods

Annual Rate ÷ Periods per Year = Rate per Period

The compound interest table lists the compound amount of $1 for various periods at various rates. To find the compound amount using the compound interest table, multiply the principal by the appropriate table value.

$$\text{Principal} \times \begin{array}{c}\text{Compound Interest} \\ \text{Table Value}\end{array} = \text{Compound Amount}$$

## PRACTICE PROBLEMS

Find the compound amount and the compound interest using the compound interest table on page 329. Round to the nearest cent.

| | Principal | Investment Terms | Compound Amount | Compound Interest |
|---|---|---|---|---|
| **1.** | $ 850 | 7 years at 8% compounded quarterly | _____ | _____ |
| **2.** | $5,600 | 3 years at 6% compounded semiannually | _____ | _____ |
| **3.** | $1,750 | 2 years at 12% compounded monthly | _____ | _____ |
| **4.** | $3,090 | 12 years at 9% compounded annually | _____ | _____ |

## FINDING THE EFFECTIVE INTEREST RATE

The stated rate, or **nominal rate,** stated by banks and other financial institutions is the annual interest rate when interest is compounded annually. The true interest rate, or **effective rate,** of interest is slightly higher than the nominal rate when interest is compounded more than once a year.

To find the effective rate of interest, first find the compound interest table value for 1 year for an investment. Then, subtract 1 from the table value and convert the remaining decimal to a percent.

### EXAMPLE

What is the effective rate of interest on savings accounts at the Atlantic Credit Union if the investment terms are 6% interest compounded quarterly?

## SOLUTION

**Step 1** Find the compounding periods and rate per period for 1 year.

|  | Periods | Compounding |
|---|---|---|
| Year | × per Year = | Periods |
| 1 | × 4 = | 4 |

6% is the nominal rate →

| Annual | Periods | Rate per |
|---|---|---|
| Rate | ÷ per Year = | Period |
| 6% | ÷ 4 = | 1.5% |

**Step 2** Find the table value for 4 periods at 1.5% using the compound interest table. Subtract 1 from the table value and convert the difference to a percent.

Table Value − 1.0000 = Effective Rate

1.0614 − 1.0000 = 0.0614 or 6.14%

## PRACTICE PROBLEMS

Find the effective rate of interest for the following investments using the compound interest table on page 329. Round to the nearest hundredth of a percent.

| Investment Terms | Effective Rate |
|---|---|
| **1.** 8% annual rate compounded semiannually | _____ |
| **2.** 12% annual rate compounded quarterly | _____ |
| **3.** 18% annual rate compounded monthly | _____ |

## FINDING PRESENT VALUE

Suppose you deposit $1,600 in a savings account today at 7% interest compounded annually. In 5 years you will have a balance of $2,244.16. The principal, $1,600, is called the **present value** and the compound amount, $2,244.16, is called the **future amount.** If an individual or a business knows an amount that will be needed in the future, the present value to be invested can be found by using a present value table. The present value table on page 332 lists the present value of $1 compounded at various interest rates for various periods. To find the present value (principal) of a future amount (compound amount), multiply the table value from the present value table by the future amount.

Present Value

Future Amount × Table Value = Present Value

## PRESENT VALUE (Present Value of $1)

| Period | 1% | 1½% | 2% | 3% | 4% | 5% | 6% | 7% | 8% | 9% | 10% | 11% | 12% | 14% |
|---|---|---|---|---|---|---|---|---|---|---|---|---|---|---|
| 1 | 0.9901 | 0.9852 | 0.9804 | 0.9709 | 0.9615 | 0.9524 | 0.9434 | 0.9346 | 0.9259 | 0.9174 | 0.9091 | 0.9009 | 0.8929 | 0.8772 |
| 2 | 0.9803 | 0.9707 | 0.9612 | 0.9426 | 0.9246 | 0.9070 | 0.8900 | 0.8734 | 0.8573 | 0.8417 | 0.8264 | 0.8116 | 0.7972 | 0.7695 |
| 3 | 0.9706 | 0.9563 | 0.9423 | 0.9151 | 0.8890 | 0.9638 | 0.8396 | 0.8163 | 0.7938 | 0.7722 | 0.7513 | 0.7312 | 0.7118 | 0.6750 |
| 4 | 0.9610 | 0.9422 | 0.9238 | 0.8885 | 0.8548 | 0.8227 | 0.7921 | 0.7629 | 0.7350 | 0.7084 | 0.6830 | 0.6587 | 0.6355 | 0.5921 |
| 5 | 0.9515 | 0.9283 | 0.9057 | 0.8626 | 0.8219 | 0.7835 | 0.7473 | 0.7130 | 0.6806 | 0.6499 | 0.6209 | 0.5935 | 0.5674 | 0.5194 |
| 6 | 0.9420 | 0.9145 | 0.8880 | 0.8375 | 0.7903 | 0.7462 | 0.7050 | 0.6663 | 0.6302 | 0.5963 | 0.5645 | 0.5346 | 0.5066 | 0.4556 |
| 7 | 0.9327 | 0.9010 | 0.8706 | 0.8131 | 0.7599 | 0.7107 | 0.6651 | 0.6227 | 0.5835 | 0.5470 | 0.5132 | 0.4817 | 0.4523 | 0.3996 |
| 8 | 0.9235 | 0.8877 | 0.8535 | 0.7894 | 0.7307 | 0.6768 | 0.6274 | 0.5820 | 0.5403 | 0.5019 | 0.4665 | 0.4339 | 0.4039 | 0.3506 |
| 9 | 0.9143 | 0.8746 | 0.8368 | 0.7664 | 0.7026 | 0.6446 | 0.5919 | 0.5439 | 0.5002 | 0.4604 | 0.4241 | 0.3909 | 0.3606 | 0.3075 |
| 10 | 0.9053 | 0.8617 | 0.8203 | 0.7441 | 0.6756 | 0.6139 | 0.5584 | 0.5083 | 0.4632 | 0.4224 | 0.3855 | 0.3522 | 0.3220 | 0.2697 |
| 11 | 0.8963 | 0.8489 | 0.8043 | 0.7224 | 0.6496 | 0.5847 | 0.5268 | 0.4751 | 0.4289 | 0.3875 | 0.3505 | 0.3173 | 0.2875 | 0.2366 |
| 12 | 0.8874 | 0.8364 | 0.7885 | 0.7014 | 0.6246 | 0.5568 | 0.4970 | 0.4440 | 0.3971 | 0.3555 | 0.3186 | 0.2858 | 0.2567 | 0.2076 |
| 13 | 0.8787 | 0.8240 | 0.7730 | 0.6810 | 0.6006 | 0.5303 | 0.4688 | 0.4150 | 0.3677 | 0.3262 | 0.2897 | 0.2575 | 0.2292 | 0.1821 |
| 14 | 0.8700 | 0.8119 | 0.7579 | 0.6611 | 0.5775 | 0.5051 | 0.4423 | 0.3878 | 0.3405 | 0.2992 | 0.2633 | 0.2320 | 0.2046 | 0.1597 |
| 15 | 0.8613 | 0.7999 | 0.7430 | 0.6419 | 0.5553 | 0.4810 | 0.4173 | 0.3624 | 0.3152 | 0.2745 | 0.2394 | 0.2090 | 0.1827 | 0.1401 |
| 16 | 0.8528 | 0.7880 | 0.7284 | 0.6232 | 0.5339 | 0.4581 | 0.3936 | 0.3387 | 0.2919 | 0.2519 | 0.2176 | 0.1883 | 0.1631 | 0.1229 |
| 17 | 0.8444 | 0.7764 | 0.7142 | 0.6050 | 0.5134 | 0.4363 | 0.3714 | 0.3166 | 0.2703 | 0.2311 | 0.1978 | 0.1696 | 0.1456 | 0.1078 |
| 18 | 0.8360 | 0.7649 | 0.7002 | 0.5874 | 0.4936 | 0.4155 | 0.3503 | 0.2959 | 0.2502 | 0.2120 | 0.1799 | 0.1528 | 0.1300 | 0.0946 |
| 19 | 0.8277 | 0.7536 | 0.6864 | 0.5703 | 0.4746 | 0.3957 | 0.3305 | 0.2765 | 0.2317 | 0.1945 | 0.1635 | 0.1377 | 0.1161 | 0.0829 |
| 20 | 0.8195 | 0.7425 | 0.6730 | 0.5537 | 0.4564 | 0.3769 | 0.3118 | 0.2584 | 0.2145 | 0.1784 | 0.1486 | 0.1240 | 0.1037 | 0.0728 |
| 21 | 0.8114 | 0.7315 | 0.6598 | 0.5375 | 0.4388 | 0.3589 | 0.2942 | 0.2415 | 0.1987 | 0.1637 | 0.1351 | 0.1117 | 0.0926 | 0.0638 |
| 22 | 0.8034 | 0.7207 | 0.6468 | 0.5219 | 0.4220 | 0.3418 | 0.2775 | 0.2257 | 0.1839 | 0.1502 | 0.1228 | 0.1007 | 0.0826 | 0.0560 |
| 23 | 0.7954 | 0.7100 | 0.6342 | 0.5067 | 0.4057 | 0.3256 | 0.2618 | 0.2109 | 0.1703 | 0.1378 | 0.1117 | 0.0907 | 0.0738 | 0.0491 |
| 24 | 0.7876 | 0.6995 | 0.6271 | 0.4919 | 0.3901 | 0.3101 | 0.2470 | 0.1971 | 0.1577 | 0.1264 | 0.1015 | 0.0817 | 0.0659 | 0.0431 |
| 25 | 0.7798 | 0.6892 | 0.6095 | 0.4776 | 0.3751 | 0.2953 | 0.2330 | 0.1842 | 0.1460 | 0.1160 | 0.0923 | 0.0736 | 0.0588 | 0.0378 |
| 26 | 0.7720 | 0.6790 | 0.5976 | 0.4637 | 0.3607 | 0.2812 | 0.2198 | 0.1722 | 0.1352 | 0.1064 | 0.0839 | 0.0663 | 0.0525 | 0.0331 |
| 27 | 0.7644 | 0.6690 | 0.5859 | 0.4502 | 0.3468 | 0.2678 | 0.2074 | 0.1609 | 0.1252 | 0.0976 | 0.0763 | 0.0597 | 0.0469 | 0.0291 |
| 28 | 0.7568 | 0.6591 | 0.5744 | 0.4371 | 0.3335 | 0.2551 | 0.1956 | 0.1504 | 0.1159 | 0.0895 | 0.0693 | 0.0538 | 0.0419 | 0.0255 |
| 29 | 0.7493 | 0.6494 | 0.5631 | 0.4243 | 0.3207 | 0.2429 | 0.1846 | 0.1406 | 0.1073 | 0.0822 | 0.0630 | 0.0485 | 0.0374 | 0.0224 |
| 30 | 0.7419 | 0.6398 | 0.5521 | 0.4120 | 0.3083 | 0.2314 | 0.1741 | 0.1314 | 0.0994 | 0.0754 | 0.0573 | 0.0437 | 0.0334 | 0.0196 |
| 35 | 0.7059 | 0.5939 | 0.5000 | 0.3554 | 0.2534 | 0.1813 | 0.1301 | 0.0937 | 0.0676 | 0.0490 | 0.0356 | 0.0259 | 0.0189 | 0.0102 |
| 40 | 0.6717 | 0.5513 | 0.4529 | 0.3066 | 0.2083 | 0.1420 | 0.0972 | 0.0668 | 0.0460 | 0.0318 | 0.0221 | 0.0154 | 0.0107 | 0.0053 |

## EXAMPLE

Victoria Barnes must pay $2,800 on a note due in 5 years. How much must she invest today (present value) at 6% interest compounded quarterly to provide the needed $2,800 in 5 years?

## SOLUTION

**Step 1**  Find the compounding periods and the rate per period.

Years × Periods per Year = Compounding Periods
    5    ×      4       =        20

Rate ÷ Periods per Year =   Rate per Period
 6% ÷       4         =        1.5%

**Step 2**  Find the present value.

Future Amount × Table Value = Present Value
   $2,800    ×    0.7425    =    $2,079

*The value for 20 periods at 1.5% on the present value table is 0.7425*

**Step 3**  Use the compound interest table to check the answer.

Present Value × Table Value =     Future Value
   $2,079    ×    1.3469    = $2,800.21 ($0.21 difference
                                 due to rounding)

*The value for 20 periods at 1.5% on the compound interest table is 1.3469*

**332**  ■  LESSON 10-2

Find the present value of the following future amounts using the present value table on page 332. Use the compound interest table to verify your answers. Round to the nearest cent.

| | Future Amount | Investment Terms | Present Value |
|---|---|---|---|
| **1.** | $ 27,800 | 3 years at 12% compounded quarterly | _____ |
| **2.** | 103,300 | 30 months at 18% compounded monthly | _____ |
| **3.** | 84,100 | 4½ years at 6% compounded semiannually | _____ |

## CALCULATING PRESENT AND FUTURE INVESTMENT VALUES

Compound interest and present value tables can be used to predict present and future values of business investments, such as real estate or stocks and bonds.

### EXAMPLE

Sinclare Investments built an office complex for $450,000 in a fast-growing area of the city. The owners predict that the property in the area will increase in value at 20% per year, compounded quarterly, for the next 3 years. The company would like to sell the complex and invest the proceeds at 8% compounded semiannually for the next 3 years. At what price should Sinclare Investments sell the building today so that their investment value will be the same as the building's value in 3 years?

### SOLUTION

**Step 1**  Use the compound interest table to find the value of the business complex in 3 years if it increases in value by 20% compounded quarterly.

*The table value for 12 periods at 5% on the compound interest table is 1.7959*

Principal × Table Value = Compound Amount

$450,000 ×    1.7959    =        $808,155 ⟵ Building value in 3 years

**Step 2**  Use the present value table to find the present value of $808,155 if invested at 8% compounded semiannually for 3 years.

*The table value for 6 periods at 4% on the present value table is 0.7903.*

Future Amount × Table Value =        Present Value

$808,155    ×    0.7903    = $638,684.896 ≈ $638,684.90

Sinclare Investments should sell the building for no less than $638,684.90 since this amount invested at 8% compounded semiannually for 3 years would equal the value of the building in 3 years.

### PRACTICE PROBLEM

Henry Fleiger wants to buy Ellen Smith's share of their business when she retires in 4 years. Her share is now worth $150,000 and increases in value at 10% per year compounded semiannually. Calculate the value of Ellen's share of the business in 4 years, and the amount Henry must invest today at 8% compounded semiannually to accumulate the necessary cash to buy her share when she retires.

## LESSON 10-2    EXERCISES

Find the compound amount and compound interest using the compound interest table on page 329. Round to the nearest cent.

| | Principal | Investment Terms | Compound Amount | Compound Interest |
|---|---|---|---|---|
| 1. | $ 5,200 | 5 years at 9% compounded annually | _____ | _____ |
| 2. | 3,500 | 3 years at 8% compounded quarterly | _____ | _____ |
| 3. | 9,400 | 6 years at 6% compounded quarterly | _____ | _____ |
| 4. | 25,100 | 13 years at 14% compounded semiannually | _____ | _____ |
| 5. | 810 | $1\frac{1}{2}$ years at 12% compounded monthly | _____ | _____ |
| 6. | 7,100 | $7\frac{1}{2}$ years at 8% compounded semiannually | _____ | _____ |
| 7. | 9,160 | 12 years at 7% compounded annually | _____ | _____ |
| 8. | 10,300 | 11 years at 6% compounded semiannually | _____ | _____ |
| 9. | 2,950 | $1\frac{3}{4}$ years at 18% compounded monthly | _____ | _____ |
| 10. | 22,800 | 5 years at 12% compounded quarterly | _____ | _____ |

Find the effective rate of interest for the following investment terms. Use the compound interest table on page 329. Round to the nearest hundredth of a percent.

| Investment Terms | Effective Rate |
|---|---|
| 11. 12% annual rate compounded quarterly | _____ |
| 12. 9% annual rate compounded annually | _____ |
| 13. 10% annual rate compounded semiannually | _____ |
| 14. 6% annual rate compounded quarterly | _____ |
| 15. 12% annual rate compounded monthly | _____ |

|  | Investment Terms | Effective Rate |
|---|---|---|
| **16.** | 12% annual rate compounded semiannually | _____ |
| **17.** | 8% annual rate compounded quarterly | _____ |
| **18.** | 7% annual rate compounded annually | _____ |
| **19.** | 18% annual rate compounded monthly | _____ |
| **20.** | 8% annual rate compounded semiannually | _____ |

Find the present value and the interest earned. Use the present value table on page 332. Use the compound interest table to verify your answers. Round to the nearest cent.

|  | Future Value | Investment Terms | Present Value | Interest Earned |
|---|---|---|---|---|
| **21.** | $ 1,100 | 3 years at 4% compounded quarterly | _____ | _____ |
| **22.** | 33,200 | 4 years at 5% compounded annually | _____ | _____ |
| **23.** | 6,800 | $3\frac{1}{2}$ years at 8% compounded semiannually | _____ | _____ |
| **24.** | 18,000 | $5\frac{1}{4}$ years at 12% compounded quarterly | _____ | _____ |
| **25.** | 4,870 | 7 years at 8% compounded quarterly | _____ | _____ |
| **26.** | 24,570 | 4 years at 5% compounded annually | _____ | _____ |
| **27.** | 6,700 | 11 years at 6% compounded semiannually | _____ | _____ |
| **28.** | 20,600 | $1\frac{1}{2}$ years at 12% compounded monthly | _____ | _____ |
| **29.** | 21,800 | 5 years at 9% compounded annually | _____ | _____ |
| **30.** | 42,900 | 30 months at 18% compounded monthly | _____ | _____ |

## BUSINESS APPLICATIONS

Round amounts to the nearest cent and percents to the nearest tenth. If necessary, use the compound interest table on page 329, the daily compound interest table on page 323, or the present value table on page 332.

**31.** Nicki Forbes deposited $4,900 in a savings account that paid 8% interest compounded quarterly. She deposited an additional $4,900 in a savings certificate paying 8% simple interest. After 6 years, how much interest did she earn (a) from the savings account and (b) from the savings certificate? (c) How much more interest did she earn from the savings account?

(a) _____

(b) _____

(c) _____

**32.** Mr. and Mrs. Vouche deposited $18,000 in a certificate of deposit offering 10% interest, compounded semiannually. After 2 years the bank changed the interest rate to 12%, compounded quarterly. Calculate (a) the balance in the account after 2 years, and (b) the balance after 4 years.

(a) _____

(b) _____

**33.** The owner of Lake Machinery will be replacing the equipment in 3 years. The owner would like to have enough money saved to cover the cost of $134,000. How much money should the owner invest today at 8% interest, compounded semiannually, to cover the cost of the equipment?

_____

**34.** The First United Bank offers a 12-month certificate of deposit at 6% interest, compounded quarterly. What is the effective rate of interest on the certificate?

_____

**35.** Jake Armon loaned his sister Marie $6,000 for 1 year at 8.5% simple interest. Marie spent $400 and then invested the remaining amount at 12% interest, compounded semiannually, for 1 year. Calculate (a) the amount Marie owes Jake, and (b) the compound amount of Marie's investment. (c) Is the compound amount of Marie's investment large enough to cover the amount due Jake?

(a) _____

(b) _____

(c) _____

**36.** In 7 years Quality Realty must repay a single-payment note with a maturity value of $48,500 on a tract of land. The company would like to pay off the note now by offering the holder of the note the present value of $48,500. If current investment rates are 6% interest, compounded quarterly, what amount should Quality Realty offer the holder of the note?

_____

**37.** Doug Stoner will need $65,000 to remodel his new auto showroom in 3 years. How much should he invest today in a certificate of deposit at 8%, compounded quarterly, to accumulate the $65,000?

_____

**38.** George Dangerly loaned his brother $15,000 to begin a new business. At the end of 5 years, his brother will repay him at the principal plus 10% interest compounded semiannually. How much will George receive?

_____

**39.** The Northstar Savings and Loan is offering 12% interest compounded monthly on their 6-month certificates of deposit. What is the effective rate of interest?

_____

As you learned in the previous lessons, a business owner or individual may invest a lump sum of money over a period of time that will earn simple or compound interest. Money also may be invested over a period of time as a series of equal deposits or payments at regular intervals that will earn compound interest at a specific rate. This method of investing is called an *annuity.*

Businesses, financial institutions, and other organizations invest in annuities to raise money to pay off bond debts, notes due, or stock dividends. They also invest in annuities to provide for future needs, such as new facilities and equipment or employee retirement benefits. Individuals may purchase annuities, such as an IRA (Individual Retirement Account) or an insurance policy, from insurance companies, financial institutions, or securities brokers.

## FINDING THE AMOUNT OF AN ORDINARY ANNUITY

An *ordinary annuity* is a series of regular payments where each payment is made at the end of the payment period. The *payment period* is the length of time between payments. Payments usually are made annually, semiannually, quarterly, or monthly. The *term of the annuity* is the length of time from the beginning of the first payment period to the end of the last payment period. The *amount of the annuity* is the sum of all payments plus their accumulated interest; this amount also is called the *cash value.*

### EXAMPLE

Each year for the next 3 years, the Richmond Rice Company will deposit $500 in an ordinary annuity account paying 12% interest compounded annually. Find the amount of the annuity at the end of 3 years.

### SOLUTION

**Step 1** Find the interest earned and the balance at the end of 1 year. Since the first deposit of $500 is made at the end of the year, no interest is earned during the first year. The balance is $500.

**Step 2** Find the balance at the end of the second year.

Previous
Balance × Rate = Interest
$500   × 0.12 = $60

First Year                    Second   Second Year
Balance  + Interest + Payment =   Balance
$500    +  $60   +  $500   =   $1,060

**Step 3** Find the balance at the end of the third year.

Previous
Balance × Rate = Interest
$1,060 × 0.12 = $127.20

Second Year                    Third    Third Year
Balance   + Interest + Payment =   Balance        Cash value at end
$1,060   + $127.20 +   $500   = $1,687.20 ←── of third year

Find the yearly compound interest and the balance at the end of each year for the following ordinary annuity. Round to the nearest cent.

Annuity payment: $520        Annual rate: 9%
Payment period: Annually      Annuity term: 3 years

|  | Interest | Balance |
|---|---|---|
| **1.** Year 1 | _____ | _____ |
| **2.** Year 2 | _____ | _____ |
| **3.** Year 3 | _____ | _____ |

## FINDING THE AMOUNT OF AN ORDINARY ANNUITY USING A TABLE

The amount of an ordinary annuity can be found more easily by using the amount of an annuity table below. The table lists the value of an annuity of $1 compounded at various rates for various time periods. To find the amount of an ordinary annuity using the annuity table, multiply the payment by the appropriate table value.

Annuity         Ordinary Annuity        Amount of
Payment ×       Table Value       = Ordinary Annuity

Before using the ordinary annuity table, find the number of annuity periods and the rate per period. The compound interest earned can be found by subtracting the sum of the payments from the amount of the annuity.

Amount of Annuity − Sum of Payments = Compound Interest

## AMOUNT OF AN ORDINARY ANNUITY

| Period | 1% | 1½% | 2% | 2½% | 3% | 4% | 5% | 6% | 8% | 10% | 12% |
|---|---|---|---|---|---|---|---|---|---|---|---|
| 1 | 1.00000 | 1.00000 | 1.00000 | 1.00000 | 1.00000 | 1.00000 | 1.00000 | 1.00000 | 1.00000 | 1.00000 | 1.00000 |
| 2 | 2.01000 | 2.01500 | 2.02000 | 2.02500 | 2.03000 | 2.04000 | 2.05000 | 2.06000 | 2.08000 | 2.10000 | 2.12000 |
| 3 | 3.03010 | 3.04522 | 3.06040 | 3.07562 | 3.09090 | 3.12160 | 3.15250 | 3.18360 | 3.24640 | 3.31000 | 3.37440 |
| 4 | 4.06040 | 4.09090 | 4.12161 | 4.15252 | 4.18363 | 4.24646 | 4.31013 | 4.37462 | 4.50611 | 4.64100 | 4.77933 |
| 5 | 5.10101 | 5.15227 | 5.20404 | 5.25633 | 5.30914 | 5.41632 | 5.52563 | 5.63709 | 5.86660 | 6.10510 | 6.35285 |
| 6 | 6.15202 | 6.22955 | 6.30812 | 6.38774 | 6.46841 | 6.63298 | 6.80191 | 6.97532 | 7.33593 | 7.71561 | 8.11519 |
| 7 | 7.21354 | 7.32299 | 7.43428 | 7.54743 | 7.66246 | 7.89829 | 8.14201 | 8.39384 | 8.92280 | 9.48717 | 10.08901 |
| 8 | 8.28567 | 8.43284 | 8.58297 | 8.73612 | 8.89234 | 9.21423 | 9.54911 | 9.89747 | 10.63663 | 11.43589 | 12.29969 |
| 9 | 9.36853 | 9.55933 | 9.75463 | 9.95452 | 10.15911 | 10.58280 | 11.02656 | 11.49132 | 12.48756 | 13.57948 | 14.77566 |
| 10 | 10.46221 | 10.70272 | 10.94972 | 11.20338 | 11.46388 | 12.00611 | 12.57789 | 13.18079 | 14.48656 | 15.93742 | 17.54874 |
| 11 | 11.56683 | 11.86326 | 12.16872 | 12.48347 | 12.80780 | 13.48635 | 14.20679 | 14.97164 | 16.64549 | 18.53117 | 20.65458 |
| 12 | 12.68250 | 13.04121 | 13.41209 | 13.79555 | 14.19203 | 15.02581 | 15.91713 | 16.86994 | 18.97713 | 21.38428 | 24.13313 |
| 13 | 13.80933 | 14.23683 | 14.68033 | 15.14044 | 15.61779 | 16.62684 | 17.71298 | 18.88214 | 21.49530 | 24.52271 | 28.02911 |
| 14 | 14.94742 | 15.45038 | 15.97394 | 16.51895 | 17.08632 | 18.29191 | 19.59863 | 21.01507 | 24.21492 | 27.97498 | 32.39260 |
| 15 | 16.09690 | 16.68214 | 17.29342 | 17.93193 | 18.59891 | 20.02359 | 21.57856 | 23.27597 | 27.15211 | 31.77248 | 37.27971 |
| 16 | 17.25786 | 17.93237 | 18.63929 | 19.38022 | 20.15688 | 21.82453 | 23.65749 | 25.67253 | 30.32428 | 35.94973 | 42.75328 |
| 17 | 18.43044 | 19.20136 | 20.01207 | 20.86473 | 21.76159 | 23.69751 | 25.84037 | 28.21288 | 33.75023 | 40.54470 | 48.88367 |
| 18 | 19.61475 | 20.48938 | 21.41231 | 22.38635 | 23.41444 | 25.64541 | 28.13238 | 30.90565 | 37.45024 | 45.59917 | 55.74971 |
| 19 | 20.81090 | 21.79672 | 22.84056 | 23.94601 | 25.11687 | 27.67123 | 30.53900 | 33.75999 | 41.44626 | 51.15909 | 63.43968 |
| 20 | 22.01900 | 23.12367 | 24.29737 | 25.54466 | 26.87037 | 29.77808 | 33.06595 | 36.78559 | 45.76196 | 57.27500 | 72.05244 |
| 21 | 23.23919 | 24.47052 | 25.78332 | 27.18327 | 28.67649 | 31.96920 | 35.71925 | 39.99273 | 50.42292 | 64.00250 | 81.69874 |
| 22 | 24.47159 | 25.83758 | 27.29898 | 28.86286 | 30.53678 | 34.24797 | 38.50521 | 43.39229 | 55.45676 | 71.40275 | 92.50258 |
| 23 | 25.71630 | 27.22514 | 28.84496 | 30.58443 | 32.45288 | 36.61789 | 41.43048 | 46.99583 | 60.89330 | 79.54302 | 104.60289 |
| 24 | 26.97346 | 28.63352 | 30.42186 | 32.34904 | 34.42647 | 39.08260 | 44.50200 | 50.81558 | 66.76476 | 88.49733 | 118.15524 |
| 25 | 28.24320 | 30.06302 | 32.03030 | 34.15776 | 36.45926 | 41.64591 | 47.72710 | 54.86451 | 73.10594 | 98.34706 | 133.33387 |

Lenny Stubing purchased an ordinary annuity from an investment broker at 12% interest compounded semiannually. If his semiannual deposit is $630, what is the amount of the annuity and the interest earned at the end of 8 years? Round to the nearest cent.

**SOLUTION**

**Step 1**  Find the rate per period.

Annual      Periods      Rate per
  Rate  ÷ per Year =   Period
  12%  ÷      2     =    6%

**Step 2**  Find the amount of the annuity.

Annuity                              Amount of
Payment × Table Value = Ordinary Annuity
  $630    ×   25.67253   =    $16,173.69

*The table value for 16 periods at 6% on the amount of an annuity table is 25.67253*

**Step 3**  Find the interest earned.

Amount of        Total
  Annuity   −  Payments =   Interest
$16,173.69 − $10,080 = $6,093.69

*$630 × 16*

**PRACTICE PROBLEMS**

Find the amount of the ordinary annuity and the interest earned. Round to the nearest cent. Use the amount of an annuity table on page 340.

| | Annuity Payment | Payment Period | Term of Annuity | Rate | Amount of Annuity | Interest |
|---|---|---|---|---|---|---|
| 1. | $3,220 | Annually | 15 years | 12% | _____ | _____ |
| 2. | $1,830 | Semiannually | 8 years | 6% | _____ | _____ |

**FINDING THE AMOUNT OF AN ANNUITY DUE**

An ordinary annuity is a series of equal payments made at the *end* of the payment period. An *annuity due* is a series of equal payments made at the *beginning* of the payment period. Some common annuities due are leases, life insurance premiums, and savings account deposits.

To find the *value of an annuity due* using the amount of an annuity table on page 340, follow these steps.

1. Find the rate per compounding period.
2. Find the number of compounding periods, then add 1. Since the payment is made at the beginning of the compounding period, there is one additional compounding period.
3. Find the table value, then subtract 1. Although extra interest is earned, an extra payment is not made.
4. Multiply the payment by the revised table value.

The effect is the same as starting an ordinary annuity one period earlier.

## EXAMPLE

For the past 6 years, Carmen Luzner has invested $300 at the beginning of each quarter of the year in a savings account paying 6% interest compounded quarterly. Find the balance in her account and the interest earned. Round to the nearest cent.

## SOLUTION

**Step 1**  Find the rate per compounding period.

Annual   Payment   Rate per
Rate ÷ per Year = Period
6%   ÷    4    =  1.5%

**Step 2**  Find the number of compounding periods, then add 1.

           Payments   Compounding
Years × per Year =   Periods
6   ×   4    =    24

24 + 1 = 25 (compounding periods for an annuity due)

**Step 3**  Locate the table value on the amount of an annuity table for 25 periods at 1.5%; then, subtract 1 from the table value.

30.06302 − 1 = 29.06302 (table value for an annuity due)

**Step 4**  Find the amount of the annuity due.

Annuity       Table       Amount of
Payment × Value − 1 = Annuity Due
$300   × 29.06302 ≈  $8,718.91

**Step 5.**  Find the amount of interest earned.

Amount of       Sum of
Annuity − Payments = Interest
$300 × 24   $8,718.91 − $7,200.00 = $1,518.91

## PRACTICE PROBLEMS

Find the amount of the annuity due and the interest earned. Round to the nearest cent. Use the amount of an annuity table on page 340.

| | Annuity Payment | Payment Period | Rate | Term of Annuity | Amount of Annuity Due | Interest |
|---|---|---|---|---|---|---|
| **1.** | $ 919 | Semiannually | 16% | 9 years | _____ | _____ |
| **2.** | 2,300 | Annually | 6% | 20 years | _____ | _____ |

## PRESENT VALUE OF AN ORDINARY ANNUITY

There are various ways in which business owners and individuals invest money to meet financial obligations. For example, if a business owner must pay $6,000 a month for the next 2 years for equipment maintenance, the owner may decide to deposit a lump sum of money today in an account earning compound interest and withdraw $6,000 at the end of each month for the next 2 years to pay the expense. The lump sum deposited today is called the *present value of an annuity.* The owner must invest the present value of an annuity to be able to withdraw payments of $6,000 per month for the next 2 years.

If the amount of the annuity payment is known, the present value of the annuity can be found by using the present value of an annuity table below. To find the present value of an ordinary annuity, multiply the payment by the appropriate table value found on the present value of an annuity table:

$$\text{Annuity Payment} \times \text{Table Value} = \text{Present Value of an Annuity}$$

### PRESENT VALUE OF AN ANNUITY

| Period | 1% | 1½% | 2% | 2½% | 3% | 4% | 5% | 6% | 8% | 10% | 12% |
|--------|------|------|------|------|------|------|------|------|------|------|------|
| 1 | 0.99010 | 0.98522 | 0.98039 | 0.97561 | 0.97087 | 0.96154 | 0.95238 | 0.94340 | 0.92593 | 0.90909 | 0.89286 |
| 2 | 1.97040 | 1.95588 | 1.94156 | 1.92742 | 1.91347 | 1.88609 | 1.85941 | 1.83339 | 1.78326 | 1.73554 | 1.69005 |
| 3 | 2.94099 | 2.91220 | 2.88388 | 2.85602 | 2.82861 | 2.77509 | 2.72325 | 2.67301 | 2.57710 | 2.48685 | 2.40183 |
| 4 | 3.90197 | 3.85438 | 3.80773 | 3.76197 | 3.71710 | 3.62990 | 3.54595 | 3.46511 | 3.31213 | 3.16987 | 3.03735 |
| 5 | 4.85343 | 4.78264 | 4.71346 | 4.64583 | 4.57971 | 4.45182 | 4.32948 | 4.21236 | 3.99271 | 3.79079 | 3.60478 |
| 6 | 5.79548 | 5.69719 | 5.60143 | 5.50813 | 5.41719 | 5.24214 | 5.07569 | 4.91732 | 4.62288 | 4.35526 | 4.11141 |
| 7 | 6.72819 | 6.59821 | 6.47199 | 6.34939 | 6.23028 | 6.00205 | 5.78637 | 5.58238 | 5.20637 | 4.86842 | 4.56376 |
| 8 | 7.65168 | 7.48593 | 7.32548 | 7.17014 | 7.01969 | 6.73274 | 6.46321 | 6.20979 | 5.74664 | 5.33493 | 4.96764 |
| 9 | 8.56602 | 8.36052 | 8.16224 | 7.97087 | 7.78611 | 7.43533 | 7.10782 | 6.80169 | 6.24689 | 5.75902 | 5.32825 |
| 10 | 9.47130 | 9.22218 | 8.98259 | 8.75206 | 8.53020 | 8.11090 | 7.72173 | 7.36009 | 6.71008 | 6.14457 | 5.65022 |
| 11 | 10.36763 | 10.07112 | 9.78685 | 9.51421 | 9.25262 | 8.76048 | 8.30641 | 7.88687 | 7.13896 | 6.49506 | 5.93770 |
| 12 | 11.25508 | 10.90751 | 10.57534 | 10.25776 | 9.95400 | 9.38507 | 8.86325 | 8.38384 | 7.53608 | 6.81369 | 6.19437 |
| 13 | 12.13374 | 11.73153 | 11.34837 | 10.98318 | 10.63496 | 9.98565 | 9.39357 | 8.85268 | 7.90378 | 7.10336 | 6.42355 |
| 14 | 13.00370 | 12.54338 | 12.10625 | 11.69091 | 11.29607 | 10.56312 | 9.89864 | 9.29498 | 8.24424 | 7.36669 | 6.62817 |
| 15 | 13.86505 | 13.34323 | 12.84926 | 12.38138 | 11.93794 | 11.11839 | 10.39766 | 9.71225 | 8.55948 | 7.60608 | 6.81086 |
| 16 | 14.71787 | 14.13126 | 13.57771 | 13.05500 | 12.56110 | 11.65230 | 10.83777 | 10.10590 | 8.85137 | 7.82371 | 6.97399 |
| 17 | 15.56225 | 14.90765 | 14.29187 | 13.71220 | 13.16612 | 12.16567 | 11.27407 | 10.47726 | 9.12164 | 8.02155 | 7.11963 |
| 18 | 16.39827 | 15.67256 | 14.99203 | 14.35336 | 13.75351 | 12.65930 | 11.68959 | 10.82760 | 9.37189 | 8.20141 | 7.24967 |
| 19 | 17.22601 | 16.42617 | 15.67846 | 14.97889 | 14.32380 | 13.13394 | 12.08532 | 11.15812 | 9.60360 | 8.36492 | 7.36578 |
| 20 | 18.04555 | 17.16864 | 16.35143 | 15.58916 | 14.87747 | 13.59033 | 12.46221 | 11.46992 | 9.81815 | 8.51356 | 7.46944 |
| 21 | 18.85698 | 17.90014 | 17.01121 | 16.18455 | 15.41502 | 14.02916 | 12.82115 | 11.76408 | 10.01680 | 8.64869 | 7.56200 |
| 22 | 19.66038 | 18.62082 | 17.65805 | 16.76541 | 15.93692 | 14.45112 | 13.16300 | 12.04158 | 10.20074 | 8.77154 | 7.64465 |
| 23 | 20.45582 | 19.33086 | 18.29220 | 17.33211 | 16.44361 | 14.85684 | 13.48857 | 12.30338 | 10.37106 | 8.88322 | 7.71843 |
| 24 | 21.24339 | 20.03041 | 18.91393 | 17.88499 | 16.93554 | 15.24696 | 13.79864 | 12.55036 | 10.52876 | 8.98474 | 7.78432 |
| 25 | 22.02316 | 20.71961 | 19.52346 | 18.42438 | 17.41315 | 15.62208 | 14.09394 | 12.78336 | 10.67478 | 9.07704 | 7.84314 |

### EXAMPLE

Karl Klingman wants to deposit an amount of money today that will yield an annuity paying $2,300 semiannually for the next 9 years. All withdrawals will be made at the end of the period. If money can be invested now at 10% interest compounded semiannually, find the present value of the ordinary annuity. Round to the nearest cent.

### SOLUTION

Find the present value of the annuity.

18 compounded periods
(9 years × 2 periods)
5% per period (10% ÷ 2)

| Annuity Payment | × | Table Value | = | Present Value of an Annuity |
|---|---|---|---|---|
| $2,300 | × | 11.68959 | = | $26,886.06 |

The table value for 18 periods at 5% on the present value of an annuity table is 11.68959

If Mr. Klingman invests $26,886.06 in an account today, he will be able to withdraw $2,300 from the account every 6 months for the next 9 years.

As you have learned, a future amount can be accumulated by depositing a lump sum of money today or by making payments over a period of time into an annuity account. The present value of an annuity is the lump sum that can be deposited today that will grow to the same total as would the periodic payments of an annuity.

Find the present value of the following ordinary annuities. Round to the nearest cent. Use the present value of an annuity table on page 343.

| | Expected Annuity Payments | Payment Period | Term of Annuity | Rate | Present Value of Annuity |
|---|---|---|---|---|---|
| 1. | $4,900 | Quarterly | 4 years | 12% | _____ |
| 2. | $1,890 | Monthly | 2 years | 18% | _____ |
| 3. | $7,900 | Semiannually | 6 years | 10% | _____ |

## EQUIVALENT CASH VALUE

Because the value of business investments increases or decreases over a period of time, often business owners must calculate the present value of investments before making investment decisions. The present value is called the *equivalent cash value* and can be found using present value tables or present value of an annuity tables.

### EXAMPLE 1

Mary Ann Johnson wishes to sell her business. Company A offered her $265,000 in cash. Company B offered her an $84,000 down payment with payments of $13,500 at the end of each quarter for 4 years. If money can be invested at 12% compounded quarterly, which offer should Mary Ann accept?

### SOLUTION

**Step 1** Find the present value of the $13,500 quarterly ordinary annuity payments.

Annuity    Table    Present Value
Payment × Value = of an Annuity
$13,500 × 12.5611 = $169,574.85

The table value for 16 periods (4 years × 4) at 3% (12% ÷ 4) on the present value of an annuity table is 12.56110

**Step 2** Find the equivalent cash value of Company B's offer.

              Present Value    Equivalent
Down Payment + of an Annuity = Cash Value
   $84,000    +   $169,574.85 = $253,574.85

**Step 3** Compare offers from Company A and Company B.

Mary Ann should accept the cash offer of $265,000 from Company A since the equivalent cash value of Company B's offer is less.

### EXAMPLE 2

In 7 years George Cunningham plans to retire and move to Florida. Find the lump sum that he should deposit today to accumulate enough money to receive annual annuity payments of $30,000 for 10 years when he retires. Assume all investments yield 12% compounded annually.

## SOLUTION

**Step 1**  Find the present value of the ordinary annuity payment of $30,000.

The table value for 10 periods at 12% on the present value of an annuity table is 5.65022

| Annuity Payment | × | Table Value | = | Present Value of an Annuity |
|---|---|---|---|---|
| $30,000 | × | 5.65022 | = | $169,506.60 |

A deposit of $169,506.60 will yield annual payments of $30,000 for 10 years

**Step 2**  Since the annuity payments do not begin for 7 years, find the present value of $169,506.60 using the present value table on page 343.

The table value for 7 periods at 12% on the present value table is 0.4523

| Future Amount | × | Table Value | = | Present Value |
|---|---|---|---|---|
| $169,506.60 | × | 0.4523 | = | $76,667.835 ≈ $76,667.84 |

A deposit of $76,667.84 will mature to $169,506.60 in 7 years. At that time the money can be deposited into an annuity account that pays $30,000 a year.

## PRACTICE PROBLEM

Daniel LaRosa wants to sell his business to his daughter and work as an employee until he retires in 6 years. How much of the proceeds from the sale of his business must he deposit today to accumulate enough money to receive annual annuity payments of $20,000 for 12 years upon his retirement? Assume money can earn 12% interest compounded semiannually.

## FINDING THE SINKING FUND PAYMENT

An additional way for a business owner or individual to invest funds for future use is to establish a ***sinking fund.*** For example, the owners of a company may know that they will need $750,000 to purchase new equipment in 5 years. They may decide to open a sinking fund account into which periodic payments are made. These periodic payments plus the compound interest earned on the payments produce the necessary amount of $750,000 needed in 5 years. A sinking fund is usually set up by a company to meet future financial needs, such as building new facilities, replacing worn equipment, or retiring a bond debt.

A sinking fund usually requires payments at the end of each payment period into an ordinary annuity. For an *ordinary annuity,* the amount of the payment is known and the future amount of the annuity must be found. For a *sinking fund,* the future amount of the annuity is known and the amount of the payment must be found.

The sinking fund payment can be found by using the sinking fund table on page 346. Multiply the appropriate table value by the future amount needed to find the amount of the sinking fund payment.

| Future Amount | × | Sinking Fund Table Value | = | Sinking Fund Payment |
|---|---|---|---|---|

---

ANSWER FOR PRACTICE PROBLEM — $124,750.57

# SINKING FUND

| Period | 1% | 1½% | 2% | 2½% | 3% | 4% | 5% | 6% | 8% | 10% | 12% |
|--------|------|------|------|------|------|------|------|------|------|------|------|
| 1 | 1.00000 | 1.00000 | 1.00000 | 1.00000 | 1.00000 | 1.00000 | 1.00000 | 1.00000 | 1.00000 | 1.00000 | 1.00000 |
| 2 | 0.49751 | 0.49628 | 0.49505 | 0.49383 | 0.49261 | 0.49020 | 0.48780 | 0.48544 | 0.48077 | 0.47619 | 0.47170 |
| 3 | 0.33002 | 0.32838 | 0.32675 | 0.32514 | 0.32353 | 0.32035 | 0.31721 | 0.31411 | 0.30803 | 0.30211 | 0.29635 |
| 4 | 0.24628 | 0.24444 | 0.24262 | 0.24082 | 0.23903 | 0.23549 | 0.23201 | 0.22859 | 0.22192 | 0.21547 | 0.20923 |
| 5 | 0.19604 | 0.19409 | 0.19216 | 0.19025 | 0.18835 | 0.18463 | 0.18097 | 0.17740 | 0.17046 | 0.16380 | 0.15741 |
| 6 | 0.16255 | 0.16053 | 0.15853 | 0.15655 | 0.15460 | 0.15076 | 0.14702 | 0.14336 | 0.13632 | 0.12961 | 0.12323 |
| 7 | 0.13863 | 0.13656 | 0.13451 | 0.13250 | 0.13051 | 0.12661 | 0.12282 | 0.11914 | 0.11207 | 0.10541 | 0.09912 |
| 8 | 0.12069 | 0.11858 | 0.11651 | 0.11447 | 0.11246 | 0.10853 | 0.10472 | 0.10104 | 0.09401 | 0.08744 | 0.08130 |
| 9 | 0.10674 | 0.10461 | 0.10252 | 0.10046 | 0.09843 | 0.09449 | 0.09069 | 0.08702 | 0.08008 | 0.07364 | 0.06768 |
| 10 | 0.09558 | 0.09343 | 0.09133 | 0.08926 | 0.08723 | 0.08329 | 0.07950 | 0.07587 | 0.06903 | 0.06275 | 0.05698 |
| 11 | 0.08645 | 0.08429 | 0.08218 | 0.08011 | 0.07080 | 0.07415 | 0.07039 | 0.06679 | 0.06008 | 0.05396 | 0.04842 |
| 12 | 0.07885 | 0.07668 | 0.07456 | 0.07249 | 0.07046 | 0.06655 | 0.06283 | 0.05928 | 0.05270 | 0.04676 | 0.04144 |
| 13 | 0.07241 | 0.07024 | 0.06812 | 0.06605 | 0.06403 | 0.06014 | 0.05646 | 0.05296 | 0.04652 | 0.04078 | 0.03568 |
| 14 | 0.06690 | 0.06472 | 0.06260 | 0.06054 | 0.05853 | 0.05467 | 0.05102 | 0.04758 | 0.04130 | 0.03575 | 0.03087 |
| 15 | 0.06212 | 0.05994 | 0.05783 | 0.05577 | 0.05377 | 0.04994 | 0.04634 | 0.04296 | 0.03683 | 0.03147 | 0.02682 |
| 16 | 0.05794 | 0.05577 | 0.05365 | 0.05160 | 0.04961 | 0.04582 | 0.04227 | 0.03895 | 0.03298 | 0.02782 | 0.02339 |
| 17 | 0.05426 | 0.05208 | 0.04997 | 0.04793 | 0.04595 | 0.04220 | 0.03870 | 0.03544 | 0.02963 | 0.02466 | 0.02046 |
| 18 | 0.05098 | 0.04881 | 0.04670 | 0.04467 | 0.04271 | 0.03899 | 0.03555 | 0.03236 | 0.02670 | 0.02193 | 0.01794 |
| 19 | 0.04805 | 0.04588 | 0.04378 | 0.04176 | 0.03981 | 0.03614 | 0.03275 | 0.02962 | 0.02413 | 0.01955 | 0.01576 |
| 20 | 0.04542 | 0.04325 | 0.04116 | 0.03915 | 0.03722 | 0.03358 | 0.03024 | 0.02718 | 0.02185 | 0.01746 | 0.01388 |
| 21 | 0.04303 | 0.04087 | 0.03878 | 0.03679 | 0.03487 | 0.03128 | 0.02800 | 0.02500 | 0.01983 | 0.01562 | 0.01224 |
| 22 | 0.04086 | 0.03870 | 0.03663 | 0.03465 | 0.03275 | 0.02920 | 0.02597 | 0.02305 | 0.01803 | 0.01401 | 0.01081 |
| 23 | 0.03889 | 0.03673 | 0.03467 | 0.03270 | 0.03081 | 0.02731 | 0.02414 | 0.02128 | 0.01642 | 0.01257 | 0.00956 |
| 24 | 0.03707 | 0.03492 | 0.03287 | 0.03091 | 0.02905 | 0.02559 | 0.02247 | 0.01968 | 0.01498 | 0.01130 | 0.00846 |
| 25 | 0.03541 | 0.03326 | 0.03122 | 0.02928 | 0.02743 | 0.02401 | 0.02095 | 0.01823 | 0.01368 | 0.01017 | 0.00750 |

## EXAMPLE

The Hampton County School District has issued $285,000 worth of bonds that must be paid off in 4 years. The school board decides to set up a sinking fund to accumulate the $285,000. The rate of interest on the sinking fund is 10% compounded annually. Find the amount of the annual payment. Use the sinking fund table on this page.

## SOLUTION

Find the amount of the sinking fund payment.

Future    Table    Sinking Fund
Amount × Value = Payment

$285,000 × 0.21547 = $61,408.95

The table value for 4 periods at 10% on the sinking fund table is 0.21547

The school district will accumulate $285,000 in 4 years by making annual payments of $61,408.95 in a sinking fund.

## PRACTICE PROBLEMS

Find the amount of each sinking fund payment. Round to the nearest cent. Use the sinking fund table above.

| | Future Amount | Payment Period | Term of Sinking Fund | Rate | Sinking Fund Payment |
|---|---|---|---|---|---|
| 1. | $ 42,000 | Annually | 11 years | 12% | _____ |
| 2. | 835,000 | Semiannually | 6 years | 10% | _____ |
| 3. | 97,500 | Quarterly | 3 years | 8% | _____ |
| 4. | 8,250 | Annually | 14 years | 6% | _____ |

## LESSON 10-3    EXERCISES

Find the yearly compound interest and the balance at the end of each year for the following ordinary annuities. Round answers to the nearest cent.

1. Annuity payment: $1,700
   Payment period: Annually
   Annual rate: 10.5%
   Annuity term: 3 years

   Year 1: Interest $_____        Balance $_____

   Year 2: Interest $_____        Balance $_____

   Year 3: Interest $_____        Balance $_____

2. Annuity payment: $2,500
   Payment period: Annually
   Annual rate: 9%
   Annuity term: 4 years

   Year 1: Interest $_____        Balance $_____

   Year 2: Interest $_____        Balance $_____

   Year 3: Interest $_____        Balance $_____

   Year 4: Interest $_____        Balance $_____

3. Annuity payment: $956
   Payment period: Annually
   Annual rate: 7.5%
   Annuity term: 5 years

   Year 1: Interest $_____        Balance $_____

   Year 2: Interest $_____        Balance $_____

   Year 3: Interest $_____        Balance $_____

   Year 4: Interest $_____        Balance $_____

   Year 5: Interest $_____        Balance $_____

Find the amount of the ordinary annuity and the interest earned. Round to the nearest cent. Use the amount of an annuity table on page 340.

| | Annuity Payment | Payment Period | Rate | Term of Annuity | Amount of Annuity | Interest Earned |
|---|---|---|---|---|---|---|
| 4. | $ 744 | Semiannually | 12% | 10 years | _____ | _____ |
| 5. | 1,350 | Annually | 10% | 25 years | _____ | _____ |
| 6. | 310 | Monthly | 18% | 2 years | _____ | _____ |

Find the amount of the annuity due and the interest earned. Round to the nearest cent. Use the amount of an annuity table on page 340.

| | Annuity Payment | Payment Period | Rate | Term of Annuity | Amount of Annuity Due | Interest Earned |
|---|---|---|---|---|---|---|
| 7. | $1,700 | Quarterly | 12% | 5 years | _____ | _____ |
| 8. | 925 | Annually | 5% | 12 years | _____ | _____ |
| 9. | 540 | Monthly | 12% | 2 years | _____ | _____ |

Find the amount of the ordinary annuity and the amount of the annuity due. Find the difference in the interest earned. Round amounts to the nearest cent. Use the amount of an annuity table on page 340.

| | Annuity Payment | Payment Period | Term of Annuity | Rate of Investment | Amount of Ordinary Annuity | Amount of Annuity Due | Difference in Interest Earned |
|---|---|---|---|---|---|---|---|
| 10. | $ 895 | Annually | 17 years | 10% | _____ | _____ | _____ |
| 11. | 340 | Quarterly | 6 years | 16% | _____ | _____ | _____ |
| 12. | 4,000 | Semiannually | 9 years | 12% | _____ | _____ | _____ |

Find the present value of each ordinary annuity. Round answers to the nearest cent. Use the present value of an annuity table on page 343.

| | Amount of Expected Payments | Payment Period | Rate | Term of Annuity | Present Value of Annuity |
|---|---|---|---|---|---|
| 13. | $1,250 | Monthly | 18% | 2 years | _____ |
| 14. | 2,800 | Annually | 8% | 22 years | _____ |
| 15. | 760 | Quarterly | 16% | 4 years | _____ |

Find the amount of the sinking fund payment. Round to the nearest cent. Use the sinking fund table on page 346.

| | Future Amount | Payment Period | Term of Sinking Fund | Rate of Investment | Sinking Fund Payment |
|---|---|---|---|---|---|
| 16. | $63,200 | Annually | 15 years | 5% | _____ |
| 17. | 10,700 | Quarterly | 6 years | 6% | _____ |
| 18. | 47,800 | Semiannually | 11 years | 6% | _____ |

## BUSINESS APPLICATIONS

Round amounts to the nearest cent.

19. When Oliver Bond graduated from college 10 years ago, he began depositing $415 every 6 months in an ordinary annuity account paying 10% interest, compounded semiannually. Calculate (a) the amount of the ordinary annuity and (b) the amount of interest earned.

(a) _____

(b) _____

20. At the end of each year for 5 years the Allied Metals Company made annual deposits of $10,000 in a savings account yielding 8% interest compounded annually. For the next 4 years the company made no deposits into the account. Calculate (a) the amount of the ordinary annuity and (b) the total interest earned at the end of 9 years.

(a) _____

(b) _____

21. Starting this month, Elwood Myers will receive monthly retirement checks of $2,100 for the next 20 years. His employer can invest money at 12% compounded annually. Determine the amount the employer must invest today to be able to make the monthly payments to Elwood.

_____

22. Rachel Young plans to invest $1,900 at the end of each year for the next 6 years in an annuity to save for her daughter's college education. If the annuity pays 6% interest compounded quarterly, how much will she have saved for her daughter?

_____

23. The United Delivery Service will pay $15,200 at the end of each year for the next 4 years to pay the balance due on the purchase of their trucks. Money can be invested at 6% compounded annually. What lump sum should the company deposit today to made the annual annuity payments?

_____

**24.** Finwick Brothers, Inc., took out a note for $230,000 including interest due in 5 years. The current rate of interest on investments is 12% compounded quarterly. The company set up a sinking fund to pay the note. Calculate (a) the amount of the quarterly deposit and (b) the amount of interest earned.

(a) _____

(b) _____

**25.** Alicia Hamburg can invest money at 10% interest compounded semiannually. How much more interest will Alicia earn if she makes semiannual payments of $620 for 9 years in an annuity due account rather than an ordinary annuity account?

_____

**26.** The owner of the Cash-N-Carry supermarket makes deposits of $3,000 at the end of each month in a savings account paying 12% interest compounded monthly. She discontinues making payments after 2 years. What is the balance in her account at the end of 4 years?

_____

**27.** The Allied Motors Corporation has a bond debt of $219,800 that is due to mature in 7 years. The corporate treasurer has recommended that a sinking fund be set up to pay off the bond. The funds can be invested at 8% interest compounded semiannually. What will be the amount of the semiannual payment to the sinking fund?

_____

**28.** Mary Anne Lykes has made quarterly payments of $542 in a savings account for the past 4 years at 6% interest compounded quarterly. Calculate (a) the cash value of the annuity due and (b) the interest earned.

(a)

(b) _____

**29.** Dr. Martin Stoner is a physician who plans to retire in 9 years. What lump sum should he invest today that will provide him with $35,000 at the end of each year for 10 years after he retires? Assume he can invest money at 10% interest compounded annually.

_____

**30.** Raymond Goetz deposits $2,000 at the end of each year in an IRA paying 6% interest compounded annually. (a) Calculate the cash value of the annuity in 20 years. (b) Determine what lump sum of money he could deposit in an account paying 6% interest compounded annually to accumulate the same amount in 20 years.

(a)

(b) _____

**31.** Robert Johnson is ordered by the court to pay child support of $4,800 at the end of each 6-month period for his 16-year-old daughter. Calculate the amount he must deposit today at 8% interest compounded semiannually to make the payments for 2 years until her eighteenth birthday.

_____

**32.** Glenda Gleason has been offered $125,000 cash for her business or $20,000 down and $18,000 at the end of each year for 8 years. If money can be invested at 10% compounded annually, which is the better offer?

_____

**33.** According to their rental contract, Lincoln Builders will pay $9,000 at the end of each month for equipment rentals for the next 2 years. If the owners can invest money at 18% compounded monthly, calculate the lump sum they can deposit today to make the payments for the length of the contract.

_____

**34.** California Plastics is selling its manufacturing plant. Company A has offered $75,000 down and $50,000 a year for the next 5 years. Company B has offered $29,000 down and $60,000 a year for the next 5 years. If money can be invested at 6% compounded annually, calculate the (a) equivalent cash value of each offer. (b) Determine which offer California Plastics should accept.

(a) _____

(b) _____

## CHAPTER 10 REVIEW

### BUSINESS APPLICATIONS

If necessary, use the daily compound interest table on page 323, the compound interest table on page 329, the present value table on page 332, the amount of an annuity table on page 340, the present value of an annuity table on page 343, or the sinking fund table on page 346. Round amounts to the nearest cent.

1. On November 15 Maxine Levitte deposited $1,380 in a savings account that earns 6.25% interest compounded daily. Thirty days later, Maxine deposited an additional $475. Sixty days after her last deposit, Maxine withdrew $815. Find the balance in her account 90 days after her first deposit of November 15.

_____

2. Find (a) the compound amount and (b) the compound interest for $14,700 invested for 5 years at 16% compounded quarterly.

(a) _____

(b) _____

3. Find (a) the compound amount and (b) the compound interest for $106,800 invested for 7 years at 14% compounded semiannually.

(a) _____

(b) _____

4. Find (a) the present value and (b) the compound interest for $63,500 invested for 3 years at 8% compounded quarterly.

(a) _____

(b) _____

**5.** Find (a) the present value and (b) the compound interest for $129,300 invested for 15 years at 10% compounded semiannually.

(a) _____

(b) _____

**6.** Find (a) the present value and (b) the compound interest for $57,200 invested for 20 years at 5% interest compounded annually.

(a) _____

(b) _____

**7.** Find (a) the amount of the ordinary annuity and (b) the compound interest for quarterly payments of $739 for 6 years at 12% compounded quarterly.

(a) _____

(b) _____

**8.** Find (a) the amount of the ordinary annuity and (b) the compound interest for semiannual payments of $3,450 for 9 years at 10% compounded semiannually.

(a) _____

(b) _____

**9.** Find (a) the amount of the annuity due and (b) the compound interest for quarterly payments of $2,600 for 5 years at 8% compounded quarterly.

(a) _____

(b) _____

**10.** Find (a) the amount of the annuity due and (b) the compound interest for annual payments of $8,350 for 15 years at 6% compounded annually.

(a) _____

(b) _____

**11.** Find the present value of an ordinary annuity with quarterly payments of $13,100 for 3 years at 16% compounded quarterly.

_____

**12.** Find the present value of an ordinary annuity with semiannual payments of $6,810 for 11 years at 6% compounded semiannually.

_____

**13.** Find the annual sinking fund payment for a future amount of $173,200 if investment terms are 8% compounded annually for 16 years.

_____

**14.** Find the quarterly sinking fund payment for a future amount of $37,800 if investment terms are 8% compounded quarterly for 5 years.

_____

**15.** The Maxwell Insurance Company plans to build new corporate offices in $6\frac{1}{2}$ years. They will need a total of $567,000 for the construction. If money can be invested at 8% compounded semiannually, calculate the lump sum they should invest today to have the needed construction money.

_____

**16.** Mrs. Kelly plans to retire in 12 years. If she deposits $5,000 at the end of each year in an investment account at 12% interest compounded annually, will she accumulate enough money in 12 years to receive annual annuity payments of $20,000 for 10 years?

_____

**17.** Tom Melon invested $9,800 in his brother's business. After 5 years his investment has increased at a rate of 14% compounded semiannually. What is Tom's investment now worth?

_____

**18.** At the beginning of each year, the Hanover Insurance Company collects $1.3 million from insurance premiums. If they invest this amount at 8% compounded quarterly, what is the amount of the annuity due at the end of 5 years?

_____

**19.** Stouville Company has $12,000 to invest for 1 year (365 days). Meredith National Bank offers a passbook savings account that pays 5.50% interest compounded daily. Meredith also has available a 1-year savings certificate that pays 6.00% simple interest. Which investment will pay Stouville Company the greater interest and how much greater is it?

_____

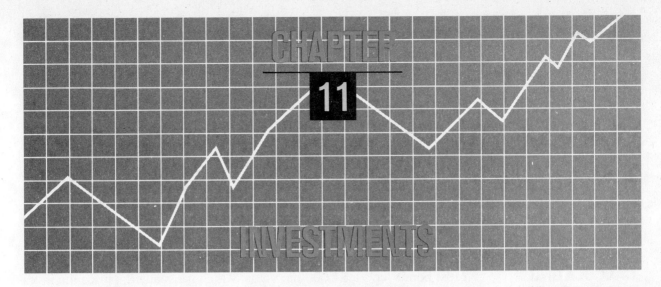

# CHAPTER
## 11

# INVESTMENTS

## LEARNING OBJECTIVES

1. Calculate the price on bonds based on market value.
2. Determine accrued interest on bond purchases and sales.
3. Calculate current yield and yield to maturity on bonds.
4. Calculate buy and sell transactions on common and preferred stocks.
5. Calculate dividends on stock.
6. Calculate buy and sell transactions on mutual funds.
7. Determine rental income.

An investment is money placed in a savings account or used to buy savings certificates, bonds, stocks, mutual funds, or real estate, with the hope that it will earn additional income and increase in value.

Often, businesses have excess funds that are not needed for current, day-to-day, operations. These funds can be invested for safekeeping and to earn interest or a return until they are needed. Businesses also invest money to accumulate funds for major purchases of equipment, buildings, or land in the future or to build a reserve to fall back on during periods of decreased business activity that might occur in the future.

With all investments, the potential for return and the degree of risk are directly related. Often if the possibility exists for a high return, a higher degree of risk may be involved. Conversely, investment opportunities that offer a lower rate of return often involve a lesser risk factor.

This chapter will discuss investments, such as bonds, stocks, mutual funds, and real estate. It will present methods used to determine investment costs, investment earnings, and gain or loss upon sale of investments.

From an investor's viewpoint, a **bond** is an investment that pays interest. Corporations and states and local governments, such as cities and counties, borrow money by selling bonds to investors. The investor is called the **bondholder.**

The corporation or government agrees to pay interest on the bond and repay the amount borrowed, called the **par value** or **face value,** at a specified time in the future. The repayment date, called the **maturity date,** may be 10, 20, 30, 40, or more years from the date it is issued.

**CALCULATING BOND PRICES**

Most bonds have a $1,000 par value, which is usually the price the investor pays when buying the bond from the issuing corporation or government. Often, however, investors do not keep the bond until the maturity date but sell it to another investor.

A bond's price may fluctuate above or below the $1,000 par value during its lifetime. The amount the investor actually pays is called the **market value.** If the market value is more than the par value, the bond is purchased at a **premium.** If purchased for less than the par value, the bond is purchased at a **discount.**

The bond market price quotation is stated as a whole or mixed number, such as 103 or $85\frac{1}{4}$, which represents the percent of par value. To find the bond's market value, convert the price quotation (percent) to a decimal and multiply it by the bond's $1,000 par value.

### EXAMPLE

Redinius Supply Company is considering the purchase of two $1,000 par value bonds with price quotations of 102 and $86\frac{3}{4}$. What is the market value of each?

### SOLUTION

Price Quotation
(Converted to a decimal number) × Par Value = Market Value

| 1.02 | × $1,000 | = $1,020.00 ← Premium |
| 0.8675 | × $1,000 | = $ 867.50 ← Discount |

### PRACTICE PROBLEMS

Calculate the market value of each of the following bonds. Identify whether each bond is being purchased at a premium or a discount. Par value is $1,000.

| | Price Quotation | Market Value | Premium or Discount |
|---|---|---|---|
| **1.** | $102\frac{1}{4}$ | _____ | _____ |
| **2.** | $92\frac{1}{8}$ | _____ | _____ |

|     | Price Quotation | Market Value | Premium or Discount |
| --- | --- | --- | --- |
| **3.** | $64\frac{3}{8}$ | _____ | _____ |
| **4.** | $105\frac{3}{4}$ | _____ | _____ |
| **5.** | $98\frac{7}{8}$ | _____ | _____ |

## READING BOND QUOTATIONS

Bonds are traded on organized marketplaces called *bond exchanges.* Market information like that shown below is available in newspapers and financial publications.

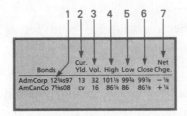

1. Adams Corporation is shown in abbreviated form as AdmCorp. The $12\frac{3}{4}$s means that the bond pays $12\frac{3}{4}\%$ interest on the par value of $1,000. The 97 stands for the bond's maturity date, 1997. In the quotation for AmCanCo, the maturity date is 2008.
2. The 13 under *Cur. Yld.* for AdmCorp means that an investor will receive 13% return on the bond if it is purchased at the day's final price quotation. The "cv" for AmCanCo means that this is a **convertible bond**—the bond can be converted into a certain number of shares of the company's common stock.
3. The 32 under *Vol.* for AdmCorp indicates the number of bonds sold on this day.
4. The $101\frac{1}{8}$ under *High* means the **highest market price** for the day was 101.125% of the par value or $1,011.25.
5. The $99\frac{3}{4}$ under *Low* means the **lowest market price** for the day was 99.75% of the par value or $997.50.
6. The $99\frac{7}{8}$ under *Close* means that the **closing market price** was 99.875% of the par value or $998.75.
7. The $-\frac{1}{8}$ under *Net Chg.* shows that the last sale was $\frac{1}{8}\%$ or $1.25 lower than yesterday's last sale.

ANSWERS FOR PRACTICE PROBLEMS

**Starting on page 358:**

|     | Market Value | Premium or Discount |
| --- | --- | --- |
| (1) | $1,022.50 | premium |
| (2) | 921.25 | discount |
| (3) | 643.75 | discount |
| (4) | 1,057.50 | premium |
| (5) | 988.75 | discount |

## PRACTICE PROBLEMS

Use the following bond quotations to answer the questions below.

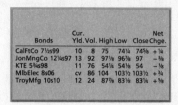

| Bonds | Cur. Yld. | Vol. | High | Low | Close | Net Chge. |
|---|---|---|---|---|---|---|
| CalFtCo 7½s99 | 10 | 8 | 75 | 74¼ | 74⅝ | +¼ |
| JonMngCo 12¼s97 | 13 | 92 | 97⅛ | 96⅞ | 97 | −⅜ |
| KTE 5¾s98 | 11 | 76 | 54¼ | 54⅛ | 54 | −⅛ |
| MlbElec 8s06 | cv | 86 | 104 | 103½ | 103½ | +¾ |
| TroyMfg 10s10 | 12 | 24 | 87⅞ | 83⅛ | 83¼ | +⅝ |

**1.** What was the highest price paid for a CalFtCo bond on this day? __________

**2.** What was the lowest price paid for a KTE bond on this day? __________

**3.** What is the annual rate of interest on a TroyMfg bond? __________

**4.** How many JonMngCo bonds sold this day? __________

**5.** How much higher or lower than yesterday did the last MlbElec bond sell today? __________

**6.** In what year does the TroyMfg bond come due? __________

## CALCULATING INTEREST EARNED

Interest on bonds usually is paid twice a year. Simple interest paid on bonds is calculated by using the interest formula, $I = P \times R \times T$. The principal is par value, *not* market value.

### EXAMPLE

Wylie Corporation purchased a $1,000 par value $8\frac{1}{2}$ percent bond at 94. What is the semiannual interest earned on this bond?

### SOLUTION

| Principal × | Rate × | Time = | Semiannual Interest |
|---|---|---|---|
| $1,000 × | 0.085 × | $\frac{1}{2}$ = | $42.50 |

ANSWERS FOR PRACTICE PROBLEMS

(1) $750
(2) $541.25
(3) 10.00%
(4) 92
(5) $7.50 higher
(6) 2010

Calculate the total semiannual interest earned on each of the following investments in bonds.

| Bond | Par Value | Rate | Number of Bonds | Total Semiannual Interest |
|------|-----------|------|-----------------|---------------------------|
| **1.** GileCo | $1,000 | 9.25% | 5 | _____ |
| **2.** Mixell | 1,000 | 11.00% | 10 | _____ |

## ACCRUED INTEREST ON BOND PURCHASES

**Accrued interest** is accumulated interest on the bond since the last semi-annual interest payment. Since bonds are often purchased between interest payment dates, the buyer pays the seller the accrued interest since the last interest payment on the bond. The buyer, then, will receive the next full semiannual interest payment. When finding the number of days since the last payment of accrued interest, assume each month has 30 days.

When calculating the amount of accrued interest use a 360-day year. Accrued interest is added to total cost of the bonds to determine total amount due.

### EXAMPLE

K-Company owns a $1,000 par value 10% Century bond that pays semi-annual interest on May 1 and November 1. Sheldon Company purchased the bond from K-Company on August 21. Calculate the accrued interest that Sheldon Company should pay to K-Company. Round to the nearest cent.

### SOLUTION

**Step 1** Calculate the days of accrued interest.

90 days (May 1 to August 1)
+20 days (August 1 to August 21)
110 days

**Step 2** Calculate the accrued interest.

Par Value × Interest × Time = Accrued Interest
$1,000 × 0.1 × $\frac{110}{360}$ = $30.56

### PRACTICE PROBLEMS

Calculate the accrued interest on each of the following corporation bond purchases. Use 30-day months and a 360-day year. Round to the nearest cent.

| | Par Value | Rate | Semiannual Interest Pay Dates | Bond Purchase Date | Days of Accrued Interest | Accrued Interest |
|---|-----------|------|-------------------------------|--------------------|--------------------------|------------------|
| **1.** | $1,000 | 10% | March 1, Sept. 1 | June 16 | _____ | _____ |
| **2.** | 1,000 | 9% | April 15, Oct 16 | June 25 | _____ | _____ |

Top: **(1)** $231.25  **(2)** $550.00

Bottom:

| | Days of Accrued Interest | Accrued Interest |
|---|--------------------------|------------------|
| **(1)** | 105 | $29.17 |
| **(2)** | 69 | 17.25 |

## BUYING BONDS

Investors buy and sell bonds through a **broker,** whose company is associated with a bond exchange. A **brokerage fee,** charged both when bonds are bought and sold, increases the investor's **total cost** at the time of purchase.

The total cost of buying a bond between interest periods includes the market price plus the accrued interest plus the brokerage fee.

Market Value + Accrued Interest + Brokerage Fee = Total Cost of Bond

### EXAMPLE

Sixty days after the last interest date, Richard Graham bought five $1,000 par value bonds earning 12% at 105. The brokerage fee or commission was $8.00 per bond. What was Mr. Graham's total cost for the bonds?

### SOLUTION

**Step 1** Calculate the market price for the five bonds.

Par Value × Price Quotation = Market Value

$1,000 × 5 ⟶ $5,000 × 1.05 = $5,250

**Step 2** Calculate the accrued interest and brokerage fee.

Number of Days
Since Last
Par Value × Rate × Interest Payment = Accrued Interest
$5,000 × 0.12 × $\frac{60}{360}$ = $100

Number of    Brokerage Fee
Bonds   ×    per Bond   = Brokerage Fee
5    ×    $8    =    $40

**Step 3** Calculate the total cost for bonds.

Market Value + Accrued Interest + Brokerage Fee = Total Cost
$5,250 + $100 + $40 = $5,390

## ACCRUED INTEREST ON BOND SALES

When an individual sells a bond, he or she receives the **net proceeds.** The net proceeds is the market value of the bond plus the accrued interest minus the brokerage fee. When the net proceeds is greater than the original total cost of the bond there is a gain on the investment. When the net proceeds is less than the original total cost, there is a loss on the investment.

Market Value + Accrued Interest − Brokerage Fee = Net Proceeds

### EXAMPLE

Several years later, Richard Graham sold the five bonds described in the preceding example at $106\frac{1}{4}$ each. This was transacted 30 days after the last interest date. The brokerage fee was $8.75 per bond. Find the net proceeds from selling the bonds and the gain or loss on the investment.

**SOLUTION**

**Step 1**   Calculate the market price for the five bonds.

Par Value × Price Quotation = Market Value

$1,000 × 5 ⟶ $5,000   ×   1.0625   =   $5,312.50

**Step 2**   Calculate the accrued interest.

$$\text{Par Value} \times \text{Rate} \times \frac{\text{Number of Days Since Last Interest Payment}}{} = \text{Accrued Interest}$$

$5,000   × 0.12 ×   $\frac{30}{360}$   =   $50

**Step 3**   Calculate the brokerage fee.

Number of   Brokerage Fee
   Bonds   ×   per Bond   = Brokerage Fee
   5   ×   $8.75   =   $43.75

**Step 4**   Calculate the net proceeds.

Market Value + Accrued Interest − Brokerage Fee = Net Proceeds
   $5,312.50   +   $50   −   $43.75   =   $5,318.75

**Step 5**   Find the gain or loss on investment.

Total Cost − Net Proceeds =   Loss
$5,390.00 −   $5,318.75   = $71.25

**PRACTICE PROBLEMS**

Find the total cost of the following 12%, $1,000 par value bonds purchased by Ripka Company.

| Bond | Price Quotation | Number of Bonds Purchased | Brokerage Fee per Bond | Days Since Last Interest Payment | Total Cost |
|------|------|------|------|------|------|
| 1. BTTL | $90\frac{1}{2}$ | 5 | $8 | 65 | _____ |
| 2. CAMDEN | $101\frac{1}{4}$ | 8 | 7 | 32 | _____ |

Ripka Company sold the bonds as shown below. Calculate the net proceeds from each sale and the gain or loss.

| Bond | Price Quotation | Number of Bonds Sold | Brokerage Fee per Bond | Days Since Last Interest Payment | Net Proceeds | Gain or Loss |
|------|------|------|------|------|------|------|
| 3. BTTL | $101\frac{1}{4}$ | 5 | $10 | 100 | _____ | _____ |
| 4. CAMDEN | 92 | 8 | 8 | 35 | _____ | _____ |

## LESSON 11-1    EXERCISES

Calculate the market value of each of the following bonds. Identify whether each bond is being purchased at a premium or at a discount.

|  | Price Quotation | Market Value | Premium or Discount |
|---|---|---|---|
| **1.** | 101 | _____ | _____ |
| **2.** | 98 | _____ | _____ |
| **3.** | $85\frac{1}{2}$ | _____ | _____ |

Use the following bond quotations to answer the questions below.

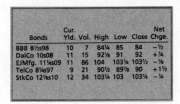

| Bonds | Cur. Yld. | Vol. | High | Low | Close | Net Chge. |
|---|---|---|---|---|---|---|
| BBB 8½s98 | 10 | 7 | 84¼ | 85 | 84 | −½ |
| DalCo 10s08 | 11 | 15 | 92⅛ | 91 | 92 | +¼ |
| EJMfg. 11¼s09 | 11 | 86 | 104 | 103¼ | 103½ | −⅛ |
| TelCo 8¼s97 | 9 | 21 | 90½ | 89⅞ | 90 | +1½ |
| StkCo 12⅛s10 | 12 | 34 | 103¼ | 103 | 103¼ | −¼ |

**4.** What was the highest price paid for a TelCo bond on this day?

_____

**5.** In what year does the DalCo bond come due? _____

**6.** What is the annual rate of interest on the EJMfg. bond? _____

**7.** What was the lowest price paid for a StkCo bond on this day?

_____

**8.** How many BBB bonds sold on this day? _____

**9.** At what price did the last StkCo bond sell on this day? _____

**10.** How much higher or lower than yesterday did the last TelCo bond sell

today? _____

Calculate the total semiannual interest earned on the following bonds.

| Bond | Par Value | Price Quotation | Rate | Number of Bonds Owned | Total Semiannual Interest |
|---|---|---|---|---|---|
| **11.** DDCo | $1,000 | $87\frac{1}{2}$ | 10.00% | 1 | _____ |
| **12.** EarlCo | 1,000 | 99 | 10.50% | 1 | _____ |

Calculate the accrued interest on each of the following corporate bond purchases. Use 30-day months and a 360-day year.

| Par Value | Rate | Semiannual Interest Pay Dates | Bond Purchase Date | Days of Accrued Interest | Accrued Interest |
|---|---|---|---|---|---|
| **13.** $1,000 | 10.00% | Feb. 1, Aug. 1 | May 25 | _____ | _____ |
| **14.** 1,000 | 9.00% | April 1, Oct. 1 | Aug 13 | _____ | _____ |

Find the total cost of the following $1,000 par value bonds purchased.

| Bond | Price Quotation | Number Purchased | Brokerage Fee per Bond | Accrued Interest | Total Cost |
|------|-----------------|------------------|------------------------|------------------|------------|
| 15. CartCo | 105 | 6 | $10 | $100 | _____ |
| 16. GTL | 98 | 20 | 8 | 250 | _____ |

The bonds were sold as shown below. Calculate the net proceeds.

| Bond | Price Quotation | Number Purchased | Brokerage Fee per Bond | Accrued Interest | Net Proceeds |
|------|-----------------|------------------|------------------------|------------------|--------------|
| 17. CartCo | $101\frac{1}{2}$ | 6 | $10 | $150 | _____ |
| 18. GTL | 104 | 20 | 8 | 200 | _____ |

List the net proceeds and the total cost from each investment in bonds, and determine gain or loss. Identify the gain (G) or loss (L).

| Bond | Net Proceeds | Total Cost | Gain or Loss |
|------|--------------|------------|--------------|
| 19. CartCo | _____ | _____ | _____ |
| 20. GTL | _____ | _____ | _____ |

## BUSINESS APPLICATIONS

21. Struthers Brothers Clothing purchased eight $1,000 par value, 10% bonds at $74\frac{1}{8}$, and paid a $10 brokerage fee per bond. Later, they sold the eight bonds at $87\frac{1}{4}$ and paid a $12 brokerage fee per bond. There is no accrued interest on the purchase or sale. What was (a) total cost, (b) net proceeds, and (c) gain or loss on the bonds?

(a) _____

(b) _____

(c) _____

22. Ryder Company owns 20 Butterworth Corporation $1,000 par value, $9\frac{1}{8}$% bonds it purchased at $98\frac{1}{2}$. What is the total semiannual interest?

_____

23. On May 25 Lewis Company purchased a $1,000 par value, 12% bond, which pays semiannual interest on February 1 and August 1, at $105\frac{1}{2}$ and paid a $15 brokerage fee. What is the total cash payment owed for the bond and the accrued interest?

_____

A corporation is financed by selling shares of *stock.* An investor who buys stock, called a *stockholder* or *shareholder,* owns a part of the corporation. Two classes of stock are *common stock* and *preferred stock.* Common stock is the most popular class and most widely owned. Preferred stock holders receive preferential treatment regarding distribution of the company's earnings and sharing in the company's assets if the company goes out of business.

An investor in stocks can make money by selling the shares at a profit and by receiving a part of the corporation's earnings, called *dividends.*

**READING STOCK QUOTATIONS**

Stocks are traded on organized marketplaces, such as the New York Stock Exchange. An investor buys or sells stocks through a broker whose company is affiliated with these exchanges.

Each day's stock transactions, such as the quotations below, are printed in newspapers and financial publications. The amounts quoted represent prices in dollars and fractions of a dollar. To determine a stock's price, called the *market value,* convert the amount to a whole number and a decimal and add a dollar sign. For instance, Calco had a high quote of $40\frac{1}{4}$, which converted to dollars would be $40.25 per share.

The market prices or stock quotations are found in the financial section of most major newspapers.

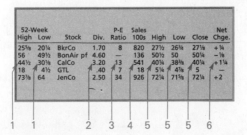

| 52-Week | | Stock | Div. | P-E Ratio | Sales 100s | High | Low | Close | Net Chge. |
|---|---|---|---|---|---|---|---|---|---|
| High | Low | | | | | | | | |
| 25⅜ | 20¼ | BkrCo | 1.70 | 8 | 820 | 27½ | 26¾ | 27⅛ | + ¼ |
| 56 | 49½ | BonAir pf | 4.60 | — | 136 | 50½ | 50 | 50¼ | – ⅛ |
| 44½ | 30⅛ | CalCo | 3.20 | 13 | 541 | 40¼ | 38⅜ | 40¼ | + 1¼ |
| 18 | 4½ | GTL | .40 | 7 | 18 | 5¼ | 4⅞ | 5 | — |
| 73⅞ | 64 | JenCo | 2.50 | 34 | 926 | 72¼ | 71⅝ | 72¼ | + 2 |

1   1                2   3   4   5   5   5   6

1. Caledonia Corporation stocks are shown in abbreviated form as CALCo. The *highest* price paid for a share of Caledonia Corporation stock in the last 52 weeks is $44.50; the *lowest* price paid is $30.125.
2. It is estimated that Caledonia Corporation will pay annual *dividends* of $3.20 per share this year.
3. The P-E, **price-earnings ratio,** shows the relationship of the stock's selling price to the company's earnings per share for the past 12 months.
4. The *sales 100s* indicates the number of shares sold this day in hundreds; 54,100 shares were sold ($541 \times 100 = 54,100$).
5. The *highest* price paid for a share during the day was $40.25, the *lowest* was $38.375. The last sale of the day was $40.25 per share.
6. Today's closing price was $1.25 higher than yesterday's closing price.

## PRACTICE PROBLEMS

Refer to the stock price quotations above to answer the following.

1. What is the highest price GTL sold at in the last 52 weeks? _____

2. How many shares of JenCo were sold on this day? _____

3. At what price did the last share of BkrCo stock sell on this day? _____

4. What is the estimated amount of dividends per share that will be paid on GTL stock this year? _____

5. What is the lowest price paid for a share of BonAir preferred in the last 52 weeks? _____

6. What was the last sale price of JenCo yesterday? _____

7. What is the lowest price at which GTL sold on this day? _____

8. How many times earnings did JenCo sell on this day? _____

## BUYING AND SELLING STOCKS

Investors who trade in stocks through a brokerage firm must pay a **brokerage fee** when stocks are bought and sold. Each brokerage firm sets its own fee, based on the number of shares traded and the price per share. The brokerage fee is added to market value to determine investor's cost.

### EXAMPLE

Kendall College purchased 1,000 shares of Sterling Electric at $40\frac{1}{2}$. The brokerage fee percent is 1.1%. What was the cost of buying the stock?

### SOLUTION

Calculate the market value and the total cost.

|  | Total Market Value | + | Brokerage Fee | = | Total Cost |  |
|---|---|---|---|---|---|---|
| $40.50 \times 1,000 \longrightarrow$ | $40,500.00 | + | $445.50 | = | $40,945.50 | $40,500 \times 1.1\%$ |

The total selling price minus the brokerage fee is the investor's **net proceeds.** When the net proceeds is greater than the original cost of the stock, there is a gain on the investment. When the net proceeds is less than the original total cost, there is a loss on the investment.

## EXAMPLE

Later, Kendall College sold the 1,000 shares of Sterling Electric stock at $46\frac{1}{4}$ and paid a $484.20 brokerage fee. What was the net proceeds from selling the stock and the gain or loss on the sale?

## SOLUTION

**Step 1**  Calculate net proceeds.

Total Market   Brokerage
Price      −     Fee    = Net Proceeds

$46.25 × 1,000 →$46,250.00  −  $484.20  =  $45,765.80

**Step 2**  Calculate gain or loss.

Net Proceeds − Total Cost =   Gain
$45,765.80  −  $40,945.50 = $4,820.30

## PRACTICE PROBLEMS

Calculate the total cost of the following stocks purchased by **All Seasons Insurance Company.**

| Stock | Number of Shares Purchased | Price per Share | Brokerage Fee | Total Cost |
|---|---|---|---|---|
| 1. AnCo | 1,000 | $20\frac{1}{2}$ | $328.40 | _____ |
| 2. BenCo | 500 | $47\frac{1}{4}$ | 314.23 | _____ |

Calculate the net proceeds from the following stocks sold by **All Seasons Insurance Company.**

| Stock | Number of Shares Sold | Price per Share | Brokerage Fee | Net Proceeds |
|---|---|---|---|---|
| 3. AnCo | 1,000 | $32\frac{1}{4}$ | $412.84 | _____ |
| 4. BenCo | 500 | $48\frac{1}{2}$ | 322.35 | _____ |

List the net proceeds and total cost from All Seasons Insurance Company's investments in stocks and determine the gain or loss. Identify the gain (G) or loss (L).

| Stock | Net Proceeds | Total Cost | Gain or Loss |
|---|---|---|---|
| 5. AnCo | _____ | _____ | _____ |
| 6. BenCo | _____ | _____ | _____ |

## CALCULATING DIVIDENDS EARNED

**Dividends** are a portion of a corporation's earnings distributed to stockholders. Corporations are not required to pay dividends, but most do since it helps maintain stockholder satisfaction. Members of the **board of directors,** who are elected by the stockholders, determine if a dividend will be paid and how much it will be. Dividends may be paid in cash or in additional shares of stock called **stock dividend.** Dividends are ordinarily paid quarterly; however, the board of directors may decide to pay dividends more or less frequently.

If a corporation has both common stock and preferred stock, the preferred stock dividend must be paid before the common stock can be paid. Also, if the corporation should go out of business, preferred stockholders have a claim on the distribution of corporation assets before common stockholders. For this reason, preferred stock is considered to be a safer investment than the common stock of the same company.

## PREFERRED STOCK DIVIDENDS

Preferred stock has a **par value** or **stated value,** such as $50 or $100, that is assigned when issued by the corporation. The par value or stated value does not necessarily reflect the stock's market value.

Preferred stock also has a specified **dividend rate** stated as a percent of its par value. To find the dividend, multiply the par value by the dividend rate by the time period.

---

### EXAMPLE

Simpson Industries purchased 500 shares of $100 par value, 6% preferred stock. What is the total dividend Simpson Industries will receive per quarter?

### SOLUTION

**Step 1**  Calculate the quarterly dividend per share.

Par Value × Dividend Rate × Time = Quarterly Dividend
  $100   ×   0.06%   ×   $\frac{1}{4}$ =    $1.50

**Step 2**  Calculate the total quarterly dividend.

Quarterly Dividend
   per Share      × Number of Shares = Total Quarterly Dividend
    $1.50      ×      500      =      $750.00

---

### PRACTICE PROBLEMS

Calculate the total quarterly dividend for each of the following investments in preferred stock.

| Stock | Par Value | Dividend Rate | Number of Shares Owned | Total Quarterly Dividend |
|-------|-----------|---------------|------------------------|--------------------------|
| 1. DXN | $100 | 8.00% | 225 | _____ |
| 2. Eaton | 50 | 10.00% | 350 | _____ |

---

## COMMON STOCK DIVIDENDS

Common stock does not have a specified dividend rate like preferred stock. Instead, the common stock dividend, usually based on quarterly earnings, is declared as a dollar-and-cents amount by the board of directors. The amount may vary from one quarter to another. To find the total annual dividend, multiply the number of shares by the sum of the quarterly dividends per share.

### EXAMPLE

Windsor Amusement Company owns 575 shares of Mankato Corporation common stock. Last year Mankato Corporation paid quarterly dividends of $1.15, $0.95, $0.80, and $1.05. What were the total dividends Windsor Amusement Company received for the year?

### SOLUTION

**Step 1**  Calculate the total annual dividends per share.

$$\underset{\text{Quarterly Dividends per Share}}{\$1.15 + \$0.95 + \$0.80 + \$1.05} = \underset{\substack{\text{Total Annual} \\ \text{Dividends per Share}}}{\$3.95}$$

**Step 2**  Calculate the total annual dividends.

$$\underset{\substack{\text{Total Annual} \\ \text{Dividends per Share}}}{\$3.95} \times \underset{\substack{\text{Number} \\ \text{of Shares}}}{575} = \underset{\substack{\text{Total} \\ \text{Annual Dividends}}}{\$2,271.25}$$

### PRACTICE PROBLEMS

Calculate the total annual dividends for each of the following investments in common stock.

| | Company | Quarterly Dividends per Share | Number of Shares | Total Annual Dividends |
|---|---|---|---|---|
| **1.** | CSLA | $1.16, $1.32, $1.04, $1.22 | 70 | _____ |
| **2.** | LaCroix | $0.15, $0.14, $0.15, $0.18 | 455 | _____ |
| **3.** | Nolan | $0.86, $0.92, $0.94, $0.98 | 125 | _____ |

## LESSON 11-2    EXERCISES

Refer to the stock price quotation below to answer the following questions. State money amounts as dollars and cents.

| 52-Week | | Stock | Div. | P-E Ratio | Sales 100s | High | Low | Close | Net Chge. |
|---------|---|-------|------|-----------|------------|------|-----|-------|-----------|
| High | Low | | | | | | | | |
| 88¼ | 62½ | GraenOil | 6.16 | 12 | 812 | 74 | 72½ | 72½ | −¼ |
| 7 | 5¼ | JNO | .13 | 5 | 16 | 5½ | 5¼ | 5¼ | — |
| 36⅛ | 24½ | RussCo | 3.75 | 15 | 982 | 35 | 34⅛ | 35 | +½ |

1. How many shares of JNO stock sold on this day? _____

2. What is the highest price at which a share of RussCo stock sold on this day? _____

3. What is the lowest price at which Graen Oil sold per share in the last 52 weeks? _____

4. At what price per share was the last sale of JNO stock on this day?

   _____

5. At what price did the last share of RussCo stock sell yesterday?

   _____

Calculate the total cost of the following stocks purchased by Dexter Manufacturing Company.

| | Stock | Number of Shares Purchased | Price per Share | Brokerage Fee | Total Cost |
|---|-------|---------------------------|-----------------|---------------|------------|
| 6. | Aladn | 300 | $15\frac{7}{8}$ | $ 91.48 | _____ |
| 7. | Forst | 175 | $46\frac{1}{2}$ | 105.72 | _____ |
| 8. | NESCO | 200 | 25 | 132.60 | _____ |
| 9. | SuperStm | 215 | $18\frac{1}{4}$ | 88.76 | _____ |
| 10. | TideCo | 550 | $2\frac{1}{8}$ | 61.27 | _____ |

Calculate the net proceeds from the following investments in stocks sold by Dexter Manufacturing Company.

| | Stock | Number of Shares Sold | Price per Share | Brokerage Fee | Net Proceeds |
|---|-------|----------------------|-----------------|---------------|--------------|
| 11. | Aladn | 300 | $28\frac{1}{2}$ | $142.64 | _____ |
| 12. | Forst | 175 | $34\frac{3}{4}$ | 92.28 | _____ |
| 13. | NESCO | 200 | $27\frac{1}{4}$ | 103.85 | _____ |

| Stock | Number of Shares Sold | Price per Share | Brokerage Fee | Net Proceeds |
|-------|------|------|--------|------|
| **14.** SuperStm | 215 | 30 | 115.07 | _____ |
| **15.** TideCo | 550 | $5\frac{1}{2}$ | 89.60 | _____ |

List the net proceeds from Problems 11 to 15 and the total costs from Problems 6 to 10 from Dexter Manufacturing Company's investments in stocks and determine the gain or loss. Identify the gain (G) or loss (L).

| Stock | Net Proceeds | Total Cost | Gain or Loss |
|-------|------|------|------|
| **16.** Aladn | _____ | _____ | _____ |
| **17.** Forst | _____ | _____ | _____ |
| **18.** NESCO | _____ | _____ | _____ |
| **19.** SuperStm | _____ | _____ | _____ |
| **20.** TideCo | _____ | _____ | _____ |

Calculate the total quarterly dividends for each of the following investments in preferred stock.

| Stock | Par Value | Dividend Rate | Number of Shares Owned | Total Quarterly Dividend |
|-------|------|------|------|------|
| **21.** LilCo | $100 | 9.00% | 50 | _____ |
| **22.** OrchCo | 50 | 6.00% | 125 | _____ |
| **23.** NortInd | 100 | 12.00% | 45 | _____ |
| **24.** Taylor | 75 | 8.00% | 200 | _____ |
| **25.** YORK | 50 | 10.00% | 60 | _____ |

Calculate the total annual dividends for each of the following investments in common stock.

| Stock | Quarterly Dividends per Share | Number of Shares | Total Annual Dividend |
|-------|------|------|------|
| **26.** JenCo | $1.15, $1.20, $1.15, $1.05 | 420 | _____ |
| **27.** MRA | 0.42, 0.36, 0.38, 0.30 | 3,050 | _____ |
| **28.** Pace | 0.97, 1.01, 1.05, 1.10 | 25 | _____ |
| **29.** RobCo | 0.10, 0.08, 0.06, 0.14 | 4,500 | _____ |
| **30.** UtdCal | 0.70, 0.60, 0.65, 0.75 | 125 | _____ |

## BUSINESS APPLICATIONS

**31.** Mar, Inc., purchased 45 shares of Bell Corporation common stock at $64\frac{1}{2}$ and paid a $52.67 brokerage fee. Mar also purchased 35 shares of Bell Corporation preferred stock at 62 and paid a $46.80 brokerage fee. What was Mar's total cost for these investments?

_____

**32.** Claridge Enterprises purchased 375 shares of TZY Corporation common stock at $43\frac{1}{2}$ and paid a $247.65 brokerage fee. Later, Claridge Enterprises sold the stock at $52\frac{1}{4}$ and paid a $304.16 brokerage fee. What was Claridge Enterprises' (a) total cost of buying the stock, (b) net proceeds from selling the stock, and (c) gain or loss on the sale?

**(a)** _____

**(b)** _____

**(c)** _____

**33.** Tyndal Management Services purchased 150 shares of Baxter Manufacturing Company common stock at $36\frac{7}{8}$, paying a $90.86 brokerage fee, and 225 shares of Baxter common stock at $41\frac{1}{2}$, paying a $149.74 brokerage fee. Tyndal sold all of the shares at $48\frac{1}{4}$, paying a $290.50 brokerage fee. What was the gain or loss on the sale?

_____

**34.** Gillette Music Company owns 80 shares of $100 par value 9% preferred stock. Gillette purchased the stock at $91\frac{1}{4}$ and paid a $110.15 brokerage fee. (a) What is the quarterly dividend per share? (b) What is the total annual dividend? Round to two decimal places.

**(a)** _____

**(b)** _____

**35.** Rice Department Store purchased 250 shares of Calgary Mining Exploration common stock at $5\frac{1}{4}$ and paid a $55.98 brokerage fee. Last year, Calgary paid quarterly dividends of $0.10, $0.08, $0.14, and $0.06. What was the total annual dividends earned by Rice?

_____

**36.** Several years ago Jameston College purchased 400 shares of Spencer Corporation common stock at $52\frac{1}{8}$ and paid a $256.18 brokerage fee. Later, Jameston College sold 300 shares at $59\frac{1}{4}$ and paid a $244.16 brokerage fee. Recently, Jameston College sold the remaining 100 shares at $48\frac{3}{4}$ and paid a $72.40 brokerage fee. What was Jameston College's gain or loss on the 400 shares?

_____

**37.** Oakdale Publishing Company purchased 50 shares of Treasure Village $100 par value, 7% preferred stock at 78 and paid a $100.00 brokerage fee. (a) What was Oakdale's total annual dividend? (b) How much more or less interest would Oakdale have earned, compared to the annual dividends received, if the total cost had been invested for one year in a savings certificate that pays 8% simple interest?

**(a)** _____

**(b)** _____

**38.** Segel Company purchased 175 shares of Halsne Corporation common stock at $32\frac{1}{2}$ and paid a $89.40 brokerage fee. Segel received quarterly dividends of $0.70, $0.60, $0.50, and $0.75. Exactly 1 year after buying the stock, Segel sold it at $34\frac{1}{4}$ and paid a $107.42 brokerage fee. What was Segel's total gain or loss from owning this stock, including the dividends earned and the gain or loss on the sale?

_____

A **mutual fund company** is an investment company that pools a large number of investors' money to buy stocks and bonds. A mutual fund company might have investments in as many as 100 companies at any one time. By owning shares in the mutual fund company, an investor indirectly holds an interest in all of these companies.

Mutual fund companies earn money by receiving interest and dividends on their investments and by selling the investments at a profit.

**READING MUTUAL FUND QUOTATIONS**

A mutual fund company often has a number of different funds designed to meet different investor needs. One fund might consist of investments in bonds, another in common stocks, another in preferred stocks, and a fund combining all three.

Mutual fund price quotations like the ones shown below are published in newspapers and financial publications.

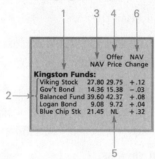

1. The mutual fund company is shown in abbreviated form.
2. Different investment programs offered by Kingston Mutual Fund Company.
3. The **net asset value** (NAV) is the price at which investors *sell* their shares, shown in dollar and cents, such as $27.80. The mutual fund company calculates the NAV by dividing the program's total investment value by the number of shares owned by investors.
4. The **offer price** is the price at which the investor *buys* shown in dollars and cents. It is the NAV plus a sales commission charge.
5. The Blue Chip Stock is a **no-load fund.** No sales commission is charged so that the offer price is the same as the NAV, $21.45.
6. The change from yesterday's to today's NAV is shown in dollars and cents, such as +$0.12.

### PRACTICE PROBLEMS

Refer to the price quotations for Kingston Mutual Funds shown above to answer these questions.

1. At what price would an investor buy a share of the Viking stock fund?

   _____

2. At what price would an investor sell a share of the Logan bond fund?

   _____

3. How much higher or lower did a share of the Balanced fund sell today than yesterday? _____

---

**ANSWERS FOR PRACTICE PROBLEMS**

(1) $29.75   (2) $9.08   (3) $0.08 lower

The mutual fund company sells its shares to investors and buys them back. The transaction may be handled by a broker. Usually, investors buy many mutual fund shares at a time. To find the total cost of a mutual fund, multiply the offer price by the number of shares purchased. When selling mutual funds, the net proceeds is found by multiplying the NAV price by the number of shares sold. As with other investments, the gain or loss on the sale of mutual fund shares is the difference between the net proceeds and the total cost.

### EXAMPLE

Giles and Company purchased 100 shares of Capital Stock mutual fund when the NAV was $12.45 and the offer price was $13.32. They sold the 100 shares of Capital Stock mutual fund when the NAV was $16.48 and the offer price was $17.63. What was Giles' total cost, net proceeds, and gain or loss on the sale?

### SOLUTION

**Step 1** Find the total cost.

Offer Price × Number of Shares = Total Cost
$13.32 × 100 = $1,332

**Step 2** Find the net proceeds.

NAV Price × Number of Shares = Net Proceeds
$16.48 × 100 = $1,648

**Step 3** Find the gain or loss.

Net Proceeds − Total Cost = Gain on Sale
$1,648 − $1,332 = $316

### PRACTICE PROBLEMS

Calculate the total cost, net proceeds, and gain or loss for the following transactions. Identify the gain (G) or loss (L).

| | Mutual Fund | Number of Shares | Purchase (Offer Price) | Sale (NAV) | Total Cost | Net Proceeds | Gain or Loss |
|---|---|---|---|---|---|---|---|
| 1. | Growth Plus | 200 | $33.34 | $38.20 | _____ | _____ | _____ |
| 2. | Global | 85 | 10.00 | 7.14 | _____ | _____ | _____ |
| 3. | Income | 150 | 26.48 | 27.15 | _____ | _____ | _____ |
| 4. | Hy Yield | 325 | 15.44 | 21.05 | _____ | _____ | _____ |

## LESSON 11-3   EXERCISES

Use the following mutual fund quotations to answer the questions below.

|  | NAV | Offer Price | NAV Change |
|---|---|---|---|
| **Talbott Funds:** | | | |
| Phoenix | 23.60 | 25.25 | + .30 |
| Equity | 14.08 | 15.07 | − .04 |
| Select | 34.72 | 37.15 | + .47 |
| Income | 9.64 | 10.31 | + .13 |
| Am. Growth | 16.80 | NL | + .09 |

1. How many different investment programs does the Talbott Mutual Fund Company offer? _____

2. At what price would an investor buy a share of the Income fund?

   _____

3. At what price would an investor sell a share of the Phoenix fund?

   _____

4. How much higher or lower did a share of the Select fund sell at today than yesterday? _____

5. At what price per share would an investor buy a share of the American Growth fund? _____

Use the price quotations shown above for Talbott Mutual Fund Company to calculate the price per share and the total cost or net proceeds for the following transactions.

| | Mutual Fund | Number of Shares | Transaction | Price per Share | Total Cost or Net Proceeds |
|---|---|---|---|---|---|
| 6. | Select | 75 | Purchased | _____ | _____ |
| 7. | Am Growth | 50 | Sold | _____ | _____ |
| 8. | Income | 160 | Purchased | _____ | _____ |
| 9. | Equity | 250 | Purchased | _____ | _____ |
| 10. | Phoenix | 350 | Sold | _____ | _____ |

Calculate the gain or loss on the following mutual fund transactions. Identify the gain (G) or loss (L).

| | Number of Shares | Purchase | | Sale | | Gain or Loss |
|---|---|---|---|---|---|---|
| | | NAV | Offer Price | NAV | Offer Price | |
| 11. | 375 | $13.60 | $14.55 | $17.80 | $19.05 | _____ |
| 12. | 180 | 8.95 | 9.58 | 22.40 | 23.97 | _____ |
| 13. | 290 | 24.72 | 26.45 | 20.16 | 21.57 | _____ |

| | Number of Shares | Purchase | | Sale | | Gain or Loss |
|---|---|---|---|---|---|---|
| | | NAV | Offer Price | NAV | Offer Price | |
| **14.** | 315 | 38.16 | 40.83 | 39.80 | 42.59 | _____ |
| **15.** | 1,000 | 4.75 | 5.08 | 23.65 | 25.31 | _____ |

## BUSINESS APPLICATIONS

**16.** Stribley Engineering purchased 75 shares of Cosmos mutual fund when the NAV was $16.90 and the offer price $18.08, 100 shares when the NAV was $19.40 and the offer price $20.79, and 200 shares when the NAV was $24.60 and the offer price $26.32. What was Stribley's total cost for these mutual fund shares?

_____

**17.** Bayless Boot Works purchased 500 shares of Holiday mutual fund when the NAV was $26.84 and the offer price was $28.73. Bayless sold 300 shares when the NAV was $36.10 and the offer price $38.63 and the remaining 200 shares when the NAV was $39.20 and the offer price $41.94. What was Bayless Boat Works' total gain or loss?

_____

**18.** Compet Company purchased 450 shares of Jennings mutual fund, a no-load fund, when the NAV was $46.18. Later, Compet sold the 450 shares when the NAV was $53.24. What was Compet's gain or loss on the sale of these shares?

_____

**19.** Jantzen Wholesale Company purchased 200 shares of Chanhassen mutual fund when the NAV was $20.05 and the offer price was $21.45, 300 shares when the NAV was $17.98 and the offer price was $19.20, and 500 shares when the NAV was $16.35 and the offer price was $17.50. What was Jantzen's average cost per share?

_____

*Real estate investment* consists of buying a house, apartment building, warehouse, office building, commercial building, agricultural land, or other form of real property and renting it to others. The investor is called the **landlord,** the renter the **tenant.**

## RENTAL INCOME

*Rental income* represents total rent received from tenants. Tenants of apartments and houses usually pay monthly rent. Tenants of buildings used for business purposes might make monthly, quarterly, semiannual, or annual rent payments.

Landlords of certain types of property, such as an apartment building, anticipate property not being fully rented for the entire year. Therefore, they deduct a vacancy allowance when estimating expected annual rental income. Usually the vacancy allowance is a small percentage of the annual rental income. To find the annual rental income, multiply the monthly rent per unit by the number of units by 12 months. The annual rental income minus the vacancy allowance equals the estimated annual rental income.

### EXAMPLE 1

Damon Brothers, Inc., owns a 16-unit apartment building. Last year all 16 apartments were rented for the entire year at $650 per month per apartment. What was Damon Brothers' annual rental income?

### SOLUTION

| Monthly Rental Income per Unit | | Number of Units | | Months in a Year | | Annual Rental Income |
|---|---|---|---|---|---|---|
| $650 | × | 16 | × | 12 | = | $124,800 |

### EXAMPLE 2

This year Damon Brothers intends to rent for the same amount per apartment but estimates a 6% vacancy rate. What is Damon Brothers' estimated annual rental income?

### SOLUTION

**Step 1** Calculate the vacancy allowance.

| Annual Rental Income | × | Vacancy Allowance Rate | = | Vacancy Allowance |
|---|---|---|---|---|
| $124,800 | × | 0.06 | = | $7,488 |

**Step 2** Calculate the estimated annual rental income.

| Annual Rental Income | − | Vacancy Allowance | = | Estimated Annual Rental Income |
|---|---|---|---|---|
| $124,800 | − | $7,488 | = | $117,312 |

Sometimes a *percentage lease* is used for retail buildings, whereby the tenant pays a set annual amount plus a percent of annual net sales. Tenants of agricultural land often pay a certain amount per acre rented or a portion of harvested crops.

---

### EXAMPLE 1

Geoff Oil Company rents to Winslow Clothiers on a percentage lease at $4,000 per month plus 2% of Winslow's annual net sales. Winslow's annual net sales were $1,550,400. What was Geoff Oil Company's annual rental income?

### SOLUTION

**Step 1** Calculate the percentage of net sales.

$$
\begin{array}{ccc}
 & & \text{Percentage} \\
\text{Net Sales} & \times \text{ Net Sales Rate} = & \text{of Net Sales} \\
\$1,550,400 \times & 0.02 & = \quad \$31,008
\end{array}
$$

**Step 2** Calculate the annual rental income.

$$
\begin{array}{cccc}
\text{Annual} & \text{Percentage} & & \text{Annual} \\
\text{Fixed Rent} + & \text{of Net Sales} = & & \text{Rental Income}
\end{array}
$$

$\$4,000 \times 12 \longrightarrow \$48,000 \ + \ \$31,008 \ = \ \$79,008$

---

### EXAMPLE 2

Munson Construction owns a 240-acre farm that it rents to its tenant for the year at $125 per acre. Munson receives semiannual rent payments. What is Munson's semiannual rental income?

### SOLUTION

$$
\begin{array}{ccccc}
\text{Annual Rent} & \text{Number} & \text{Semiannual} & & \\
\text{per Acre} & \times \text{ of Acres} \times & \text{Time} & = & \text{Rental Income} \\
\$125 & \times \quad 240 \quad \times & \frac{1}{2} & = & \$15,000
\end{array}
$$

---

### PRACTICE PROBLEMS

Calculate the vacancy allowance, if any, and the annual gross rental income earned for each of the following real estate investment properties.

| Property | Number of Rental Units | Rental Unit/ Rental Period | Rent per Unit | Vacancy Allowance Rate | Vacancy Allowance | Annual Gross Rental Income |
|---|---|---|---|---|---|---|
| 1. House | 1 | Monthly | $ 775 | — | _____ | _____ |
| 2. Apartment building | 24 | Monthly | 625 | 4% | _____ | _____ |
| 3. Farm | 160 Acres | Per acre | 115 | — | _____ | _____ |
| 4. Warehouse | 1 | Semiannually | 4,800 | — | _____ | _____ |
| 5. Office building | 32 | Monthly | 995 | 10% | _____ | _____ |

Calculate the annual rental income for the following percentage leases.

| | Monthly Rent | Annual Net Sales | Percent of Annual Net Sales | Annual Rental Income |
|---|---|---|---|---|
| 6. | $3,000 | $1,250,000 | 1% | _____ |
| 7. | 0 | 4,650,000 | 3% | _____ |
| 8. | 1,500 | 246,500 | 3% | _____ |

## CASH FLOW

*Cash flow* is the difference between cash received and cash paid out over a specified time period such as a month, a quarter, or a year. *Positive cash flow* is when more cash is received than paid out. If more cash is paid out than received, it is a *negative cash flow.*

Normal expenditures incurred by a real estate investor include real estate taxes, property insurance, maintenance and repairs, and utilities. Often, an investor can buy real estate by making a down payment of perhaps 25% of the purchase price and borrowing the rest. Therefore, a loan could be a major cash expenditure.

### EXAMPLE

Chang, Inc., owns an apartment building from which it received $78,250 rental income last year. Chang's expenditures were as follows: loan payment, $42,170; real estate taxes, $12,124; property insurance, $3,400; and maintenance and repairs, $8,090. What was Chang's positive or negative cash flow for the year?

### SOLUTION

| Annual Rental Income | − | Loan Payment | − | Real Estate Taxes | − | Property Insurance | − | Maintenance and Repairs | = | Positive Cash Flow |
|---|---|---|---|---|---|---|---|---|---|---|
| $78,250 | − | $42,170 | − | $12,124 | − | $3,400 | − | $8,090 | = | $12,466 |

### PRACTICE PROBLEMS

Calculate the positive (+) or negative (−) cash flow for each of the following real estate investments.

| | Annual Rental Income | Loan Payment | Real Estate Taxes | Property Insurance | Maintenance and Repairs | Cash Flow (+ or −) |
|---|---|---|---|---|---|---|
| 1. | $10,200 | $ 7,820 | $ 1,640 | $ 378 | $ 805 | _____ |
| 2. | 38,416 | 0 | 4,483 | 914 | 1,725 | _____ |

**Starting on page 382:**

| | Vacancy Allowance | Annual Gross Rental Income |
|---|---|---|
| (1) | $ 0 | $ 9,300 |
| (2) | 7,200 | 172,800 |
| (3) | 0 | 18,400 |
| (4) | 0 | 9,600 |
| (5) | 38,208 | 343,872 |
| (6) | $48,500 | (7) $139,500 | (8) $25,395 |

| | Annual Rental Income | Loan Payment | Real Estate Taxes | Property Insurance | Maintenance and Repairs | Cash Flow (+ or −) |
|---|---|---|---|---|---|---|
| 3. | 26,905 | 18,050 | 3,860 | 726 | 4,537 | _____ |
| 4. | 79,540 | 51,310 | 13,276 | 2,850 | 3,509 | _____ |

## CALCULATING NET INCOME OR NET LOSS

A real estate investor has **net income** when income exceeds expenses. When expenses exceed income, there is a **net loss.**

**Expenses** are expired costs, such as real estate taxes, insurance, maintenance and repairs, interest on a loan, labor, and utilities. **Depreciation,** the gradual decrease in value of an asset because of wear, tear, obsolescence, and other factors, is another expense that an investor can record. The Internal Revenue Service (IRS) sets guidelines for the amount of depreciation that can be claimed each year and allows depreciation for buildings, equipment, appliances, and furnishings, but not land. Recognizing depreciation expense is very favorable for the real estate investor, since it does not require a cash outlay but does reduce net income and income taxes. Since depreciation expense may be a large amount, it is possible for an investor to show a net loss for income tax purposes but to also have a positive cash flow.

Not all expenditures are expenses. With a loan payment that consists partly of principal and partly of interest, only the interest is an expense.

### EXAMPLE

Last year King Auto Company had the following income and expenses from an apartment building: rental income, $249,600; interest expense, $148,684; depreciation expense, $63,475; real estate taxes, $27,400; property insurance, $10,340; and maintenance and repairs, $8,216. What was King Auto's net income or net loss for the year?

### SOLUTION

| Annual Rental Income | − | Interest Expense | − | Depreciation Expense | − | Real Estate Taxes | − | Property Insurance | − | Maintenance and Repairs | = | Net Loss |
|---|---|---|---|---|---|---|---|---|---|---|---|---|
| $249,600 | − | $148,684 | − | $63,475 | − | $27,400 | − | $10,340 | − | $8,216 | = | $8,515 |

### PRACTICE PROBLEMS

Calculate the net income (NI) or net loss (NL) from each of the following real estate investments.

| | Annual Rental Income | Interest Expense | Depreciation Expense | Real Estate Taxes | Property Insurance | Maintenance and Repairs | Net Income or Net Loss |
|---|---|---|---|---|---|---|---|
| 1. | $82,800 | $52,624 | $34,015 | $14,528 | $4,510 | $2,176 | _____ |
| 2. | 11,580 | 5,620 | 4,285 | 2,150 | 593 | 1,430 | _____ |

## CALCULATING RETURN ON INVESTMENT

A real estate investor's **return on investment** is calculated by dividing the annual net income by the investor's total cash investment to purchase the property. The **total cash investment** includes cash paid to buy the property or the cash down payment plus any costs incurred to buy the property. These costs, called **closing costs,** might include legal fees, property appraisal fees, a credit report on the buyer, and lender's fees and are expressed as a percent.

### EXAMPLE

Carlyle Industries purchased an office building for $346,500 cash and incurred closing costs of $2,750. Last year Carlyle's net income from the property was $18,200. What rate of return did Carlyle earn from their investment? Round to the nearest hundredth of a percent.

### SOLUTION

**Step 1**  Calculate the total cash investment.

Cash Price + Closing Costs = Total Cash Investment

$346,500 + $2,750 = $349,250

**Step 2**  Calculate the return on investment.

Annual Net Income ÷ Total Cash Investment = Return on Investment

$18,200 ÷ $349,250 = 0.0521 or 5.21%

### PRACTICE PROBLEMS

Calculate the rate of return on the following real estate investments. Round answers to two decimal places.

| | Annual Net Income | Cash Investment | Closing Costs | Return on Investment |
|---|---|---|---|---|
| 1. | $ 9,678 | $128,600 | $3,480 | _____ |
| 2. | 6,219 | 48,150 | 1,635 | _____ |
| 3. | 35,874 | 347,940 | 8,678 | _____ |
| 4. | 905 | 15,285 | 2,150 | _____ |
| 5. | 4,832 | 52,560 | 3,110 | _____ |

## LESSON 11-4    EXERCISES

Calculate the vacancy allowance and the annual gross rental income earned from each of the following real estate investment properties.

| Property | Number of Rental Units | Rental Unit/ Rental Period | Rent per Unit | Vacancy Allowance Rate | Vacancy Allowance | Annual Rental Income |
|---|---|---|---|---|---|---|
| 1. Apartment building | 32 | Monthly | $ 850 | 5% | _____ | _____ |
| 2. House | 1 | Monthly | 925 | — | _____ | _____ |
| 3. Warehouse | 1 | Semiannually | 4,250 | — | _____ | _____ |
| 4. Farm | 240 acres | Per acre | 145 | — | _____ | _____ |
| 5. Office building | 48 | Monthly | 725 | 8% | _____ | _____ |

Calculate the annual gross rental income for each of the following percentage leases for retail buildings. The landlord receives a monthly rent plus a percent of annual net sales.

| | Monthly Rent | Annual Net Sales | Percent of Annual Net Sales | Annual Rental Income |
|---|---|---|---|---|
| 6. | $1,000 | $ 480,000 | 1.5% | _____ |
| 7. | 3,750 | 1,250,000 | 1.0% | _____ |
| 8. | 0 | 874,600 | 4.0% | _____ |
| 9. | 2,100 | 2,460,200 | 1.5% | _____ |
| 10. | 4,800 | 3,760,500 | 2.0% | _____ |

Calculate the positive (+) or negative (−) cash flow for each of the following real estate investments.

| | Annual Rental Income | Loan Payment | Real Estate Taxes | Property Insurance | Maintenance and Repairs | Cash Flow (+ or −) |
|---|---|---|---|---|---|---|
| 11. | $115,200 | $87,461 | $46,000 | $17,430 | $12,415 | _____ |
| 12. | 9,900 | 5,592 | 1,876 | 624 | 385 | _____ |
| 13. | 72,480 | 16,788 | 5,078 | 815 | 1,820 | _____ |
| 14. | 21,500 | 16,874 | 5,400 | 935 | 350 | _____ |
| 15. | 26,845 | 18,205 | 4,360 | 438 | 797 | _____ |

Calculate the net income (NI) or net loss (NL) from each of the following real estate investments.

| | Annual Rental Income | Interest | Depreciation Expense | Real Estate Taxes | Property Insurance | Maintenance and Repairs | Net Income or Net Loss |
|---|---|---|---|---|---|---|---|
| 16. | $ 86,390 | $ 28,490 | $ 46,852 | $21,815 | $ 1,473 | $ 4,360 | _____ |
| 17. | 32,475 | 18,315 | 0 | 3,628 | 347 | 0 | _____ |
| 18. | 55,200 | 21,480 | 16,728 | 12,905 | 3,916 | 2,472 | _____ |
| 19. | 286,495 | 120,057 | 274,371 | 79,714 | 26,805 | 15,305 | _____ |
| 20. | 33,360 | 8,436 | 14,318 | 4,115 | 924 | 815 | _____ |

Calculate the rate of return on the following real estate investments. Round your answers off at two decimal places.

| | Annual Net Income | Cash Investment | Closing Costs | Return on Investment |
|---|---|---|---|---|
| 21. | $18,029 | $147,350 | $2,897 | _____ |
| 22. | 6,440 | 82,460 | 3,416 | _____ |
| 23. | 10,938 | 208,741 | 8,738 | _____ |
| 24. | 9,307 | 28,432 | 1,518 | _____ |
| 25. | 17,382 | 120,291 | 5,730 | _____ |

## BUSINESS APPLICATIONS

26. DeBaca Investment Company owns a 16-unit apartment building. DeBaca rents 8 units at $650 per month, 4 units at $700 per month, 2 units at $800 per month, and 2 units at $1,200 per month. What is DeBaca's annual rental income?

27. Chambers, Inc., owns a retail building that it rents to the Clothes Horse clothing store for $3,850 per month. Last year the Clothes Horse had net sales of $689,850. How much more or less would Chambers have received in rent if it would have had a percentage lease where it received $2,000 per month plus 4% of net sales?

**28.** Alverez Appliance Store owns a 240-acre farm that it rents to its tenant for a portion of the crops that the tenant grows. Last year Alverez received 1,685 bushels of soybeans that it sold for $5.20 per bushel, 4,210 bushels of corn that it sold for $1.30 per bushel, and 385 bushels of wheat that it sold for $1.20 per bushel. What was Alverez's average annual rent income per acre?

**29.** Forsythe Investment Company owns a 32-unit apartment building and rents each apartment at $650 per month. Last year there was a vacancy rate of 18%. Forsythe estimates that if it hires a full-time property manager for $30,000 a year, it can raise its rent to $700 per month per unit and cut the vacancy rate to 5%. How much will Forsythe gain or lose by hiring the property manager compared to last year's results?

**30.** Livingstone Medical Center owns a 16-unit apartment building from which it received rental income of $115,200 last year. Expenses were: interest, $42,621; depreciation, $64,580; real estate taxes, $24,987; property insurance, $3,840; and repairs, $7,258. What was Livingstone's net income or net loss for the year?

**31.** Southwest Office Supply, Inc., owns an 8-unit office building that it rents for $700 per unit per month. Last year Southwest's costs and expenses were as follows: loan payment, $34,873; real estate taxes, $6,478; property insurance, $1,945; and repairs, $2,080. What was Southwest's positive or negative cash flow for the year?

**32.** Hermann Publishing Company made a $49,850 down payment and paid $5,750 closing costs to purchase an 8-unit apartment building. Last year the rental income was $69,600 and expenses were $62,965. What was Hermann's rate of return on their investment? Round to two decimal places.

_____

**33.** R & R Investment Company purchased an 8-unit apartment building. R & R estimates the following costs and expenses for next year: loan payment, $38,540; real estate taxes, $9,680; property insurance, $2,068; and repairs, $2,000. R & R wants a positive cash flow of $8,000 for the year. Assuming there will be no vacancies, at what amount should R & R rent each apartment per month to achieve its goal?

_____

**34.** Milford, Inc., owns a retail building from which it received $40,575 gross rental income last year. Costs and expenses were as follows: loan payment, $18,285 (principal was $4,169; interest was $14,116); depreciation, $24,378; real estate taxes, $8,450; property insurance, $4,216; and repairs, $2,645. (a) What was Milford's positive or negative cash flow for the year? (b) What was Milford's net income or net loss for the year?

**(a)** _____

**(b)** _____

## CHAPTER 11 REVIEW

### CALCULATING BOND PRICES

Calculate the market value of each of the following bonds. Identify whether the bond is being purchased at a premium or a discount.

|  | Price Quotation | Market Value | Premium or Discount |
|---|---|---|---|
| **1.** | 86 | _____ | _____ |
| **2.** | $102\frac{1}{2}$ | _____ | _____ |

### READING BOND QUOTATIONS

Use the following bond quotations to answer the questions below.

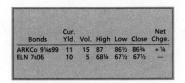

| Bonds | Cur. Yld. | Vol. | High | Low | Close | Net Chge. |
|---|---|---|---|---|---|---|
| ARKCo 9¼s99 | 11 | 15 | 87 | 86½ | 86¾ | + ¼ |
| ELN 7s06 | 10 | 5 | 68¼ | 67½ | 67½ | — |

**3.** What is the interest rate paid on the ARKCo bond? _____

**4.** What was the highest price paid for an ELN bond? _____

**5.** In what year does the ELN bond come due? _____

**6.** At what price was the last ARKCo bond sold? _____

### CALCULATING BOND INTEREST EARNED

Calculate the total semiannual interest earned on each investment.

|  | Bond | Par Value | Market Value | Interest Rate | Bonds Owned | Semiannual Interest |
|---|---|---|---|---|---|---|
| **7.** | JolTrk | $1,000 | $82\frac{1}{4}$ | 7.50% | 5 | _____ |
| **8.** | KDB | 1,000 | $42\frac{1}{8}$ | 5.25% | 2 | _____ |

### ACCRUED INTEREST ON BONDS

Calculate the accrued interest on each of the following corporate bond purchases. Use 30-day months and a 360-day year. Round to the nearest cent.

|  | Bond's Par Value | Bond's Interest Rate | Semiannual Interest Payment Dates | Bond Purchase Date | Days of Accrued Interest | Accrued Interest |
|---|---|---|---|---|---|---|
| **9.** | $1,000 | 12.00% | April 1, Oct. 1 | July 16 | _____ | _____ |
| **10.** | 1,000 | 8.50% | May 15, Nov. 15 | Aug. 25 | _____ | _____ |

## BUYING BONDS

Find the total cost of the following $1,000 par value bonds purchased by Levitt Company.

| Bond | Price Quotation per Bond | Number of Bonds Purchased | Brokerage Fee per Bond | Accrued Interest | Total Cost |
|------|------|------|------|------|------|
| 11. BKR | $96\frac{1}{2}$ | 4 | $10 | $125.00 | _____ |
| 12. SalCo | $105\frac{1}{4}$ | 5 | 12 | 200.00 | _____ |

## SELLING BONDS

Levitt Company sold the bonds shown below. Calculate the net proceeds from each sale.

| Bond | Price Quotation per Bond | Number of Bonds Purchased | Brokerage Fee per Bond | Accrued Interest | Net Proceeds |
|------|------|------|------|------|------|
| 13. BKR | $101\frac{1}{4}$ | 4 | $12 | $250.00 | _____ |
| 14. SalCo | $93\frac{3}{8}$ | 5 | 10 | 100.00 | _____ |

## GAIN OR LOSS ON BONDS

Refer to Problems 11 to 14 to determine the net proceeds and the total cost from each of Levitt Company's investments in bonds. Then determine Levitt's gain (G) or loss (L).

| Bond | Net Proceeds | Total Cost | Gain or Loss |
|------|------|------|------|
| 15. BKR | _____ | _____ | _____ |
| 16. SalCo | _____ | _____ | _____ |

## READING STOCK QUOTATIONS

Refer to the stock price quotations below to answer the following questions. State amounts as dollars and cents.

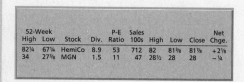

| 52-Week High | Low | Stock | Div. | P-E Ratio | Sales 100s | High | Low | Close | Net Chge. |
|------|------|------|------|------|------|------|------|------|------|
| 82¼ | 67¼ | HemiCo | 8.9 | 53 | 712 | 82 | 81⅜ | 81⅞ | +2⅛ |
| 34 | 27⅜ | MGN | 1.5 | 11 | 47 | 28½ | 28 | 28 | −¼ |

17. What is the highest price a share of MGN stock sold? _____

18. What is the lowest price a share of HemiCo stock sold? _____

19. How many shares of HemiCo stock sold on this day? _____

20. How much higher or lower was today's last sale of MGN stock than yesterday's last sale? _____

## BUYING STOCKS

Calculate the total cost of the following stocks purchased by Hanover Company.

| Stock | Number of Shares Purchased | Price per Share | Brokerage Fee | Total Cost |
|---|---|---|---|---|
| 21. CalCo | 180 | $24\frac{1}{8}$ | $ 73.85 | _____ |
| 22. RDMill | 560 | $39\frac{1}{2}$ | 308.19 | _____ |

## SELLING STOCKS

Calculate the net proceeds from the following investments in stock sold by Hanover Company.

| Stock | Number of Shares Sold | Price per Share | Brokerage Fee | Net Proceeds |
|---|---|---|---|---|
| 23. CalCo | 180 | $46\frac{1}{2}$ | $124.25 | _____ |
| 24. RDMill | 560 | $31\frac{5}{8}$ | 257.38 | _____ |

## GAIN OR LOSS ON STOCKS

Refer to Problems 21 to 24 to list the net proceeds and the total cost from Hanover Company's investment in stocks. Then determine the gain (G) or loss (L).

| Stock | Net Proceeds | Total Cost | Gain or Loss |
|---|---|---|---|
| 25. CalCo | _____ | _____ | _____ |
| 26. RDMill | _____ | _____ | _____ |

## PREFERRED STOCK DIVIDENDS

Calculate the total quarterly dividends for these preferred stocks.

| Stock | Par Value | Dividend Rate | Number of Shares Owned | Total Quarterly Dividend |
|---|---|---|---|---|
| 27. DRA | $100 | 9.50% | 230 | _____ |
| 28. GrAmCo | 50 | 8.00% | 125 | _____ |

## COMMON STOCK DIVIDENDS

Calculate the total annual dividends for these common stocks.

| Company | Quarterly Dividends per Share | Number of Shares | Total Annual Dividend |
|---|---|---|---|
| 29. LockCo | $0.87, $0.86, $0.70, $0.81 | 180 | _____ |
| 30. Milo | $0.13, $0.16, $0.18, $0.20 | 4,260 | _____ |

## READING MUTUAL FUND QUOTATIONS

Use the following mutual fund quotations to answer the questions below.

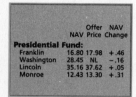

**31.** At what price would an investor buy a share of the Franklin fund?

_____

**32.** At what price would an investor sell a share of the Lincoln fund?

_____

**33.** How much higher or lower did a share of the Monroe fund sell today
than yesterday _____

**34.** At what price would an investor buy a share of the Washington fund?

_____


## BUYING AND SELLING MUTUAL FUNDS

Use the price quotations shown below for the Presidential Fund to calculate the gain (G) or loss (L) on the following mutual fund transactions.

|  | Mutual Fund | Number of Shares | Purchase | | Sale | | Gain or Loss |
|---|---|---|---|---|---|---|---|
|  |  |  | NAV | Offer Price | NAV | Offer Price |  |
| **35.** | Lincoln | 400 | $12.16 | $13.01 | $18.48 | $19.77 | _____ |
| **36.** | Monroe | 160 | 8.05 | 8.61 | 5.12 | 5.48 | _____ |


## REAL ESTATE—GROSS RENTAL INCOME

Calculate the vacancy allowance and the annual gross rental income
earned from each of the following real estate investment properties.

| Property | Number of Rental Units | Rental Period | Rent per Unit | Vacancy Allowance Rate | Vacancy Allowance | Annual Rental Income |
|---|---|---|---|---|---|---|
| **37.** Apartment building | 16 | Monthly | $ 700 | 5% | _____ | _____ |
| **38.** House | 1 | Monthly | 920 | 8% | _____ | _____ |

Calculate the annual rental income from each of the following percentage leases for retail buildings. The landlord receives a monthly rent plus a percent of annual net sales.

| | Monthly Rent | Annual Net Sales | Percent of Annual Net Sales | Annual Gross Rental Income |
|---|---|---|---|---|
| **39.** | $2,400 | $1,365,800 | 1.5% | _____ |
| **40.** | 1,500 | 652,600 | 1.0% | _____ |

## REAL ESTATE—CASH FLOW

Calculate the positive (+) or negative (−) cash flow for each of the following real estate investments.

| | Annual Rental Income | Loan Payment | Real Estate Taxes | Property Insurance | Repairs | Cash Flow (+ or −) |
|---|---|---|---|---|---|---|
| **41.** | $26,950 | $12,814 | $8,798 | $1,362 | $3,481 | _____ |
| **42.** | 8,620 | 0 | 2,465 | 915 | 876 | _____ |

## REAL ESTATE—NET INCOME OR NET LOSS

Calculate the net income (NI) or net loss (NL) from each of the following real estate investments.

| | Annual Rental Income | Interest | Depreciation Expense | Real Estate Taxes | Property Insurance | Repairs | Net Income or Net Loss |
|---|---|---|---|---|---|---|---|
| **43.** | $43,105 | $26,413 | $34,182 | $14,346 | $4,720 | $2,841 | _____ |
| **44.** | 18,724 | 3,214 | 5,789 | 4,109 | 1,560 | 1,205 | _____ |

## REAL ESTATE—RETURN ON INVESTMENT

Calculate the rate of return on the following real estate investments. Round your answers to two decimal places.

| | Annual Net Income | Cash Investment | Closing Costs | Return on Investment |
|---|---|---|---|---|
| **45.** | $ 8,715 | $104,370 | $ 3,685 | _____ |
| **46.** | 35,804 | 451,980 | 10,614 | _____ |

## BUSINESS APPLICATIONS

**47.** Vaske Equipment Company purchased 250 shares of Bosworth Corporation common stock at $28\frac{1}{4}$ and paid a $124.86 brokerage fee. Vaske received quarterly dividends of $0.62, $0.50, $0.46, and $0.65. Exactly 1 year after buying the stock, Vaske sold it at $35\frac{3}{8}$ and paid a $146.25 brokerage fee. What was (a) the quarterly dividends earned; (b) the total gain or loss from owning this stock, including the dividends earned and the gain or loss on the sale?

(a) _____

(b) _____

**48.** Oakes Electric Company purchased 1,200 shares of Butell Mutual Fund when the NAV was $18.42 and the offer price was $19.71. Oakes sold 700 shares when the NAV was $24.60 and the offer price was $26.32 and sold the remaining 500 shares when the NAV was $19.23 and the offer price was $20.58. What was Oake's gain or loss on the sale of its mutual fund shares?

_____

**49.** Logan Enterprises owns an apartment building from which it received $173,650 rental income last year. Costs and expenses were as follows: loan payment, $86,314 ($17,418 was principal, $68,896 was interest); depreciation, $73,240; real estate taxes, $44,780; property insurance, $16,481; and repairs, $9,256. What was Logan's (a) positive or negative cash flow for the year, and (b) net income or net loss for the year?

(a) _____

(b) _____

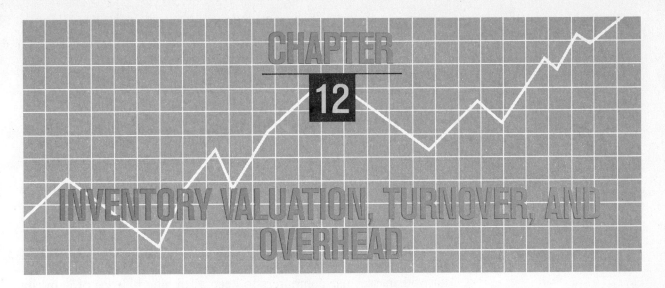

# CHAPTER 12
# INVENTORY VALUATION, TURNOVER, AND OVERHEAD

## LEARNING OBJECTIVES

1. Use perpetual and periodic inventory systems to record purchases and sales of merchandise inventory.
2. Calculate the cost of goods available for sale and the cost of goods sold.
3. Calculate the ending inventory valuation using the specific identification, average cost, FIFO, and LIFO methods.
4. Estimate ending inventory value using the retail method.
5. Calculate the inventory turnover ratio at retail and at cost.
6. Allocate overhead expenses based on floor space and sales.

Merchandise offered for resale by a business is referred to as **merchandise inventory,** or simply **inventory.** Most businesses have a large investment in their merchandise inventory. Therefore, much management attention is directed toward recording and regulating the amount and value of merchandise inventory in stock.

## LESSON 12-1    PERPETUAL AND PERIODIC INVENTORY SYSTEMS

**PERPETUAL INVENTORY SYSTEM**

In a **perpetual inventory system,** inventory cards often are used to record the amount of merchandise purchased and sold. The balance on hand at any given time is always shown on the perpetual **inventory card.** The inventory card on page 398 shows the stock number, date, unit cost, units purchased, units sold, and running balance of quantity in stock.

To find the **running balance** on an inventory card, add the units purchased to the preceding balance or subtract the units sold from the preceding balance. There are 19 items currently on hand according to the inventory card shown below. When the balance falls below the reorder point, more stock must be ordered.

**INVENTORY CARD**

| Stock No.: 10051-D | | | | | |
|---|---|---|---|---|---|
| Item: Dishwasher | **Maximum: 70** **Reorder Point: 30** | | | | |
| Order Date | Purchase Order No. | Unit Cost | Units Purchased | Units Sold | Balance |
| 1/8 | Beginning inventory | $475 | | | 50 |
| 2/15 | | | | 33 | 17 |
| 5/3 | 442-009 | 479 | 46 | | 63 |
| 7/23 | | | | 47 | 16 |
| 9/30 | 245-008 | 481 | 40 | | 56 |
| 10/24 | | | | 37 | 19 |

Many businesses use computers to keep a running balance of items in stock. A computerized inventory system will automatically record the date, unit cost, and quantity purchased and sold. This information is recorded by computer optical scanners that read uniform product codes (UPC) found on labels of most goods. Many businesses have increased inventory accuracy and reduced labor costs by installing a computerized inventory system.

### PRACTICE PROBLEMS

Find the running balance on the following inventory card.

**INVENTORY CARD**

| Stock No.: K9-321 | | | | | |
|---|---|---|---|---|---|
| Item: Color television | **Maximum: 102** **Reorder Point: 40** | | | | |
| Order Date | Purchase Order No. | Unit Cost | Units Purchased | Units Sold | Balance |
| 1/5 | Beginning inventory | $539 | | | 59 |
| 1/13 | | | | 21 | _____ |
| 2/3 | | 545 | 55 | | _____ |
| 2/19 | | | | 57 | _____ |
| 2/29 | | 589 | 63 | | _____ |

## PERIODIC INVENTORY SYSTEM

Periodically, a *physical inventory* is taken and compared with the inventory shown on the inventory cards or computer printouts. A physical inventory is an actual count of items on hand. An inventory taken at regular intervals, usually monthly, quarterly, semiannually, or yearly, is called a *periodic inventory.*

ANSWERS FOR PRACTICE PROBLEMS

38, 93, 36, 99

An inventory is taken to determine the cost of goods sold and to place a value on goods still in stock. The **cost of goods sold** is found by subtracting the ending inventory value from the cost of goods available for sale. The **cost of goods available for sale** is the sum of the beginning inventory value plus the cost of purchases.

$$\begin{array}{c}\text{Beginning} \\ \text{Inventory Value}\end{array} + \text{Purchases} = \begin{array}{c}\text{Cost of Goods} \\ \text{Available for Sale}\end{array}$$

$$\begin{array}{c}\text{Cost of Goods} \\ \text{Available for Sale}\end{array} - \begin{array}{c}\text{Ending} \\ \text{Inventory Value}\end{array} = \begin{array}{c}\text{Cost of Goods} \\ \text{Sold}\end{array}$$

---

### EXAMPLE

Calculate the cost of goods available for sale and the cost of goods sold for the dishwasher, model 10051-D, shown on the inventory card on page 398. The ending inventory value is $11,480.

### SOLUTION

**Step 1** Find the extensions of the beginning inventory and the purchases.

| Date | Description | Unit Cost | × | Quantity | = | Extension |
|------|-------------|-----------|---|----------|---|-----------|
| 1/8 | Beginning inventory | $475 | × | 50 | = | $23,750 |
| 5/3 | Purchase | 479 | × | 46 | = | 22,034 |
| 9/30 | Purchase | 481 | × | 40 | = | 19,240 |

**Step 2** Find the cost of goods available for sale.

$$\begin{array}{c}\text{Beginning} \\ \text{Inventory Value} +\end{array} \quad \text{Purchases} \quad = \begin{array}{c}\text{Cost of Goods} \\ \text{Available for Sale}\end{array}$$
$$\$23,750 \quad + \$22,034 + \$19,240 = \quad \$65,024$$

**Step 3** Find the cost of goods sold.

$$\begin{array}{c}\text{Cost of Goods} \\ \text{Available for Sale} -\end{array} \begin{array}{c}\text{Ending} \\ \text{Inventory Value} =\end{array} \begin{array}{c}\text{Cost of Goods} \\ \text{Sold}\end{array}$$
$$\$65,024 \quad - \quad \$9,139 \quad = \quad \$55,885$$

---

### PRACTICE PROBLEMS

Calculate the cost of goods available for sale and the cost of goods sold.

| | Beginning Inventory | Purchases | Ending Inventory | Cost of Goods Available | Cost of Goods Sold |
|---|---------------------|-----------|------------------|-------------------------|--------------------|
| 1. | $10,390 | $ 52,500 | $19,100 | _____ | _____ |
| 2. | 27,150 | 64,400 | 33,800 | _____ | _____ |
| 3. | 83,300 | 100,300 | 41,030 | _____ | _____ |

**4.** Calculate (a) the extensions, (b) the cost of goods available for sale, and (c) the cost of goods sold. The ending inventory value is $12,350.

| Date | Description | Unit Cost | Quantity | Extension |
|------|-------------|-----------|----------|-----------|
| 1/1 | Beginning inventory | $69.71 | 151 | **a.** _____ |
| 1/12 | Purchase | 72.15 | 160 | **a.** _____ |
| 1/25 | Purchase | 72.98 | 203 | **a.** _____ |

**b.** Cost of goods available for sale:  $_____

**c.** Cost of goods sold:  $_____

---

ANSWERS FOR
PRACTICE PROBLEMS

**Starting on page 399:**

| | Cost of Goods Available | Cost of Goods Sold |
|---|---|---|
| (1) | $ 62,890 | $ 43,790 |
| (2) | 91,550 | 57,750 |
| (3) | 183,600 | 142,570 |

(4) (a) $10,526.21, $11,544.00, $14,814.94   (b) 36,885.15   (c) 24,555.15

## LESSON 12-1    EXERCISES

Calculate the cost of goods available for sale and the cost of goods sold.

| | Beginning Inventory | Purchases | Ending Inventory | Cost of Goods Available | Cost of Goods Sold |
|---|---|---|---|---|---|
| 1. | $ 9,849 | $44,730 | $23,420 | _____ | _____ |
| 2. | 80,535 | 94,500 | 66,990 | _____ | _____ |
| 3. | 1,470 | 3,969 | 1,890 | _____ | _____ |
| 4. | 12,688 | 11,405 | 11,601 | _____ | _____ |
| 5. | 26,160 | 70,850 | 40,330 | _____ | _____ |

Find the extensions, the cost of goods available for sale, and the cost of goods sold.

6. The ending inventory value is $520,500.

| Date | Description | Unit Cost | Quantity | Extension |
|---|---|---|---|---|
| 7/2 | Beginning inventory | $362 | 1,200 | _____ |
| 7/9 | Purchase | 367 | 900 | _____ |
| 7/16 | Purchase | 370 | 750 | _____ |
| 7/23 | Purchase | 385 | 1,100 | _____ |
| 7/30 | Purchase | 401 | 500 | _____ |

Cost of goods available for sale: _____

Cost of goods sold: _____

7. The ending inventory value is $39,935.

| Date | Description | Unit Cost | Quantity | Extension |
|---|---|---|---|---|
| 3/1 | Beginning inventory | $11.95 | 2,700 | _____ |
| 3/8 | Purchase | 11.15 | 3,180 | _____ |
| 3/17 | Purchase | 12.20 | 1,070 | _____ |
| 3/24 | Purchase | 13.00 | 950 | _____ |
| 3/31 | Purchase | 13.29 | 1,000 | |

Cost of goods available for sale: _____

Cost of goods sold: _____

## BUSINESS APPLICATIONS

**8.** The beginning inventory of Elk's Hunting Supplies for the month of September is $245,600. Purchases for the month total $100,800 and the ending inventory value is $96,540. Find (a) the cost of goods available for sale and (b) the cost of goods sold.

(a) _____

(b) _____

**9.** On September 1 Williamson's Lighting Center had a beginning inventory of $1,290,000. During the month the owner made purchases of $230,000, $50,800, and $72,300. The inventory value on September 30 was $456,000. Find (a) the cost of goods available for sale and (b) the cost of goods sold.

(a) _____

(b) _____

Enter the running balance on the following inventory cards. Then, calculate (a) the cost of goods available for sale and (b) the cost of goods sold.

**10.** The ending inventory value is $317.

### INVENTORY CARD

**Stock No.:** 945—TR
**Item:** Pocket calculator

| Order Date | Unit Cost | Units Purchased | Units Sold | Balance |
|---|---|---|---|---|
| 1/5 | $ 8.25 | Beginning balance | | 100 |
| 2/2 | | | 46 | _____ |
| 3/15 | 8.79 | 57 | | _____ |
| 3/21 | | | 98 | _____ |
| 4/28 | 9.05 | 105 | | _____ |
| 5/20 | 9.72 | 84 | | _____ |
| 6/10 | | | 153 | _____ |
| 6/15 | 10.10 | 132 | | _____ |
| 7/30 | | | 148 | _____ |

(a) _____

(b) _____

**THE SPECIFIC IDENTIFICATION INVENTORY VALUATION METHOD**

Many businesses that sell a small volume of high-priced items such as automobiles or jewelry can easily identify the original purchase cost of items in stock. The **specific identification** inventory valuation method is used when each item in stock is valued at its actual purchase cost. The value of ending inventory is the sum of the values of the remaining items in stock.

### EXAMPLE

Business Systems Plus opened for business on January 1. At that time the owner purchased the following computer components:

20 Model X-L Computers for $1,560 each
24 Model T-7 Computers for $1,912 each
27 Model R2-5 Computers for $2,393 each

On January 31 the owner had in stock 7 model X-Ls, 10 model T-7s, and 9 model R2-5s. Use the specific identification inventory method to find the cost of goods available for sale, the ending inventory value, and the cost of goods sold.

### SOLUTION

**Step 1**  Find the cost of goods available for sale.

| Units | Quantity in Stock, Jan. 1 × | Unit Cost | = Extension |
|---|---|---|---|
| Model X-L | 20 | × $1,560 = | $31,200 |
| Model T-7 | 24 | × 1,912 = | 45,888 |
| Model R2-5 | 27 | × 2,393 = | 64,611 |
| | | | $141,699    Cost of goods available for sale |

**Step 2**  Find the ending inventory value.

| Units | Quantity in Stock, Jan. 31 × | Unit Cost | = Extension |
|---|---|---|---|
| Model X-L | 7 | × $1,560 = | $10,920 |
| Model T-7 | 10 | × 1,912 = | 19,120 |
| Model R2-5 | 9 | × 2,393 = | 21,537 |
| | | | $51,577    Ending inventory value |

**Step 3**  Find the cost of goods sold.

| Cost of Goods Available for Sale | − | Ending Inventory Value | = | Cost of Goods Sold |
|---|---|---|---|---|
| $141,699 | − | $51,577 | = | $90,122 |

Use the specific identification method to calculate the cost of goods available for sale, the ending inventory value, and the cost of goods sold.

| Units | Quantity in Stock, March 1 | Quantity in Stock, March 31 | Unit Cost |
|---|---|---|---|
| Recliner | 37 | 11 | $219 |
| Rolltop desk | 26 | 8 | 349 |
| Sleeper sofa | 43 | 19 | 695 |

1. Cost of goods available for sale: $_____

2. Inventory value, March 31: $_____

3. Cost of goods sold: $_____

## THE AVERAGE COST INVENTORY VALUATION METHOD

When a wide variety of items are bought and sold in large volumes at various prices throughout the year, the cost of each individual item is not easily identified. Therefore, other inventory valuation methods are used to find the ending inventory value.

Companies that sell in bulk products such as grain and oil may use the average cost method. The **average cost method** uses an average unit cost to value ending inventory. To calculate the average unit cost, divide the cost of goods available for sale by the number of units available for sale. To find the ending inventory value, multiply the number of units in stock by the average unit cost.

$$\frac{\text{Cost of Goods Available for Sale}}{\text{Units Available for Sale}} = \text{Average Unit Cost}$$

$$\text{Number of Units in Stock} \times \text{Average Unit Cost} = \text{Ending Inventory Value}$$

### EXAMPLE

The inventory card below shows the number of classical music albums in stock from January through June at Sharp's Music City. On June 30 the ending inventory is 7. Use the average cost method to find the average unit cost and the ending inventory value.

### SOLUTION

**Step 1** Find the number of units available for sale and the cost of goods available for sale.

| Date | Description | Quantity | Unit Cost | Extension |
|---|---|---|---|---|
| 1/2 | Beginning inventory | 12 | $16.50 | $198.00 |
| 3/10 | Purchase | 10 | 16.80 | 168.00 |
| 4/21 | Purchase | 15 | 17.00 | 255.00 |
| 6/30 | Purchase | 8 | 17.10 | 136.80 |
| | | 45 Units available | | $757.80 Cost of goods available |

**(1)** 47,062 **(2)** 18,406 **(3)** 28,656

**Step 2** Find the average unit cost.

$$\text{Average Unit Cost} = \frac{\text{Cost of Goods Available for Sale}}{\text{Units Available for Sale}} = \frac{\$757.80}{45} = \$16.84$$

**Step 3** Find the ending inventory value.

| Number of Units in Stock | × Average Unit Cost = | Ending Inventory Value |
|---|---|---|
| 7 | × $16.84 = | $117.88 |

PRACTICE PROBLEMS

Calculate the average unit cost and the ending inventory value using the average cost method.

| | Units Available | Cost of Goods Available | Number of Units in Stock | Average Unit Cost | Ending Inventory Value |
|---|---|---|---|---|---|
| 1. | 1,340 | $69,881 | 563 | _____ | _____ |
| 2. | 3,590 | 76,467 | 986 | _____ | _____ |
| 3. | 12,900 | 16,125 | 1,400 | _____ | _____ |

4. Use the average cost method to calculate (a) the cost of goods available for sale, (b) the average unit cost, and (c) the ending inventory for the following item. The ending inventory is 503 units.

| Date | Description | Quantity | Unit Cost | Extension |
|---|---|---|---|---|
| 3/1 | Beginning inventory | 1,020 | $2.15 | _____ |
| 3/19 | Purchase | 710 | 2.27 | _____ |
| 3/31 | Purchase | 535 | 3.10 | _____ |
| | | _____ | | _____ (total) |

a. Cost of goods available for sale: $_____

b. Average unit cost: $_____

c. Ending inventory value: $_____

ANSWERS FOR
PRACTICE PROBLEMS

| | Average Unit Cost | Ending Inventory Value |
|---|---|---|
| (1) | $52.15 | $29,560.45 |
| (2) | 21.30 | 21,001.80 |
| (3) | 1.25 | 1,750.00 |

(4)(a) 5,463.20  (b) 2.41  (c) 1,212.23

## THE FIRST-IN, FIRST-OUT (FIFO) INVENTORY VALUATION METHOD

The *first-in, first-out (FIFO)* inventory valuation method assumes that the first items purchased are the first items sold. Therefore, the value of the ending inventory is based on the cost of the last items purchased.

### EXAMPLE

The following information from an inventory card shows the number of picture frames purchased during the month of November at Remark's Art Supplies and Crafts. On November 30, there were 53 frames in stock. Use the FIFO method to find the ending inventory valuation.

| Date | Description | Quantity | Unit Cost |
|------|-------------|----------|-----------|
| November 1 | Beginning inventory | 38 | $21.60 |
| November 11 | Purchase | 19 | 22.03 |
| November 19 | Purchase | 21 | 22.57 |
| November 29 | Purchase | 35 | 23.10 |

### SOLUTION

The value of the 53 units in stock is made up of the last 35 units purchased on November 29 at $23.10 each plus 18 of the 21 units purchased on November 19 at $22.57 each. To find the ending inventory valuation, total the extensions of the last 53 units purchased.

| Units | | Unit Cost | | Extension |
|-------|---|-----------|---|-----------|
| 35 | × | $23.10 | = | $ 808.50 |
| 18 | × | 22.57 | = | 406.26 |
| 53 Total units | | | | $1,214.76 Ending inventory value |

### PRACTICE PROBLEM

Calculate the ending inventory value for the following items using the FIFO inventory method.

The ending inventory is 32 units.

| Date | Description | Quantity | Unit Cost |
|------|-------------|----------|-----------|
| 5/1 | Beginning inventory | 12 | $132 |
| 5/18 | Purchase | 23 | 137 |
| 5/30 | Purchase | 27 | 142 |

Ending inventory value: $_____

## THE LAST-IN, FIRST-OUT (LIFO) INVENTORY VALUATION METHOD

The *last-in, first-out (LIFO)* method is the opposite of the FIFO method. The LIFO method assumes that the last items purchased are the first items sold. Therefore, the ending inventory valuation is based on the cost of the first items purchased.

The FIFO and LIFO methods make no attempt to identify specific items sold or retained in inventory; rather, they identify a flow of costs for accounting purposes.

ANSWER FOR
PRACTICE PROBLEM

4,519

## EXAMPLE

The following information from an inventory card shows the number of compact disk players purchased during the first quarter of the year at Summit Appliance and Stereo. On March 31 there were 31 disk players in stock. Use the LIFO method to find the ending inventory valuation.

| Date | Description | Quantity | Unit Cost |
|------|-------------|----------|-----------|
| 1/2 | Beginning inventory | 14 | $345 |
| 2/27 | Purchase | 24 | 352 |
| 3/5 | Purchase | 18 | 361 |
| 3/30 | Purchase | 29 | 366 |

## SOLUTION

The value of the 31 units in stock is made up of the first 14 units in stock on January 1 at $345 each plus 17 of the 24 units purchased on February 27 at $352 each. To find the ending inventory value, total the extensions of the first 31 units purchased.

| Units | × Unit Cost | = Extensions |
|-------|-------------|--------------|
| 14 | × $345 | = $ 4,830 |
| 17 | × $352 | = 5,984 |
| 31 Total units | | $10,814 Ending inventory value |

## PRACTICE PROBLEM

Calculate the ending inventory value using the LIFO inventory method.

The ending inventory is 70 units.

| Date | Description | Quantity | Unit Cost |
|------|-------------|----------|-----------|
| 10/1 | Beginning inventory | 63 | $ 9.45 |
| 10/16 | Purchase | 105 | $10.20 |
| 10/31 | Purchase | 51 | $10.75 |

Ending inventory value: $_____

## ESTIMATING INVENTORY VALUATION

Most businesses take one physical inventory a year, usually at the end of the annual accounting period. Thus, when the value of inventory is needed to prepare financial statements on a monthly or quarterly basis, the value of inventory at cost must be estimated.

Many retailers use the **retail inventory method** to estimate their inventory. With this method the ending inventory value at retail is calculated and then converted to an estimated ending inventory value at cost.

To find the **ending inventory value at retail,** subtract net sales from merchandise available for sale at retail. To find the estimated inventory value at cost, multiply the inventory at retail by a ratio of goods available for sale at cost to goods available for sale at retail.

$$\text{Goods Available for Sale at Retail} - \text{Net Sales} = \text{Inventory Value at Retail}$$

$$\frac{\text{Goods Available for Sale at Cost}}{\text{Goods Available for Sale at Retail}} \times \text{Inventory Value at Retail} = \text{Estimated Inventory Value at Cost}$$

## EXAMPLE

On October 1 Silk Creations had a beginning inventory value of $28,000 at cost and $42,100 at retail. During the month, purchases were $56,000 at cost and $61,000 at retail. Net sales were $64,000. Calculate the estimated ending inventory at cost using the retail method.

## SOLUTION

**Step 1**  Calculate the goods available for sale at cost and at retail.

Beginning Inventory + Purchases = Goods Available for Sale

| | | |
|---|---|---|
| $28,000 | + $56,000 = | $84,000 (at cost) |
| $42,100 | + $61,000 = | $103,100 (at retail) |

**Step 2**  Calculate the ending inventory value at retail.

$$\text{Goods Available for Sale at Retail} - \text{Net Sales} = \text{Ending Inventory Value at Retail}$$
$$\$103,100 - \$64,000 = \$39,100$$

**Step 3**  Calculate estimated inventory value at cost. Round to the nearest cent.

$$\frac{\text{Goods Available for Sale at Cost}}{\text{Goods Available for Sale at Retail}} \times \frac{\text{Inventory}}{\text{Value at Retail}} = \frac{\text{Estimated Inventory}}{\text{Value at Cost}}$$

$$\frac{\$84,000}{\$103,100} \times \$39,100 = \$31,856.45$$

## PRACTICE PROBLEMS

Calculate the inventory value at retail and the estimated ending inventory value at cost using the retail inventory method. Round to the nearest cent.

| | Net Sales | Goods Available for Sale at Cost | Goods Available for Sale at Retail | Inventory Value at Retail | Estimated Inventory Value at Cost |
|---|---|---|---|---|---|
| **1.** | $ 73,000 | $ 94,050 | $132,600 | _____ | _____ |
| **2.** | 89,000 | 106,100 | 165,200 | _____ | _____ |
| **3.** | 228,700 | 184,800 | 319,200 | _____ | _____ |

## LESSON 12-2    EXERCISES

Use the specific identification inventory evaluation method to calculate the cost of goods available for sale, the total ending inventory value, and the cost of goods sold.

**1.** LEISURE-TIME OUTDOOR FURNITURE INVENTORY SUMMARY

| Units | Quantity in Stock, May 1 | Quantity in Stock, May 31 | Unit Cost |
|---|---|---|---|
| Lawn chair | 98 | 34 | $ 49.90 |
| Picnic table | 32 | 16 | 159.90 |
| Swing | 18 | 8 | 189.95 |
| Bench | 23 | 13 | 99.95 |

**a.** Cost of goods available for sale: _____

**b.** Ending inventory value: _____

**c.** Cost of goods sold: _____

**2.** HARTFORD APPLIANCE COMPANY INVENTORY

| Units | Quantity in Stock, April 1 | Quantity in Stock, April 30 | Unit Cost |
|---|---|---|---|
| Model C-365 | 65 | 15 | $260.85 |
| Model C-356 | 84 | 27 | 275.68 |
| Model DA-63 | 61 | 16 | 219.34 |
| Model DA-64 | 47 | 9 | 214.87 |

**a.** Cost of goods available for sale: _____

**b.** Ending inventory value: _____

**c.** Cost of goods sold: _____

Calculate the average unit cost and ending inventory value using the average cost method.

| | Number of Units Available | Cost of Goods Available for Sale | Number of Units in Stock | Average Unit Cost | Ending Inventory Value |
|---|---|---|---|---|---|
| **3.** | 22,400 | $ 45,248 | 985 | _____ | _____ |
| **4.** | 430 | 29,025 | 107 | _____ | _____ |
| **5.** | 2,780 | 119,540 | 560 | _____ | _____ |
| **6.** | 10,800 | 130,680 | 2,700 | _____ | _____ |
| **7.** | 910 | 455 | 312 | _____ | _____ |
| **8.** | 8,790 | 968,658 | 2,300 | _____ | _____ |

Calculate the ending inventory value using the average cost, FIFO, and LIFO inventory valuation methods. Round the average unit cost to the nearest cent.

**9.** Product: Men's suit #349
Ending Inventory: 34 units

| Date | Description | Quantity | Unit Cost |
|------|-------------|----------|-----------|
| 8/1 | Beginning inventory | 27 | $167.59 |
| 8/15 | Purchase | 30 | 171.84 |
| 8/30 | Purchase | 8 | 172.95 |

**a.** Average cost ending inventory value: _____

**b.** FIFO ending inventory value: _____

**c.** LIFO ending inventory value: _____

**10.** Product: Candy bar #12-78
Ending Inventory: 578 units

| Date | Description | Quantity | Unit Cost |
|------|-------------|----------|-----------|
| 10/1 | Beginning inventory | 450 | $1.05 |
| 10/18 | Purchase | 875 | 1.29 |
| 10/29 | Purchase | 556 | 1.25 |

**a.** Average cost ending inventory value: _____

**b.** FIFO ending inventory value: _____

**c.** LIFO ending inventory value: _____

Calculate the inventory value at retail and the estimated inventory value at cost using the retail inventory method. Round to the nearest cent.

| | Net Sales | Goods Available for Sale at Cost | Goods Available for Sale at Retail | Inventory Value at Retail | Estimated Inventory Value at Cost |
|------|-----------|------------|------------|------------|------------|
| **11.** | $173,600 | $141,200 | $ 207,600 | _____ | _____ |
| **12.** | 59,800 | 65,800 | 89,700 | _____ | _____ |
| **13.** | 617,000 | 796,650 | 1,400,100 | _____ | _____ |

## BUSINESS APPLICATIONS

Round to the nearest cent unless otherwise directed.

**14.** In May the Children's World toy store had a beginning inventory of 9 skateboards at a cost of $24.90 each. During the month the store purchased 10 at $25 each, 12 at $25.80 each, and sold 22. Find (a) the average unit cost and (b) the ending inventory value using the average cost method.

(a) _____

(b) _____

**15.** On April 1 Diamond Jim's Jewelry had the following watches in stock:

15 diamond watches for $303 each
20 sports watches for $219 each
37 digital watches for $76 each
19 children's watches for $27 each

During the month they sold 9 diamond watches, 13 sports watches, 29 digital watches, and 11 children's watches. Find (a) the cost of goods available for sale, (b) the ending inventory value using the specific identification method, and (c) the cost of goods sold.

(a) _____

(b) _____

(c) _____

**16.** Dilling's Sports Supply purchased the following number of tennis rackets during the year:

January: 163 rackets at $98 each
February: 94 rackets at $105 each
July: 71 rackets at $95 each
October: 105 rackets at $110 each

At the end of the year, there were 172 tennis rackets in stock. Use the LIFO inventory valuation method to calculate (a) the cost of goods available for sale, (b) the ending inventory value, and (c) the cost of goods sold.

(a) _____

(b) _____

(c) _____

**17.** On May 1 Delmar Food's beginning inventory in the produce department was $36,500 at cost and $45,000 at retail. During the month, purchases were $37,100 at cost and $47,000 at retail. Net sales were $57,000 for the month. Calculate (a) the inventory value at retail and (b) the estimated inventory value at cost using the retail method.

(a) _____

(b) _____

**18.** On April 1 the McGregory Department Store's beginning inventory in the furniture department was $123,300 at cost and $285,000 at retail. Purchases during the month totaled $165,600 at cost and $110,400 at retail. Sales for the month were $232,600. Calculate (a) the inventory value at retail and (b) the estimated inventory at cost using the retail method. Round to the nearest cent.

(a) _____

(b) _____

## CALCULATING INVENTORY TURNOVER AT RETAIL

The *inventory turnover* is the number of times a business replaces its stock of merchandise during a given time period, usually 1 year. For example, an inventory turnover of 3 means that the inventory was replaced on the average of three times during the year. Inventory turnover varies widely depending on the type of business. For example, the turnover for expensive jewelry may be a 4, while the turnover for fresh vegetables may be 240.

Businesses that value inventory at retail find the inventory turnover at retail. The *inventory turnover at retail* is expressed as a ratio of sales to average inventory value at retail.

$$\frac{\text{Sales}}{\text{Average Inventory Value at Retail}} = \text{Inventory Turnover Ratio at Retail}$$

The *average inventory value* is found by dividing the sum of the inventory values at retail by the number of times inventory is taken per year. For example, if inventory is taken semiannually, divide the sum of the beginning, middle, and ending inventory values by 3 to find the average inventory value. Likewise, if inventory is taken at the beginning of the year and at the end of each month, divide the sum of the inventory values by 13.

### EXAMPLE

Total yearly sales for the Pet Parade are $1,566,000. The beginning inventory value at retail is $473,300, and the ending inventory value at retail is $236,700. Find the inventory turnover ratio at retail. Round to the nearest hundredth.

### SOLUTION

**Step 1** Find the average inventory value at retail.

$$\text{Average Inventory Value at Retail} = \frac{\text{Sum of Inventory Values}}{\text{Number of Inventories}}$$

$$= \frac{(\$473,300 + \$236,700)}{2} = \$355,000$$

**Step 2** Find the inventory turnover ratio at retail. Round to the nearest hundredth.

$$\text{Inventory Turnover Ratio at Retail} = \frac{\text{Sales}}{\text{Average Inventory Value}} = \frac{\$1,566,000}{\$355,000} = 4.411 \approx 4.41$$

### PRACTICE PROBLEMS

Calculate (a) the average inventory value and (b) the inventory turnover ratio at retail. Round to the nearest hundredth.

1. Beginning and ending inventory values at retail are $7,200 and $4,020. Sales are $16,000.
   a. Average inventory: $_____
   b. Turnover ratio: _____

**2.** Quarterly inventory values at retail are $73,200, $56,310, $42,700, $29,200, and $61,300. Net sales are $445,600.
  **a.** Average inventory: $_____
  **b.** Turnover ratio: _____

**3.** Semiannual inventory values at retail are $103,200, $94,300, and $86,999. Net sales are $351,250.
  **a.** Average inventory: $_____
  **b.** Turnover ratio: _____

## CALCULATING INVENTORY TURNOVER AT COST

Businesses that value inventory at cost, such as wholesalers, usually find it more convenient to calculate inventory turnover at cost. The *inventory turnover at cost* is expressed as a ratio of cost of goods sold to average inventory value at cost.

$$\frac{\text{Cost of Goods Sold}}{\text{Average Inventory Value}} = \text{Inventory Turnover Ratio at Cost}$$

Inventory turnover ratio at retail is usually lower than the inventory turnover ratio at cost due to theft, markdown, spoilage, product deterioration, and so on.

### EXAMPLE

Cost of goods sold at Bell-Wick Manufacturers is $9,350,600 for the year. The semiannual inventory values at cost are $4,340,000, $4,400,000, and $3,300,000. Calculate the inventory turnover ratio at cost. Round ratio to the nearest hundredth.

### SOLUTION

**Step 1**  Find the average inventory value at cost.

$$\text{Average Inventory Value at Cost} = \frac{\text{Sum of Inventory Values}}{\text{Number of Inventories}} \quad \$4,340,000 + \$4,400,000 + \$3,300,000$$

$$= \frac{\$12,040,000}{3} = \$4,013,333.33$$

**Step 2**  Find the inventory turnover ratio at cost. Round to the nearest hundredth.

$$\text{Inventory Turnover Ratio at Cost} = \frac{\text{Cost of Goods Sold}}{\text{Average Inventory Value at Cost}} = \frac{\$9,350,600}{\$4,013,333.33} = 2.329 \approx 2.33$$

## PRACTICE PROBLEMS

Calculate (a) the average inventory value at cost and (b) the turnover ratio at cost. Round ratios to the nearest hundredth.

1. Beginning and ending inventory values at cost are $16,100 and $15,030. Cost of goods sold is $66,900.
   a. Average inventory value: $_____
   b. Turnover ratio: _____
2. Semiannual inventory values at cost are $45,300, $67,200, and $50,799. Cost of goods sold is $81,650.
   a. Average inventory value: $_____
   b. Turnover ratio: _____
3. Quarterly inventory values at cost are $71,300, $60,000, $49,900, $55,300, and $63,600. Cost of goods sold is $648,216.
   a. Average inventory value: $_____
   b. Turnover ratio: _____

## ALLOCATING OVERHEAD TO DEPARTMENTS

Expenses, such as wages and supplies, that can be identified as being incurred by a specific department of the business are called **direct expenses.** Other expenses, such as rent, insurance, and utilities, that cannot be identified as being incurred by a specific department are called **indirect expenses** or **overhead.** Businesses that prepare departmental budgets or that determine net income or loss by department distribute overhead to the departments through a process called **allocation.**

The two most common bases for allocating overhead are the amount of floor space in each department and the amount of sales in each department. The method used depends on which will make the fairest allocation of overhead. To find the **overhead allocated to each department** multiply the total overhead by a ratio of floor space or sales per department to total floor space or total sales.

$$\frac{\text{Department's Floor Space or Sales}}{\text{Total Floor Space or Sales}} \times \text{Overhead} = \frac{\text{Overhead}}{\text{per Department}}$$

### EXAMPLE

The total overhead for Elite Manufacturers is $72,000. The company allocates its overhead based on floor space occupied by each department. Department A occupies 4,800 square feet, Department B occupies 4,200 square feet, and Department C occupies 3,000 square feet. Find the amount of overhead allocated to each department.

### SOLUTION

Step 1   Find the total square feet of floor space.

Department A + Department B + Department C = Total Square Feet
   4,800      +      4,200    +     3,000    =      12,000

**Step 2** Calculate the overhead per department.

|  | Square Feet Per Department | ÷ | Total Square Feet | × Overhead | = | Overhead per Department |
|---|---|---|---|---|---|---|
| Dept. A | 4,800 | ÷ | 12,000 | × $72,000 | = | $28,800 |
| Dept. B | 4,200 | ÷ | 12,000 | × $72,000 | = | 25,200 |
| Dept. C | 3,000 | ÷ | 12,000 | × $72,000 | = | 18,000 |
|  | 12,000 |  |  |  |  | $72,000 |

## PRACTICE PROBLEM

Calculate the amount of overhead allocated to each department based on floor space. Total overhead is $62,000.

| Department | Floor Space (Square Feet) | Overhead per Department |
|---|---|---|
| Sales | 5,760 | **(b)** _____ |
| Purchasing | 5,040 | **(c)** _____ |
| Advertising | 3,960 | **(d)** _____ |
| Marketing | 3,240 | **(e)** _____ |
| Total square feet: | **(a)** _____ |  |

## LESSON 12-3     EXERCISES

Calculate the average inventory value and the average turnover ratio at retail. Round ratios to the nearest tenth.

| Inventory Date | Inventory Values at Retail | Average Inventory | Net Sales | Turnover Ratio |
|---|---|---|---|---|
| **1.** January 1 | $218,300 | | | |
| June 30 | 175,000 | | | |
| December 31 | 98,298 | _____ | $ 655,464 | _____ |
| **2.** January 1 | 48,700 | | | |
| March 31 | 57,400 | | | |
| June 30 | 49,700 | | | |
| September 30 | 25,200 | | | |
| December 31 | 67,500 | _____ | 1,653,012 | _____ |

Calculate the average inventory value at cost and the average turnover ratio at cost. Round ratios to the nearest tenth.

| Inventory Date | Inventory Values at Cost | Cost of Goods Sold | Average Inventory | Inventory Turnover Ratio at Cost |
|---|---|---|---|---|
| **3.** January 1 | $21,000 | | | |
| June 30 | 15,500 | | | |
| December 31 | 25,000 | $1,270,100 | _____ | _____ |
| **4.** January 1 | 98,100 | | | |
| March 31 | 72,000 | | | |
| June 30 | 51,400 | | | |
| September 30 | 71,300 | | | |
| December 31 | 65,600 | 3,619,840 | _____ | _____ |

Find the overhead ratio and the amount of overhead allocated to each department. Total columns. Round to the nearest hundredth.

**5.** Business: Renaldo's Fashion Design, Inc.
Overhead Expense: $59,700 allocated based on floor space

| Department | Floor Space | Overhead per Department |
|---|---|---|
| Design | 8,640 sq ft | _____ |
| Sales | 5,280 sq ft | _____ |
| Advertising | 5,760 sq ft | _____ |
| Marketing | 4,320 sq ft | _____ |
| Totals | _____ | _____ |

**6.** Business: Business Plus
   Overhead Expense: $21,300 allocated based on sales

| Department | Sales | Overhead per Department |
|---|---|---|
| Computers | $160,860 | _____ |
| Calculators | 110,304 | _____ |
| Software | 128,688 | _____ |
| Miscellaneous Supplies | 59,748 | _____ |
| Totals | _____ | _____ |

## BUSINESS APPLICATIONS

**7.** The Radcliff Company has a beginning inventory at retail of $70,245, ending inventory at retail of $84,420, and sales of $308,700. Find the inventory turnover ratio at retail. Round to the nearest tenth.

**8.** The McDaniel's Company has semiannual inventory values at cost of $92,350, $105,600, and $84,500. The cost of goods sold is $1,486,500. Find the inventory turnover ratio at cost. Round to the nearest tenth.

**9.** The amount of overhead expense allocated to the sales department based on floor space is $11,515. If the sales department occupies 35% of the total floor space, what is the total amount of overhead expense?

**10.** Shirewood Bakery had beginning and ending inventory values at cost of $12,340 and $9,380. The cost of goods sold is $460,225. Find the inventory turnover ratio at cost. Round to the nearest tenth.

**11.** The Bensen Company allocates their overhead expenses of $9,800 based on sales. Sales in the motor division total $47,484. Total sales for the company are $131,900. What is the amount of overhead allocated to the motor division?

## CHAPTER 12 REVIEW

Calculate the cost of goods available for sale and the cost of goods sold. Round amounts to the nearest cent.

| | Beginning Inventory | Purchases | Ending Inventory | Cost of Goods Available | Cost of Goods Sold |
|---|---|---|---|---|---|
| 1. | $14,773 | $67,095 | $35,130 | _____ | _____ |
| 2. | 2,205 | 5,954 | 2,835 | _____ | _____ |

Use the specific identification inventory valuation method to find the cost of goods available for sale, the ending inventory value, and the cost of goods sold.

### THE RACER'S EDGE CYCLE SHOP
### INVENTORY SUMMARY

| Units | Quantity in Stock, January 1 | Quantity in Stock, January 31 | Unit Cost |
|---|---|---|---|
| Model #82V | 47 | 6 | $689 |
| Model #39D | 39 | 7 | $439 |
| Model #73D | 94 | 15 | $395 |

**3.** Cost of goods available for sale: _____

**4.** Ending inventory value: _____

**5.** Cost of goods sold: _____

Calculate the ending inventory valuation using the average cost method, FIFO method, and LIFO method. Round amounts to the nearest cent.

Product: Accent lamps #679-00-D
Ending Inventory: 2,030 units

| Date | Description | Quantity | Unit Cost |
|---|---|---|---|
| 11/3 | Beginning inventory | 1,950 | $0.62 |
| 11/19 | Purchase | 2,000 | 0.89 |
| 11/30 | Purchase | 1,500 | 1.05 |

**6.** Average cost inventory value: _____

**7.** FIFO inventory value: _____

**8.** LIFO inventory value: _____

Calculate the inventory value at retail and the estimated ending inventory at cost using the retail inventory method. Round to the nearest cent.

| | Net Sales | Goods Available for Sale at Cost | Goods Available for Sale at Retail | Inventory Value at Retail | Estimated Inventory Value at Cost |
|---|---|---|---|---|---|
| **9.** | $ 54,750 | $ 70,538 | $ 99,450 | _____ | _____ |
| **10.** | 171,520 | 107,730 | 239,400 | _____ | _____ |

Calculate the inventory turnover ratio at retail and at cost. Round to the nearest hundredth.

| | Average Inventory at Retail | Average Inventory at Cost | Sales | Cost of Goods | Turnover at Retail | Turnover at Cost |
|---|---|---|---|---|---|---|
| **11.** | $32,100 | $19,550 | $138,244 | $85,085 | _____ | _____ |

## BUSINESS APPLICATIONS

**12.** Inventory at the Recreational Warehouse was $16,500 on January 1, $12,890 on June 30, and $14,240 on December 31. Calculate the average inventory value. Round to the nearest cent.

_____

**13.** Clemson Sports Manufacturing allocates its overhead expense of $22,400 based on total floor space of 8,000 sq ft. Find the amount of overhead allocated to the shipping department, which occupies 3,040 sq ft.

_____

## CUMULATIVE REVIEW III

Round amounts to the nearest cent and percents to two decimal places.

1. L. T. Salvadore, Inc., had sales of $5,384,718 in 19X3 and $5,986,052 in 19X4. What percent did Salvadore's sales increase or decrease from 19X3 to 19X4?

_____

2. Jay Carter is paid $10.25 per hour and is paid $1\frac{1}{2}$ times the regular rate for hours worked over 40 in a week. Last week Jay worked 47 hours. What was Jay's gross pay for the week?

_____

3. Delling Auction Company borrowed $25,500 at $9\frac{1}{2}$% simple interest for 90 days. What is interest on the loan using the exact-interest method?

_____

4. On October 6 Bingham Corporation purchased goods with a $1,750 list price and 40% trade discount. Terms were 2/10, n/30. How much does Bingham owe if payment is made on October 16?

_____

5. Central City Variety sold a customer the following items: eight at $13.98 each, five at $10.25 each, and six at $15.75 each. The entire order is subject to a 5% sales tax. What is the total amount due from the customer?

_____

**6.** Quality Printing Company can purchase an offset press for $8,995 cash. The press can also be purchased on the installment plan with $995 down and $177.96 per month for 60 months. What is the total sale price for the press?

**7.** Morton Insurance Agency's office building has a $675,500 market value. Property in the city is assessed at 80% of market value and the tax rate is $4.80 per $100. What is the annual property tax on Morton's office building?

**8.** On September 1 Claremont Gifts had a beginning inventory of $472,355. During the month, Claremont made purchases of $8,475, $12,316, and $5,428. On September 30, the inventory value was $481,509. What was the cost of goods sold?

**9.** Filmore Office Equipment Company had six model PL193 file cabinets on hand at a cost of $58.75 each and made the following purchases: March 10, four at $56.90 each; May 24, six at $55.75 each; August 12, five at $58.20 each, and October 9, three at $59.60 each. On December 31 there are nine file cabinets in inventory. What is the year-end inventory valuation using the FIFO inventory valuation method?

**10.** The Bike Shop had four model TR-100 bicycles on hand at a cost of $149.60 each and made purchases: March 5, two at $147.75; June 5, six at $154.90 each; July 15, eight at $157.80 each; and August 18, eight at $159.60 each. On December 31 there are eight model TR-100 bicycles in inventory. What is the year-end inventory valuation using the LIFO inventory valuation method?

# CHAPTER

## 13

# DEPRECIATION

### LEARNING OBJECTIVES

1. Calculate the depreciation rate, annual depreciation, and book value of an asset using straight-line, units-of-production, declining-balance, sum-of-the-years'-digits, and the ACRS depreciation methods.

2. Prepare a depreciation schedule using the straight-line, units-of-production, declining-balance, sum-of-the-years'-digits, and ACRS depreciation methods.

Business assets that are used in the operation of a business over a long period of time, such as buildings, equipment, vehicles, and office furniture, are called **fixed assets.** For each year of useful life of the fixed asset, a portion of the cost of an asset is deducted as a business expense. The deduction, called **depreciation,** reduces the value of the asset.

The **useful life** of an asset is estimated by the business owner based on industry and government guidelines and other factors, such as the use of the asset, the replacement policy, and obsolescence. However, the useful life does not necessarily relate to the real life of the asset. For example, a truck may be fully depreciated on business records after 5 years and still be in good working condition.

A business may take depreciation deductions only on assets that meet the following requirements:

1. It must be used in business or held for the production of income.
2. It must have a determinable life and that life must be longer than 1 year.
3. It must be something that wears out, decays, gets used up, becomes obsolete, or loses value from natural causes.

The amount of depreciation expense taken on an asset each year depends on the depreciation method used. The first four depreciation methods presented in this chapter—straight line, units of production, declining balance, and sum of the years' digits—are frequently used for financial accounting purposes and state income tax returns.

The fifth method, the Accelerated Cost Recovery System (ACRS), is used for calculating depreciation for federal income tax on assets purchased on or after January 1, 1981. A business may use its own judgment in selecting an appropriate depreciation method as long as IRS guidelines are followed and the method is used consistently.

## LESSON 13-1    STRAIGHT-LINE DEPRECIATION

**CALCULATING STRAIGHT-LINE DEPRECIATION**

In the *straight-line method of depreciation,* the amount of depreciation is the same for each year of the asset's useful life. The estimated value of an asset at the end of its useful life is called *salvage value, scrap value, disposal value,* or *trade-in value.* The *depreciable amount* of an asset is the difference between cost and salvage value.

Cost − Salvage Value = Depreciable Amount

To find the annual depreciation using the straight-line method, divide the depreciable amount by the years of useful life.

$$\text{Annual Depreciation} = \frac{\text{Depreciable Amount}}{\text{Useful Life}}$$

### EXAMPLE

Use the straight-line method to find the annual depreciation on a CAD/CAM computer system that cost $60,000 with a useful life of 5 years and estimated salvage value of $5,500.

### SOLUTION

**Step 1**  Find the depreciable amount.

| Cost | − Salvage Value | = Depreciable Amount |
|------|------|------|
| $60,000 − | $5,500 | = $54,500 |

**Step 2**  Find the annual depreciation.

$$\frac{\text{Annual}}{\text{Depreciation}} = \frac{\text{Depreciable Amount}}{\text{Useful Life}} = \frac{\$54,500}{5} = \$10,900$$

The annual depreciation also can be found by multiplying the depreciable amount by the depreciation rate. To find the *depreciation rate,* divide 100% by the years of useful life.

$$\text{Depreciation Rate} = \frac{100\%}{\text{Useful Life}}$$

### EXAMPLE

The Southeast Excavating Company purchased a loader that cost $122,000 with a useful life of 5 years. The trade-in value is estimated to be $19,000. Find the annual depreciation using the straight-line depreciation rate.

## SOLUTION

**Step 1** Find the depreciable amount.

Cost — Salvage Value = Depreciable Amount

$122,000 — $19,000 = $103,000

**Step 2** Find the annual depreciation rate.

$$\text{Depreciation Rate} = \frac{100\%}{\text{Useful Life}} = \frac{100\%}{5} = 20\%$$

**Step 3** Find the annual depreciation.

| Depreciable Amount | × | Depreciation Rate | = | Annual Depreciation |
|---|---|---|---|---|
| $103,000 | × | 0.2 | = | $20,600 |

## PRACTICE PROBLEMS

Find the annual depreciation using the straight-line depreciation method. Round to the nearest dollar.

| | Asset | Cost | Salvage Value | Useful Life | Annual Depreciation |
|---|---|---|---|---|---|
| **1.** | Cash register | $1,300 | $215 | 7 years | _____ |
| **2.** | Freezer case | 4,260 | 500 | 10 years | _____ |
| **3.** | Meat cutter | 325 | 50 | 5 years | _____ |

Find the annual depreciation using the straight-line depreciation rate. Round percents to the nearest tenth and amounts to the nearest dollar.

| | Asset | Cost | Salvage Value | Useful Life | Depreciation Rate | Annual Depreciation |
|---|---|---|---|---|---|---|
| **4.** | Delivery truck | $24,100 | $3,000 | 4 years | _____ | _____ |
| **5.** | Forklift | 4,200 | 610 | 5 years | _____ | _____ |
| **6.** | CB radio | 726 | 80 | 3 years | _____ | _____ |

**BOOK VALUE**

***Accumulated depreciation*** is total depreciation to date. The original cost of an asset minus accumulated depreciation is the ***book value*** of the asset.

Depreciation expense is deducted each year from book value until salvage value equals book value.

Cost — Accumulated Depreciation = Book Value

---

ANSWERS FOR
PRACTICE PROBLEMS

| | Rate | Annual Depreciation |
|---|---|---|
| **(4)** | 25% | $5,275 |
| **(5)** | 20% | 718 |
| **(6)** | 33.3% | 215 |

| | Annual Depreciation | Rate |
|---|---|---|

**(1)** $155 **(2)** $376 **(3)** $55

**EXAMPLE**

Jennings Insurance purchased a copying machine for $4,600 with a useful life of 9 years and an estimated salvage value of $600. Use the straight-line method to find the accumulated depreciation and ending book value for the third year of useful life. Round to the nearest dollar.

**SOLUTION**

**Step 1**  Find the depreciable amount.

Cost  − Salvage Value = Depreciable Amount
$4,600 −     $600     =         $4,000

**Step 2**  Find the annual depreciation (to the nearest dollar).

$$\frac{\text{Annual}}{\text{Depreciation}} = \frac{\text{Depreciable Amount}}{\text{Useful Life}} = \frac{\$4,000}{9} = \$444$$

**Step 3**  Find the accumulated depreciation after three years.

Annual        Number of    Accumulated
Depreciation ×   Years   = Depreciation
$444      ×     3     =     $1,332

**Step 4**  Find the ending book value for the third year.

Accumulated      Ending
Cost  − Depreciation = Book Value
$4,600 −    $1,332    =    $3,268

**PRACTICE PROBLEMS**

Find the accumulated depreciation and the ending book value for the third year using the straight-line method. Round to the nearest dollar.

| | Asset | Cost | Salvage Value | Useful Life | Accumulated Depreciation, Year 3 | Ending Book Value, Year 3 |
|---|---|---|---|---|---|---|
| 1. | Computer | $3,780 | $420 | 4 years | _____ | _____ |
| 2. | Cellular phone | 564 | 75 | 3 years | _____ | _____ |
| 3. | Electronic typewriter | 1,093 | 205 | 6 years | _____ | _____ |

## PREPARING A STRAIGHT-LINE DEPRECIATION SCHEDULE

The annual depreciation, the accumulated depreciation, and the book value over the useful life of an asset can be shown by preparing a **depreciation schedule.** Most businesses prepare a depreciation schedule using a computer spreadsheet program. The formulas used to compute the required amounts are entered on the spreadsheet. Then, the spreadsheet program automatically calculates these amounts.

### EXAMPLE

Prepare a straight-line depreciation schedule for an asset that cost $18,000 with a useful life of 5 years, scrap value of $4,500, and annual depreciation of $2,700.

### SOLUTION

**Step 1** Find the accumulated depreciation and ending book value for each of the 5 years.

| Year | Annual Depreciation | × | Number of Years | = | Accumulated Depreciation | Cost | − | Accumulated Depreciation | = | Ending Book Value |
|------|-----|---|---|---|--------|---------|---|--------|---|---------|
| 1 | $2,700 | × | 1 | = | $ 2,700 | $18,000 | − | $ 2,700 | = | $15,300 |
| 2 | 2,700 | × | 2 | = | 5,400 | 18,000 | − | 5,400 | = | 12,600 |
| 3 | 2,700 | × | 3 | = | 8,100 | 18,000 | − | 8,100 | = | 9,900 |
| 4 | 2,700 | × | 4 | = | 10,800 | 18,000 | − | 10,800 | = | 7,200 |
| 5 | 2,700 | × | 5 | = | 13,500 | 18,000 | − | 13,500 | = | 4,500 |

**Step 2** Prepare a straight-line depreciation schedule using the above amounts. The following depreciation schedule was prepared using a computer spreadsheet.

```
                  Depreciation Schedule

            Annual           Accumulated        Ending
  Year      Depreciation     Depreciation       Book Value

  0                             ----              $18,000
  1         $2,700           $ 2,700              15,300
  2          2,700             5,400              12,600
  3          2,700             8,100               9,900
  4          2,700            10,800               7,200
  5          2,700            13,500               4,500*

  *Ending book value equals salvage value.
```

## PRACTICE PROBLEM

Prepare a depreciation schedule using the straight line method for a tractor that cost $20,900 with a salvage value of $3,500 and an estimated useful life of 3 years. Round amounts to the nearest dollar.

### DEPRECIATION SCHEDULE

| Year | Annual Depreciation | Accumulated Depreciation | Ending Book Value |
|------|---------------------|--------------------------|-------------------|
| 0 | — | — | $20,900 |
| 1 | _____ | _____ | _____ |
| 2 | _____ | _____ | _____ |
| 3 | _____ | _____ | _____ |

## LESSON 13-1   EXERCISES

Find the annual depreciation using the straight-line depreciation method. Round to the nearest dollar.

| | Asset | Cost | Salvage Value | Useful Life | Annual Depreciation |
|---|---|---|---|---|---|
| 1. | Industrial vacuum | $ 6,400 | $ 800 | 4 years | _____ |
| 2. | Stage lighting | 9,710 | 900 | 8 years | _____ |

Find the annual depreciation using the straight-line depreciation rate. Round percents to the nearest tenth. Round amounts to the nearest dollar.

| | Cost | Salvage Value | Useful Life | Depreciation Rate | Annual Depreciation |
|---|---|---|---|---|---|
| 3. | $264,800 | $24,800 | 20 years | _____ | _____ |
| 4. | 340,000 | 46,500 | 10 years | _____ | _____ |

Find the annual depreciation and ending book value for the first year using the straight-line depreciation method. Round to the nearest dollar.

| | Cost | Salvage Value | Useful Life | Annual Depreciation | Ending Book Value |
|---|---|---|---|---|---|
| 5. | $ 6,300 | $ 1,700 | 8 years | _____ | _____ |
| 6. | 13,000 | 3,000 | 10 years | _____ | _____ |

Find the accumulated depreciation and ending book value for the third year using the straight-line depreciation method. Round to the nearest dollar.

| | Cost | Salvage Value | Useful Life | Accumulated Depreciation, Year 3 | Ending Book Value, Year 3 |
|---|---|---|---|---|---|
| 7. | $ 7,600 | $ 3,120 | 8years | _____ | _____ |
| 8. | 27,300 | 11,000 | 4 years | _____ | _____ |

Prepare a depreciation schedule using the straight-line depreciation method. Round to the nearest dollar.

9. Asset:          Office Duplicating Machine
   Cost:           $4,560
   Useful Life:    3 years
   Salvage Value:  $1,785

| Year | Annual Depreciation | Accumulated Depreciation | Ending Book Value |
|---|---|---|---|
| 0 | — | — | $4,560 |
| 1 | _____ | _____ | _____ |
| 2 | _____ | _____ | _____ |
| 3 | _____ | _____ | _____ |

## BUSINESS APPLICATIONS

Round amounts to the nearest dollar and percents to the nearest tenth.

10. An electronic display typewriter costs $1,300. The estimated useful life of the typewriter is 8 years with a salvage value of $420. Find (a) the annual rate of depreciation, (b) the annual amount of depreciation, and (c) the book value at the end of the second year using the straight-line depreciation method.

(a) _____

(b) _____

(c) _____

11. Studio lighting at the New York Photography Center cost $25,600. The lighting has a useful life of 10 years and a trade-in value of $3,000. Find the book value at the end of the fifth year using the straight-line depreciation method.

_____

12. A small-engine plane that cost $785,900 has an estimated life of 10 years and a salvage value of $30,000. Find the book value of the plane after 4 years using the straight-line depreciation method.

_____

13. A street-cleaning machine that cost $18,500 has an estimated life of 8 years and a trade-in value of $700. Find the book value of the machine after 5 years using the straight-line depreciation method.

_____

**CALCULATING UNITS-OF-PRODUCTION DEPRECIATION**

The units-of-production depreciation method can be used to calculate depreciation on equipment, vehicles, or similar assets when the amount of depreciation is based on hours used, miles driven, or number of items produced.

The units-of-production method depreciates a set amount per unit of use or production. For example, a machine may be depreciated by the number of items it produces or by the number of hours it is used.

The asset's useful life is expressed in terms of expected useful units used or produced over the lifetime of the asset. To find the depreciation rate per unit, divide the depreciable amount by the useful life.

$$\frac{\text{Depreciation Amount}}{\text{Useful Life}} = \text{Depreciation Rate per Unit}$$

To find the annual depreciation expense, multiply the number of units produced or used during the year by the depreciation rate.

Number of Units × Depreciation Rate = Annual Depreciation

**EXAMPLE**

Allied Machines purchased a drill press for $56,500 with a useful life of 90,000 hours and salvage value of $7,000. During the first year the drill press was used 22,500 hours. Find the depreciable amount, the depreciation rate per unit, and the annual depreciation expense using the units-of-production method.

**SOLUTION**

**Step 1**  Find the depreciable amount.

Cost     − Salvage Value = Depreciable Amount
$56,500 −     $7,000     =        $49,500

**Step 2**  Find the depreciation rate.

$$\frac{\text{Depreciation}}{\text{Rate}} = \frac{\text{Depreciable Amount}}{\text{Useful Life}} = \frac{\$49,500}{90,000 \text{ hours}} = \$0.55 \text{ per hour}$$

**Step 3**  Find the annual depreciation expense.

Number of    Depreciation      Annual
   Units   ×     Rate     = Depreciation
  22,500   ×    $0.55     =    $12,375

## PRACTICE PROBLEMS

Use the units-of-production depreciation method to find the depreciation rate (round to the nearest cent) and the annual depreciation (round to the nearest dollar).

| | Cost | Salvage Value | Useful Life | Yearly Units Used/Produced | Depreciation Rate | Annual Depreciation |
|---|---|---|---|---|---|---|
| 1. | $ 90,150 | $10,000 | 35,000 miles | 1,460 miles | _____ | _____ |
| 2. | 164,875 | 40,000 | 18,500 hours | 2,100 hours | _____ | _____ |
| 3. | 121,520 | 27,000 | 278,000 units | 55,600 units | _____ | _____ |

## PREPARING A UNITS-OF-PRODUCTION DEPRECIATION SCHEDULE

The units-of-production depreciation schedule is prepared by first calculating the depreciation rate per unit. Then, the yearly depreciation, accumulated depreciation, and book value are calculated and entered on the depreciation schedule. The depreciation expense taken each year depends on the number of units produced or used during the year.

### EXAMPLE

Prepare a units-of-production depreciation schedule for a van that costs $18,600. It has an estimated life of 100,000 miles of use and a salvage value of $1,600. The van was driven the following number of miles per year:

Year 1   18,500 miles

Year 2   25,200 miles

Year 3   20,500 miles

Year 4   19,000 miles

Year 5   16,800 miles

### SOLUTION

**Step 1**   Calculate the depreciable amount.

Cost   − Salvage Value = Depreciable Amount

$18,600 −     $1,600     =         $17,000

**Step 2**   Calculate the depreciation rate.

$$\text{Depreciation Rate} = \frac{\text{Depreciable Amount}}{\text{Useful Life}} = \frac{\$17,000}{100,000 \text{ miles}} = \$0.17 \text{ per mile}$$

**Step 3** Calculate the annual depreciation, accumulated depreciation, and ending book value for each year.

| Year | Units Used × | Depreciation Rate | = Annual Depreciation | Cost | − Accumulated Depreciation | = Ending Book Value |
|------|------|------|------|------|------|------|
| 1 | 18,500 × | $0.17 | = $3,145 | $18,600 − | $ 3,145 | = $15,455 |
| 2 | 25,200 × | 0.17 | = 4,284 | 18,600 − | 7,429 | = 11,171 |
| 3 | 20,500 × | 0.17 | = 3,485 | 18,600 − | 10,914 | = 7,686 |
| 4 | 19,000 × | 0.17 | = 3,230 | 18,600 − | 14,144 | = 4,456 |
| 5 | 16,800 × | 0.17 | = 2,856 | 18,600 − | 17,000 | = 1,600 |

**Step 4** Prepare a depreciation schedule. The following depreciation schedule was prepared using a computer spreadsheet.

```
                  Depreciation Schedule


          Units     Annual        Accumulated
   Year   Used      Depreciation  Depreciation    Book Value

    0                                              $18,600
    1     18,500    $3,145         $ 3,145          15,455
    2     25,200     4,284           7,429          11,171
    3     20,500     3,485          10,914           7,686
    4     19,000     3,230          14,144           4,456
    5     16,800     2,856          17,000           1,600
```

## PRACTICE PROBLEM

Browden Construction purchased a pickup truck for $27,000 with an estimated life of 80,000 miles and a salvage value of $3,000. Prepare a depreciation schedule using the units-of-production depreciation method.

### DEPRECIATION SCHEDULE

| Year | Miles Driven | Depreciation Rate | Annual Depreciation | Accumulated Depreciation | Book Value |
|------|------|------|------|------|------|
| 0 | — | — | — | — | $27,000 |
| 1 | 20,500 | ____ | ____ | ____ | ____ |
| 2 | 25,700 | ____ | ____ | ____ | ____ |
| 3 | 24,400 | ____ | ____ | ____ | ____ |

## LESSON 13-2    EXERCISES

Find the amount of depreciation for the following assets using the units-of-production method. Round the amount of depreciation to the nearest dollar.

| | Asset | Cost | Salvage Value | Useful Life | Units Used | Amount of Depreciation |
|---|---|---|---|---|---|---|
| 1. | Truck | $ 29,000 | $ 4,200 | 80,000 miles | 16,000 miles | _____ |
| 2. | Sound system | 12,000 | 2,200 | 14,000 hours | 2,500 hours | _____ |
| 3. | Tape duplicator | 1,800 | 500 | 5,000 hours | 985 hours | _____ |
| 4. | Plane | 156,000 | 16,000 | 125,000 miles | 24,500 miles | _____ |
| 5. | Equipment | 24,900 | 4,900 | 250,000 units | 23,700 units | _____ |

Calculate the depreciation rate per unit, the annual depreciation, the accumulated depreciation, and the ending book value for the following assets using the units-of-production method. Round depreciation rates to the nearest cent. Round all other amounts to the nearest dollar.

6. Cost:            $1,950
   Salvage Value:   None
   Useful Life:     15,000 hours

| Year | Units Used | Depreciation Rate | Annual Depreciation | Accumulated Depreciation | Ending Book Value |
|---|---|---|---|---|---|
| 1 | 3,200 | _____ | _____ | _____ | _____ |
| 2 | 1,400 | _____ | _____ | _____ | _____ |
| 3 | 1,000 | _____ | _____ | _____ | _____ |

Prepare a depreciation schedule using the units of production method for heating equipment costing $110,000. The salvage value is estimated to be $11,980 and the useful life is 26,000 hours. The equipment was in service for the following hours: year 1 for 5,300 hours, year 2 for 6,700 hours, year 3 for 4,300 hours, year 4 for 4,800 hours, and year 5 for 4,900 hours. Round the depreciation rate to the nearest cent. Round all other amounts to the nearest dollar.

| 7. Year | Depreciation Rate | Annual Depreciation | Accumulated Depreciation | Ending Book Value |
|---|---|---|---|---|
| 0 | — | — | — | $110,000 |
| 1 | _____ | _____ | _____ | _____ |
| 2 | _____ | _____ | _____ | _____ |
| 3 | _____ | _____ | _____ | _____ |
| 4 | _____ | _____ | _____ | _____ |
| 5 | _____ | _____ | _____ | _____ |

## BUSINESS APPLICATIONS

8. Find the yearly depreciation for an industrial mold costing $32,500 and producing 36,500 units. The useful life of the mold is 200,000 units and the scrap value is $12,500.

_____

9. The Banks Vending Service depreciate their delivery trucks using the units-of-production method. The cost of each truck is $16,500, the useful life is 100,000 miles, and the trade-in value is $3,500. Find (a) the annual depreciation to the nearest dollar and (b) ending book value for a delivery truck that was driven 21,300 miles during the first year of operation.

(a) _____

(b) _____

10. Jawarski Septic Service purchased a sewerpump truck for $43,000. The truck has an estimated useful life of 120,000 miles and a salvage value of $7,000. The truck was driven 42,300 miles in the first year. Calculate (a) the depreciation rate per mile, (b) the annual depreciation expense, and (c) the ending book value.

(a) _____

(b) _____

(c) _____

11. The useful life of a frontloader is 80,000 hours, the salvage value is $15,000, and the depreciation expense is $0.81 per hour of use. Find the cost of the frontloader.

_____

**CALCULATING
DECLINING-
BALANCE
DEPRECIATION**

The *declining-balance depreciation method* is an accelerated depreciation method. An accelerated depreciation method produces larger depreciation deductions during the early years and progressively smaller deductions in later years. The declining-balance method will produce a book value similar to the market value of such assets as automobiles, which depreciate the most during the first year.

The 200% declining-balance rate, also called *double declining balance,* is twice the straight-line depreciation rate or 200% of the straight-line rate. The 150% declining-balance rate is $1\frac{1}{2}$ times the straight-line depreciation rate or 150% of the straight-line rate.

$$200\% \text{ Declining-Balance Rate} = \frac{100\%}{\text{Useful Life}} \times 2$$

$$150\% \text{ Declining-Balance Rate} = \frac{100\%}{\text{Useful Life}} \times 1.5$$

Salvage value is *not* deducted from original cost when using the declining-balance method. Instead, the original cost is multiplied by the depreciation rate to find the depreciation for the first year. The book value for the succeeding years cannot drop below the salvage value.

To find the annual depreciation deduction using the declining-balance method, multiply the ending book value for the previous year by the declining-balance rate:

Book Value × Declining-Balance Rate = Annual Depreciation

**EXAMPLE**

On January 1 Wilson Dairy, Inc., constructed a milking barn for $350,000. The estimated life of the barn is 20 years and the salvage value is $10,000. Find the annual depreciation and ending book value for the first year using the 150% declining-balance depreciation method.

**SOLUTION**

**Step 1**  Find the 150% declining-balance rate. Round to the nearest tenth.

$$\frac{100\%}{\text{Useful Life}} = \frac{100\%}{20} \times 1.5 = 7.5\% \text{ Declining-balance rate}$$

**Step 2**  Find the annual depreciation and ending book value for the first year.

| Book Value | × | Depreciation Rate | = | Annual Depreciation |
|---|---|---|---|---|
| $350,000 | × | 0.075 | = | $26,250 |

↑
For the first year,
the book value is
the cost of the asset.

| Cost | − | Annual Depreciation | = | Ending Book Value |
|---|---|---|---|---|
| $350,000 | − | $26,250 | = | $323,750 |

## PRACTICE PROBLEMS

Use the 200% declining-balance method to find the declining-balance rate, the annual depreciation, and the ending book value for the first year. Round percents to the nearest tenth. Round amounts to the nearest dollar.

|  | Cost | Salvage Value | Useful Life | 200% Declining-Balance Rate | Annual Depreciation | Ending Book Value |
|---|---|---|---|---|---|---|
| 1. | $23,400 | $4,000 | 4 | _____ | _____ | _____ |
| 2. | 10,700 | 1,000 | 7 | _____ | _____ | _____ |
| 3. | 8,200 | 500 | 3 | _____ | _____ | _____ |
| 4. | 56,000 | 5,000 | 10 | _____ | _____ | _____ |

**PREPARING A DECLINING-BALANCE DEPRECIATION SCHEDULE**

When preparing a depreciation schedule using the declining-balance method, remember to multiply the book value, which decreases each year, by the declining-balance rate, which remains the same. The amount of depreciation in the final year may have to be adjusted so that the book value is never less than the salvage value.

---

### EXAMPLE

The Victory Remodeling Company purchased a diesel flatbed truck for $17,500 with a salvage value of $2,200. Prepare a depreciation schedule for the 5-year useful life of the truck using the 200% declining-balance method. Round to the nearest dollar.

### SOLUTION

**Step 1**  Find the 200% declining-balance rate.

$$\frac{100\%}{\text{Useful Life}} = \frac{100\%}{5} \times 2 = 40\% \text{ Declining-balance rate.}$$

**Step 2**  Find the annual depreciation, accumulated depreciation, and ending book value for the 5 years of useful life.

| Year | Book Value × | Depreciation Rate | = Annual Depreciation | Cost | − Accumulated Depreciation | = Ending Book Value |
|---|---|---|---|---|---|---|
| 1 | $17,500 × | 0.4 | = $7,000 | $17,500 − | $ 7,000 | = $10,500 |
| 2 | 10,500 × | 0.4 | = 4,200 | 17,500 − | 11,200 | = 6,300 |
| 3 | 6,300 × | 0.4 | = 2,520 | 17,500 − | 13,720 | = 3,780 |
| 4 | 3,780 × | 0.4 | = 1,512 | 17,500 − | 15,232 | = 2,268 |
| 5 | 2,268 × | 0.4 | = 907 | 17,500 − | 16,139 | = 1,361 |

---

ANSWERS FOR PRACTICE PROBLEMS

| 200% Declining Balance Rate | Annual Depreciation | Ending Book Value |
|---|---|---|
| (1) 50% | $11,700 | $11,700 |
| (2) 28.6% | 3,060 | 7,640 |
| (3) 66.7% | 5,469 | 2,731 |
| (4) 20% | 11,200 | 44,800 |

**Step 3**   Adjust the fifth-year depreciation amounts and the ending book value so that it equals the salvage value.

Change the fifth-year depreciation.

$2,268 − $2,200 = $68

Change the fifth-year accumulated depreciation and the ending book value.

$$\begin{matrix} & \text{Accumulated} & \text{Ending} \\ \text{Cost} & -\text{Depreciation} & =\text{Book Value} \\ \$17,500 & -\quad\$15,300 & =\quad\$2,200 \end{matrix}$$

$15,232 + $68

**Step 4**   Prepare a depreciation schedule. The following depreciation schedule was prepared using a computer spreadsheet program.

```
                    Depreciation Schedule


           Rate of         Annual        Accumulated     Ending
  Year  Depreciation    Depreciation    Depreciation   Book Value

   0                                                     $17,500
   1        0.4           $7,000         $ 7,000          10,500
   2        0.4            4,200          11,200           6,300
   3        0.4            2,520          13,720           3,750
   4        0.4            1,512          15,232           2,268
   5        0.4               68*         15,300*          2,200*

*Adjusted
```

## PRACTICE PROBLEM

Prepare a depreciation schedule for a diesel tractor truck that cost $120,000 with a salvage value of $10,000 and an estimated life of 3 years using the 200% declining-balance method. Round amounts to the nearest dollar. Round the depreciation rate to the nearest tenth of a percent.

| Year | Depreciation Rate | Annual Depreciation | Accumulated Depreciation | Ending Book Value |
|------|------|------|------|------|
| 1 | _____ | _____ | _____ | _____ |
| 2 | _____ | _____ | _____ | _____ |
| 3 | _____ | _____ | _____ | _____ |

ANSWERS FOR PRACTICE PROBLEM

| Year | Depreciation Rate | Annual Depreciation | Accumulated Depreciation | Ending Book Value |
|------|------|------|------|------|
| 1 | 66.7% | $80,040 | $80,040 | $39,960 |
| 2 | 66.7% | 26,653 | 106,693 | 13,307 |
| 3 | 66.7% | 3,307 (adjusted) | 110,000 (adjusted) | 10,000 (adjusted) |

## LESSON 13-3    EXERCISES

Find the depreciation rate, annual depreciation, and ending book value for the first year using the 200% declining-balance depreciation method. Round percents to the nearest tenth. Round amounts to the nearest dollar.

| | Cost | Useful Life | Depreciation Rate | Annual Depreciation | Book Value |
|---|---|---|---|---|---|
| **1.** | $318,000 | 20 years | _____ | _____ | _____ |
| **2.** | 52,000 | 5 years | _____ | _____ | _____ |
| **3.** | 4,200 | 15 years | _____ | _____ | _____ |

Find the rate, annual depreciation, and ending book value for each of the first three years using the declining-balance depreciation method. Round amounts to the nearest dollar. Round percents to the nearest tenth.

**4.** Asset:            Tractor

   Cost:             $16,400

   Salvage Value:  $5,000

   Useful Life:     5 years

   Rate:            200% declining balance _____

   Year 1:  Depreciation: _____    Book Value: _____

   Year 2:  Depreciation: _____    Book Value: _____

   Year 3:  Depreciation: _____    Book Value: _____

**5.** Asset:            Cabinets and Files

   Cost:             $8,300

   Salvage Value:  $1,900

   Useful Life:     7 years

   Rate:            200% declining balance _____ %

   Year 1:  Depreciation: _____    Book Value: _____

   Year 2:  Depreciation: _____    Book Value: _____

   Year 3:  Depreciation: _____    Book Value: _____

## BUSINESS APPLICATIONS

Round amounts to the nearest dollar and percents to the nearest tenth.

**6.** Handi-Craft, Inc., purchased new display shelves at a cost of $9,570. The shelves have a useful life of 7 years and an estimated scrap value of $1,300. Find the book value for the end of the third year using the 200% declining balance method.

**7.** A drilling rig has a purchase price of $92,000, a useful life of 10 years, and a salvage value of $6,500. Find the annual depreciation and the ending book value for the first, second, and third year using the 200% declining-balance method.

| Year | Annual Depreciation | Book Value |
|------|---------------------|------------|
| 1 | _____ | _____ |
| 2 | _____ | _____ |
| 3 | _____ | _____ |

**8.** Dr. Robbins purchased an X-ray machine for his office costing $134,000. The machine has an estimated life of 8 years and a trade-in value of $15,000. Find (a) the rate of depreciation, (b) the annual depreciation for the first year, (c) the accumulated depreciation for the second year, and (d) the ending book value for the third year using the 200% declining-balance method.

(a) _____

(b) _____

(c) _____

(d) _____

**9.** An automatic watering system purchased by the Western Garden Nursery cost $4,900, including installation. The system has a useful life of 5 years and a salvage value of $650. Prepare a depreciation schedule for the watering using the 200% declining-balance method.

| Year | Depreciation Rate | Annual Depreciation | Accumulated Depreciation | Book Value |
|------|-------------------|---------------------|--------------------------|------------|
| 0 | — | — | — | $4,900 |
| 1 | _____ | _____ | _____ | _____ |
| 2 | _____ | _____ | _____ | _____ |
| 3 | _____ | _____ | _____ | _____ |
| 4 | _____ | _____ | _____ | _____ |
| 5 | _____ | _____ | _____ | _____ |

Another accelerated depreciation method is the ***sum-of-the-years'-digits depreciation method.*** Annual depreciation is found by multiplying the depreciable amount (cost minus salvage value) by the depreciation rate, expressed as a fraction. Each year the fraction becomes smaller, which gives a lower depreciation expense:

Depreciable       Depreciation          Annual
  Amount  × Rate (fraction) = Depreciation

The numerator of the fraction is the number of years of remaining useful life of the asset. The denominator is the sum of the years of useful life. For example, if an asset has a useful life of 5 years, the fraction for the first year is $\frac{5}{15}$. The numerator of 5 is the number of years of useful life, and the denominator of 15 is the sum of the digits $1 + 2 + 3 + 4 + 5$. The following is the depreciation rate for each year for an asset with a useful life of 5 years:

| Year 1 | Year 2 | Year 3 | Year 4 | Year 5 |
|--------|--------|--------|--------|--------|
| $\frac{5}{15}$ | $\frac{4}{15}$ | $\frac{3}{15}$ | $\frac{2}{15}$ | $\frac{1}{15}$ |

The denominator of the depreciation rate fraction also can be found by using the sum-of-the-years'-digits formula, in which $n$ is the number of years of useful life:

$$\text{Sum-of-the-Years' Digits} = \frac{n(n + 1)}{2}$$

### EXAMPLE 1

Use the sum-of-the-years'-digits depreciation method to find each year's depreciation rate fraction for an asset with a useful life of 9 years.

### SOLUTION

**Step 1**   Find the denominator of the depreciation rate fraction.

Sum-of-the-Years' Digits         = Denominator
$1 + 2 + 3 + 4 + 5 + 6 + 7 + 8 + 9 =$         45

*or*

$$\frac{n(n + 1)}{2} = \frac{9(9 + 1)}{2} \qquad = \qquad 45$$

**Step 2**   Express the depreciation rates as fractions.

| Year 1 | Year 2 | Year 3 | Year 4 | Year 5 | Year 6 | Year 7 | Year 8 | Year 9 |
|--------|--------|--------|--------|--------|--------|--------|--------|--------|
| $\frac{9}{45}$ | $\frac{8}{45}$ | $\frac{7}{45}$ | $\frac{6}{45}$ | $\frac{5}{45}$ | $\frac{4}{45}$ | $\frac{3}{45}$ | $\frac{2}{45}$ | $\frac{1}{45}$ |

## EXAMPLE 2

The Woodland Apartments built a utility storage shed at a cost of $18,400 with an estimated useful life of 12 years and a salvage value of $2,500. Find the depreciation rate, annual depreciation, and ending book value for the first year. Round depreciation to the nearest dollar.

**SOLUTION**

**Step 1**   Express the depreciation rate as a fraction.

Sum-of-the-Years' Digits = Denominator

$$\frac{n(n + 1)}{2} = \frac{12(12 + 1)}{2} = \quad 78$$

First-year depreciation rate fraction is $\frac{12}{78}$.

**Step 2**   Find the depreciable amount.

Cost   − Salvage Value = Depreciable Amount
$18,400 −    $2,500   =      $15,900

**Step 3**   Find the first-year depreciation.

Depreciable   Depreciation       First-Year
  Amount  ×    Rate   =   Depreciation
  $15,900  ×    $\frac{12}{78}$   = $2,446.15 ≈ $2,446

**Step 4**   Find the ending book value.

         First-Year      Ending
  Cost  − Depreciation = Book Value
$18,400 −    $2,446   =   $15,954

## PRACTICE PROBLEMS

Express the sum-of-the-years'-digits depreciation rate as a fraction for each year of life for the following assets.

**1.** Computer with 5 years' useful life.

| Year 1 | Year 2 | Year 3 | Year 4 | Year 5 |
|--------|--------|--------|--------|--------|
|        |        |        |        |        |

**2.** Intercom system with 7 years' useful life.

| Year 1 | Year 2 | Year 3 | Year 4 | Year 5 |
|--------|--------|--------|--------|--------|
|        |        |        |        |        |

| Year 6 | Year 7 |
|--------|--------|
|        |        |

**3.** Office furniture with 9 years' useful life.

| Year 1 | Year 2 | Year 3 | Year 4 | Year 5 |
|--------|--------|--------|--------|--------|
|        |        |        |        |        |

| Year 6 | Year 7 | Year 8 | Year 9 |
|--------|--------|--------|--------|
|        |        |        |        |

Find the first year's depreciation for the following assets using the sum-of-the-years'-digits method. Round to the nearest dollar.

| | Cost | Useful Life | Salvage Value | First-Year Depreciation |
|---|---|---|---|---|
| 4. | $ 950 | 3 years | $ 50 | _____ |
| 5. | 11,300 | 15 years | 1,000 | _____ |
| 6. | 27,800 | 20 years | 3,000 | _____ |

## PREPARING A SUM-OF-THE-YEARS'-DIGITS DEPRECIATION SCHEDULE

When preparing a depreciation schedule using the sum-of-the-years'-digits method, multiply the depreciation fraction, which becomes smaller each year, by the depreciable amount, not the book value.

### EXAMPLE

The Bates Company purchased packaging equipment for $17,150. The useful life is estimated to be 4 years, at which time it will have a trade-in value of $2,500. Prepare a depreciation schedule using the sum-of-the-years'-digits method. Round to the nearest dollar.

### SOLUTION

**Step 1** Find the depreciable amount.

Cost   − Salvage Value = Depreciable Amount
$17,150 −   $2,500   =   $14,650

**Step 2** Find the depreciation rate for each year.

Sum-of-the-Years' Digits = Denominator
  1 + 2 + 3 + 4   =   10

Depreciation rates expressed as fractions

Year 1   Year 2   Year 3   Year 4

$\frac{4}{10}$      $\frac{3}{10}$      $\frac{2}{10}$      $\frac{1}{10}$

**Step 3** Find the annual depreciation, accumulated depreciation, and ending book value for each year.

| Year | Depreciable Amount | × | Depreciation Rate | = Depreciation (Annual) | Cost | − Accumulated Depreciation | = Ending Book Value |
|---|---|---|---|---|---|---|---|
| 1 | $14,650 | × | 4/10 | = $5,860 | $17,150 − | $ 5,860 | = $11,290 |
| 2 | 14,650 | × | 3/10 | = 4,395 | 17,150 − | 10,255 | = 6,895 |
| 3 | 14,650 | × | 2/10 | = 2,930 | 17,150 − | 13,185 | = 3,965 |
| 4 | 14,650 | × | 1/10 | = 1,465 | 17,150 − | 14,650 | = 2,500 |

**Step 4** Prepare a depreciation schedule. The following depreciation schedule was prepared using a computer spreadsheet program.

```
                    Depreciation  Schedule


         Depreciation    Annual          Accumulated      Ending
  Year   Rate            Depreciation    Depreciation     Book Value

   0                                                      $17,150
   1       4/10          $5,860          $ 5,860           11,290
   2       3/10           4,395           10,255            6,895
   3       2/10           2,930           13,185            3,965
   4       1/10           1,465           11,650            2,500
```

## PRACTICE PROBLEM

Prepare a sum-of-the-years'-digits depreciation schedule for automatic canning machinery that cost $14,900 with a trade-in value of $2,100 and an estimated life of 5 years.

| Year | Depreciation Rate | Annual Depreciation | Accumulated Depreciation | Ending Book Value |
|---|---|---|---|---|
| 0 | — | — | — | $14,900 |
| 1 | _____ | _____ | _____ | _____ |
| 2 | _____ | _____ | _____ | _____ |
| 3 | _____ | _____ | _____ | _____ |
| 4 | _____ | _____ | _____ | _____ |
| 5 | _____ | _____ | _____ | _____ |

ANSWERS FOR PRACTICE PROBLEM

| Year | Depreciation Rate | Annual Depreciation | Accumulated Depreciation | Ending Book Value |
|---|---|---|---|---|
| 1 | $\frac{5}{15}$ | $4,267 | $ 4,267 | $10,633 |
| 2 | $\frac{4}{15}$ | 3,413 | 7,680 | 7,220 |
| 3 | $\frac{3}{15}$ | 2,560 | 10,240 | 4,660 |
| 4 | $\frac{2}{15}$ | 1,707 | 11,947 | 2,953 |
| 5 | $\frac{1}{15}$ | 853 | 12,800 | 2,100 |

## LESSON 13-4    EXERCISES

Find the annual depreciation for the year indicated using the sum-of-the-years'-digits depreciation method. Round to the nearest dollar.

1. Asset:           Pickup Truck
   Cost:            $6,434
   Salvage Value:   $650
   Useful Life:     7 years
   Year 5:   Depreciation: _____

2. Asset:           D-9 Dozer
   Cost:            $289,900
   Salvage Value:   $43,000
   Useful Life:     10 years
   Year 6:   Depreciation: _____

3. Asset:           Concrete Saw
   Cost:            $2,330
   Salvage Value:   $195
   Useful Life:     3 years
   Year 2:   Depreciation: _____

4. Asset:           Generator
   Cost:            $560
   Salvage Value:   None
   Useful Life:     5 years
   Year 4:   Depreciation: _____

Find the annual depreciation and ending book value for the first and second year using the sum-of-the-years'-digits depreciation method.

5. Asset:           Cement Mixer
   Cost:            $640
   Salvage Value:   None
   Useful Life:     9 years
   Year 1:   Depreciation: _____    Book Value: _____
   Year 2:   Depreciation: _____    Book Value: _____

Prepare a depreciation schedule for the following assets using the sum-of-the-years'-digits method. Round to the nearest dollar.

6. Asset:           Drill Press
   Cost:            $2,900
   Salvage Value:   $250
   Useful Life:     5 years

| Year | Depreciation Rate | Annual Depreciation | Accumulated Depreciation | Book Value |
|------|-------------------|---------------------|--------------------------|------------|
| 0    | —                 | —                   | —                        | $2,900     |
| 1    | _____           | _____             | _____                  | _____    |
| 2    | _____           | _____             | _____                  | _____    |
| 3    | _____           | _____             | _____                  | _____    |
| 4    | _____           | _____             | _____                  | _____    |
| 5    | _____           | _____             | _____                  | _____    |

## BUSINESS APPLICATIONS

**7.** Little Roman's Pizza purchased dining-room tables and chairs costing $4,700 with a trade-in value of $400 and an estimated life of 7 years. Find (a) the depreciation rate fraction for the fifth year and (b) the annual depreciation for the fifth year using the sum-of-the-years'-digits method.

(a) _____

(b) _____

**8.** Find the book value at the end of the fourth year using the sum-of-the-years'-digits method for wheel alignment equipment costing $6,700. The equipment has an estimated life of 11 years and a salvage value of $800.

_____

**9.** Sparkling Pools, Inc., built a display pool and deck costing $17,400 with a scrap value of $3,400 and a useful life of 8 years. Find (a) the depreciation rate and (b) the annual depreciation for the sixth year using the sum-of-the-years'-digits method.

(a) _____

(b) _____

**10.** Fab Nails and Hair Care purchased manicure tables and chairs costing $980 with no trade-in value and an estimated life of 6 years. Find the annual depreciation for (a) the third year, (b) fourth year, and (c) fifth year.

(a) _____

(b) _____

(c) _____

## LESSON 13-5  ACCELERATED COST RECOVERY SYSTEM (ACRS)

**CALCULATING ACRS DEPRECIATION**

Another accelerated depreciation method called the ***accelerated cost recovery system (ACRS)*** was established by the Economic Recovery Tax Act of 1981. The ACRS must be used on assets purchased on or after January 1, 1981. Assets purchased before 1981 must be depreciated according to the method used at the time of purchase, unless the IRS gives special permission to change.

The ACRS method of depreciation is used primarily for calculating depreciation expenses for federal income tax purposes. Most businesses use one of the previous depreciation methods discussed when preparing financial records or state income tax returns.

Since no salvage value is considered when using ACRS, the depreciation is based on the original cost of the asset. Instead of taking depreciation deductions over the useful life of the asset, depreciation deductions are taken over a specific ***recovery period.*** The recovery period is shorter than the useful life for most assets; therefore, annual depreciation is increased. The large amount of depreciation expense in the early years of an asset's life lowers the amount of taxable income and reduces the amount of taxes paid.

Under ACRS, assets are classified as having a 3-, 5-, 10-, 15-, 18-, and 19-year life, or recovery period. To find the annual depreciation, multiply the cost of the asset by the recovery rate found in a table similar to the following ACRS depreciation rates table. Congress frequently changes the recovery periods and recovery rates. Therefore, check the latest IRS tax guidelines for the correct ACRS procedures for the current year.

### ACRS DEPRECIATION RATES

| Recovery Periods | Recovery Rate |
|---|---|
| 3-year property: | |
|   1st year | 25% |
|   2d year | 38% |
|   3d year | 37% |
| 5-year property: | |
|   1st year | 15% |
|   2d year | 22% |
|   3d through 5th year | 21% |
| 10-year property: | |
|   1st year | 8% |
|   2d year | 14% |
|   3d year | 12% |
|   4th through 6th year | 10% |
|   7th through 10th year | 9% |
| 18-year property: | |
|   1st through 2d year | 9% |
|   3d year | 8% |
|   4th through 5th year | 7% |
|   6th year | 6% |
|   7th through 12th year | 5% |
|   13th through 18th year | 4% |

**EXAMPLE**

On March 15 the J & J Wrecking Company purchased a safe including installation for $3,750. The safe is a 5-year property according to ACRS guidelines. Find the annual depreciation and ending book value for the first and second year using ACRS.

**SOLUTION**

**Step 1** Find the recovery rate for the first year by locating the 5-year property category in the table; then read down to the first year and across to the recovery rate, 15%.

**Step 2** Find the first year's depreciation and ending book value. Round to the nearest dollar.

|  | Recovery |  |  |
|---|---|---|---|
| Cost | × Rate | = Annual Depreciation |  |
| $3,750 × | 0.15 | = | $562.50 ≈ $563 |

|  | Accumulated |  |
|---|---|---|
| Cost | − Depreciation | = Book Value |
| $3,750 − | $563 | = $3,187 |

**Step 3** Find the second year's depreciation and ending book value. The recovery rate for a 5-year property, second year, is 22%.

|  | Recovery |  |  |
|---|---|---|---|
| Cost | × Rate | = Annual Depreciation |  |
| $3,750 × | 0.22 | = | $825 |

|  | Accumulated |  |
|---|---|---|
| Cost | − Depreciation | = Book Value |
| $3,750 − | ($1,388) | = $2,362 |

$563 + $825

**PRACTICE PROBLEMS**

Find the recovery rate, the annual depreciation, and the ending book value for the first year using the ACRS method of depreciation. Use the table on page 449. Round to the nearest dollar.

| | Cost | Recovery Period | Recovery Rate | Annual Depreciation | Ending Book Value |
|---|---|---|---|---|---|
| **1.** | $ 730 | 5 years | _____ | _____ | _____ |
| **2.** | 1,850 | 10 years | _____ | _____ | _____ |
| **3.** | 11,800 | 3 years | _____ | _____ | _____ |

Find the accumulated depreciation and the ending book value for the third year using the ACRS method of depreciation.

| | Cost | Recovery Period | Accumulated Depreciation | Ending Book Value |
|---|---|---|---|---|
| **4.** | $40,600 | 5 years | _____ | _____ |
| **5.** | 81,300 | 18 years | _____ | _____ |

## CALCULATING DEPRECIATION FOR A PORTION OF A YEAR

For the 3-, 5-, and 10-year property, the full first-year percentage applies no matter when in the tax year the property is placed in service. The percentages for 15-, 18-, and 19-year real property depend on when the asset is placed in service during the tax year. The following table shows the recovery rate for each month of the tax year for 18-year recovery property. Each month of the tax year is assigned a depreciation rate. To find the annual depreciation, multiply the cost by the rate assigned to the month the asset was placed in service. Use the rates assigned to that month for each year in the recovery period.

### 18-YEAR REAL PROPERTY (Placed in Service after June 22, 1984)

| Year | Month Placed in Service | | | | | | | | | | | |
|---|---|---|---|---|---|---|---|---|---|---|---|---|
| | 1 | 2 | 3 | 4 | 5 | 6 | 7 | 8 | 9 | 10 | 11 | 12 |
| 1st | 9% | 9% | 8% | 7% | 6% | 5% | 4% | 4% | 3% | 2% | 1% | 0.4% |
| 2d | 9% | 9% | 9% | 9% | 9% | 9% | 9% | 9% | 9% | 10% | 10% | 10% |
| 3d | 8% | 8% | 8% | 8% | 8% | 8% | 8% | 8% | 9% | 9% | 9% | 9% |
| 4th | 7% | 7% | 7% | 7% | 7% | 8% | 8% | 8% | 8% | 8% | 8% | 8% |
| 5th | 7% | 7% | 7% | 7% | 7% | 7% | 7% | 7% | 7% | 7% | 7% | 7% |
| 6th | 6% | 6% | 6% | 6% | 6% | 6% | 6% | 6% | 6% | 6% | 6% | 6% |

### EXAMPLE

On June 3 Gladwell Insurance moved into its new office building costing $1,500,000. According to the ACRS guidelines the building is classified as 18-year property. Find the annual depreciation deduction on the building for the first and second year using ACRS.

### SOLUTION

**Step 1** Find the recovery rate for the first year by reading down the Year column and across to the 6th month of the year (June). The recovery rate is 5%. Find the depreciation for the first year.

| | Cost | × | Recovery Rate | = Depreciation |
|---|---|---|---|---|
| | $1,500,000 | × | 0.05 | = $75,000 |

**Step 2** The recovery rate for the second year, 6th month, is 9%. Find the depreciation for the second year.

$$\begin{array}{ccc} & \text{Recovery} & \\ \text{Cost} \times & \text{Rate} & = \text{Depreciation} \\ \$1,500,000 \times & 0.09 & = \$135,000 \end{array}$$

## PRACTICE PROBLEMS

Use the ACRS to find the recovery rate, the annual depreciation, and the ending book value for the first year for property assigned to the 18-year recovery period. Use the table on page 451.

| | Cost | Month Placed in Service | Recovery Rate | Annual Depreciation | Book Value |
|---|---|---|---|---|---|
| 1. | $ 4,600 | April | _____ | _____ | _____ |
| 2. | 17,800 | August | _____ | _____ | _____ |

Use ACRS to find the accumulated depreciation and the ending book value for the third year for property assigned to the 18-year recovery period.

| | Cost | Month Placed in Service | Accumulated Depreciation | Ending Book Value |
|---|---|---|---|---|
| 3. | $ 29,300 | October | _____ | _____ |
| 4. | 119,000 | February | _____ | _____ |

## PREPARING AN ACRS DEPRECIATION SCHEDULE

To find the annual depreciation when preparing a depreciation schedule using the ACRS method, multiply the cost by the appropriate rate for each successive year. There is no need to estimate the salvage value or the useful life when using the ACRS method.

### EXAMPLE

Oregon Investment Association purchased office furniture costing $24,700. The furniture is assigned to the 5-year property class. Prepare a depreciation schedule for the recovery period using the ACRS method.

ANSWERS FOR
PRACTICE PROBLEMS

| | Recovery Rate | Annual Depreciation | Book Value |
|---|---|---|---|
| (1) | 7% | $322 | $ 4,278 |
| (2) | 4% | 712 | 17,088 |

| | Accumulated Depreciated | Ending Book Value |
|---|---|---|
| (3) | $ 6,153 | $23,147 |
| (4) | 30,940 | 88,060 |

## SOLUTION

**Step 1** Find the annual depreciation, accumulated depreciation, and book value for each recovery year using the table on page 451.

| Cost | × | Recovery Rate | = | Annual Depreciation | | Cost | − | Accumulated Depreciation | = | Book Value |
|---|---|---|---|---|---|---|---|---|---|---|
| $24,700 | × | 0.15 | = | $3,705 | | $24,700 | − | $ 3,705 | = | $20,995 |
| 24,700 | × | 0.22 | = | 5,434 | | 24,700 | − | 9,139 | = | 15,561 |
| 24,700 | × | 0.21 | = | 5,187 | | 24,700 | − | 14,326 | = | 10,374 |
| 24,700 | × | 0.21 | = | 5,187 | | 24,700 | − | 19,513 | = | 5,187 |
| 24,700 | × | 0.21 | = | 5,187 | | 24,700 | − | 24,700 | = | 0 |

**Step 2** Prepare a depreciation schedule. The following depreciation schedule was prepared using a computer spreadsheet.

```
                   Depreciation Schedule


            Recovery    Annual          Accumulated    Ending
    Year    Rate        Depreciation    Depreciation   Book Value


      0                                                $24,700
      1      15%        $3,705          $ 3,705         20,995
      2      22%         5,434            9,139         15,561
      3      21%         5,187           14,326         10,374
      4      21%         5,187           19,513          5,187
      5      21%         5,187           24,700          -0-
```

## PRACTICE PROBLEM

Unicom Insurance purchased electronic office equipment costing $107,800. According to ACRS, the equipment is in the 5-year property class. Prepare a depreciation schedule for the recovery period using ACRS.

| Year | Recovery Rate | Annual Depreciation | Accumulated Depreciation | Ending Book Value |
|---|---|---|---|---|
| 0 | — | — | — | $107,800 |
| 1 | _____ | _____ | _____ | _____ |
| 2 | _____ | _____ | _____ | _____ |
| 3 | _____ | _____ | _____ | _____ |
| 4 | _____ | _____ | _____ | _____ |
| 5 | _____ | _____ | _____ | _____ |

ANSWERS FOR PRACTICE PROBLEM

| Year | Recovery Rate | Annual Depreciation | Accumulated Depreciation | Ending Book Value |
|---|---|---|---|---|
| 1 | 15% | 16,170 | 16,170 | 91,630 |
| 2 | 22% | 23,716 | 39,886 | 67,914 |
| 3 | 21% | 22,638 | 62,524 | 45,276 |
| 4 | 21% | 22,638 | 85,162 | 22,638 |
| 5 | 21% | 22,638 | 107,800 | 0 |

## LESSON 13-5　　EXERCISES

Find the recovery rate, the annual depreciation, and the ending book value
for the first year using ACRS. Use the table on page 449. Round to the
nearest dollar.

| | Cost | Recovery Period | Recovery Rate | Annual Depreciation | Ending Book Value |
|---|---|---|---|---|---|
| 1. | $ 954 | 3 years | _____ | _____ | _____ |
| 2. | 2,030 | 10 years | _____ | _____ | _____ |
| 3. | 15,400 | 5 years | _____ | _____ | _____ |

Find the accumulated depreciation and the ending book value for the third
year using ACRS.

| | Cost | Recovery Period | Accumulated Depreciation | Ending Book Value |
|---|---|---|---|---|
| 4. | $ 1,670 | 5 years | _____ | _____ |
| 5. | 31,300 | 10 years | _____ | _____ |
| 6. | 241,800 | 3 years | _____ | _____ |

Find the recovery rate, the annual depreciation, and the ending book value
for the first year for the following property assigned to the 18-year recovery period.

| | Cost | Month Placed in Service | Recovery Rate | Annual Depreciation | Ending Book Value |
|---|---|---|---|---|---|
| 7. | $ 13,700 | January | _____ | _____ | _____ |
| 8. | 28,900 | September | _____ | _____ | _____ |
| 9. | 38,200 | March | _____ | _____ | _____ |

Find the accumulated depreciation and the ending book value for the third
year for the following property assigned to the 18-year recovery period.

| | Cost | Month Placed in Service | Accumulated Depreciation | Ending Book Value |
|---|---|---|---|---|
| 10. | $ 31,200 | February | _____ | _____ |
| 11. | 72,700 | November | _____ | _____ |
| 12. | 115,700 | May | _____ | _____ |

## BUSINESS APPLICATIONS

Use the tables on pages 449 and 451 to solve the following.

13. James Brothers and Associates purchased paper recycling equipment that cost $1,560,000. The equipment is assigned to the 5-year recovery period. Find (a) the recovery rate and (b) the annual depreciation for the fourth year using the ARCS.

(a) _____

(b) _____

14. During the month of July, New Mountain Realty purchased an apartment building that cost $2,350,000. The building is assigned to the 18-year recovery period. Find the recovery rate and the annual depreciation for the 4th, 5th, and 6th year.

| Year | Recovery Rate | Annual Depreciation |
|------|------|------|
| 4 | _____ | _____ |
| 5 | _____ | _____ |
| 6 | _____ | _____ |

15. Prepare a depreciation schedule using the ACRS method of depreciation for a pickup truck that cost $14,700. The truck is assigned to the 3-year recovery period.

| Year | Recovery Rate | Annual Depreciation | Accumulated Depreciation | Ending Book Value |
|------|------|------|------|------|
| 0 | — | — | — | $14,700 |
| 1 | _____ | _____ | _____ | _____ |
| 2 | _____ | _____ | _____ | _____ |
| 3 | _____ | _____ | _____ | _____ |

16. Kastle Kleaners purchased new dry cleaning equipment for a cost of $41,700. The equipment is in the 5-year property class. Prepare a depreciation schedule using the ACRS method of depreciation.

| Year | Recovery Rate | Annual Depreciation | Accumulated Depreciation | Ending Book Value |
|------|------|------|------|------|
| 0 | — | — | — | $41,700 |
| 1 | _____ | _____ | _____ | _____ |
| 2 | _____ | _____ | _____ | _____ |
| 3 | _____ | _____ | _____ | _____ |
| 4 | _____ | _____ | _____ | _____ |
| 5 | _____ | _____ | _____ | _____ |

## CHAPTER 13 REVIEW

Use the tables on pages 449 and 451 to solve the following. Round percents to the nearest tenths and amounts to the nearest dollar.

1. Use the straight-line depreciation method to find (a) the annual depreciation and (b) the ending book value for the first year for a generator that costs $1,550, has an estimated life of 8 years, and a trade-in value of $150.

(a) _____

(b) _____

2. Find the accumulated depreciation at the end of 5 years for a forklift that cost $8,700 with a trade-in value of $1,300 and an estimated life of 10 years. Use the straight-line depreciation method.

_____

3. United Metals purchased a melting furnace for $112,700. The furnace has a scrap value of $15,000 and an estimated useful life of 10 years. Find the book value at the end of 3 years using the 200% declining-balance method.

_____

4. Jamestown Manufacturers purchased industrial sorting equipment for $43,500. The trade-in value is $16,800, and the estimated life is 890,000 units sorted. Use the units-of-production method to find (a) the first year's depreciation and (b) the ending book value if 210,000 units were sorted the first year.

(a) _____

(b) _____

5. The Orange City Country Club purchased a utility truck that cost $15,600 with a trade-in value of $4,400 and an estimated life of 80,000 miles. The truck was driven 17,500 miles the first year and 13,200 miles the second year. Find (a) the accumulated depreciation and (b) the ending book value for the second year using the units-of-production method of depreciation.

(a) _____

(b) _____

6. The Izell Company purchased a networked computer system costing $52,000 including installation. The company estimated that the life of the system is 5 years with a scrap value of $5,500. Find the book value at the end of the first year using (a) the 200% declining-balance method and (b) the sum-of-the-years'-digits method.

(a) _____

(b) _____

7. The River Market purchased a computerized checkout system costing $129,000. The system has an estimated life of 6 years with a salvage value of $10,000. Prepare a depreciation schedule using the sum-of-the-years'-digits method. Round to the nearest dollar.

| Year | Depreciation Rate | Annual Depreciation | Accumulated Depreciation | Ending Book Value |
|------|------------------|--------------------|-------------------------|-------------------|
| 0 | — | — | — | $129,000 |
| 1 | _____ | _____ | _____ | _____ |
| 2 | _____ | _____ | _____ | _____ |
| 3 | _____ | _____ | _____ | _____ |
| 4 | _____ | _____ | _____ | _____ |
| 5 | _____ | _____ | _____ | _____ |
| 6 | _____ | _____ | _____ | _____ |

8. The Home Decorating Depot purchased a glass display case for $1,090. The case has a useful life of 8 years and no salvage value. Depreciation is calculated by the sum-of-the-years'-digits method. For the fifth year of useful life, find (a) the depreciation rate fraction and (b) the annual depreciation amount. Round to the nearest dollar.

(a) _____

(b) _____

9. Wang-Chow Engineering purchased new office furniture at a cost of $5,600. The furniture is in the 5-year property class. Find (a) the accumulated depreciation and (b) the ending book value for the second year using the ACRS method of depreciation.

(a) _____

(b) _____

**10.** In August, the Livingston Financial Consultants purchased an office building that cost $573,400. The building is in the 18-year recovery class. Find the recovery rate and the annual depreciation for the fourth, fifth, and sixth year using the ACRS method of depreciation.

| Year | Recovery Rate | Annual Depreciation |
|------|---------------|---------------------|
| 4 | _____ | _____ |
| 5 | _____ | _____ |
| 6 | _____ | _____ |

**11.** Finwick Distributors purchased a company van that cost $18,300. The van is assigned to the 3-year recovery class. Find (a) the accumulated depreciation and (b) the ending book value for the second year using ACRS.

(a) _____

(b) _____

**12.** Oak Hill Park purchased a manufactured home that cost $57,600 to be used as their model center. The home is in the 10-year property class. Prepare an ACRS depreciation schedule for a manufactured home.

| Year | Recovery Rate | Annual Depreciation | Accumulated Depreciation | Ending Book Value |
|------|---------------|---------------------|--------------------------|-------------------|
| 0 | — | — | — | $57,600 |
| 1 | _____ | _____ | _____ | _____ |
| 2 | _____ | _____ | _____ | _____ |
| 3 | _____ | _____ | _____ | _____ |
| 4 | _____ | _____ | _____ | _____ |
| 5 | _____ | _____ | _____ | _____ |
| 6 | _____ | _____ | _____ | _____ |
| 7 | _____ | _____ | _____ | _____ |
| 8 | _____ | _____ | _____ | _____ |
| 9 | _____ | _____ | _____ | _____ |
| 10 | _____ | _____ | _____ | _____ |

## LEARNING OBJECTIVES

**1.** Understand and complete an income statement and balance sheet.

**2.** Prepare vertical and horizontal analyses of the income statement and balance sheet.

**3.** Analyze various elements of the income statement and balance sheet employing current ratios, the acid-test ratio, and accounts-receivable turnover ratios.

**4.** Prepare percent analyses regarding the aging of receivables, return on total assets, and return on owners' equity.

The ***income statement*** and ***balance sheet*** are ***financial statements*** that show a company's progress and financial condition.

Business managers analyze these statements to determine the company's performance and to plan where efforts should be directed to increase profitability and improve the company's financial standing.

Financial statement information is presented in terms of dollar amounts. Since an amount by itself has little significance and because it is difficult to develop an accurate relationship by comparing dollar amounts, financial statements are generally analyzed in terms of ratios and percents.

Many businesses use computers to prepare financial statements and to analyze financial statement amounts and relationships. Still, business managers and employees must understand the financial statements and be able to interpret the results of the analysis.

In this chapter you will learn how income statements and balance sheets are prepared. You will also learn how to use the data contained on these financial statements to analyze the business performance.

## LESSON 14-1    INCOME STATEMENT ANALYSIS

The ***income statement*** summarizes a company's income, costs, and expenses for a certain period of time such as a month, quarter, or year. If income exceeds costs and expenses, the result is called ***net income.*** If costs and expenses exceed income, there is a ***net loss.***

## UNDERSTANDING THE INCOME STATEMENT

The specific information shown on an income statement will vary somewhat from one type of business to another. An income statement for a **merchandising business,** like a retailer or wholesaler, shows income, cost of goods sold, and expenses. An income statement for a **service business,** like a real estate agency or a consulting business, shows just income and expenses.

```
                    Ellerbrook Fashions
1                    Income Statement
               For the Month Ended June 30, 19—
2   Revenue:
      Sales ....................................  $34,965
      Less: Sales Returns and Allowances ......      820
        Net Sales ............................             $34,145
    Cost of Goods Sold:
      Merchandise Inventory, June 1 ...........   74,625
      Add: Purchases..........................    21,015
3     Goods Available for Sale ...............    95,640
      Less: Merchandise Inventory, June 30 ....   77,390
        Cost of Goods Sold ...................              18,250
4→  Gross Profit on Sales.....................             $15,895
    Operating Expenses:
      Salaries..................................   4,260
      Rent ....................................    3,400
      Advertising .............................    1,715
      Utilities.................................     930
5     Employer's Taxes ........................      515
      Depreciation—Equipment ................       305
      Bad Debts ...............................      235
      Insurance ...............................      135
      Supplies ................................      160
      Miscellaneous ...........................      205
    Total Operating Expenses..................              11,860
6→  Net Income ...............................             $ 4,035
```

Ellerbrook Fashions' income statement for the month ended June 30, 19—, is shown above. The following information is provided in the income statement.

1. The three-line heading shows the name of the business, the type of financial statement, and the time period covered.
2. The term **revenue** means the same as *income.* Total sales were $34,965. Merchandise was returned to Ellerbrook, or price allowances were granted, of $820. The difference between these two amounts is **net sales,** $34,145.
3. Add beginning inventory and purchases to find merchandise available for sale. Subtract ending inventory from merchandise available for sale to find the cost of goods sold.
4. Gross profit, $15,895, is the difference between net sales and the cost of goods sold.
5. The total operating expenses are $11,860.
6. Total operating expenses are deducted from gross profit on sales to find the net income of $4,035.

## PRACTICE PROBLEM

Complete the income statement for Jenny's Fabric Shop for the month ended July 31: Sales, $8,900; Sales Returns and Allowances, $300; Merchandise Inventory, July 1, $19,000; Purchases in July, $6,500; Merchandise Inventory, July 31, $21,100; Rent, $1,000; Salaries, $980; Advertising, $450; Utilities, $120; Employer's Taxes, $100; Depreciation—Equipment, $95; Insurance, $80; Supplies Expense, $60, and Miscellaneous Expense, $130.

**1.** _____

**2.** _____

**3.** _____

Revenue:

    Sales . . . . . . . . . . . . . . . . . . . . . . . . .   _____

    Less: Sales Returns and Allowances . . .   _____

**4.**     Net Sales . . . . . . . . . . . . . . . . . . .   _____

Cost of Goods Sold

    Merchandise Inventory, July 1 . . . . . . . .   _____

    Add: Purchases . . . . . . . . . . . . . . . . . . . .   _____

**5.**     Goods Available for Sale . . . . . . . . . . . . .   _____

**6.**     Less: Merchandise Inventory, July 31 . .   _____

**7.**     Cost of Goods Sold . . . . . . . . . . . . . .   _____

**8.** Gross Profit on Sales . . . . . . . . . . . . . . . . . . .   _____

Operating Expenses:

    Rent . . . . . . . . . . . . . . . . . . . . . . . . . . . . .   _____

    Salaries . . . . . . . . . . . . . . . . . . . . . . . . . .   _____

    Advertising . . . . . . . . . . . . . . . . . . . . . . . .   _____

    Utilities . . . . . . . . . . . . . . . . . . . . . . . . . . .   _____

    Employer's Taxes . . . . . . . . . . . . . . . . . . . .   _____

    Depreciation—Equipment . . . . . . . . . . . .   _____

    Insurance . . . . . . . . . . . . . . . . . . . . . . .   _____

    Supplies . . . . . . . . . . . . . . . . . . . . . . . . .   _____

    Miscellaneous . . . . . . . . . . . . . . . . . . . . . .   _____

**9.** Total Operating Expenses . . . . . . . . . . . . .   _____

**10.** Net Income . . . . . . . . . . . . . . . . . . . . . . .   =========

---

## VERTICAL ANALYSIS

In a *vertical analysis* of the income statement, each item on the statement is compared to and expressed as a percent of net sales. Divide each income and expense amount by net sales to find the percent of net sales.

$$\text{Percent of Net Sales} = \frac{\text{Item Amount}}{\text{Net Sales}}$$

Vertical analysis helps management recognize which cost or expense items may be disproportionately high or low in comparison to the others. The results are also useful when comparing the company's performance to industry averages and to other standards of performance.

### EXAMPLE

In March, Ling's Entertainment Center had net sales of $36,400 and an advertising expense of $3,025. What percent was advertising expense of net sales? Round the percent to the nearest hundredth.

### SOLUTION

$$\frac{\text{Percent of}}{\text{Net Sales}} = \frac{\text{Advertising Expense}}{\text{Net Sales}} = \frac{\$3,025}{\$36,400} \approx 0.08310 \text{ or } 8.31\%$$

### PRACTICE PROBLEMS

Selected items from the income statement for Tammy's Hobby Shop are shown below. What percent is each item of net sales? Round percents to the nearest hundredth.

| Income Statement Item | Amount | Percent of Net Sales |
|---|---|---|
| 1. Sales | $10,365 | _____ |
| 2. Less: Sales returns and allowances | 520 | _____ |
| Net Sales | 9,845 | 100.00% |
| 3. Cost of goods sold | 5,460 | _____ |
| 4. Total operating expenses | 3,155 | _____ |
| 5. Net income | 1,230 | _____ |

## HORIZONTAL ANALYSIS

A *horizontal analysis* compares the present period's income statement to a previous period's income statement. This analysis identifies the amount and percent of increase or decrease in each income statement item from one period to the next. The amounts on the previous income statement are always the base, or 100%, to which the most recent period's data are compared. To perform a horizontal analysis, first find the amount of increase or decrease, then find the percent of increase or decrease.

$$\text{Percent of Increase or Decrease} = \frac{\text{Amount of Increase or Decrease}}{\text{Previous Period Amount}}$$

The horizontal analysis helps management pinpoint why net income has increased or decreased.

## EXAMPLE

Carmichael Company's net sales were $650,500 in 19X1 and $602,350 in 19X2. What was the amount and percent of increase or decrease from 19X1 to 19X2? Compute each percent to the nearest hundredth.

## SOLUTION

**Step 1** Calculate the amount of increase or decrease.

$$\begin{array}{ccc} 19X1 & 19X2 & \text{Amount} \\ \text{Sales} & - & \text{Sales} & = \text{of Decrease} \end{array}$$

$$\$650,500 - \$602,350 = \$48,150$$

**Step 2** Calculate the percent of decrease.

$$\frac{\text{Percent}}{\text{of Decrease}} = \frac{\text{Amount of Decrease}}{19X1 \text{ Sales}} = \frac{\$48,150}{\$650,500} \approx 0.0740 \text{ or } 7.40\%$$

Divide by previous amount.

## PRACTICE PROBLEMS

A portion of the comparative income statement for March and April is shown below for Mendota Industries. Calculate the amount and percent of increase (+) or decrease (−) from March to April. Round each percent to the nearest hundredth.

Mendota Industries
Comparative Income Statement
For the Months Ended March and April, 19—

| | March | April | Increase or Decrease Amount | Percent |
|---|---|---|---|---|
| 1. Sales | $5,260 | $5,895 | $_____ | _____ |
| 2. Less: Sales Returns and Allowances | 375 | 210 | _____ | _____ |
| 3. Net Sales | 4,885 | 5,685 | _____ | _____ |
| 4. Cost of Goods Sold | 2,565 | 3,085 | _____ | _____ |
| 5. Gross Profit | 2,320 | 2,600 | _____ | _____ |
| 6. Total Operating Expenses | 1,570 | 1,460 | _____ | _____ |
| 7. Net Income | 750 | 1,140 | _____ | _____ |

ANSWERS FOR
PRACTICE PROBLEMS

| | Amount | Percent |
|---|---|---|
| (1) | +$635 | +12.07% |
| (2) | −165 | −44.00% |
| (3) | +800 | +16.38% |
| (4) | +520 | +20.27% |
| (5) | +280 | +12.07% |
| (6) | −110 | −7.01% |
| (7) | +390 | +52.00% |

## COMPARING ACTUAL PERFORMANCE TO BUDGETED AMOUNTS

A *budget* is a plan for the company's future income, costs, expenses, and net income. An analysis shows the amount and percent that actual amounts are over or under budgeted amounts. The budgeted amount is the base, or 100%, to which the actual performance is compared.

$$\text{Percent Over or Under Budget} = \frac{\text{Amount Actual Performance Is Over or Under Budget}}{\text{Budgeted Amount}}$$

### EXAMPLE

Susuki Company projected sales of $49,375 for July. Actual sales were $51,215. What amount and percent were actual sales over or under the budgeted amount? Round the percent to the nearest hundredth.

### SOLUTION

**Step 1**  Calculate the amount over or under the budget.

Actual Amount − Budgeted Amount = Amount Over Budget

$51,215  −  $49,375  =  $1,840

**Step 2**  Calculate the percent actual performance is over the budgeted amount.

$$\frac{\text{Percent Over}}{\text{Budget}} = \frac{\text{Amount Over Budget}}{\text{Budgeted Amount}} = \frac{\$1,840}{\$49,375} \approx 0.03726 \text{ or } 3.73\%$$

### PRACTICE PROBLEMS

A portion of Crosby Real Estate Company's actual income statement and budget for 19X4 are shown below. What amount and percent is each item on the actual income statement over (+) or under (−) the budget?

Crosby Real Estate Company
Income Statement
For the Year Ended December 31, 19X4

|  | Actual | Budgeted | Over or Under Amount | Over or Under Percent |
|---|---|---|---|---|
| Revenue: | | | | |
| 1.  Sales Commission Earned | $145,216 | $150,000 | $_____ | _____ |
| Expenses: | | | | |
| 2.  Rent | 16,200 | 14,400 | _____ | _____ |
| 3.  Office Salaries | 34,250 | 35,600 | _____ | _____ |
| 4.  Advertising | 24,750 | 20,000 | _____ | _____ |
| 5.  Total Operating Expenses | 96,200 | 95,000 | _____ | _____ |
| 6.  Net Income | 49,016 | 55,000 | _____ | _____ |

## LESSON 14-1    EXERCISES

1. Complete the income statement for Bellnote Music Store for the month ending September 30: sales, $35,965; sales returns and allowances, $1,420; merchandise inventory, September 1, $82,145; purchases in September, $15,650; merchandise inventory, September 30, $76,410; rent, $3,000; salaries, $2,800; advertising, $975; utilities, $310; depreciation—equipment, $275; insurance, $210; bad debts, $150; supplies expense, $125; and miscellaneous expense, $100.

_____

_____

_____

Revenue:

   Sales . . . . . . . . . . . . . . . . . . . . . . . . . . . . . . . . .   _____

   Less: Sales Returns and Allowances   _____

      Net Sales . . . . . . . . . . . . . . . . . . . . . . . . . . .      _____

Cost of Goods Sold:

   Merchandise Inventory, September 1 . . . . . .   _____

   Add: Purchases . . . . . . . . . . . . . . . . . . . . . . . . .   _____

   Merchandise Available for Sale . . . . . . . . . .   _____

   Less: Merchandise Inventory, September 30   _____

      Cost of Goods Sold . . . . . . . . . . . . . . . . . . .      _____

Gross Profit on Sales . . . . . . . . . . . . . . . . . . . . .      _____

Operating Expenses:

   Rent . . . . . . . . . . . . . . . . . . . . . . . . . . . . . . . . .   _____

   Salaries . . . . . . . . . . . . . . . . . . . . . . . . . . . . . . .   _____

   Advertising . . . . . . . . . . . . . . . . . . . . . . . . . . . .   _____

   Utilities . . . . . . . . . . . . . . . . . . . . . . . . . . . . . . .   _____

   Depreciation—Equipment . . . . . . . . . . . . . . .   _____

   Insurance . . . . . . . . . . . . . . . . . . . . . . . . . . . . .   _____

   Bad Debts . . . . . . . . . . . . . . . . . . . . . . . . . . . .   _____

   Supplies . . . . . . . . . . . . . . . . . . . . . . . . . . . . . .   _____

   Miscellaneous . . . . . . . . . . . . . . . . . . . . . . . . .   _____

Total Operating Expenses . . . . . . . . . . . . . . . . .      _____

Net Income . . . . . . . . . . . . . . . . . . . . . . . . . . . . .      ========

**2.** The comparative income statement shown below is for Urbana Motor-cycle Sales for July and August. Calculate the amount and percent of increase (+) or decrease (−) from July to August. Round each percent to the nearest hundredth.

Urbana Motorcycle Sales
Comparative Income Statement
For the Months Ended July and August, 19—

| | July | August | Increase or Decrease Amount | Percent |
|---|---|---|---|---|
| Revenue: | | | | |
| Sales .............................. | $ 94,315 | $ 89,740 | _____ | _____ |
| Less: Sales Returns and Allowances. | 2,470 | 1,835 | _____ | _____ |
| Net Sales ......................... | 91,845 | 87,905 | _____ | _____ |
| Cost of Goods Sold: | | | | |
| Beginning Inventory ............... | 126,315 | 133,880 | _____ | _____ |
| Add: Purchases ................... | 63,905 | 47,105 | _____ | _____ |
| Merchandise Available for Sale ..... | 190,220 | 180,985 | _____ | _____ |
| Less: Ending Inventory ............ | 133,880 | 129,160 | _____ | _____ |
| Cost of Goods Sold ............... | 56,340 | 51,825 | _____ | _____ |
| Gross Profit ...................... | 35,505 | 36,080 | _____ | _____ |
| Operating Expenses: | | | | |
| Salaries ........................... | 14,386 | 12,508 | _____ | _____ |
| Rent ............................. | 4,200 | 4,500 | _____ | _____ |
| Advertising........................ | 1,650 | 1,375 | _____ | _____ |
| Depreciation—Equipment.......... | 480 | 480 | _____ | _____ |
| Supplies .......................... | 216 | 205 | _____ | _____ |
| Miscellaneous ..................... | 348 | 376 | _____ | _____ |
| Total Operating Expenses ........... | 21,280 | 19,444 | _____ | _____ |
| Net Income ........................ | $ 14,225 | $ 16,636 | _____ | _____ |

**3.** Hudson Department Store's actual income statement and budget for 19X6 are shown below. What amount and percent is each item on the actual income statement over (+) or under (−) the budget? Round each percent off to the nearest hundredth.

Hudson Department Store
Income Statement
For the Year Ended December 31, 19X6

| | Actual | Budgeted | Over or Under Amount | Percent |
|---|---|---|---|---|
| Revenue: | | | | |
| Sales ............................ | $962,835 | $950,750 | _____ | _____ |
| Less: Sales Returns and Allowances | 8,418 | 9,500 | _____ | _____ |
| Net Sales ......................... | 954,417 | 941,250 | _____ | _____ |
| Cost of Goods Sold: | | | | |
| Merchandise Inventory, Jan. 1 ........ | 368,425 | 375,000 | | |
| Add: Purchases .................... | 489,320 | 475,000 | _____ | _____ |
| Merchandise Available for Sale ....... | 857,745 | 850,000 | _____ | _____ |
| Less: Merchandise Inventory, Dec. 31 . | 365,140 | 370,000 | _____ | _____ |
| Cost of Goods Sold ................ | 492,605 | 480,000 | _____ | _____ |
| Gross Profit ....................... | 461,812 | 461,250 | _____ | _____ |
| Operating Expenses: | | | | |
| Salaries .......................... | 224,600 | 210,000 | _____ | _____ |
| Depreciation—Building .............. | 68,000 | 68,000 | _____ | _____ |
| Advertising........................ | 42,560 | 48,000 | _____ | _____ |
| Depreciation—Equipment............ | 13,500 | 13,500 | _____ | _____ |
| Utilities.......................... | 18,572 | 16,200 | _____ | _____ |
| Real Estate Taxes .................. | 28,306 | 26,150 | _____ | _____ |
| Bad Debts ........................ | 18,390 | 15,000 | _____ | _____ |
| Supplies ......................... | 13,070 | 12,000 | _____ | _____ |
| Miscellaneous ..................... | 4,400 | 5,000 | _____ | _____ |
| Total Operating Expenses ............. | 431,398 | 413,850 | _____ | _____ |
| Net Income ......................... | $ 30,414 | $ 47,400 | _____ | _____ |

## BUSINESS APPLICATIONS

Round percents to the nearest hundredth.

4. Last year Atkinson Insurance Agency had a revenue of $216,486 and the following operating expenses: Salaries, $89,500, Rent, $30,000; Advertising, $18,000; Depreciation—Equipment, $12,650; Employer's Taxes, $10,740; Utilities, $7,470; Supplies, $3,675; and Miscellaneous, $1,365. What was Atkinson's net income or net loss?

_____

5. Last year the following amounts were listed on Harmon and Company's income statement: Net Sales, $624,816; Cost of Goods Sold, $378,419; and Total Operating Expenses, $153,706. What percent of net sales were (a) cost of goods sold, (b) gross profit, (c) total operating expenses, and (d) net income?

(a) _____

(b) _____

(c) _____

(d) _____

6. Net sales for Lopez Manufacturing Company were as follows for three consecutive years: 19X4, $1,476,397; 19X5, $1,807,916; and 19X6, $1,532,681. What was the percent of increase or decrease in net sales from (a) 19X4 to 19X5 and (b) 19X5 to 19X6?

(a) _____

(b) _____

The **balance sheet,** which shows what a business owns, how much it owes to others, and the difference between these two amounts, is divided into three sections: assets, liabilities, and owners' equity. An **asset** is anything of value that is owned by a business, such as cash, merchandise inventory, buildings, and land. A **liability** is an amount owed to another, called a **creditor.** Liabilities result from buying on credit or from obtaining a bank loan. **Owners' equity,** the difference between total assets and total liabilities, shows the amount of assets which the owner may claim after all liabilities are paid. Owners' equity may also be called **net worth, capital,** and **stockholders' equity.** On any balance sheet, total assets must equal total liabilities and owners' equity, as shown in the **accounting equation:**

Assets = Liabilities + Owners' Equity

## UNDERSTANDING THE BALANCE SHEET

The balance sheet may be prepared in **account form,** with the assets listed on the left and the liabilities and owners' equity listed on the right, or in **report form,** with all items listed vertically. A report form balance sheet is shown below.

Sports World
Balance Sheet
December 31, 19—

### Assets

| | | |
|---|---:|---:|
| **Current Assets:** | | |
| Cash | $ 14,815 | |
| Accounts Receivable | 3,960 | |
| Merchandise Inventory | 148,560 | |
| Supplies | 2,015 | |
| Total Current Assets | | $169,350 |
| **Fixed Assets:** | | |
| Equipment | 21,700 | |
| Building | 125,060 | |
| Land | 65,000 | |
| Total Fixed Assets | | 211,760 |
| Total Assets | | $381,110 |

### Liabilities

| | | |
|---|---:|---:|
| **Current Liabilities:** | | |
| Notes Payable | $24,600 | |
| Accounts Payable | 37,350 | |
| Total Current Liabilities | | 61,950 |
| **Long-Term Liabilities:** | | |
| Notes Payable | 38,050 | |
| Mortgage—Notes Payable | 84,320 | |
| Total Long-Term Liabilities | | 122,370 |
| Total Liabilities | | $184,320 |

### Owners' Equity

| | | |
|---|---|---:|
| David M. Olson, Capital | | 196,790 |
| Total Liabilities and Capital | | $381,110 |

1. The three-line heading shows the name of the business, the type of financial statement, and the date of preparation.
2. The assets section is divided into two categories. **Current assets** includes cash and other liquid assets that will be converted into cash or consumed within one year. **Fixed assets** are those that will be used in the business for a long time. Total current assets and total fixed assets equal **total assets.**
3. The liability section is divided into two categories: current liabilities and long-term liabilities. While **current liabilities** are those due within a year, **long-term liabilities** are due more than 1 year from the balance sheet date. Total current liabilities and total long-term liabilities equal **total liabilities.**
4. The owners' equity section is the financial worth that the owner, David M. Olson, has in the business. Owners' equity is calculated by deducting total liabilities from total assets.

## PRACTICE PROBLEM

Complete the balance sheet on December 31, 19X8, for Holiday Gift Shop, which is owned by Carla Ellis: Cash, $2,460; Accounts Receivable, $980; Merchandise Inventory, $75,290; Supplies, $310; Equipment, $12,250; Building, $80,500; Land, $46,000; Notes Payable (current portion), $1,200; Accounts Payable, $7,250; Notes Payable (long-term portion), $14,300; and Mortgage Notes Payable, $74,500.

1. _____
2. _____
3. _____

### Assets

Current Assets:
    Cash . . . . . . . . . . . . . . . . . . . . . . . . . . . . .   _____
    Accounts Receivable. . . . . . . . . . . . . . . . .   _____
    Merchandise Inventory. . . . . . . . . . . . . . .   _____
    Supplies . . . . . . . . . . . . . . . . . . . . . . . . . .   _____
4.     Total Current Assets. . . . . . . . . . . . . . . . _____
Fixed Assets:
    Equipment . . . . . . . . . . . . . . . . . . . . . . . . .   _____
    Building . . . . . . . . . . . . . . . . . . . . . . . . . . .   _____
    Land . . . . . . . . . . . . . . . . . . . . . . . . . . . . .   _____
5.     Total Fixed Assets. . . . . . . . . . . . . . . . . _____
6. Total Assets . . . . . . . . . . . . . . . . . . . . . . . . _____

### Liabilities

Current Liabilities:
    Notes Payable . . . . . . . . . . . . . . . _____
    Accounts Payable . . . . . . . . . . . . . _____
7.     Total Current Liabilities . . . . . . . . . . . . . _____
Long-Term Liabilities:
    Notes Payable . . . . . . . . . . . . . . . . _____
    Mortgage Notes Payable . . . . . . . _____
8.     Total Long-Term Liabilities. . . . . . . . . . _____
9. Total Liabilities . . . . . . . . . . . . . . . . . . . . . . _____

### Owners' Equity

10. Carla Ellis, Capital . . . . . . . . . . . . . . . . . . . _____
11. Total Liabilities and Owners' Equity. . . . . . _____

A *vertical analysis* of the balance sheet shows the percent that each individual item is of the total of that section. Each asset amount is compared to the total assets. Each liability and owners' equity amount is compared to the total liabilities and owners' equity.

$$\text{Percent of Total Assets} = \frac{\text{Individual Asset Amount}}{\text{Total Assets}}$$

$$\text{Percent of Total Liabilities and Owner's Equity} = \frac{\text{Individual Liability or Owners' Equity Amount}}{\text{Total Liabilities and Owners' Equity}}$$

The individual percents in each section of the balance sheet must total 100%. Therefore, it may be necessary to round an item's percent up or down after your calculations are completed.

Management scrutinizes these percent relationships to determine if any individual item is disproportionately higher or lower when compared with other balance sheet amounts.

### EXAMPLE

The April 30 balance sheet for Kramer Upholstery shows cash of $15,438 and total assets of $124,500. What percent is cash of the total assets? Round percent to the nearest tenth.

### SOLUTION

$$\frac{\text{Percent of}}{\text{Total Assets}} = \frac{\text{Cash Amount}}{\text{Total Assets}} = \frac{\$15,438}{\$124,500} \approx 0.124 \text{ or } 12.4\%$$

### PRACTICE PROBLEMS

Assets from the balance sheet for Chan's Jewelery are shown below. What percent is each amount of the total assets? Round percents to the nearest tenth.

| | Assets | Amount | Percent of Total |
|---|---|---|---|
| 1. | Cash | $ 35,725 | _____ |
| 2. | Accounts receivable | 16,005 | _____ |
| 3. | Merchandise inventory | 169,765 | _____ |
| 4. | Supplies | 2,286 | _____ |
| 5. | Store equipment | 45,442 | _____ |
| 6. | Office equipment | 16,577 | _____ |
| | Total Assets | $285,800 | 100.0% |

ANSWERS FOR PRACTICE PROBLEMS

**On this page:** (1) 12.5% (2) 5.6% (3) 59.4% (4) 0.8% (5) 15.9% (6) 5.8%

**From page 472:**
(1) Holiday Gift Shop (2) Balance Sheet (3) December 31, 19X8
(4) $79,040 (5) $138,750 (6) $217,790 (7) $8,450 (8) $88,800
(9) $97,250 (10) $120,540 (11) $217,790

## HORIZONTAL ANALYSIS

A *horizontal analysis* compares the present period's balance sheet to a previous period's balance sheet. This analysis identifies the amount and percent of increase or decrease in each balance sheet item from one period to the next. The previous period is always the base, or 100%, to which the most recent data are compared.

$$\text{Percent of Increase or Decrease} = \frac{\text{Amount of Increase or Decrease}}{\text{Previous Period's Amount}}$$

Business managers use this analysis to determine if any balance sheet amounts have changed drastically from one period to the next, thus improving or weakening the company's financial condition.

### EXAMPLE

Wilson Company's December 31, 19X3 and 19X4, balance sheets show cash as follows: 19X3, $36,215; and 19X4, $21,460. What was the amount and percent of increase or decrease in cash from 19X3 to 19X4? Round the percent to the nearest hundredth.

### SOLUTION

**Step 1**  Calculate the amount of increase or decrease.

19X3 Cash − 19X4 Cash = Amount of Decrease

$36,215  −  $21,460  =  $14,755

**Step 2**  Calculate the percent of decrease.

$$\text{Percent of Decrease} = \frac{\text{Amount of Decrease}}{\text{19X3 Cash}}$$

$$= \frac{\$14,755}{\$36,215} \approx 0.40742 \text{ or } 40.74\%$$

↑
Divide by
previous amount

### PRACTICE PROBLEMS

A portion of the comparative balance sheet for August and September is shown below for Huxley Appliance Center. What is the amount and percent of increase (+) or decrease (−) from August to September? Round each percent to the nearest hundredth.

| | | | | Increase or Decrease | |
|---|---|---|---|---|---|
| | Assets | August | September | Amount | Percent |
| 1. | Cash | $ 14,785 | $ 16,315 | $_____ | _____ |
| 2. | Accounts receivable | 21,406 | 13,820 | _____ | _____ |
| 3. | Merchandise inventory | 265,418 | 278,450 | _____ | _____ |

## COMPARING CURRENT DATA TO PRIOR PERIOD AVERAGES

A comparison of the current period's balance sheet with averages from prior periods often yields a more meaningful analysis than a comparison to the previous period only. This is because the previous period's amounts may not provide an accurate picture of the company's progress through the years.

The amount of increase or decrease from the average of prior periods to the current period is divided by the average to yield the percent of increase or decrease from the average.

$$\text{Percent Above or Below Average of Prior Periods} = \frac{\text{Amount of Increase or Decrease}}{\text{Average of Prior Periods}}$$

### EXAMPLE

Balance sheets for Caseys, Inc., show the following amounts for cash on December 31 for the past 3 years: 19X2, $39,376; 19X3, $42,516; and 19X4, $48,712. What is the amount and percent of increase or decrease from the average of 19X2 and 19X3 to 19X4? Round the percent to the nearest hundredth.

### SOLUTION

**Step 1** Calculate the total of prior period amounts.

$$\begin{array}{ccc} & & \text{Total of} \\ \text{19X2 Amount} + \text{19X3 Amount} = & \text{Prior Periods} \\ \$39,376 \quad + \quad \$42,516 \quad = & \$81,892 \end{array}$$

**Step 2** Calculate the average of prior period amounts.

$$\text{Average of Prior Periods} = \frac{\text{Total of Prior Periods}}{\text{Number of Periods}} = \frac{\$81,892}{2} = \$40,946$$

**Step 3** Calculate the amount of increase or decrease from the prior periods' average to the current period.

$$\begin{array}{ccc} \text{Current} & \text{Average of} & \text{Amount} \\ \text{Period Amount} - \text{Prior Periods} = & \text{of Increase} \\ \$48,712 \quad - \quad \$40,946 \quad = & \$7,766 \end{array}$$

**Step 4** Calculate the percent of increase over the average of the prior periods.

$$\text{Percent of Increase} = \frac{\text{Amount of Increase}}{\text{Average of Prior Periods}} = \frac{\$7,766}{\$40,946} \approx 0.18966 \text{ or } 18.97\%$$

## PRACTICE PROBLEMS

Balance sheet data for Lakeshore Industries for years 19X5, 19X6, and 19X7 are shown below. Calculate the average of 19X5 and 19X6 and the amount and percent of increase (+) or decrease (−) from the average to 19X7 data. Round percents to the nearest hundredth.

| Current Assets | 19X5 | 19X6 | 19X7 | Average of 19X5, 19X6 | Increase or Decrease Amount | Percent |
|---|---|---|---|---|---|---|
| 1. Cash | $ 15,600 | $ 12,350 | $ 18,240 | $_____ | $_____ | _____ |
| 2. Accounts receivable | 82,500 | 78,278 | 43,910 | _____ | _____ | _____ |
| 3. Merchandise inventory | 476,828 | 432,744 | 518,090 | _____ | _____ | _____ |
| 4. Prepaid insurance | 8,640 | 7,850 | 9,264 | _____ | _____ | _____ |
| 5. Supplies | 4,216 | 5,738 | 3,072 | _____ | _____ | _____ |

## LESSON 14-2    EXERCISES

1. Complete the balance sheet on December 31, 19X1, for Holly's Card Shop: Cash, $3,855; Accounts Receivable, $438; Merchandise Inventory, $68,949; Equipment, $12,150; Building and Land, $85,370; Notes Payable (current portion), $1,850; Accounts Payable, $6,280; Notes Payable (long-term portion), $6,890; and Mortgage Notes Payable, $72,580.

_____

_____

_____

### Assets

Current Assets:

Cash . . . . . . . . . . . . . . . . . . . . . . . . . . . .    _____

Accounts Receivable . . . . . . . . . . . . . . .    _____

Merchandise Inventory . . . . . . . . . . . . .    _____

    Total Current Assets . . . . . . . . . . . . . .    _____

Fixed Assets:

Equipment . . . . . . . . . . . . . . . . . . . . . . .    _____

Buildings and Land . . . . . . . . . . . . . . . .    _____

    Total Fixed Assets . . . . . . . . . . . . . . .    _____

Total Assets . . . . . . . . . . . . . . . . . . . . . . .    _____

### Liabilities

Current Liabilities:

Notes Payable . . . . . . . . . . . . . . _____

Accounts Payable . . . . . . . . . . . . _____

    Total Current Liabilities . . . . . . . . . . . .    _____

Long-Term Liabilities:

Notes Payable . . . . . . . . . . . . . . _____

Mortgage Notes Payable . . . . . . _____

    Total Long-Term Liabilities . . . . . . . . .    _____

Total Liabilities . . . . . . . . . . . . . . . . . . . . .    _____

### Owners' Equity

Holly Seger, Capital . . . . . . . . . . . . . . . . . . .    _____

Total Liabilities and Owners' Equity . . . . .    _____

**2.** The June 30, 19X8, balance sheet for Akron Office Supply is shown below. What percent is each asset item of the total assets? What percent is each liability and owners' equity item of the total liabilities and owners' equity? Round percents to the nearest hundredth.

Akron Office Supply
Balance Sheet
June 30, 19X8

| | Amount | Percent of Totals |
|---|---|---|
| **Assets** | | |
| Current Assets: | | |
| Cash .................................... | $ 39,293 | _____ |
| Accounts Receivable .................. | 59,107 | _____ |
| Merchandise Inventory ............... | 240,114 | _____ |
| Supplies ............................. | 10,912 | _____ |
| Total Current Assets ............... | 349,426 | _____ |
| Fixed Assets: | | |
| Equipment .......................... | 30,774 | _____ |
| Buildings and Land .................. | 98,400 | _____ |
| Total Fixed Assets ................. | 129,174 | _____ |
| Total Assets ......................... | $478,600 | 100.00% |
| **Liabilities** | | |
| Current Liabilities: | | |
| Notes Payable ....................... | $ 15,076 | _____ |
| Accounts Payable .................... | 90,264 | _____ |
| Total Current Liabilities ........... | 105,340 | _____ |
| Long-Term Liabilities: | | |
| Notes Payable ....................... | 9,955 | _____ |
| Mortgage Notes Payable ............. | 57,385 | _____ |
| Total Long-Term Liabilities ......... | 67,340 | _____ |
| **Owners' Equity** | | |
| Charles Furgeson, Capital .............. | 305,920 | _____ |
| Total Liabilities and Owners' Equity ..... | $478,600 | 100.00% |

**3.** The comparative balance sheet for December 31, 19X7 and 19X8, is shown below for the Ski Shop. What is the amount and percent of increase (+) or decrease (−) from 19X7 to 19X8? Round each percent to the nearest hundredth.

Ski Shop
Comparative Balance Sheet
December 31, 19X7 and 19X8

| | 19X7 | 19X8 | Increase or Decrease Amount | Percent |
|---|---|---|---|---|
| **Assets** | | | | |
| Current Assets: | | | | |
| Cash | $ 5,672 | $ 6,715 | _____ | _____ |
| Accounts Receivable | 10,416 | 3,205 | _____ | _____ |
| Notes Receivable | 1,570 | 2,150 | _____ | _____ |
| Merchandise Inventory | 48,360 | 57,918 | _____ | _____ |
| Prepaid Insurance | 2,450 | 2,675 | _____ | _____ |
| Total Current Assets | 68,468 | 72,663 | _____ | _____ |
| Fixed Assets: | | | | |
| Equipment | 27,550 | 29,706 | _____ | _____ |
| Vehicles | 38,520 | 32,646 | _____ | _____ |
| Buildings and Land | 122,380 | 135,670 | _____ | _____ |
| Total Fixed Assets | 188,450 | 198,022 | _____ | _____ |
| Total Assets | $256,918 | $270,685 | _____ | _____ |
| **Liabilities** | | | | |
| Current Liabilities: | | | | |
| Notes Payable | $ 15,934 | $ 12,810 | _____ | _____ |
| Accounts Payable | 12,470 | 14,860 | _____ | _____ |
| Total Current Liabilities | 28,404 | 27,670 | _____ | _____ |
| Long-Term Liabilities: | | | | |
| Notes Payable | 8,460 | 7,960 | _____ | _____ |
| Mortgage Notes Payable | 18,470 | 35,765 | _____ | _____ |
| Total Long-Term Liabilities | 26,930 | 43,725 | _____ | _____ |
| Total Liabilities | 55,334 | 71,395 | _____ | _____ |
| **Owners' Equity** | | | | |
| Michelle Nolan, Capital | 201,584 | 199,290 | _____ | _____ |
| Total Liabilities and Capital | $256,918 | $270,685 | _____ | _____ |

## BUSINESS APPLICATIONS

Round percents to the nearest hundredth.

4. The August 31 balance sheet for Hargrove Industries shows the following: Cash, $43,261; Accounts Receivable, $87,295; Merchandise Inventory, $479,708; Supplies, $14,637; Equipment, $147,216; Vehicles, $86,480; Buildings, $578,249; Land, $188,920; Notes Payable, $38,764; Accounts Payable, $146,093; and Mortgage Notes Payable, $499,762. What is the owners' equity?

_____

5. Balance sheets dated December 31 for the past 7 years for Valdez Insurance Agency show accounts receivable of the following amounts: 19X1, $43,760; 19X2, $38,216; 19X3, $51,725; 19X4, $73,985; 19X5, $88,705; 19X6, $109,842; and 19X7, $121,684. What percent is the 19X7 accounts receivable above or below the average accounts receivable for years 19X1 to 19X6?

_____

6. The December 31, 19X3, balance sheet for Hy-Tyme Video showed owners' equity of $38,450. In the following 4 years, Hy-Tyme's owners' equity increased or decreased the following percents of the previous year's amount: 19X4, decreased 12%; 19X5, decreased 8%; 19X6, increased 28%; and 19X7, increased 78%. What was Hy-Tyme's owners' equity on the balance sheet for (a) 19X4, (b) 19X5, (c) 19X6, and (d) 19X7? Round all amounts to the nearest cent.

(a) _____

(b) _____

(c) _____

(d) _____

A *ratio* expresses the relative value of two amounts. For instance, if one amount is 5 times as large as another, like $20 and $4, their relationship can be expressed as 5 to 1 or 5:1.

Ratios are used to analyze elements of a financial statement and to analyze business performance. Acceptable ratios vary with the type of business, economic conditions, and management's general business philosophy.

## CURRENT RATIO

The current ratio shows the relationship of the company's total current assets to total current liabilities:

$$\text{Current Ratio} = \frac{\text{Total Current Assets}}{\text{Total Current Liabilities}}$$

The current ratio, also called the **working-capital ratio,** indicates the company's ability to pay current debts from its cash and other assets such as accounts receivable that will be converted to cash within the coming year. A current ratio of 2:1 is often considered to be the minimum acceptable level.

## EXAMPLE

A recent balance sheet for Roanoke Mills shows total current assets of $476,824 and total current liabilities of $216,109. What is the current ratio? Round to the nearest hundredth.

## SOLUTION

$$\text{Current Ratio} = \frac{\text{Total Current Asset}}{\text{Total Current Liabilities}}$$
$$= \frac{\$476,824}{\$216,109} \approx 2.206 \text{ or } 2.21:1 \quad \leftarrow \quad \text{Express quotient as a ratio of 2.21 to 1}$$

## PRACTICE PROBLEMS

Calculate the current ratio for each of the companies listed below. Round to the nearest hundredth.

| Company | Total Current Assets | Total Current Liabilities | Current Ratio |
|---|---|---|---|
| 1. Aqua-Lite | $768,415 | $396,818 | _____ |
| 2. Smith Systems | 106,732 | 31,690 | _____ |
| 3. Perkins' Fashions | 347,908 | 148,834 | _____ |
| 4. Carmel, Inc. | 87,176 | 104,642 | _____ |
| 5. Tulane Office Supply | 349,763 | 121,825 | _____ |

ANSWERS FOR PRACTICE PROBLEMS     (1) 1.94:1    (2) 3.37:1    (3) 2.34:1    (4) .83:1    (5) 2.87:1

## ACID-TEST RATIO

The **acid-test ratio** shows the relationship of the quick assets to the total current liabilities. **Quick assets** are cash and any other assets that may be converted into cash in a very short time period, such as marketable securities (short-term investments), accounts receivable, and notes receivable.

$$\text{Acid-Test Ratio} = \frac{\text{Quick Assets}}{\text{Total Current Liabilities}} \quad \leftarrow \text{Cash, marketable securities, receivables}$$

The acid-test ratio, also called the **quick ratio,** shows the company's ability to pay off current liabilities quickly if that becomes necessary or advisable. An acid-test ratio of 1:1 is often considered to be the minimum acceptable level.

### EXAMPLE

Pixler Company's balance sheet shows the following amounts: Cash, $12,350; Marketable Securities, $15,755; Notes Receivable, $18,960; Accounts Receivable, $27,580; and Total Current Liabilities, $62,146. What is the acid-test ratio? Round the answer to the nearest hundredth.

### SOLUTION

**Step 1**  Total the quick assets.

| | Marketable | Notes | Accounts | |
| Cash | + Securities | + Receivable | + Receivable | = Quick Assets |
| $12,350 + | $15,755 + | $18,960 + | $27,580 = | $74,645 |

**Step 2**  Find the acid-test ratio.

$$\text{Acid-Test Ratio} = \frac{\text{Quick Assets}}{\text{Total Current Liabilities}}$$
$$= \frac{\$74,645}{\$62,146} \approx 1.2011 \text{ or } 1.2:1$$

### PRACTICE PROBLEMS

Balance sheet data is shown below for several companies. Calculate the acid-test ratio for each. Round answers to the nearest hundredth.

| | Cash | Marketable Securities | Notes Receivable | Accounts Receivable | Total Current Liabilities | Acid-Test Ratio |
|---|---|---|---|---|---|---|
| 1. | $ 4,682 | $ 0 | $ 5,690 | $13,824 | $ 30,568 | _____ |
| 2. | 39,744 | 86,514 | 0 | 98,247 | 184,695 | _____ |
| 3. | 56,875 | 16,075 | 10,815 | 85,449 | 150,432 | _____ |
| 4. | 18,605 | 0 | 0 | 48,725 | 85,762 | _____ |
| 5. | 24,628 | 8,615 | 0 | 52,907 | 61,140 | _____ |

## ACCOUNTS RECEIVABLE TURNOVER RATIO

The *accounts receivable turnover ratio* measures how quickly accounts receivable from credit sales are turned into cash. This analysis compares net credit sales to average accounts receivable.

$$\text{Accounts Receivable Turnover Ratio} = \frac{\text{Net Credit Sales}}{\text{Average Accounts Receivable}}$$

The average accounts receivable is calculated by adding the accounts receivable amounts from two or more balance sheets and then dividing by the total number of amounts added.

The accounts receivable turnover ratio is used to analyze the company's credit granting and collection practices. Acceptable levels of performance vary from one industry to another, but a high accounts receivable turnover rate is desirable.

### EXAMPLE

The accounts receivable amounts shown on several balance sheets for Hahn Industries in 19X9 are as follows: January 1, $57,284; March 31, $53,723; June 30, $43,011; September 30, $40,842; and December 31, $62,410. Net credit sales for the year ending December 31, 19X9, were $286,854. What was the accounts receivable turnover ratio? Round the answer to the nearest hundredth.

### SOLUTION

**Step 1**  Total the five accounts receivable amounts.

| Jan. 1 Accounts Receivable | Mar. 31 Accounts Receivable | June 30 Accounts Receivable | Sept. 30 Accounts Receivable | Dec. 31 Accounts Receivable | Total Accounts Receivable |
|---|---|---|---|---|---|
| $57,284 + | $53,723 + | $43,011 + | $40,842 + | $62,410 = | $257,270 |

**Step 2**  Calculate the average accounts receivable.

$$\text{Average Accounts Receivable} = \frac{\text{Total Accounts Receivable}}{\text{Number of Amounts}}$$

$$= \frac{\$257,270}{5} = \$51,454$$

**Step 3**  Calculate the accounts receivable turnover ratio.

$$\text{Accounts Receivable Turnover Ratio} = \frac{\text{Net Credit Sales}}{\text{Average Accounts Receivable}}$$

$$= \frac{\$286,854}{\$51,454} \approx 5.574 \text{ or } 5.57:1$$

## PRACTICE PROBLEMS

Calculate the average accounts receivable and the accounts receivable turnover ratio for each of the following companies. Round the accounts receivable turnover ratio to the nearest hundredth.

| | Jan. 1 Accts. Rec. | Mar. 31 Accts. Rec. | June 30 Accts. Rec. | Sept. 30 Accts. Rec. | Dec. 31 Accts. Rec. | Average Accts. Rec. | Net Credit Sales | Accts. Rec. Turnover Ratio |
|---|---|---|---|---|---|---|---|---|
| 1. | $16,346 | $12,160 | $ 8,924 | $11,842 | $18,103 | $_____ | $ 92,176 | _____ |
| 2. | 49,315 | — | 27,406 | — | 48,712 | _____ | 173,096 | _____ |
| 3. | 26,340 | 15,707 | 16,804 | 12,923 | 30,416 | _____ | 67,272 | _____ |
| 4. | 41,620 | 40,743 | 34,832 | 36,217 | 58,388 | _____ | 374,016 | _____ |
| 5. | 67,418 | 53,237 | — | 41,819 | 76,510 | _____ | 606,172 | _____ |

## LESSON 14-3    EXERCISES

Calculate the current ratio for each of the companies listed below. Round answers to the nearest hundredth.

| Company | Total Current Assets | Total Current Liabilities | Current Ratio |
|---|---|---|---|
| 1. Foremost | $ 208,416 | $ 93,748 | _____ |
| 2. Geary's | 866,339 | 641,096 | _____ |
| 3. Judd, Inc. | 1,842,113 | 607,521 | _____ |
| 4. Kabelco | 586,499 | 691,327 | _____ |
| 5. Lystodd, Inc. | 580,341 | 216,740 | _____ |

Balance sheet data are shown below for several companies. Calculate the acid-test ratio for each. Round answers to the nearest hundredth.

| | Cash | Marketable Securities | Notes Receivable | Accounts Receivable | Total Current Liabilities | Acid-Test Ratio |
|---|---|---|---|---|---|---|
| 6. | $ 8,412 | $ 3,614 | $ 2,870 | $25,481 | $ 36,359 | _____ |
| 7. | 23,975 | 0 | 8,325 | 47,418 | 52,728 | _____ |
| 8. | 15,405 | 27,850 | 5,615 | 89,314 | 130,211 | _____ |
| 9. | 37,814 | 5,000 | 3,525 | 92,872 | 173,012 | _____ |
| 10. | 6,372 | 0 | 0 | 14,781 | 5,962 | _____ |

### BUSINESS APPLICATIONS

Round answers to the nearest hundredth.

11. The December 31 balance sheet for Nu-Energy, Inc., shows the following amounts: Cash, $14,387; Marketable Securities, $35,618; Notes Receivable, $8,000; Accounts Receivable, $49,934; Merchandise Inventory, $187,462; Supplies, $614; Total Current Assets, $296,015; and Total Current Liabilities, $127,416. What is (a) the current ratio, and (b) the acid-test ratio?

(a) _____

(b) _____

12. The December 31 balance sheet for Chambers, Inc., shows the following amounts: Cash, $22,850; Marketable Securities, $2,500; Notes Receivable, $10,500; Accounts Receivable, $68,970; Merchandise Inventory, $57,600; Total Current Assets, $162,420; and Total Current Liabilities, $72,508. Business consultant Scott Michaels states that for a business like Chambers, the current ratio should be 2.50 : 1 and the acid-test ratio should be 1.25 : 1. How many dollars above or below the standard described by Mr. Michaels are (a) the total current assets used in calculating the current ratio and (b) the quick assets used in calculating the acid-test ratio?

(a) _____

(b) _____

13. Last year five balance sheets prepared by Loftis, Inc., showed the following amounts for accounts receivable: $48,204, $52,371, $36,972, $41,065, and $58,263. For the year, net cash sales were $714,964 and net credit sales were $289,764. (a) What percent, rounded to the nearest tenth, were net credit sales of the total net sales? (b) What was the accounts receivable turnover rate?

(a) _____

(b) _____

14. The December 31, 19X3, balance sheet for McLaughlin Enterprises shows total current assets of $876,492 and total current liabilities of $610,568. (a) What was the current ratio, rounded to the nearest tenth? (b) Next December 19X4, McLaughlin estimates total current liabilities will be $547,310. What should total current assets be to achieve a current ratio of 2.20 : 1?

(a) _____

(b) _____

A *percent analysis* is often made by businesses to estimate the amount of accounts receivable that actually will be collected and to determine the owners' rate of return on their investment. Analyzing business performance in terms of percent relationships is often more meaningful than only analyzing dollar amounts.

**AGING THE RECEIVABLES**

Most businesses that extend credit to customers find that eventually some of the customers will not pay the amount due. *Aging the receivables* is a method of analyzing the company's accounts receivables, based on their age, to estimate the amount that will be uncollectible. The term *age* refers to the length of time an account is overdue.

When analyzing the company's current accounts receivable, it is not known which specific customers will not pay. Experience shows, however, that the longer an account is overdue, the more difficult it will be to collect. Therefore, this analysis begins by categorizing receivable amounts into time periods based upon the date from which credit was granted. Using the company's collection experience as a guideline, a percent is applied to the amount in each time period to estimate the amount that will be uncollectible. The estimated net value of accounts receivable is calculated by deducting the estimated total uncollectible accounts receivable from the total accounts receivable.

The process of aging the accounts receivable helps give management a clearer picture of the true value of the receivables. The results often indicate whether changes are warranted in the company's credit granting and collection procedures.

## EXAMPLE

The December 31 balance sheet for Buy-Rite Office Supply showed accounts receivable of $71,180. Buy-Rite listed the receivables in time periods and estimated the percent of each category that will be uncollectible, based on their previous collection experience. Calculate total estimated accounts receivable that will be uncollectible and the estimated net value of the accounts receivable.

## SOLUTION

**Step 1**  Multiply each amount by the percent to estimate the amount uncollectible at each age.

**Step 2**  Total the amounts to find the total estimated uncollectible accounts receivable.

| Age of Receivables | Accounts Receivable Amount | × | Estimated Percent Uncollectible | = | Estimated Amount Uncollectible | |
|---|---|---|---|---|---|---|
| Not due | $62,600 | × | 2% | = | $1,252.00 | |
| 0 to 30 days past due | 3,250 | × | 5% | = | 162.50 | |
| 31 to 60 days past due | 2,270 | × | 10% | = | 227.00 | Step 1 |
| 61 to 90 days past due | 1,415 | × | 20% | = | 283.00 | |
| 91 to 180 days past due | 970 | × | 30% | = | 291.00 | |
| Over 180 days past due | 675 | × | 70% | = | 472.50 | |
| Total Accounts Receivable | $71,180 | | | | | |
| Total Estimated Uncollectible Accounts Receivable: | | | | | $2,688.00 | ←Step 2 |

**Step 3** Calculate the estimated net value of the accounts receivable.

| Total Accounts Receivable | − | Estimated Uncollectible Accounts Receivable | = | Estimated Net Value of Accounts Receivable |
|---|---|---|---|---|
| $71,180 | − | $2,688 | = | $68,492 |

## PRACTICE PROBLEMS

On December 31 McGuire Industries had total accounts receivable of $51,305. McGuire categorized the receivables by age and estimated the percent that will be uncollectible as shown below. What is the estimated amount of each category that will be uncollectible, total estimated amount uncollectible, and estimated net value of the accounts receivable?

| Age of Receivables | Accounts Receivable Amount | Estimated Percent Uncollectible | Estimated Amount Uncollectible |
|---|---|---|---|
| 1. Not due | $42,500 | 1% | $_____ |
| 2.  0 to  60 days past due | 4,550 | 2% | _____ |
| 3. 61 to 180 days past due | 2,680 | 5% | _____ |
| 4. Over  180 days past due | 1,575 | 40% | _____ |
| Total Accounts Receivable | $51,305 | | |

**5.** Total Estimated Amount Uncollectible:  $_____

**6.** Estimated Net Value of Accounts Receivable:  $_____

## RETURN ON TOTAL ASSETS

The return on total assets is a percent analysis that shows how productive the assets have been in generating net income. It is calculated by dividing the total of net income plus interest expense by the average total assets.

$$\text{Return on Total Assets} = \frac{\text{Net Income} + \text{Interest Expense}}{\text{Average Total Assets}}$$

In preparing the income statement, interest expense is deducted, along with the other expenses, before arriving at net income. Since interest expense is considered a cost of financing rather than operating the business, it is added to net income when calculating the return on total assets.

The *average total assets* is calculated by adding the total assets from all balance sheets prepared during the time period being analyzed and dividing by the number of amounts added.

This analysis shows how efficiently management has utilized the company's assets to earn a profit. Acceptable rates vary from industry to industry and with current economic conditions.

## EXAMPLE

Last year Arment Wholesale Company had net income of $87,255 and interest expense of $9,645. Balance sheets prepared during the year show total assets of the following amounts: January 1, $946,276; June 30, $978,042; and December 31, $997,109. What was the rate of return on the total assets? Round answer to the nearest hundredth.

## SOLUTION

**Step 1**   Find total assets.

January 1  June 30  December 31
Total Assets + Total Assets + Total Assets = Total Assets
$946,276  +  $978,042  +  $997,109  = $2,921,427

**Step 2**   Calculate the average total assets.

$$\text{Average Total Assets} = \frac{\text{Total Assets}}{\text{Number of Amounts}} = \frac{\$2,921,427}{3} = \$973,809$$

**Step 3**   Add net income and interest expense.

Net Income + Interest Expense = Net Income and Interest Expense
$87,255  +  $9,645  =  $96,900

**Step 4**   Calculate the rate of return on total assets.

$$\text{Return on Total Assets} = \frac{\text{Net Income and Interest Expense}}{\text{Average Total Assets}}$$

$$= \frac{\$96,900}{\$973,809} \approx 0.09950 \text{ or } 9.95\%$$

## PRACTICE PROBLEMS

Calculate the return on total assets for the following companies. Note that the number of balance sheets prepared in a year varies from company to company. Round percents to the nearest hundredth.

| | Net Income | Interest Expense | Jan. 1 | March 31 | June 30 | Sept. 30 | Dec. 31 | Return on Assets |
|---|---|---|---|---|---|---|---|---|
| | | | Total Assets (Balance Sheet Dates) | | | | | |
| 1. | $28,294 | $12,470 | $374,213 | $360,548 | $357,142 | $362,169 | $370,418 | _____ |
| 2. | 58,724 | 16,350 | 518,077 | 540,392 | 560,781 | 584,316 | 592,344 | _____ |
| 3. | 14,208 | 2,526 | 185,417 | — | — | — | 213,705 | _____ |
| 4. | 93,086 | 0 | 692,182 | — | 727,051 | — | 759,862 | _____ |
| 5. | 10,382 | 21,474 | 426,315 | 428,804 | 415,621 | 405,476 | 392,464 | _____ |

LESSON 14-4   ■   **489**

## RETURN ON OWNERS' EQUITY

The **return on owners' equity** shows the rate of return the owners of the business earned on their investment. It is calculated by dividing net income by the average owners' equity:

$$\text{Return on Owners' Equity} = \frac{\text{Net Income}}{\text{Average Owners' Equity}}$$

The **average owners' equity** is calculated by adding the owners' equity amounts from the balance sheets prepared during the accounting period and then dividing by the number of amounts added.

What is considered to be an acceptable rate of return varies from industry to industry and with the rate of return available on other investment alternatives. Owners or investors in a business, however, find a high rate of return on owners' equity desirable.

### EXAMPLE

Mesaba Company's income statement for the year ended December 31, 19X8, shows net income of $93,742. The owners' equity amounts taken from balance sheets prepared during the year were as follows: January 1, $846,370; March 31, $850,721; June 30, $875,413; September 30, $864,312; and December 31, $905,849. What was the return on owners' equity? Round the percent to the nearest hundredth.

### SOLUTION

**Step 1**  Add the owners' equity amounts.

| Jan. 1 Owners' Equity | | March 31 Owners' Equity | | June 30 Owners' Equity | | Sept. 30 Owners' Equity | | Dec. 31 Owners' Equity | | Total Owners' Equity |
|---|---|---|---|---|---|---|---|---|---|---|
| $846,370 | + | $850,721 | + | $875,413 | + | $864,312 | + | $905,849 | = | $4,342,665 |

**Step 2**  Calculate the average owners' equity.

$$\text{Average Owners' Equity} = \frac{\text{Total Owners' Equity}}{\text{Number of Amounts}}$$

$$= \frac{\$4,342,665}{5} = \$868,533$$

**Step 3**  Calculate the return on owners' equity.

$$\text{Return on Owners' Equity} = \frac{\text{Net Income}}{\text{Average Owners' Equity}}$$

$$= \frac{\$93,742}{\$868,533} \approx 0.10793 \text{ or } 10.79\%$$

## PRACTICE PROBLEMS

Calculate the return on owners' equity for the following companies. Round your percents to the nearest hundredth.

| | Net Income | Owners' Equity (Balance Sheet Dates) | | | | | Return on Owners' Equity |
|---|---|---|---|---|---|---|---|
| | | Jan. 1 | March 31 | June 30 | Sept. 30 | Dec. 31 | |
| 1. | $27,492 | $376,915 | $380,217 | $384,503 | $390,436 | $395,784 | _____ |
| 2. | 12,486 | 197,824 | — | 188,978 | — | 184,647 | _____ |
| 3. | 79,138 | 498,744 | — | — | — | 526,306 | _____ |
| 4. | 98,936 | 582,780 | 592,113 | 603,804 | 615,735 | 628,463 | _____ |
| 5. | 41,703 | 614,092 | — | 592,814 | — | 580,431 | _____ |

## LESSON 14-4    EXERCISES

Accounts receivable data are shown below for two companies. For each, calculate the estimated amount of each category that will be uncollectible, the total estimated amount uncollectible, and the estimated net value of the accounts receivable.

**1.**

| Age of Receivables | Accounts Receivable Amount | Estimated Percent Uncollectible | Estimated Amount Uncollectible |
|---|---|---|---|
| Not due | $27,200 | 2% | _____ |
| 0 to 30 days past due | 2,600 | 4% | _____ |
| 31 to 60 days past due | 980 | 10% | _____ |
| 61 to 90 days past due | 610 | 20% | _____ |
| 91 to 180 days past due | 440 | 30% | _____ |
| Over 180 days past due | 274 | 50% | _____ |
| Total Accounts Receivable | $32,104 | | |

Total Estimated Amount Uncollectible:                    _____

Estimated Net Value of Accounts Receivable:         _____

**2.**

| Age of Receivables | Accounts Receivable Amount | Estimated Percent Uncollectible | Estimated Amount Uncollectible |
|---|---|---|---|
| Not due | $53,300 | 2% | _____ |
| 0 to 30 days past due | 3,460 | 5% | _____ |
| 31 to 60 days past due | 1,650 | 10% | _____ |
| 61 to 90 days past due | 1,100 | 20% | _____ |
| 91 to 180 days past due | 950 | 30% | _____ |
| Over 180 days past due | 790 | 70% | _____ |
| Total Accounts Receivable | $61,250 | | |

Total Estimated Amount Uncollectible:                    _____

Estimated Net Value of Accounts Receivable:         _____

**3.** Calculate the return on total assets for the following companies. Round percents to the nearest hundredth.

| | | Total Assets (Balance Sheet Dates) | | | | | |
|---|---|---|---|---|---|---|---|
| Net Income | Interest Expense | Jan. 1 | March 31 | June 30 | Sept. 30 | Dec. 31 | Return on Assets |
| $16,248 | $ 3,750 | $167,215 | $ — | $174,395 | $ — | $180,342 | _____ |
| 72,605 | 13,500 | 685,107 | 642,916 | 690,349 | 706,515 | 746,763 | _____ |
| 21,454 | 6,304 | 278,644 | — | — | — | 315,912 | _____ |
| 16,341 | 12,913 | 468,930 | 452,871 | 437,304 | 424,462 | 418,433 | _____ |
| 87,412 | 36,950 | 976,436 | — | 962,916 | — | 980,428 | _____ |
| 41,854 | 0 | 347,850 | — | — | — | 379,352 | _____ |
| 15,681 | 11,614 | 571,285 | — | 540,312 | — | 536,762 | _____ |
| 52,650 | 12,000 | 618,044 | 624,376 | 620,900 | 625,318 | 632,732 | _____ |
| 19,545 | 2,640 | 207,641 | — | 240,716 | — | 271,403 | _____ |

**4.** Calculate the return on owners' equity for the following companies. Round percents to the nearest hundredth.

| | Owners' Equity (Balance Sheet Dates) | | | | | |
|---|---|---|---|---|---|---|
| Net Income | Jan. 1 | March 31 | June 30 | Sept. 30 | Dec. 31 | Return on Owners' Equity |
| $15,823 | $143,752 | $ — | $148,745 | $ — | $153,603 | _____ |
| 26,276 | 94,352 | 99,347 | 104,073 | 110,853 | 116,805 | _____ |
| 54,383 | 738,964 | — | — | — | 715,806 | _____ |
| 72,720 | 248,907 | 256,312 | 260,311 | 271,467 | 275,803 | _____ |
| 14,361 | 371,631 | — | 368,642 | — | 367,519 | _____ |
| 72,467 | 925,367 | — | — | — | 915,841 | _____ |
| 65,890 | 526,876 | 532,819 | 540,005 | 546,192 | 565,378 | _____ |
| 55,654 | 403,914 | — | 390,470 | — | 405,733 | _____ |
| 8,762 | 297,420 | — | — | — | 280,412 | _____ |
| 51,306 | 473,046 | 480,113 | 481,217 | 490,744 | 491,805 | _____ |

## BUSINESS APPLICATIONS

Round percents to the nearest hundredth.

5. The June 30 balance sheet for Central Distributing Company shows accounts receivable of $106,660. Central's management classified the receivables into the following groups and estimated the percent of each group that will be uncollectible. What percent of the total accounts receivable does Central Distributing expect to collect?

| Age of Receivables | Accounts Receivable Amount | Estimated Percent Uncollectible |
|---|---|---|
| Not due | $87,450 | 2% |
| 0 to 90 days past due | 12,760 | 15% |
| Over 90 days past due | 6,450 | 40% |

6. Last year the Baxter Company had net income of $46,372 and interest expense of $12,600. Baxter's January 1 and December 31 balance sheets showed total assets of $732,716 and $755,048, and owners' equity of $481,976 and $490,712. What was Baxter's (a) return on total assets, and (b) return on owners' equity?

(a) _____

(b) _____

7. The January 1, June 30, and December 31 balance sheets for Stafford, Inc., shows the following owners' equity amounts: $684,720, $699,404, and $728,473. The return on owners' equity was 12.56%. What was Stafford's net income for the year?

8. In 19X2, City-Wide Furniture had net income of $54,362. City-Wide's January 1 and December 31 balance sheets showed owners' equity of $689,750 and $694,346. In 19X3, City-Wide had net income of $84,716. City-Wide's January 1 and December 31 balance sheets showed owners' equity of $694,346 and $748,312. How much more or less was the 19X3 return on owners' equity than the 19X2 rate?

9. The December 31, 19X4, balance sheet for Mandrell Fashions showed total accounts receivable of $95,430. Mandrell classified the receivables into groups and estimated the percent of each group that would be uncollectible as follows: not due, $66,400, 1.5% uncollectible; 0 to 90 days overdue, $24,650, 12% uncollectible; and over 90 days overdue, $4,380, 50% uncollectible. Four years later Mandrell analyzed what had happened to these receivables and found that $90,348 had been collected and the rest had been written off the books. How many dollars more or less did Mandrell actually collect of the accounts receivable than was estimated in 19X4?

## CHAPTER 14 REVIEW

### UNDERSTANDING THE INCOME STATEMENT

The following information was included on Arkansas Millwork's income statement for the quarter ending March 31: Sales, $48,360; Sales Returns and Allowances, $1,710; Merchandise Inventory, January 1, $67,345; Net Purchases, $15,780; Merchandise Inventory, December 31, $51,875; and Total Operating Expenses, $8,735. What was (1) the net sales, (2) cost of goods sold, (3) gross profit, and (4) net income or net loss?

1. _____   2. _____   3. _____   4. _____

### VERTICAL ANALYSIS—INCOME STATEMENT

Calculate the percent that each of the following income statement items is of net sales. Round your percents to the nearest hundredth.

| Item | Amount | Percent |
|------|--------|---------|
| **5.** Sales | $56,240 | _____ |
| **6.** Sales returns and allowances | 1,850 | _____ |
| Net Sales | 54,390 | 100.00% |
| **7.** Cost of goods sold | 31,250 | _____ |
| **8.** Gross profit | 23,140 | _____ |
| **9.** Total operating expenses | 17,122 | _____ |
| **10.** Net income | 6,018 | _____ |

### HORIZONTAL ANALYSIS—INCOME STATEMENT

Amounts from the comparative income statement for Brighton Shoe Company for August and September are shown below. Calculate the amount and percent of increase (+) or decrease (−) from August to September. Round each percent to the nearest hundredth.

| Item | August | September | Increase or Decrease Amount | Percent |
|------|--------|-----------|--------|---------|
| **11.** Net sales | $63,740 | $64,718 | _____ | _____ |
| **12.** Cost of goods sold | 39,450 | 37,115 | _____ | _____ |
| **13.** Total operating expenses | 20,621 | 25,306 | _____ | _____ |
| **14.** Net income | 3,669 | 2,297 | _____ | _____ |

### COMPARING ACTUAL PERFORMANCE TO BUDGETED AMOUNTS—INCOME STATEMENT

Amounts from Hyline Service's actual income statement and budget for 19X9 are shown at the top of the next page. What amount and percent is each item on the actual income statement over (+) or under (−) the budget? Round percents to the nearest hundredth.

| Item | Actual | Budget | Over or Under | |
|------|--------|--------|--------|---------|
| | | | Amount | Percent |
| **15.** Net sales | $332,650 | $340,000 | _____ | _____ |
| **16.** Gross profit | 153,185 | 150,000 | _____ | _____ |
| **17.** Total operating expenses | 88,064 | 90,000 | _____ | _____ |
| **18.** Net income | 65,121 | 60,000 | _____ | _____ |

## UNDERSTANDING THE BALANCE SHEET

The following data were included on the Beck Agency's December 31 balance sheet: Total Current Assets, $145,295; Total Fixed Assets, $85,360; Total Current Liabilities, $80,350; and Total Long-Term Liabilities, $53,718. What were (19) the total assets, (20) the total liabilities, (21) owners' equity, and (22) total liabilities and owners' equity?

**19.** _____     **20.** _____     **21.** _____     **22.** _____

## VERTICAL ANALYSIS—BALANCE SHEET

Several amounts from the assets section of the September 30 balance sheet for Wexler, Inc., are shown below. What percent is each item of the total assets? Round each percent to the nearest tenth.

| Item | Amount | Percent of Total Assets |
|------|--------|-------------------------|
| **23.** Cash | $ 23,416 | _____ |
| **24.** Merchandise inventory | 90,652 | _____ |
| **25.** Equipment | 12,618 | _____ |
| **26.** Building | 70,255 | _____ |
| Total assets | 245,546 | 100.0% |

## HORIZONTAL ANALYSIS—BALANCE SHEET

Amounts from the December 31, 19X4 and 19X5, balance sheets for Campbell Bros. are shown below. What is the amount and percent of increase (+) or decrease (−) from 19X4 to 19X5? Round each percent to the nearest hundredth.

| Item | 19X4 | 19X5 | Increase or Decrease | |
|------|------|------|--------|---------|
| | | | Amount | Percent |
| **27.** Cash | $ 37,250 | $ 39,475 | _____ | _____ |
| **28.** Total current assets | 246,435 | 258,718 | _____ | _____ |
| **29.** Accounts payable | 42,093 | 38,470 | _____ | _____ |
| **30.** Total liabilities | 149,355 | 126,104 | _____ | _____ |

## COMPARING CURRENT DATA TO PRIOR PERIOD AVERAGES— BALANCE SHEET

Balance sheet data for Sanchez Real Estate for 19X1, 19X2, is 19X3 are shown below. Calculate the average of years 19X1 and 19X2 and the amount and percent of increase (+) or decrease (−) from the average to 19X3 data. Round percents to the nearest hundredth.

| Item | 19X1 | 19X2 | 19X3 | Average, 19X1, 19X2 | Increase or Decrease Amount | Percent |
|------|------|------|------|---------------------|------------------------------|---------|
| **31.** Cash | $ 26,140 | $ 22,716 | $ 30,475 | _____ | _____ | _____ |
| **32.** Merch. inventory | 184,370 | 170,208 | 165,808 | _____ | _____ | _____ |
| **33.** Notes payable | 40,200 | 34,620 | 26,454 | _____ | _____ | _____ |
| **34.** Accounts payable | 104,765 | 110,381 | 140,637 | _____ | _____ | _____ |

## CURRENT RATIO

Calculate the current ratio for each of the companies listed below. Round answers to the nearest hundredth.

| Company | Current Assets | Current Liabilities | Current Ratio |
|---------|---------------|---------------------|---------------|
| **35.** Harker | $476,318 | $214,709 | _____ |
| **36.** Imoway | 279,873 | 105,114 | _____ |
| **37.** Klinger | 310,925 | 392,064 | _____ |
| **38.** Miles | 629,470 | 138,416 | _____ |

## ACID-TEST RATIO

Balance sheet data are shown below for several companies. Calculate the acid-test ratio for each. Round answers to the nearest hundredth.

| | Cash | Marketable Securities | Notes Receivable | Accounts Receivable | Total Current Liabilities | Acid-Test Ratio |
|---|------|----------------------|------------------|---------------------|---------------------------|-----------------|
| **39.** | $21,055 | $12,615 | $ 8,500 | $ 74,318 | $115,386 | _____ |
| **40.** | 8,216 | 0 | 0 | 35,892 | 32,605 | _____ |
| **41.** | 84,379 | 40,000 | 15,650 | 143,681 | 260,540 | _____ |
| **42.** | 16,494 | 1,500 | 3,875 | 64,790 | 137,734 | _____ |

## ACCOUNTS RECEIVABLE TURNOVER RATIO

Calculate the average accounts receivable and the accounts receivable turnover ratio for each of the following companies. Round the accounts receivable turnover ratio to the nearest hundredth.

| | Jan. 1 Accts. Rec. | Mar. 31 Accts. Rec. | June 30 Accts. Rec. | Sept. 30 Accts. Rec. | Dec. 31 Accts. Rec. | Average Accts. Rec. | Net Credit Sales | Accts. Rec. Turnover Ratio |
|---|---|---|---|---|---|---|---|---|
| 43. | $32,715 | $36,804 | $41,816 | $33,803 | $34,317 | _____ | $296,413 | _____ |
| 44. | 64,816 | — | — | — | 72,304 | _____ | 146,998 | _____ |
| 45. | 71,463 | 85,728 | 62,740 | 86,323 | 94,606 | _____ | 937,089 | _____ |
| 46. | 45,624 | — | 33,076 | — | 51,222 | _____ | 368,191 | _____ |

## AGING THE RECEIVABLES

Accounts receivable data are shown below for Heartland Gifts. Calculate the estimated amount of each category that will be uncollectible, total estimated amount uncollectible, and estimated net value of the accounts receivable.

| | Age of Receivables | Accounts Receivable Amount | Estimated Percent Uncollectible | Estimated Amount Uncollectible |
|---|---|---|---|---|
| 47. | Not due | $46,750 | 2% | _____ |
| 48. | 0 to 60 days past due | 15,825 | 4% | _____ |
| 49. | 61 to 120 days past due | 6,380 | 15% | _____ |
| 50. | 120 to 180 days past due | 3,740 | 30% | _____ |
| 51. | Over 180 days past due | 8,470 | 70% | _____ |
| | Total Accounts Receivable | $81,165 | | |

52. Total Estimated Amount Uncollectible: _____

53. Estimated Net Value of Accounts Receivable: _____

## RETURN ON TOTAL ASSETS

Calculate the return on total assets for the following companies. Round percents to the nearest hundredth.

| | Net Income | Interest Expense | Total Assets (Balance Sheet Dates) | | | | | Return on Assets |
|---|---|---|---|---|---|---|---|---|
| | | | Jan. 1 | March 31 | June 30 | Sept. 30 | Dec. 31 | |
| 54. | $ 12,716 | $ 5,762 | $146,218 | $ — | $150,781 | $ — | $162,313 | _____ |
| 55. | 38,736 | 14,110 | 389,472 | — | — | — | 416,306 | _____ |
| 56. | 108,473 | 26,500 | 296,408 | 315,790 | 365,058 | 420,611 | 446,943 | _____ |
| 57. | 7,316 | 4,165 | 215,782 | — | 184,320 | — | 180,125 | _____ |

## RETURN ON OWNERS' EQUITY

Calculate the return on owners' equity for the following companies. Round percents to the nearest hundredth.

| | Net Income | Owners' Equity (Balance Sheet Dates) | | | | | Return on Owners' Equity |
| | | Jan. 1 | March 31 | June 30 | Sept. 30 | Dec. 31 | |
| --- | --- | --- | --- | --- | --- | --- | --- |
| **58.** | $16,784 | $143,716 | $ — | $132,804 | $ — | $140,912 | _____ |
| **59.** | 48,466 | 287,432 | 290,896 | 298,463 | 315,780 | 312,624 | _____ |
| **60.** | 26,314 | 685,905 | — | — | — | 638,116 | _____ |
| **61.** | 35,688 | 376,482 | — | 390,931 | — | 412,661 | _____ |

## BUSINESS APPLICATIONS

**62.** The following amounts were listed on Western Wear's income statement: Net Sales, $416,385; Cost of Goods Sold, $247,392; and total operating expenses, $197,316. What was Western Wear's (a) gross profit and (b) net income or net loss?

(a) _____

(b) _____

**63.** Net income for Palmaro Engineering was as follows for three consecutive years: 19X6, $89,316; 19X7, $74,309; and 19X8, $97,347. What was the percent of increase (+) or decrease (−) in net income from (a) 19X6 to 19X7 and (b) 19X7 to 19X8? Round percents to the nearest tenth.

(a) _____

(b) _____

**64.** The March 31 balance sheet for Pzazz Enterprises shows the following amounts: Cash, $46,354; Marketable Securities, $15,000; Notes Receivable, $10,750; Accounts Receivable, $107,318; Merchandise Inventory, $441,230; Supplies, $1,245; Total Current Assets, $621,897; and Total Current Liabilities, $214,110. What were (a) the current ratio and (b) the acid-test ratio? Round to the nearest hundredth.

(a) _____

(b) _____

**65.** Last year the Wimmer Group had net income of $178,525 and interest expense of $47,350. Wimmer's January 1 and December 31 balance sheets showed total assets of $3,746,512 and $3,915,416 and total owners' equity of $2,960,347 and $3,104,613. What was Wimmer's (a) return on total assets and (b) return on owners' equity? Round percents to the nearest hundredth.

(a) _____

(b) _____

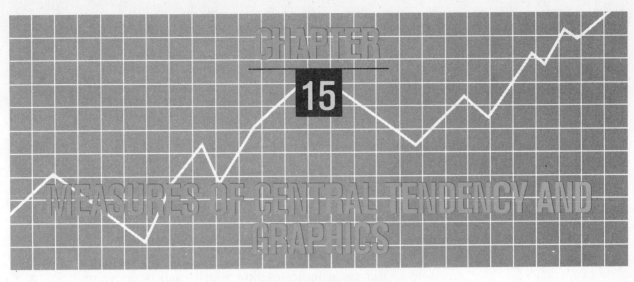

# CHAPTER
## 15

# MEASURES OF CENTRAL TENDENCY AND GRAPHICS

## LEARNING OBJECTIVES

1. To measure central tendency by calculating the mean, median, and mode.

2. To construct and derive data from bar, line, and circle graphs.

**M**easures of central tendency are statistical analyses of data that show the average of a group of amounts, the middle amount of a group of amounts, and the most frequently occurring amount in a group. These measures often reveal normal performance and indicate outstanding or substandard achievement. The results of these measures are often presented visually in the form of a graph or table.

In this chapter, you will learn how to calculate measures of central tendency and prepare graphs. Often in business, however, computer spreadsheets are used to make these calculations and the results are used to render a computer-generated graph or table as shown below.

### WEEKLY PAYROLL REGISTER
### FOR WEEK ENDING APRIL 7, 19---

| EMPLOYEE NO. | HOURS WORKED | PAY RATE | REGULAR PAY | OVERTIME PAY | GROSS EARNINGS |
|---|---|---|---|---|---|
| 101 | 40 | 8.55 | 342.00 | 0.00 | 342.00 |
| 102 | 43 | 8.40 | 336.00 | 37.80 | 373.80 |
| 103 | 38 | 9.25 | 351.50 | 0.00 | 351.50 |
| 104 | 45 | 8.70 | 348.00 | 65.25 | 413.25 |
| 105 | 39 | 8.25 | 321.75 | 0.00 | 321.75 |
| TOTALS | | | 1699.25 | 103.05 | 1802.30 |

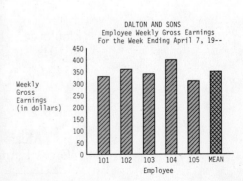

DALTON AND SONS
Employee Weekly Gross Earnings
For the Week Ending April 7, 19--

The three measures of central tendency are called the **mean,** the **median,** and the **mode.** The results of these measures help managers analyze past performance and set standards for the future.

## UNGROUPED DATA

*Ungrouped* data are a set of data with amounts listed individually. Measures of central tendency are calculated by considering each amount.

### CALCULATING THE MEAN

The **mean** is another term for **average,** calculated by adding all amounts and dividing by the number of amounts. Comparison of individual amounts to the mean indicates how much each is above or below the average performance. If relatively little variation exists between the individual amounts, the mean can provide a useful statistic. If individual amounts are widely scattered, however, the mean may not provide a realistic measure of the data.

---

### EXAMPLE

The computer spreadsheet below shows weekly payroll data for employees of Dalton and Sons. What is the mean weekly gross earnings?

```
                WEEKLY PAYROLL REGISTER
              FOR WEEK ENDING APRIL 7, 19--

   EMPLOYEE   HOURS     PAY    REGULAR   OVERTIME    GROSS
     NO.      WORKED    RATE     PAY       PAY      EARNINGS
   --------------------------------------------------------
     101        40      8.55    342.00     0.00      342.00
     102        43      8.40    336.00    37.80      373.80
     103        38      9.25    351.50     0.00      351.50
     104        45      8.70    348.00    65.25      413.25
     105        39      8.25    321.75     0.00      321.75
                                          --------------------
```

### SOLUTION

**Step 1**    Add the individual gross earnings amounts.

| Employee 101 | + | Employee 102 | + | Employee 103 | + | Employee 104 | + | Employee 105 | = | Total Gross Earnings |
|---|---|---|---|---|---|---|---|---|---|---|
| $342.00 | + | $373.80 | + | $351.50 | + | $413.25 | + | $321.75 | = | $1,802.30 |

**Step 2**    Calculate the mean.

Average earnings per employee

$$\frac{\text{Mean}}{\text{Gross Earnings}} = \frac{\text{Total Gross Earnings}}{\text{Number of Amounts}} = \frac{\$1,802.30}{5} = \$360.46$$

---

Sales by department for a period of 5 months for Claridge Department Store, rounded to the nearest hundred dollar, are shown below. What is the mean sales per department for each month? Round answers to the nearest cent.

| Month | Women's Wear | Children's Wear | Men's Wear | Home Furnishings | Jewelry | Shoes | Gifts | Mean Sales |
|---|---|---|---|---|---|---|---|---|
| 1. Feb. | $38,600 | $31,500 | $24,300 | $21,200 | $18,100 | $23,400 | $15,900 | $_____ |
| 2. March | 52,000 | 41,500 | 34,300 | 20,100 | 19,700 | 26,900 | 17,000 | _____ |
| 3. April | 51,800 | 43,400 | 35,000 | 26,800 | 16,500 | 28,000 | 14,000 | _____ |
| 4. May | 40,700 | 41,000 | 32,000 | 39,500 | 18,300 | 29,000 | 18,000 | _____ |
| 5. June | 39,000 | 40,600 | 31,100 | 43,500 | 20,000 | 27,500 | 29,300 | _____ |

## CALCULATING THE MEDIAN

The **median** is the middle number of a group of numbers arranged in rank order from highest to lowest or from lowest to highest. The median serves as a reference point for indicating where individual amounts fall within the top half or bottom half of those listed. When the database contains extremely high or low values, the median will better represent all the values in a group of numbers than will the mean. Locating the rank of the median in a group of numbers arranged in rank order can be done through the following formula:

$$\text{Rank of Middle Value} = \frac{\text{Number of Values} + 1}{2}$$

The median is the middle value when there are an uneven number of values arranged in order of size or rank order. When there are an even number of values in a series, the median is the midpoint or the average of the two middle values. Add the two middle values and divide by two to find the median.

### EXAMPLE

Sales made last month by seven salespeople at Jackson Real Estate were as follows: K. Cain, $142,560; B. Lassiter, $95,450; M. Jensen, $215,305; T. Wang, $170,080; O. Smith, $64,500; R. Dowden, $225,450; and S. Taylor, $105,000. What was the median sales amount?

### SOLUTION

**Step 1** Arrange the amounts in rank order from highest to lowest.

| Salesperson | Sales Amount | |
|---|---|---|
| R. Dowden | $225,450 | |
| M. Jensen | 215,305 | |
| T. Wang | 170,080 | |
| K. Cain | 142,560 | ← Median is the fourth value ranked highest to lowest |
| S. Taylor | 105,000 | |
| B. Lassiter | 95,450 | |
| O. Smith | 64,500 | |

**(1)** $24,714.29 **(2)** $30,214.29 **(3)** $30,785.71 **(4)** $31,214.29 **(5)** $33,000.00

**Step 2** Locate the middle amount.

$$\text{Rank of Middle Value} = \frac{\text{Number of Values} + 1}{2} = \frac{7 + 1}{2} = 4$$

**Step 3** Identify the median.

The median, or middle amount, is the fourth amount ranked highest to lowest, K. Cain's sales of $142,560.

## PRACTICE PROBLEMS

Sales made yesterday by salespeople at Clarion Entertainment center, rounded to the nearest dollar are shown below. What was the median sales amount for each salesperson? If there is no exact middle amount, express the median as an average of the amount above and the amount below the midpoint.

| Salesperson | Sales | Median |
|---|---|---|
| 1. J. Bascomb | $148, $205, $97, $258, $128 | _____ |
| 2. B. Carey | $95, $298, $315, $90, $139, $309, $210 | _____ |
| 3. S. Eaton | $106, $243, $190, $112, $134, $91 | _____ |
| 4. D. Jacobs | $95, $48, $74, $92, $108, $90, $86, $151, $32, $116, $122 | _____ |
| 5. I. Lyons | $139, $42, $216, $140, $130, $97, $116, $73, $108, $81, $262, $146 | _____ |

## CALCULATING THE MODE

The *mode,* or *modal measure,* is the number that occurs the most frequently in a group. This indicates the most commonly occurring level of performance. If two or more numbers occur an equal number of times and are the most frequently occurring numbers, each one is considered a mode. Often, there is no one number in a group that occurs more frequently than any other. In that case, there is no mode.

### EXAMPLE

The scores on preemployment tests given to 16 job applicants at McKabe Computer Services shown in order from highest to lowest, were as follows: 95, 93, 91, 91, 89, 88, 87, 87, 85, 85, 85, 82, 79, 74, 71, and 64. What is the modal test score?

### SOLUTION

Since 85 occurs most frequently, it is the mode.

ANSWERS FOR
PRACTICE PROBLEMS

(1) $148 (2) $210 (3) $123 (4) $92 (5) $123

**506** ■ LESSON 15-1

Copyright © 1989 by McGraw-Hill, Inc. All rights reserved.

## PRACTICE PROBLEMS

Find the mode for each of the following preemployment tests given by McKabe Computer Services.

| Test Scores January 15 | Test Scores March 15 | Test Scores May 15 | Test Scores July 15 | Test Scores September 15 |
|---|---|---|---|---|
| 97 | 95 | 96 | 99 | 93 |
| 94 | 94 | 94 | 94 | 92 |
| 91 | 93 | 91 | 94 | 91 |
| 91 | 90 | 90 | 91 | 91 |
| 90 | 90 | 89 | 90 | 90 |
| 88 | 89 | 86 | 89 | 87 |
| 88 | 87 | 85 | 89 | 84 |
| 88 | 86 | 80 | 89 | 84 |
| 86 | 86 | 78 | 86 | 84 |
| 83 | 83 | 74 | 84 | 83 |
| 81 | 80 | 67 | 80 | 80 |
| 79 | 78 | 61 | 76 | 76 |
| 76 | 73 | 52 | 75 | 72 |
| 62 | 64 | 51 | 75 | 64 |

Mode:  1. _____  2. _____  3. _____  4. _____  5. _____

**GROUPED DATA**

When a large number of data items are to be processed and analyzed, they are usually grouped into categories called **class intervals.** Each class interval is the same size and has an upper and lower limit or value. The difference between the upper and lower limit is the **range.**

**COMPLETING A FREQUENCY DISTRIBUTION**

To find the range of each class interval, divide the difference between the highest and lowest value in the database by the number of class intervals desired. After the class intervals have been determined, a **frequency distribution** is constructed showing the number of data items in each class interval.

---

### EXAMPLE

The costs of long-distance telephone calls made by employee Chris Davis at Schaffer Telemarketing yesterday are as follows: $1.42, $3.18, $9.65, $1.97, $14.76, $5.89, $18.46, $5.89, $3.76, $4.12, $5.19, $6.03, $26.46, $1.99, $3.06, $4.19, $12.90, $6.05, $2.26, $18.05, $10.43, $8.19, $17.32, $7.41, $2.18, $4.04, and $7.45.

**(a)** Establish the range of the class interval with a low of $0.01 and a high of $28.00 in the database.

$$\frac{\text{Highest Value} - \text{Lowest Value}}{\text{Number of Class Intervals}} = \frac{\$28.00 - \$0.01}{7} = 3.9 \text{ or } 4 \text{ (class interval range)}$$

**(b)** Count the number of scores in each interval.
**(c)** Show the frequency at each class interval.

---

ANSWERS FOR PRACTICE PROBLEMS

(1) 88  (2) 90 and 86  (3) None  (4) 89  (5) 84

## SOLUTION

| a. Class Interval | b. Number of Phone Calls (Tally) | c. Number of Phone Calls (Frequency) |
|---|---|---|
| $24.01–$28.00 | I | 1 |
| 20.01– 24.00 | | 0 |
| 16.01– 20.00 | III | 3 |
| 12.01– 16.00 | II | 2 |
| 8.01– 12.00 | III | 3 |
| 4.01– 8.00 | HHT HHT | 10 |
| 0.01– 4.00 | HHT III | 8 |

## PRACTICE PROBLEMS

Salesperson Sally Harpenau made sales of the following amounts yesterday at Bill's Souvenir Store: $2.19, $12.47, $13.80, $.99, $21.60, $4.06, $3.79, $16.29, $5.05, $3.78, $25.75, $16.47, $.98, $3.15, $10.05, $9.99, $2.15, $3.77, $31.60, $12.46, $3.89, $5.86, $7.15, $11.18, and $6.78. Establish class intervals with a range of $5.00. The highest value is $35.00, and the lowest is $0.01 in the database. Tally the individual data items. Show the frequency at each class interval.

| Sales Amount by Class Interval | Number of Sales (Tally) | Number of Sales (Frequency) |
|---|---|---|
| 1. _____ | _____ | _____ |
| 2. _____ | _____ | _____ |
| 3. _____ | _____ | _____ |
| 4. _____ | _____ | _____ |
| 5. _____ | _____ | _____ |
| 6. _____ | _____ | _____ |
| 7. _____ | _____ | _____ |

## CALCULATING THE WEIGHTED MEAN

When the mean of grouped data is calculated, it is assumed that all items of each class interval are at the midpoint of the interval. It is also assumed

---

ANSWERS FOR PRACTICE PROBLEMS

| Class Interval | Tally | Frequency |
|---|---|---|
| (1) $30.01–$35.00 | / | 1 |
| (2) 25.01– 30.00 | / | 1 |
| (3) 20.01– 25.00 | / | 1 |
| (4) 15.01– 20.00 | II | 2 |
| (5) 10.01– 15.00 | HHT | 5 |
| (6) 5.01– 10.00 | HHT | 5 |
| (7) 0.01– 5.00 | HHT HHT | 10 |

that all data items in a class interval of 95 to 99 have a middle value of 97, calculated by adding the lower limit and the upper limit and dividing by two (95 + 99 = 194; 194 ÷ 2 = 97).

The midpoint at each interval is multiplied by the frequency at that interval to determine the total for each class interval. The class interval totals are added and divided by the number of items to determine the **weighted mean.**

## EXAMPLE

Taylor Temporary Office Services administered a preemployment test to 22 job applicants. The class interval of test scores and the number of applicants scoring in each class interval were as follows: 95 to 99, 2 persons; 90 to 94, 1 person; 85 to 89, 6 persons; 80 to 84, 7 persons; and 75 to 79, 6 persons. Determine the mean score on the test. Round to the nearest hundredth.

## SOLUTION

**Step 1**  List the class interval and frequency.
**Step 2**  Calculate the midpoint at each class interval using the formula:

$$\text{Midpoint} = \frac{\text{Lower Limit} + \text{Upper Limit}}{2}$$

**Step 3**  To find the class interval totals, multiply the midpoint at each class interval by the frequency.
**Step 4**  Determine the frequency total and the sum of the class interval totals.

| Number of Applications (Frequency) | Class Interval of Test Scores | Midpoint | Class Interval | Totals |
|---|---|---|---|---|
| 2 | 95–99 | 97 | (2 × 97 = ) | 194 |
| 1 | 90–94 | 92 | (1 × 92 = ) | 92 |
| 6 | 85–89 | 87 | (6 × 87 = ) | 522 |
| 7 | 80–84 | 82 | (7 × 82 = ) | 574 |
| 6 | 75–79 | 77 | (6 × 77 = ) | 462 |
| 22 | ← Step 4 → | | | 1,844 |

**Step 5**  Calculate the mean.

$$\text{Mean Test Score} = \frac{\text{Class Interval Totals}}{\text{Number of Applicants (Frequency)}} = \frac{1,844}{22} = 83.82$$

## PRACTICE PROBLEMS

Scores by job applicants are shown at the top of the next page by class intervals. Calculate the midpoint of each class interval, the class interval totals, and the mean score.

| Number of Applicants (Frequency) | | Class Interval of Test Scores | Midpoint | Class Interval Totals |
|---|---|---|---|---|
| **1.** | 4 | 91–99 | _____ | _____ |
| **2.** | 8 | 82–90 | _____ | _____ |
| **3.** | 6 | 73–81 | _____ | _____ |
| **4.** | 5 | 64–72 | _____ | _____ |
| **5.** | 3 | 55–63 | _____ | _____ |
| **6.** | = | | | ═══════ |

**7.** Mean Test Score: _____

## CALCULATING THE MEDIAN

If the median of grouped data falls within a class interval, the class interval is the median, known as the ***median class interval.***

Sometimes, the median score falls between two class intervals. In this case, the median may be expressed as being between the two class intervals, for example, between the class interval 91 to 95 and the class interval 96 to 100. Another method is to express the median as a single score. This is calculated by adding the midpoint of the class interval above the median score and the midpoint of the class interval below the median score and dividing by two.

---

### EXAMPLE 1

Weekly earnings of employees at Rezak Manufacturing are rounded to the nearest dollar and grouped by class interval. Calculate the median.

| Number of Employees (Frequency) | Weekly Earnings (Class Interval) |
|---|---|
| 11 | $400–$450 |
| 29 | 350– 399 |
| →28 | 300– 349 |
| 9 | 250– 299 |
| 4 | 200– 249 |
| 81 | |

The forty-first amount, ranked from highest to lowest, is in the 300–349 interval.

### SOLUTION

**Step 1** Calculate the rank of the middle value.

$$\text{Middle Amount} = \frac{\text{Number of Items} + 1}{2} = \frac{81 + 1}{2} = 41$$

---

**Step 2** Identify the median class interval. The median is the 41st amount, counting from highest to lowest. The amount falls within the $300 to $349 class interval; therefore, $300 to $349 is the median.

## EXAMPLE 2

Karpinski, Inc., listed its 234 employees' median ages by class intervals while calculating the median age of the employees. The middle amount falls between the 41 to 45 class interval and the 46 to 50 class interval. What is the median age expressed as a single number?

## SOLUTION

**Step 1** Add the midpoint of the class intervals above and below the middle amount.

| Midpoint Class Interval | | Midpoint Class Interval | | Total |
|---|---|---|---|---|
| 41–45 | + | 46–50 | = | of Midpoints |
| 43 | + | 48 | = | 91 |

**Step 2** Find the median.

$$\text{Median Age} = \dfrac{\dfrac{\text{Total}}{\text{of Midpoints}}}{2} = \dfrac{91}{2} = 45.5$$

## PRACTICE PROBLEMS

Sales by representatives of Carver Home Products Company for 4 weeks are shown below by class interval. Note that a different number of salespeople made sales each week. Calculate the median sales for each week. If the median falls within a class interval, identify the median class interval. If the median falls between two class intervals, express the median as a single amount.

| Sales Amounts by Class Interval | Number of Salespeople (Frequency) | | | |
|---|---|---|---|---|
| | Week 1 | Week 2 | Week 3 | Week 4 |
| $901–$1,000 | 3 | 4 | 3 | 2 |
| 801– 900 | 2 | 7 | 5 | 9 |
| 701– 800 | 5 | 12 | 8 | 4 |
| 601– 700 | 8 | 10 | 12 | 3 |
| 501– 600 | 16 | 5 | 9 | 4 |
| 401– 500 | 12 | 13 | 6 | 11 |
| 301– 400 | 6 | 9 | 9 | 5 |
| 201– 300 | 7 | 3 | 4 | 6 |
| Median: | 1. _____ | 2. _____ | 3. _____ | 4. _____ |

Hourly wages paid to employees at four plants of Moneta Manufacturing are shown at the top of the next page by class interval. Calculate the median hourly wages paid at each branch. If the median falls within a class interval, identify the median class interval. If the median falls between two class intervals, express the median as a single amount.

| Hourly Wages by Class Interval | Number of Employees (Frequency) | | | |
|---|---|---|---|---|
| | Akron | Duluth | Memphis | Portland |
| $14.01–$15.00 | 27 | 8 | 22 | 7 |
| 13.01– 14.00 | 31 | 7 | 14 | 9 |
| 12.01– 13.00 | 46 | 12 | 16 | 22 |
| 11.01– 12.00 | 39 | 7 | 17 | 5 |
| 10.01– 11.00 | 55 | 10 | 10 | 11 |
| 9.01– 10.00 | 18 | 8 | 30 | 11 |
| 8.01– 9.00 | 26 | 7 | 38 | 3 |
| 7.01– 8.00 | 30 | 9 | 18 | 8 |
| Median Hourly Wage: | 5. _____ | 6. _____ | 7. _____ | 8. _____ |

## CALCULATING THE MODE OF GROUPED DATA

The class interval with the greatest frequency is known as the **mode,** or **modal class interval.** If two or more class intervals occur an equal number of times and are the most frequently occurring class intervals, each one is considered a modal class interval. If no class interval appears more frequently than any other, there is no mode.

### EXAMPLE

The frequency distribution of days of absence from work by employees at Hildreth Company is shown below. What is the mode?

| Number of Employees (Frequency) | Days of Absence by Class Interval |
|---|---|
| 2 | 20–24 |
| 8 | 15–19 |
| 12 | 10–14 |
| 39 | 5– 9 |
| 16 | 0– 4 |

### SOLUTION

The most frequently occurring class interval, 5–9, is the mode.

### PRACTICE PROBLEMS

Find the modal class interval of sales made by salespeople at each of the branches of Wicker Company.

| Sales by Class Interval | Number of Salespeople (Frequency) | | |
|---|---|---|---|
| | Chicago | Phoenix | Seattle |
| $901–$1,000 | 4 | 2 | 4 |
| 801– 900 | 18 | 14 | 3 |
| 701– 800 | 9 | 12 | 2 |
| 601– 700 | 21 | 10 | 7 |
| 501– 600 | 5 | 9 | 1 |
| Mode: | 1. _____ | 2. _____ | 3. _____ |

ANSWERS FOR PRACTICE PROBLEMS

**Bottom: (1)** $601–$700 **(2)** $801–$900 **(3)** $501–$700

**(8)** $12.01

**(4)** $500.50 **(5)** $11.01–$12.00 **(6)** $11.01 **(7)** $9.01–$10.00

**Starting on page 511: (1)** $501–$600 **(2)** $601–$700 **(3)** $600.50

## LESSON 15-1    EXERCISES

Sales made by salespersons at Mountain View Real Estate are shown below for the first 10 months of 19X7, rounded to the nearest dollar. What is the mean sales for the salespeople for each month? Round answers to the nearest dollar.

| Month | D. Graves | T. Hirsch | B. Ming | K. Noren | O. Smalley | Mean |
|---|---|---|---|---|---|---|
| 1. January | $ 63,615 | $104,316 | $ 91,055 | $ 44,610 | $ 86,407 | _____ |
| 2. February | 43,762 | 85,618 | 74,300 | 38,775 | 92,765 | _____ |
| 3. March | 186,785 | 145,750 | 179,315 | 54,815 | 95,380 | _____ |
| 4. April | 194,765 | 105,312 | 215,788 | 99,610 | 128,355 | _____ |
| 5. May | 215,805 | 144,760 | 380,750 | 475,890 | 205,390 | _____ |
| 6. June | 346,866 | 88,415 | 920,470 | 98,500 | 326,448 | _____ |
| 7. July | 145,690 | 70,410 | 100,550 | 350,160 | 122,058 | _____ |
| 8. August | 640,350 | 125,460 | 250,310 | 135,590 | 210,510 | _____ |
| 9. September | 421,046 | 597,850 | 95,480 | 380,416 | 650,900 | _____ |
| 10. October | 210,305 | 140,316 | 145,715 | 201,315 | 97,509 | _____ |

Shown below are expenses incurred by salespeople of Ratton Health Products, rounded to the nearest dollar. What was the median expense amount per week? If there is no exact middle amount, express the median as an average of the amount above and the amount below the middle amount.

| Week | Salespersons' Expenses | Median |
|---|---|---|
| 11. July 6 | $246, $318, $147, $406, $312, $518, $384 | _____ |
| 12. July 13 | $174, $206, $499, $308, $245 | _____ |
| 13. July 20 | $380, $402, $284, $188, $296, $216 | _____ |
| 14. July 27 | $513, $276, $412, $378 | _____ |
| 15. August 3 | $275, $416, $506, $182, $370, $397, $408 | _____ |
| 16. August 10 | $308, $472, $294, $340, $106, $579 | _____ |
| 17. August 17 | $277, $319, $594, $145, $416, $205 | _____ |
| 18. August 24 | $505, $618, $321, $406, $711 | _____ |
| 19. August 31 | $370, $162, $409, $386, $438, $256 | _____ |
| 20. September 7 | $145, $410, $477, $309, $321, $301 | _____ |

Find the modal age of each of the departments of Farley Imports.

| Employee Age: | Accounting | Management | Purchasing | Sales | Shipping |
|---|---|---|---|---|---|
| | 63 | 58 | 60 | 55 | 41 |
| | 54 | 56 | 54 | 48 | 37 |
| | 48 | 54 | 50 | 47 | 36 |
| | 47 | 54 | 38 | 46 | 36 |
| | 32 | 47 | 37 | 45 | 30 |
| | 26 | | 37 | 45 | 29 |
| | | | 28 | 45 | 26 |
| | | | 28 | 40 | 24 |
| | | | | 32 | 23 |
| | | | | 28 | 22 |
| | | | | | 19 |

Modal Age:   **21.** _____   **22.** _____   **23.** _____   **24.** _____   **25.** _____

The amounts of sales bonus earned by salespersons at National Novelties last month are as follows: $218, $305, $176, $192, $180, $122, $406, $105, $365, $195, $206, $425, $260, $275, $306, $218, $114, $138, $385, $432, $143, $216, and $205. (a) Establish class intervals with a low of $101 and a high of $450 with a range of $50 for each interval. (b) Tally individual data items. (c) Show frequency at each class interval.

| | Class Intervals | Tally | Frequency |
|---|---|---|---|
| **26.** | _____ | _____ | _____ |
| **27.** | _____ | _____ | _____ |
| **28.** | _____ | _____ | _____ |
| **29.** | _____ | _____ | _____ |
| **30.** | _____ | _____ | _____ |
| **31.** | _____ | _____ | _____ |
| **32.** | _____ | _____ | _____ |

Miles driven last month by salespeople at the Billings, Montana, and Miami, Florida, branches of Westcott Textiles are shown below and on the next page. For each branch, calculate the midpoint of each class interval, the frequency total and class interval totals, and the mean miles driven per salesperson at each branch. Round the mean to the nearest mile.

### Billings Branch

| | Number of Salespeople (Frequency) | Class Interval of Miles Driven | Midpoint | Class Interval Totals |
|---|---|---|---|---|
| **33.** | 4 | 1,501–2,000 | _____ | _____ |
| **34.** | 19 | 1,001–1,500 | _____ | _____ |
| **35.** | 6 | 501–1,000 | _____ | _____ |
| **36.** | 1 | 1– 500 | _____ | _____ |
| **37.** | = | | | = |

**38.** Mean Miles Driven: _____

## Miami Branch

| | Number of Salespeople (Frequency) | Class Interval of Miles Driven | Midpoint | Class Interval Totals |
|---|---|---|---|---|
| **39.** | 6 | 451–500 | _____ | _____ |
| **40.** | 15 | 401–450 | _____ | _____ |
| **41.** | 12 | 351–400 | _____ | _____ |
| **42.** | 9 | 301–350 | _____ | _____ |
| **43.** | 7 | 251–300 | _____ | _____ |
| **44.** | ═══ | | | ═══════ |

**45.** Mean Miles Driven: _____

Find the modal class interval for each of the following.

| Number of Job Applicants (Frequency) | Keyboarding W.P.M. by Class Interval |
|---|---|
| 0 | 91–100 |
| 1 | 81– 90 |
| 2 | 71– 80 |
| 4 | 61– 70 |
| 35 | 51– 60 |
| 28 | 41– 50 |
| 16 | 31– 40 |
| 2 | 21– 30 |

**46.** Mode: _____

| Number of Customers (Frequency) | Sales Amount by Class Interval |
|---|---|
| 6 | $46–$50 |
| 4 | 41– 45 |
| 3 | 36– 40 |
| 7 | 31– 35 |
| 12 | 26– 30 |
| 15 | 21– 25 |
| 12 | 16– 20 |
| 7 | 11– 15 |
| 12 | 6– 10 |
| 93 | 1– 5 |

**47.** Mode: _____

| Number of Sales Returns (Frequency) | Amounts by Class Interval |
|---|---|
| 4 | $901–$1,000 |
| 2 | 801– 900 |
| 6 | 701– 800 |
| 8 | 601– 700 |
| 5 | 501– 600 |
| 3 | 401– 500 |
| 2 | 301– 400 |
| 3 | 201– 300 |
| 1 | 101– 200 |

**48.** Mode: _____

| Number of Employees (Frequency) | Absences by Class Interval |
|---|---|
| 1 | 18–20 |
| 2 | 15–17 |
| 6 | 12–14 |
| 12 | 9–11 |
| 25 | 6– 8 |
| 21 | 3– 5 |
| 14 | 0– 2 |

**49.** Mode: _____

## BUSINESS APPLICATIONS

Round your answers to the nearest cent.

**50.** The number of years' employment by employees at Santana and Company is as follows: 16, 22, 4, 10, 33, 21, 2, 6, 1, 29, 12, 8, 7, 8, 18, 30, 6, 22, 5, 11, 4, 8, 1, 3, 12, 14, 7, 3, 31, 14, 3, 5, 9, 14, 2, 19, 8, 5, 13, 22, 31, 5, 4, 1, 10, 5, 16, 7, and 9. Calculate (a) the mean, (b) median, and (c) modal years of employment.

(a) _____

(b) _____

(c) _____

**51.** Monthly sales made by Pelzer Company's salespersons are shown below, grouped by $1,000 class intervals. Calculate (a) the mean, (b) median, and (c) modal sales. Space is provided for you to calculate the midpoint of each class interval, total frequency, and class interval totals.

| Number of Salespersons (Frequency) | Total Monthly Sales by Class Interval | Midpoint | Class Interval Totals |
|---|---|---|---|
| 2 | $15,001–$16,000 | _____ | _____ |
| 5 | 14,001– 15,000 | _____ | _____ |
| 4 | 13,001– 14,000 | _____ | _____ |
| 9 | 12,001– 13,000 | _____ | _____ |
| 7 | 11,001– 12,000 | _____ | _____ |
| 12 | 10,001– 11,000 | _____ | _____ |
| 3 | 9,001– 10,000 | _____ | _____ |
| 2 | 8,001– 9,000 | _____ | _____ |
| ═══ | | | ═══════════════ |

(a) _____

(b) _____

(c) _____

**52.** Sales bonuses paid to employees at Kraenbrink's last year were as follows: $425, $815, $375, $590, $960, $480, $350, $380, $250, $280, $320, $395, $250, $285, $590, $480, $210, $305, and $225. Establish class intervals with a low of $201 and a high of $1,000 with a range of $100 for each interval. Tally the individual data items. Calculate (a) the mean, (b) median, and (c) mode. Space is provided for you to list the frequency and class intervals and to calculate the midpoint of each class interval and the class interval totals.

| Number of Salespeople (Frequency) | Bonus by Class Interval | Midpoint | Class Interval Totals |
|---|---|---|---|
| _____ | _____ | _____ | _____ |
| _____ | _____ | _____ | _____ |
| _____ | _____ | _____ | _____ |
| _____ | _____ | _____ | _____ |
| _____ | _____ | _____ | _____ |
| _____ | _____ | _____ | _____ |
| _____ | _____ | _____ | _____ |
| _____ | _____ | _____ | _____ |

(a) _____

(b) _____

(c) _____

**53.** Last year, sales made by O'Leary's salespeople were as follows: $475,800, $516,750, $985,780, $621,550, $480,900, $825,300, $715,600, $390,450, $684,950, $410,650, $855,330, $905,750, $507,900, and $585,650. For next year, O'Leary's sales goal is to increase the mean sales by 15%. What is next year's mean sales goal per salesperson?

_____

**54.** Mulso Household Products Company set a sales goal of $450 per week for each salesperson. Last week, Mulso's salespeople had sales as follows: $380, $724, $515, $378, $620, $455, $410, $396, $1,016, $512, $398, $460, $395, $425, $476, $312, and $452. (a) What was the mean sales amount per salesperson? (b) How many dollars was the mean sales above or below the weekly sales goal? (c) What was the median sales amount? (d) How many salespeople reached or exceeded the sales goal? (e) How many salespeople did not reach the sales goal?

(a) _____

(b) _____

(c) _____

(d) _____

(e) _____

A *graph* is a visual device used to show business data and the relationship that exists between data items. Business executives use graphs to make visual comparisons and observations that are not easily made from large amounts of numerical data. Commonly used graphs are bar graphs, line graphs, and circle graphs. To prepare a graph, use the following steps.

1. Assemble the data to be shown. Often, dollar amounts are rounded to the nearest hundred or thousand.
2. Use a ruler, compass, protractor, and graph paper.
3. Choose a scale that matches the data shown and that will make the graph easy to interpret.
4. Label the graph at equal intervals according to the scale that you have chosen. The vertical line of bar graphs and line graphs is called the *vertical axis.* The horizontal line is called the *horizontal axis.*
5. The use of dots to show where the bars or lines should be drawn is called *plotting* the data.
6. Give the graph an informative title and, if appropriate, a subtitle.

A graph often is used as part of a report or presentation to summarize information and show relationships at a glance.

**BAR GRAPHS**

*Bar graphs* compare several data items at the same time. Since bar graphs are easily constructed from a frequency distribution using graph paper and a ruler, they are used frequently.

### EXAMPLE

The weekly payroll register for the week ending April 7, 19—, for Dalton and Sons shows employee gross earnings as follows: Employee 101, $342.00; Employee 102, $373.80; Employee 103, $351.50; Employee 104, $413.25; and Employee 105, $321.75. The mean weekly gross earnings is $360.46. Prepare a vertical-bar graph to show each employee's weekly gross earnings and the median gross earnings. Label the vertical axis with dollar amounts from $0 to $450 at $50 intervals. Give the graph an appropriate heading and label the information in the graph.

### SOLUTION

The following graph was generated by computer.

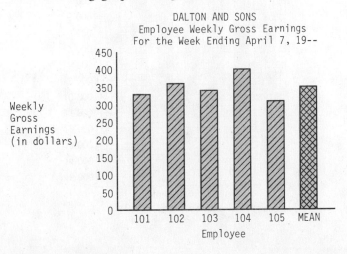

## PRACTICE PROBLEM

Sheldon Company's advertising expenditures for the first 7 months of the year were as follows: January, $8,207; February, $7,439; March, $9,892; April, $12,536; May, $10,985; June, $11,103; and July, $14,691. Present these data in a vertical-bar graph. Round the amounts to the nearest thousand dollars. Label the vertical axis from $0 to $15,000 at $1,000 intervals. Give the graph an appropriate heading and label the information in the graph.

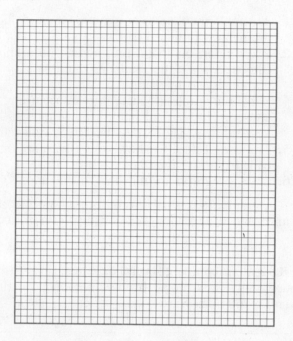

---

ANSWERS FOR
PRACTICE PROBLEMS

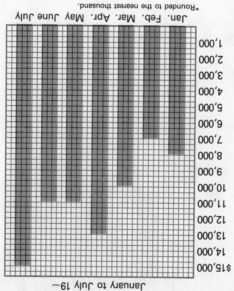

## LINE GRAPHS

*Line graphs* show the relationship between three or more items and illustrate the change over a period of time. Either a solid line or a broken line may be used. If more than one type of data is shown in a ***comparative graph,*** a solid line might be used to show one data item and a broken line can be used to show the other.

### EXAMPLE

Total sales and cash sales for the Men's Wear department of Jurgenson Clothing for the first 7 months of the year are shown below. Present these data in the form of a comparative line graph with the total sales represented by a solid line and cash sales represented by a broken line. Round amounts to the nearest thousand dollars. Label the vertical axis with dollar amounts from $15,000 to $45,000 at $5,000 intervals. Label the horizontal axis with months. Give the graph an appropriate heading and label the information in the graph.

| Month | Total Sales | Cash Sales |
|---|---|---|
| January | $41,500 | $28,400 |
| February | 24,300 | 16,800 |
| March | 34,300 | 22,600 |
| April | 35,000 | 24,000 |
| May | 32,000 | 19,000 |
| June | 31,100 | 17,200 |
| July | 28,300 | 21,000 |

### SOLUTION

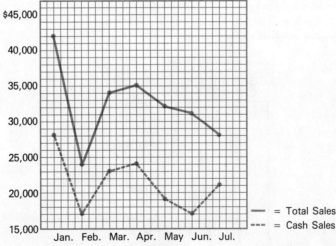

Jurgensen Clothing
A COMPARISON OF TOTAL SALES AND CASH SALES*
Men's Wear Department
January to July 19—

—— = Total Sales
--- = Cash Sales

*Rounded to the nearest thousand.

## PRACTICE PROBLEM

Net income for Adamson, Inc., was as follows for 7 consecutive years: 19X1, $123,460; 19X2, $156,250; 19X3, $84,250; 19X4, $175,390; 19X5, $65,310; 19X6, $164,050; and 19X7, $180,350. Present this information in a line graph. Round amounts to the nearest thousand dollars. Label the vertical axis by dollar amount from $50,000 to $200,000 in $10,000 intervals. Label the horizontal axis with years 19X1 to 19X7. Give the table an appropriate heading and label the information in the graph.

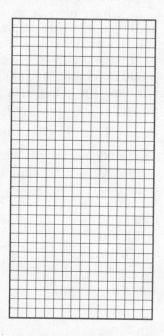

ANSWER FOR
PRACTICE PROBLEM

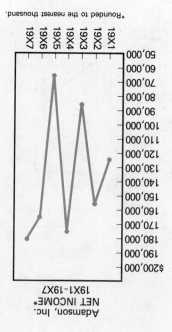

## CIRCLE GRAPHS

*Circle graphs* show the relationship among data items on a percentage or dollar basis or indicate how the whole is divided into parts. To prepare a circle graph follow these procedures.

1. Calculate the percent that each individual item is of the total.
2. Determine how many degrees of the circle should be used to represent each data item by multiplying the percent that each data item is of the total by 360 degrees.
3. Draw the circle with the aid of a compass.
4. Use a protractor to identify the degrees of the circle. Mark the point with a dot.
5. Draw a line from the center of the circle to the edge where the dot is.

### EXAMPLE

Kingsley Company's operating expenses for last year were as follows: salaries, $480,000; rent, $100,000; advertising, $80,000; depreciation, $40,000; utilities, $40,000; insurance, $20,000; and other expenses, $40,000. (a) Calculate the percent, rounded to the nearest tenth of a percent, that each expense is of the total. (b) Determine how many degrees of the circle should be used to represent each expense item. (c) Present the data in the form of a circle graph. Give the graph an appropriate heading and label the information in the graph.

### SOLUTION

**Step 1** Calculate the percent that each item is of the total and determine the number of degrees that represent each item.

| Expense | Amount | ÷ Total Expenses | = Percent of Total Expenses | × 360° = | Degrees of Circle Graph |
|---------|--------|------------------|------------------------------|----------|--------------------------|
| Salaries | $480,000 | ÷ $800,000 | = 60.0% | × 360° = | 216° |
| Rent | 100,000 | ÷ 800,000 | = 12.5% | × 360° = | 45° |
| Advertising | 80,000 | ÷ 800,000 | = 10.0% | × 360° = | 36° |
| Depreciation | 40,000 | ÷ 800,000 | = 5.0% | × 360° = | 18° |
| Utilities | 40,000 | ÷ 800,000 | = 5.0% | × 360° = | 18° |
| Insurance | 20,000 | ÷ 800,000 | = 2.5% | × 360° = | 9° |
| Other | 40,000 | ÷ 800,000 | = 5.0% | × 360° = | 18° |
| | $800,000 | | 100.0% | | 360° |

**Step 2** Draw the circle and present the data.

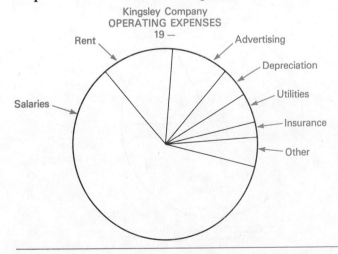

Kingsley Company
OPERATING EXPENSES
19 —

## PRACTICE PROBLEMS

Sales by department for Chelsea Department Store are listed below. Calculate the percent that each department's sales are of the total sales, rounded to the nearest tenth of a percent. Determine how many degrees of a circle graph, rounded to the nearest degree, should be used to show each department's sales. Then use the space below to present a circle graph to show this information. Give the graph an appropriate heading and label each part of the graph.

| Department | Sales Amount | Percent of Total Sales | Degrees of Circle Graph |
|---|---|---|---|
| 1. Children's Wear | $60,000 | _____ | _____ |
| 2. Women's Wear | 42,000 | _____ | _____ |
| 3. Home Furnishings | 36,000 | _____ | _____ |
| 4. Men's Wear | 32,400 | _____ | _____ |
| 5. Shoes | 27,600 | _____ | _____ |
| 6. Jewelery | 24,000 | _____ | _____ |
| 7. Gifts | 18,000 | _____ | _____ |
| 8. | $ _____ | _____ | _____ |

Draw a circle graph below.

## LESSON 15-2 EXERCISES

1. Interest rates paid on loans obtained by Holmgren Manufacturing were as follows for the past 8 years: 19X1, 10.50%; 19X2, 11.00%; 19X3, 14.25%; 19X4, 16.00%; 19X5, 17.50%; 19X6, 15.00%; 19X7, 13.50%; and 19X8, 14.25%. Present this data in a vertical-bar graph. Label the vertical axis from 10.00% to 19.00% at 0.5% intervals. Label the horizontal axis with years from 19X1 to 19X8. Give the graph an appropriate heading and label the information in the graph.

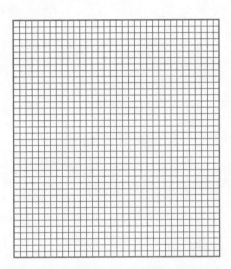

2. The percent that net income was of net sales for Kahler Wholesale Company for each of the past 10 years was as follows: 19X1, 4.5%; 19X2, 5.5%; 19X3, 7.0%; 19X4, 4.0%; 19X5, 2.5%; 19X6, 7.5%; 19X7, 12.0%; 19X8, 13.5%; 19X9, 10.5%; and 19X0, 12.5%. Present these data in a line graph. Round all percents to the nearest half percent. Label the vertical axis from 0% to 15%. Label the horizontal axis with years 19X1 and 19X0. Give the graph an appropriate heading and label the information in the graph.

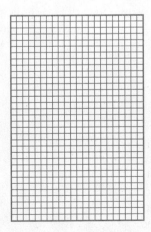

**3.** Clerken and Company's net sales and cost of goods sold for eight consecutive years are shown below. Present these data in a comparative
line graph. Round dollar amounts to the nearest $50,000. Show the net
sales as a solid line and the cost of goods sold as a broken line. Label
the vertical axis with dollar amounts from $800,000 to $2,400,000 at
$100,000 intervals. Label the horizontal axis with years from 19X1 to
19X8. Give the graph an appropriate heading and label the information in the graph.

| Year | Net Sales | Cost of Goods Sold |
|------|-----------|--------------------|
| 19X1 | $1,604,091 | $ 978,495 |
| 19X2 | 1,657,700 | 1,006,230 |
| 19X3 | 1,774,216 | 1,113,765 |
| 19X4 | 1,720,480 | 1,032,900 |
| 19X5 | 1,895,260 | 1,175,416 |
| 19X6 | 2,190,315 | 1,384,330 |
| 19X7 | 2,323,500 | 1,397,100 |
| 19X8 | 2,050,600 | 1,286,500 |

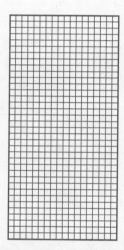

**4.** Urbandale Clothing Company's budgeted amounts for next year are
shown on the following page. Calculate the percent that each item is of
the total, rounded to the nearest tenth of a percent. Determine how
many degrees of a circle graph should be used to represent each item,
rounded to the nearest degree. Use the blank space on the next page to
present a circle graph showing Urbandale's budgeted amounts. Give
the graph an appropriate heading and label each part of the graph.

| Item | Budgeted Amount | Percent of Total | Degrees of Circle Graph |
|---|---|---|---|
| **a.** Cost of goods sold | $ 920,000 | _____ | _____ |
| **b.** Rent expense | 120,000 | _____ | _____ |
| **c.** Salary expense | 280,000 | _____ | _____ |
| **d.** Advertising expense | 80,000 | _____ | _____ |
| **e.** Other expenses | 40,000 | _____ | _____ |
| **f.** Net income | 160,000 | _____ | _____ |
| **g.** | | | |

**5.** Sales made last year by the eight branch offices of Villa Capri Health Products Company were as follows, rounded to the nearest $1,000. Calculate the percent each branch's sales is of total sales, rounded to the nearest full percent. Determine how many degrees of a circle graph should be used to show each branch's sales, rounded to the nearest degree. Use the blank space on the next page to draw a circle graph to present this information. Give the graph an appropriate heading and label each part of the graph.

| Branch | Sales Amount | Percent Total Sales | Degrees of Circle Graph |
|---|---|---|---|
| **a.** Albuquerque | $368,000 | _____ | _____ |
| **b.** Ft. Worth | 690,000 | _____ | _____ |
| **c.** Indianapolis | 552,000 | _____ | _____ |
| **d.** Louisville | 966,000 | _____ | _____ |
| **e.** New Orleans | 322,000 | _____ | _____ |
| **f.** Norfolk | 414,000 | _____ | _____ |
| **g.** Milwaukee | 782,000 | _____ | _____ |
| **h.** San Diego | 506,000 | _____ | _____ |
| **i.** | | | |

## CHAPTER 15 REVIEW

### CALCULATING THE MEAN—UNGROUPED DATA

Calculate the mean sales by department for each month for Patrick's Department Store.

| | Month | Clothing | Furniture | Gifts | Hardware | Jewelery | Mean |
|---|---|---|---|---|---|---|---|
| 1. | Sept. | $44,250 | $30,160 | $10,415 | $18,250 | $ 9,760 | _____ |
| 2. | Oct. | 56,350 | 28,410 | 12,850 | 16,910 | 14,860 | _____ |
| 3. | Nov. | 68,970 | 32,460 | 24,060 | 20,650 | 22,650 | _____ |
| 4. | Dec. | 97,386 | 40,655 | 74,920 | 31,550 | 80,380 | _____ |

### CALCULATING THE MEDIAN—UNGROUPED DATA

Individual sales made by Caltron Company's salespeople are shown below. Calculate the median sales for each salesperson. If there is no exact middle amount, express the median as an average of the amount above and the amount below the middle point.

| | Salesperson | Sales Amounts | Median |
|---|---|---|---|
| 5. | J. Dybdahl | $268, $305, $210, $315, $320, $412, $290 | _____ |
| 6. | T. Eaton | $315, $410, $308, $270, $405, $560 | _____ |
| 7. | M. Renze | $455, $310, $406, $320, $280, $510, $340 | _____ |
| 8. | S. Winfrey | $310, $450, $380, $420, $360, $460 | _____ |

### CALCULATING THE MODE—UNGROUPED DATA

Scores on three different preemployment tests given by Hollensbee Company are as follows: test no. 1: 78, 86, 79, 92, 90, 88, 86, 93, 91, 74, 72, and 70; test no. 2: 85, 90, 73, 86, 91, 93, 97, 80, 82, 77, 71, and 88; and test no. 3: 90, 78, 91, 90, 85, 77, 71, 92, 85, 86, 96, and 95. Find the modal score for each test.

**9.** Test No. 1 Mode: _____

**10.** Test No. 2 Mode: _____

**11.** Test No. 3 Mode: _____

### COMPLETING A FREQUENCY DISTRIBUTION—GROUPED DATA

**12.** Expenditures classified as miscellaneous expense at Lavine Company last month were as follows: $4.80, $5.65, $9.95, $8.14, $6.16, $6.30, $3.05, $2.78, $7.70, $5.40, $3.65, $6.15, $1.05, $.98, $4.50, $.99, and $1.65. Establish class intervals with a low of $0.01 and a high of $10.00 and a range of $1.00 for each interval. Tally the items. Then show the frequency at each class interval.

| Class Interval | Number of Expenditures (Tally) | Number of Expenditures (Frequency) |
|---|---|---|
| a. $0.01–$1.00 | _____ | _____ |
| b. 1.01– 2.00 | _____ | _____ |
| c. 2.01– 3.00 | _____ | _____ |
| d. 3.01– 4.00 | _____ | _____ |
| e. 4.01– 5.00 | _____ | _____ |
| f. 5.01– 6.00 | _____ | _____ |
| g. 6.01– 7.00 | _____ | _____ |
| h. 7.01– 8.00 | _____ | _____ |
| i. 8.01– 9.00 | _____ | _____ |
| j. 9.01–10.00 | _____ | _____ |

## CALCULATING THE MEAN—GROUPED DATA

13. The amount of travel expense spent by salespeople at Badger Specialties last month is shown below by class interval. Calculate the midpoint of each class interval, the class interval totals, and the mean expenditure per salesperson. Round the mean to the nearest cent.

| | Number of Salespeople (Frequency) | Travel Expense by Class Interval | Midpoint | Class Interval Totals |
|---|---|---|---|---|
| a. | 4 | $301–$ 500 | _____ | _____ |
| b. | 18 | 501– 700 | _____ | _____ |
| c. | 9 | 701– 900 | _____ | _____ |
| d. | 5 | 901– 1,100 | _____ | _____ |
| e. | ═══ | | | ═══ |

f. Mean Travel Expenditure: _____

## CALCULATING THE MEDIAN—GROUPED DATA

Sales made by salespersons at four branches of Pearson Company are shown below, grouped by class interval. Calculate the median sales at each branch. If the median falls within a class interval, identify the median class interval. If the median falls between two class intervals, express the median as a single amount.

| Sales by Class Interval | Cheyenne Branch (Frequency) | Fargo Branch (Frequency) | Nashville Branch (Frequency) | Tulsa Branch (Frequency) |
|---|---|---|---|---|
| $50,001–$60,000 | 3 | 2 | 12 | 14 |
| 40,001– 50,000 | 2 | 4 | 19 | 4 |
| 30,001– 40,000 | 5 | 2 | 31 | 5 |

| Sales by Class Interval | Cheyenne Branch (Frequency) | Fargo Branch (Frequency) | Nashville Branch (Frequency) | Tulsa Branch (Frequency) |
|---|---|---|---|---|
| 20,001– 30,000 | 4 | 2 | 47 | 19 |
| 10,001– 20,000 | 1 | 2 | 8 | 4 |

Median Sales: **14.** _____ **15.** _____ **16.** _____ **17.** _____

## CALCULATING THE MODE—GROUPED DATA

Find the modal class interval for each of the following.

| Number of Salespeople (Frequency) | Sales by Class Interval | Number of Miles Driven (Frequency) | Miles Driven by Class Interval | Number of Customers (Frequency) | Annual Orders by Class Interval |
|---|---|---|---|---|---|
| 8 | $175,001–$200,000 | 5 | 401–500 | 13 | 21–25 |
| 12 | 150,001– 175,000 | 24 | 301–400 | 43 | 16–20 |
| 31 | 125,001– 150,000 | 21 | 201–300 | 79 | 11–15 |
| 43 | 100,001– 125,000 | 13 | 101–200 | 146 | 6–10 |
| 10 | 75,001– 100,000 | 18 | 1–100 | 101 | 1–5 |
| 6 | 50,001– 75,000 | | | | |

**18.** Mode: _____ **19.** Mode: _____ **20.** Mode: _____

## BAR GRAPHS

**21.** July sales by department for Lendel's Emporium were as follows: bakery, $10, 680; candy, $14,250; gifts, $24,600; jewelry, $18,500; shoes, $4,350; and toys, $8,390. Prepare a vertical-bar graph to show each department's sales. Round sales amounts to the nearest thousand dollars. Give the graph an appropriate heading.

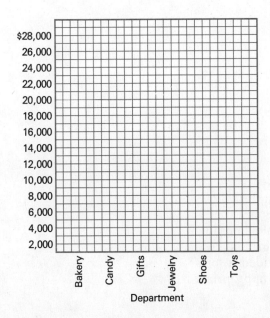

## LINE GRAPHS

**22.** Total commissions earned and total expenses for 7 consecutive years, rounded to the nearest thousand dollars, are shown below for Capital City Real Estate. Present these data as a comparative line graph. Give the graph an appropriate heading.

| Year | Total Commissions Earned | Total Expenses |
|------|--------------------------|----------------|
| 19X1 | $92,000 | $46,000 |
| 19X2 | 81,000 | 44,000 |
| 19X3 | 66,000 | 55,000 |
| 19X4 | 54,000 | 59,000 |
| 19X5 | 61,000 | 68,000 |
| 19X6 | 82,000 | 56,000 |
| 19X7 | 75,000 | 51,000 |

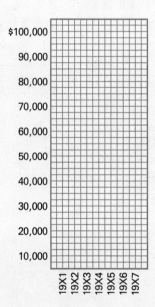

## CIRCLE GRAPHS

**23.** Sales by each branch of Faber, Inc., for 19X9 are listed on the following page. Calculate the percent that each branch's sales are of the total, rounded to the nearest tenth of a percent. Then calculate how many degrees of a circle graph, rounded to the nearest degree, should be used to show each branch's sales. Use the space provided to present a circle graph to show this data. Give the graph an appropriate heading and label each part of the graph.

| Branch | Sales Amount | Percent of Total Sales | Degrees of Circle Graph |
|---|---|---|---|
| **a.** Atlanta | $2,146,000 | _____ | _____ |
| **b.** Boston | 1,218,000 | _____ | _____ |
| **c.** Columbus | 348,000 | _____ | _____ |
| **d.** Rochester | 696,000 | _____ | _____ |
| **e.** Springfield | 1,392,000 | _____ | _____ |
| **f.** | ========= | ========= | ========= |

## BUSINESS APPLICATIONS

**24.** Sales made yesterday by Sharon DeBoise at the Cashsaver Discount Store were as follows: $2.15, $7.45, $10.18, $7.50, $.98, $1.19, $5.50, $10.40, $3.45, $5.50, $4.25, $6.15, $2.98, $1.45, $3.45, $5.60, $.99, $3.45, $7.20, $8.15, $8.10, $2.60, $1.98, $4.20, and $2.45. Calculate (a) the mean, (b) median, and (c) modal sales amounts.

(a) _____

(b) _____

(c) _____

**25.** Hourly wages paid to employees at Comstock Manufacturing are shown below by class interval. Calculate the (a) mean, (b) median, and (c) modal hourly wages. Space is provided for you to calculate the midpoint of each class interval, total frequency, and class interval totals.

| Number of Employees (Frequency) | Hourly Wages by Class Interval | Midpoint | Class Interval Totals |
|---|---|---|---|
| 84 | $13.76–$15.00 | $14.38 | $1,207.92 |
| 112 | 12.51– 13.75 | 13.13 | 1,470.56 |
| 210 | 11.26– 12.50 | 11.88 | 2,494.80 |
| 326 | 10.01– 11.25 | 10.63 | 3,465.38 |
| 62 | 8.76– 10.00 | 9.38 | 581.56 |
| 794 | | | $9,220.22 |

(a) _____

(b) _____

(c) _____

## CUMULATIVE REVIEW IV

Round amounts to the nearest cent. Round percents and ratios off to two decimal places.

1. Carlene Jacobs works an 8-hour day. As an experiment, Carlene's employer asked her to record the amount of time spent on each activity to the nearest $\frac{1}{8}$ hour. At the end of the day, Carlene had the following summary: computer entry, $2\frac{3}{4}$ hours; telephone calls, $\frac{5}{8}$ hour; processing mail, $\frac{3}{8}$ hour; billings, $1\frac{1}{4}$ hours; writing correspondence, 2 hours; and lunch, $\frac{7}{8}$ hour. How many hours of Carlene's time were unaccounted for?

_____

2. Futura Electronics had sales of $38,752 in July. Of this, 46% was from the sale of audio equipment. What was the audio equipment sales amount?

_____

3. In 19X5 the Sewing Center had sales of $876,950. In 19X6 sales increased $2\frac{1}{2}$% over 19X5 sales. What was the 19X6 sales amount?

_____

4. Melissa Meyer is paid $9.75 per hour and is paid $1\frac{1}{2}$ times the regular rate for hours worked over 40 in a week. Last week, Melissa worked 43 hours. Deductions were made as follows from Melissa's gross pay: FICA taxes, $32.58; federal income taxes, $75.00; and state income taxes, $11.25. What was Melissa's net pay for the week?

_____

5. On July 1 Cicero Company's ending bank statement was $4,618.43 and ending checkbook balance was $4,978.91. The following were found in preparing the bank reconciliation: deposit in transit, $872.90; outstanding checks, $147.35, $278.24, and $86.41; interest income, $19.17; and service charge, $18.75. What was the adjusted checkbook balance?

_____

6. Josten Distributing borrowed $1,850 at $10\frac{1}{2}$% simple interest for 150 days. What is interest on the loan using the exact-interest method?

_____

7. On February 7 the Clock Works purchased goods with a $4,850 list price and series trade discounts of 45% and 10%. Terms were 2/10, n/30. How much does the Clock Works owe if payment is made on February 17?

_____

8. Togs n' Toys purchased a shipment of stuffed animals at $8.85 each. Togs n' Toys requires a markup of 40% on the selling price. At what price should Togs n' Toys sell each of the stuffed animals?

_____

9. Crafter's Millwork can purchase a new drill press for $2,250 cash. The press can also be purchased on the installment plan with $250 down and $127.61 per month for 18 months. What is the total sale price for the drill press?

_____

10. Mikel Manufacturing offers its employees a group life insurance plan with coverage equal to $1\frac{1}{2}$ times his or her annual salary at a rate of $2.75 per $1,000 of insurance. George Harms earns $34,000 a year. What will be George's annual insurance premium?

_____

11. Petticord, Inc., purchased ten $1,000 par value $9\frac{1}{4}$% bonds at $98\frac{1}{8}$. The brokerage fee was $8 per bond and accrued interest was $16.50 per bond. What was the total cost of this investment?

_____

12. On August 1 Williby Decorating had a beginning inventory of $386,415. During the month Williby made purchases of $6,485, $10,316, and $2,426. On August 31 the inventory valuation was $369,084. What was the cost of goods sold?

13. On January 1 the Lamplighter had three model LT-65 table lamps on hand at a cost of $27.50 each. During the year, the Lamplighter made purchases of the LT-65 lamps as follows: February 15, five at $28.50 each; May 20, six at $29.00 each; August 1, four at $30.00 each; and November 5, six at $31.50 each. On December 31 there are seven of the lamps in inventory. What is the December 31 inventory using the LIFO inventory valuation method?

14. The total overhead expense for Sundall Manufacturing is $95,600. Sundall allocates its overhead based on floor space occupied by each department. Department A occupies 28,000 square feet, Department B occupies 32,000 square feet, and Department C occupies 20,000 square feet. What is the amount of overhead expense allotted to Department A?

15. The Custom Machine Shop purchased a welder for $3,750. The estimated useful life of the welder is 5 years with a salvage value of $550. The straight-line method is used to calculate depreciation. What is the welder's book value at the end of the third year?

16. Springtime Garden Service purchased a garden tiller for $4,800. The tiller has a salvage value of $400 and an estimated useful life of 8,000 hours. The tiller was used 1,600 hours during the first year. What was the garden tiller's depreciation expense for the first year? Use the units-of-production method to calculate depreciation.

**17.** Executive Management Systems purchased a computer for $4,500. The computer is a 5-year property according to ACRS guidelines. The recovery rate is 15% for the first year, 22% for the second year, and 21% for the third year. What is the computer's book value at the end of the third year, using the ACRS method?

_____

**18.** Girrard's income statement showed the following amounts for the quarter ending June 30: Sales, $62,480; Sales Returns and Allowances, $1,875; Merchandise Inventory, April 1, $147,315; Net Purchases, $45,685; Merchandise Inventory, June 30, $155,736; and Total Operating Expenses, $18,398. What was Girrard's net income or net loss?

_____

**19.** Healthco's balance sheet showed the following amounts: Cash, $39,720; Accounts Receivable, $75,908; Merchandise Inventory, $498,725; Supplies, $3,218; Building, $469,452; Land, $126,350; Accounts Payable, $124,385; and Notes Payable, $207,392. What is Healthco's owner's equity?

_____

**20.** Scores on preemployment tests given by Hudson Technologies were as follows: 88, 92, 77, 94, 83, 97, 90, 78, 66, 90, 98, 91, 82, 74, 88, 89, 94, 98, 83, 89, and 84. What was the median test score?

_____

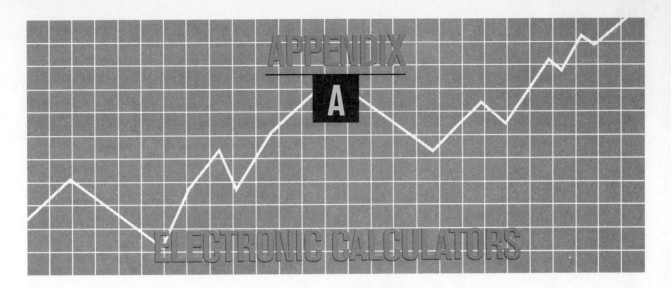

# APPENDIX A

# ELECTRONIC CALCULATORS

## LEARNING OBJECTIVES

**1.** To operate an electronic calculator using various keys to perform the four basic math operations.

**2.** To utilize the percent key and memory keys.

**3.** To perform business applications on the electronic calculator.

Using an electronic calculator to perform business mathematics calculations can increase accuracy and save a great deal of time. An inexpensive hand-held calculator can be used to perform the basic mathematical operations of addition, subtraction, multiplication, and division. The calculator's memory capabilities allow you to perform more complex business mathematics operations without reentering amounts.

**OPERATING THE CALCULATOR**

The illustration below shows features found on most hand-held calculators. The position of keys and the method of operation may vary slightly from one calculator to another. If your calculator does not operate as described in this section, refer to your calculator's operating manual for directions.

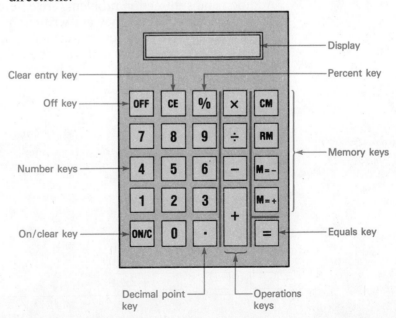

## OPERATING CONTROLS

The following operating controls are commonly found on hand-held calculators.

| ON/C | **On/clear key.** Turns on the calculator; clears all amounts from the calculator. |

**On/clear key.** Turns on the calculator; clears all amounts from the calculator.

$\boxed{\text{CE}}$    **Clear entry key.** Clears the last amount entered without clearing other amounts from the calculator.

$\boxed{+}$    **Plus key.** Adds an amount.

$\boxed{-}$    **Minus key.** Subtracts an amount.

$\boxed{+/-}$    **Change sign key.** Changes the sign of a number, either from a positive to a negative or a negative to a positive.

$\boxed{\times}$    **Multiplication key.** Enters a multiplication factor.

$\boxed{\div}$    **Dividend key.** Enters a dividend.

$\boxed{=}$    **Equals key.** Completes multiplication, division, or other calculation.

$\boxed{\%}$    **Percent key.** Multiplies an amount by a percent.

$\boxed{.}$    **Decimal point key.** Enters a decimal point.

$\boxed{M^+_=}$    **Memory equals-plus key.** Adds an amount, product, or quotient to the calculator's memory. Identified as $\boxed{M+}$ on some calculators.

$\boxed{M^-_=}$    **Memory equals-minus key.** Subtracts an amount, product, or quotient from the calculator's memory. Identified as $\boxed{M-}$ on some calculators.

$\boxed{RM}$    **Recall memory key.** Shows total stored in calculator's memory; does not clear memory.

$\boxed{CM}$    **Clear memory key.** Clears memory.

$\boxed{OFF}$    **Off key.** Turns off calculator.

## BASIC OPERATIONS

Listed below are step-by-step procedures for performing basic mathematics computations on your calculator.

Addition and subtraction problems are entered on the hand-held calculator in the same order as they are written.

When adding and subtracting decimals, enter a decimal point only when there is a decimal point in the number. The calculator will automatically place the decimal point in the answer.

### ADDITION

**EXAMPLE**

$5 + 9 + 8 =$

**SOLUTION**

Display

5 $\boxed{+}$ 9 $\boxed{+}$ 8 $\boxed{+}$ $\boxed{22}$

## SUBTRACTION

**EXAMPLE**

213 − 87 =

**SOLUTION**

Display

213 ⎡ + ⎤ ⎡ − ⎤ 87 ⎡ = ⎤ ⎡ 126 ⎤

Negative numbers can be entered first by pressing ⎡ ON/C ⎤ before pressing ⎡ − ⎤. If the answer is a negative number, a minus sign will appear after the answer on the display.

**EXAMPLE**

−49 + 8.6 + 13.06 =

**SOLUTION**

Display

⎡ − ⎤ 49 ⎡ + ⎤ 8.6 ⎡ + ⎤ 13.06 ⎡ = ⎤ ⎡ −27.34 ⎤

Negative numbers can also be entered by keying in the number and then pressing the ⎡ +/− ⎤ key. This key can change a positive number to a negative number when subtracting.

**EXAMPLE**

Find the total of checks written and the checking account balance. Beginning balance: $1,320.00. Checks: $320.40, $19.85, $102.10, $595.13.

**SOLUTION**

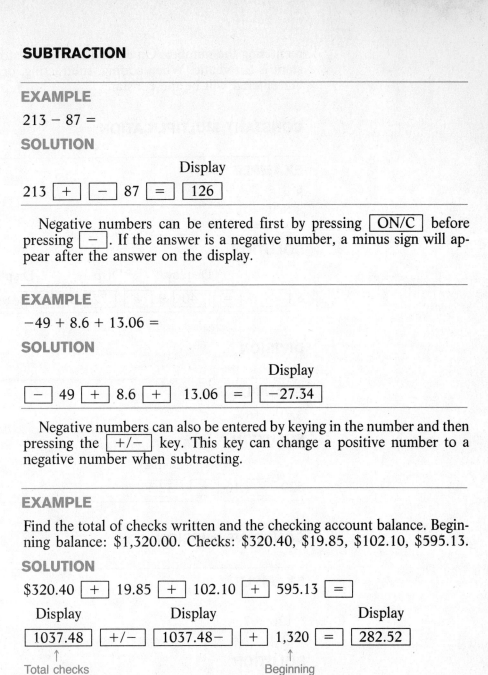

$320.40 ⎡ + ⎤ 19.85 ⎡ + ⎤ 102.10 ⎡ + ⎤ 595.13 ⎡ = ⎤

Display          Display          Display

⎡ 1037.48 ⎤ ⎡ +/− ⎤ ⎡ 1037.48− ⎤ ⎡ + ⎤ 1,320 ⎡ = ⎤ ⎡ 282.52 ⎤
↑                                      ↑
Total checks                        Beginning
                                    checkbook
                                    balance

## MULTIPLICATION

**EXAMPLE**

128.45 × 15 =

**SOLUTION**

Display

128.45 ⎡ × ⎤ 15 ⎡ = ⎤ ⎡ 1926.75 ⎤

In many business problems, it is useful to be able to add, subtract, multiply, or divide a number used in several problems without constantly reentering the number. On a hand-held calculator, operation with a constant is automatic. When adding, subtracting, or dividing, the first number entered will be the constant.

## CONSTANT MULTIPLICATION

### EXAMPLE

$8 \times 5 =$
$8 \times 9 =$
$8 \times 6 =$

### SOLUTION

Display      Display      Display

8 $\boxed{\times}$ 5 $\boxed{=}$ $\boxed{40}$ 9 $\boxed{=}$ $\boxed{72}$ 6 $\boxed{=}$ $\boxed{48}$

## DIVISION

### EXAMPLE

$520 \div 16 =$

### SOLUTION

Display

520 $\boxed{\div}$ 16 $\boxed{=}$ $\boxed{32.5}$

## CONSTANT DIVISION

### EXAMPLE

   $35 \div 7 =$
  $126 \div 7 =$
$185.5 \div 7 =$

### SOLUTION

Display      Display      Display

35 $\boxed{\div}$ 7 $\boxed{=}$ $\boxed{5}$ 126 $\boxed{=}$ $\boxed{18}$ 185.5 $\boxed{=}$ $\boxed{26.5}$

## MULTISTEP OPERATIONS

Problems that have a series of multiplication or division operations can be solved on a calculator. Combinations of multiplication and division operations also can be solved on a calculator.

### EXAMPLE 1

$2 \times 3 \times 8 =$

### SOLUTION

Display

2 $\boxed{\times}$ 3 $\boxed{\times}$ 8 $\boxed{=}$ $\boxed{48}$

## EXAMPLE 2

$8 \times 2 \div 5 =$

### SOLUTION

Display

$8$ $\boxed{\times}$ $2$ $\boxed{\div}$ $5$ $\boxed{=}$ $\boxed{3.2}$

## EXAMPLE 3

$136 \div 8 \times 4 =$

### SOLUTION

Display

$136$ $\boxed{\div}$ $8$ $\boxed{\times}$ $4$ $\boxed{=}$ $\boxed{68}$

### PRACTICE PROBLEMS

1. $346 + 109 + 872 =$
2. $\$1,489 + \$3,148 + \$621 =$
3. $208 - 149 =$
4. $\$2,168.41 - \$173.15 =$
5. $406.15 \times 3.2 =$
6. $\$5,289.60 \times 24 =$
7. $\$35.25 \times 7 =$
8. $\$35.25 \times 18 =$
9. $\$35.25 \times 12 =$
10. $145 \div 25 =$
11. $387.64 \div 5.5 =$
12. $\$31,900.75 \div 8.75 =$
13. $480 \div 15 =$
14. $10,275 \div 15 =$
15. $3,120 \div 15 =$
16. $15 \times 3 \times 9 =$
17. $35 \times 4 \div 5 =$
18. $420 \div 15 \times 9 =$
19. $19 \times 3 \times 26 =$
20. $108 \times 32 \div 75 =$

## PERCENT CALCULATIONS

Use $\boxed{\%}$ to find the percent of a number. By pressing $\boxed{\%}$ instead of $\boxed{=}$, it is not necessary to convert the percent to a decimal first.

### FINDING THE PART (PERCENTAGE)

#### EXAMPLE

What is 15% of $750?

#### SOLUTION

Display

$750$ $\boxed{\times}$ $15$ $\boxed{\%}$ $\boxed{112.5}$

ANSWERS FOR
PRACTICE PROBLEMS

(1) 1,327 (2) $5,258 (3) 59 (4) $1,995.26 (5) 1,299.68
(6) $126,950.40 (7) $246.75 (8) $634.50 (9) $423.00 (10) 5.8
(11) 70.48 (12) $3,645.80 (13) 32 (14) 685 (15) 208 (16) 405
(17) 28 (18) 252 (19) 1,482 (20) 46.08

## FINDING THE RATE (PERCENT)

**EXAMPLE**

$450 is what percent of $1,200?

**SOLUTION**

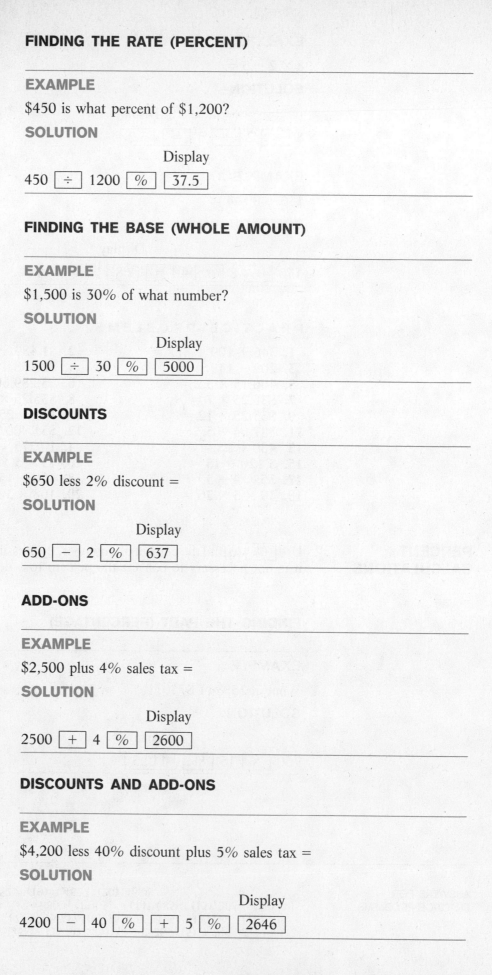

Display

450 ÷ 1200 % 37.5

## FINDING THE BASE (WHOLE AMOUNT)

**EXAMPLE**

$1,500 is 30% of what number?

**SOLUTION**

Display

1500 ÷ 30 % 5000

## DISCOUNTS

**EXAMPLE**

$650 less 2% discount =

**SOLUTION**

Display

650 − 2 % 637

## ADD-ONS

**EXAMPLE**

$2,500 plus 4% sales tax =

**SOLUTION**

Display

2500 + 4 % 2600

## DISCOUNTS AND ADD-ONS

**EXAMPLE**

$4,200 less 40% discount plus 5% sales tax =

**SOLUTION**

Display

4200 − 40 % + 5 % 2646

Round money answers to the nearest cent. Round percent answers to the nearest hundredth.

1. What is 25% of $475?
2. What is 37.5% of $1,600?
3. What is 105% of $245?
4. $125 is what percent of $550?
5. $24.50 is what percent of $187.50?
6. 14.5 is what percent of 80?
7. 105 is 14% of what number?
8. $560 is 35% of what amount?
9. $7 is 4% of what amount?
10. $111 is 37% of what amount?
11. What is $450 less 4% of $450?
12. What is $735.50 less 2% of $735.50?
13. What is $298.60 less 3% of $298.60?
14. What is $360 plus 12% of $360?
15. What is $89.80 plus 5% of $89.80?
16. What is $1,475 plus 60% of $1,475?
17. What is $1,750 less 40% discount plus 5% sales tax?
18. What is $800 less 35% discount plus 6% sales tax?
19. What is $2,890 less 48% discount plus 4% sales tax?
20. What is $3,500 less 44% discount plus 5% sales tax?

## MEMORY

The calculator's memory can be used to accumulate products, quotients, and individual amounts. Products, quotients, or individual amounts can be either added to or subtracted from the calculator's memory.

The **memory** feature allows you to do a part of a problem, store the answer, and then recall the stored answer later. This feature is useful when a previous answer is needed to solve another part of a problem. The memory on the calculator usually consists of four keys:

$\boxed{M^+_=}$    For adding a value to memory

$\boxed{M^-_=}$    For subtracting a value from memory

$\boxed{RM}$    For recalling the value in memory

$\boxed{CM}$    For clearing the memory

More complex operations that combine addition and subtraction with multiplication and division can be done using the memory. Before entering values, use $\boxed{CM}$ to clear the memory. When the memory keys are being used, an "M" appears on the calculator's display.

### EXAMPLE 1

($125 × 8) + ($206 × 13) =

### SOLUTION

Display

$\boxed{CM}$ 125 $\boxed{×}$ 8 $\boxed{M^+_=}$ 206 $\boxed{×}$ 13 $\boxed{M^+_=}$ $\boxed{RM}$ $\boxed{3678M}$

## EXAMPLE 2

$(175 \times 18) - (928 \div 16) =$

**SOLUTION**

Display

[CM] 175 [×] 18 [M⁺⁼] 928 [÷] 16 [M⁼] [RM] [3092M]

## EXAMPLE 3

$(464 \div 29) + 98 - (17 \times 4) - 20 =$

**SOLUTION**

Display

[CM] 464 [÷] 29 [M⁺⁼] 98 [M⁺⁼] 17 [×] 4 [M⁼] 20 [M⁼] [RM] [26M]

## EXAMPLE 4

$(46 \times 15) + (32 \times 8)$ less 20% of memory total =

**SOLUTION**

Display

[CM] 46 [×] 15 [M⁺⁼] 32 [×] 8 [M⁺⁼] [RM] [−] 20 [%] [756.8M]

### PRACTICE PROBLEMS

Round all money answers to the nearest cent.

1. $(45 \times 6) + (38 \times 9) =$
2. $(13.2 \times 4) - (1.9 \times 7) =$
3. $(127 \times 8) + (560 \div 35) =$
4. $(47 \times 6) + (364 \div 26) - (18 \times 3) =$
5. $(104 \times 13) + 356 + (540 \div 12) - 78 =$
6. $(\$12.25 \times 30) - \$4.50 + (\$8.98 \times 15) + \$9.75 =$
7. $(2,145 \div 33) + 105 - (192 \div 8) - 62 =$
8. $3,476 + (29 \times 16) - 1,321 + (4,920 \div 205) =$
9. $(\$12.75 \times 16) + (\$13.98 \times 15) - 20\%$ of memory total =
10. $(\$28.75 \times 45) + (\$17.75 \times 20) + 4\%$ of memory total =

## BUSINESS APPLICATIONS

Business mathematics calculations can be performed with a calculator by applying the procedures shown on the preceding pages. Often, more than one procedure can be used to obtain the correct answer. The solutions in this section show one of those acceptable procedures.

### GROSS PAY

**EXAMPLE**

Jena Loots is paid $9.75 per hour and is paid time and a half for hours worked over 40 in a week. Last week, Jena worked 46 hours. What was Jena's gross pay?

ANSWERS FOR
PRACTICE PROBLEMS

(1) 612 (2) 39.5 (3) 1,032 (4) 242 (5) 1.675 (6) $507.45 (7) 84 (8) 2,643 (9) $330.96 (10) $1,714.70

## SOLUTION

Display

| CM | 9.75 | × | 40 | M⁺₌ | 9.75 | × | 1.5 | × | 6 | M⁺₌ | RM | 477.75M |

### PRACTICE PROBLEMS

Time and a half is paid for all hours over 40 in a week. Calculate gross pay.

| | Regular Rate | Hours Worked | Gross Pay |
|---|---|---|---|
| 1. | $10.50 | 45 | $_____ |
| 2. | 8.25 | 42 | _____ |
| 3. | 10.95 | 44 | _____ |

## NET PAY

### EXAMPLE

Juan Lopez is paid $10.80 per hour and is paid time and a half for hours worked over 40 in a week. Last week Juan worked 45 hours and had the following deductions from his gross pay; FICA tax, 7.51% of gross pay; federal income tax, $64; and state income tax, $9.60. What was Juan's net pay?

### SOLUTION

Display

| CM | 10.80 | × | 40 | M⁺₌ | 10.80 | × | 1.5 | × | 5 | M⁺₌ | RM | 513 |

Display

| × | .0751 | M₌ | 64 | M₌ | 9.60 | M₌ | RM | 400.8737M |

### PRACTICE PROBLEMS

Time and a half is paid for all hours worked over 40 in a week. All gross pay is subject to the 7.51% FICA tax rate. Calculate net pay. Round to the nearest cent.

| | Regular Rate | Hours Worked | FICA Rate | Federal Income Tax | State Income Tax | Net Pay |
|---|---|---|---|---|---|---|
| 1. | $11.00 | 46 | 7.51% | $67 | $10.05 | $_____ |
| 2. | 9.85 | 45 | 7.51% | 62 | 9.30 | _____ |
| 3. | 10.55 | 44 | 7.51% | 65 | 9.75 | _____ |

## SIMPLE INTEREST

### EXAMPLE

Hardy Enterprises borrowed $5,000 at 12.5% simple interest to be repaid in 250 days. What is the interest on the loan and the total amount to be repaid?

### SOLUTION

Interest → Display

Amount to be repaid → Display

| CM | 5000 | $M^{\pm}$ | × | .125 | × | 250 | ÷ | 365 | 428.0821 | $M^{\pm}$ | RM | 5428.0821M |

### PRACTICE PROBLEMS

Calculate the amount of simple interest and total amount to be repaid for the following loans. Use a 365-day year.

| | Principal | Interest Rate | Days in Loan Period | Simple Interest | Total to Be Repaid |
|---|---|---|---|---|---|
| 1. | $3,750 | 13.0% | 200 | $_____ | $_____ |
| 2. | 9,500 | 11.5% | 90 | _____ | _____ |
| 3. | 1,800 | 10.5% | 300 | _____ | _____ |

## TRADE DISCOUNTS

### EXAMPLE

Bortz Company purchased goods with a $2,500 list price and trade discounts of 30%, 20%, and 10%. What is Bortz's cost for the goods?

### SOLUTION

Display

| 2500 | − | 30 | % | − | 20 | % | − | 10 | % | 1260 |

### PRACTICE PROBLEMS

Calculate the cost for each of the following purchases.

| | List Price | Series Trade Discounts | Cost |
|---|---|---|---|
| 1. | $3,000 | 40%, 10%, 5% | $_____ |
| 2. | 8,750 | 30%, 20%, 10% | _____ |
| 3. | 2,655 | 40%, 10%, 10% | _____ |

## CASH DISCOUNTS

### EXAMPLE

Bortz Company purchased goods at a cost of $1,260. Terms are 2/10, n/30 and Bortz paid within the discount period. What is the cash discount, and the cash price?

### SOLUTION

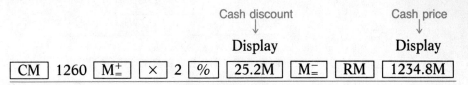

### PRACTICE PROBLEMS

Calculate the cash discount and cash price for the following purchases, all of which were paid within the cash discount period.

| | Net Price | Terms of Payment | Cash Discount | Cash Price |
|---|---|---|---|---|
| 1. | $2,000 | 2/10, n/30 | $_____ | $_____ |
| 2. | 1,975 | 3/10, n/30 | _____ | _____ |
| 3. | 8,650 | 4/10, n/60 | _____ | _____ |

## MARKUP

### EXAMPLE

Grems Sporting Goods Company purchased an exercise machine for $275 and marked it up 75% on cost. What is the selling price?

### SOLUTION

Display

275 [ + ] 75 [ % ] [ 481.25 ]

### PRACTICE PROBLEMS

Calculate the selling price for each of the following items. Markup is based on cost.

| | Cost | Markup on Cost | Selling Price |
|---|---|---|---|
| 1. | $450 | 60% | $_____ |
| 2. | 198 | 80% | _____ |
| 3. | 375 | 75% | _____ |

**ANSWERS FOR PRACTICE PROBLEMS**

**Top:**

| | Cash Discount | Cash Price |
|---|---|---|
| (1) | $ 40.00 | $1,960.00 |
| (2) | 59.25 | 1,915.75 |
| (3) | 346.00 | 8,304.00 |

**Bottom:** (1) $720.00   (2) $356.40   (3) $656.25

## MARKDOWN

### EXAMPLE

Hargate Clothing Store has a coat priced at $450. The coat is to be marked down 40% for a sale. What is the selling price?

### SOLUTION

Display

450 $\boxed{-}$ 40 $\boxed{\%}$ $\boxed{270}$

### PRACTICE PROBLEMS

Calculate the sales price for each of the following products.

| | Retail Price | Markdown | Sales Price |
|---|---|---|---|
| 1. | $398 | 25% | $_____ |
| 2. | 150 | 60% | _____ |
| 3. | 295 | 45% | _____ |

## INVOICE EXTENSIONS

### EXAMPLE

The following items were listed on an invoice for goods purchased by Calmar Company: 3 at $29.50 each; 8 at $19.95 each; and 6 at $32.45 each. Calmar received a 2% cash discount and must pay freight charges of $24.65. How much does Calmar owe on the invoice?

### SOLUTION

$\boxed{CM}$ 3 $\boxed{\times}$ 29.50 $\boxed{M^+_=}$ 8 $\boxed{\times}$ 19.95 $\boxed{M^+_=}$ 6 $\boxed{\times}$ 32.45 $\boxed{M^+_=}$

Display                    Display

$\boxed{RM}$ $\boxed{442.8}$ $\boxed{\times}$ 2 $\boxed{\%}$ $\boxed{M^-_=}$ 24.65 $\boxed{M^+_=}$ $\boxed{RM}$ $\boxed{458.594M}$

### PRACTICE PROBLEMS

Calculate the total amount due on each of the following invoices. Round answers to the nearest cent.

| Cost of Goods | Cash Discount | Freight Charges | Amount Due |
|---|---|---|---|
| 1. 6 @ $14.50<br>9 @ $12.98<br>5 @ $16.45 | 2% | $15.80 | $_____ |
| 2. 12 @ $5.65<br>25 @ $9.98<br>15 @ $7.40 | 3% | $12.30 | $_____ |

ANSWERS FOR PRACTICE PROBLEMS

**Top: (1)** $298.50    **(2)** $60.00    **(3)** $162.25

| | Cost of Goods | Cash Discount | Freight Charges | Amount Due |
|---|---|---|---|---|
| **3.** | 24 @ $16.95 | | | |
| | 18 @ $9.98 | | | |
| | 40 @ $12.25 | 2% | $29.75 | $_____ |

## SALES SLIP EXTENSIONS

### EXAMPLE

Salesperson Julie Bannon sold the following items to a customer: 5 at $1.98 each; 6 at $5.95 each; and 2 at $13.75 each. A 6% sales tax applies to the total sale. What is the total amount due from the customer?

### SOLUTION

5 $\boxed{\times}$ 1.98 $\boxed{M^{\pm}_{=}}$ 6 $\boxed{\times}$ 5.95 $\boxed{M^{\pm}_{=}}$ 2 $\boxed{\times}$ 13.75 $\boxed{M^{\pm}_{=}}$ $\boxed{RM}$

Display                    Display

$\boxed{73.1}$ $\boxed{+}$ 6 $\boxed{\%}$ $\boxed{77.486M}$

### PRACTICE PROBLEMS

Calculate total amount due from each of the following retail sales. Round answers to the nearest cent.

| | Items Sold | Sales Tax | Amount Due |
|---|---|---|---|
| **1.** | 6 @ $9.25 | | |
| | 3 @ $3.50 | | |
| | 5 @ $12.75 | 5% | $_____ |
| **2.** | 9 @ $4.25 | | |
| | 2 @ $1.75 | | |
| | 5 @ $6.80 | 6% | $_____ |
| **3.** | 4 @ $12.60 | | |
| | 6 @ $9.18 | | |
| | 8 @ $2.78 | 4% | $_____ |

## GAIN OR LOSS ON INVESTMENTS

### EXAMPLE

Marly Corporation sold 125 shares of HemiCo common stock at $37.50 per share and paid a $328 brokerage fee. Marly had purchased the stock at $22.00 per share and had paid a $192 brokerage fee. What was Marly's gain or loss on the sale of the stock?

---

**SOLUTION**

$\boxed{\text{CM}}$ 125 $\boxed{\times}$ 37.5 $\boxed{\text{M}^+_=}$ 328 $\boxed{\text{M}^-_=}$ 125 $\boxed{\times}$ 22 $\boxed{\text{M}^-_=}$ 192

Display←Gain

$\boxed{\text{M}^-_=}$ $\boxed{\text{RM}}$ $\boxed{\text{1417.5M}}$

---

## PRACTICE PROBLEMS

Calculate the gain or loss on each of the following common stock investments.

| | No. of Shares | Price per Share | Transaction | Brokerage Fee | Gain or Loss (Specify) |
|---|---|---|---|---|---|
| **1.** | 150 | $15.50 | Purchased | $162 | |
| | 150 | 28.75 | Sold | 297 | $_____ |
| **2.** | 300 | $35.25 | Purchased | $630 | |
| | 300 | 26.75 | Sold | 435 | $_____ |
| **3.** | 225 | $10.50 | Purchased | $165 | |
| | 225 | 29.25 | Sold | 418 | $_____ |

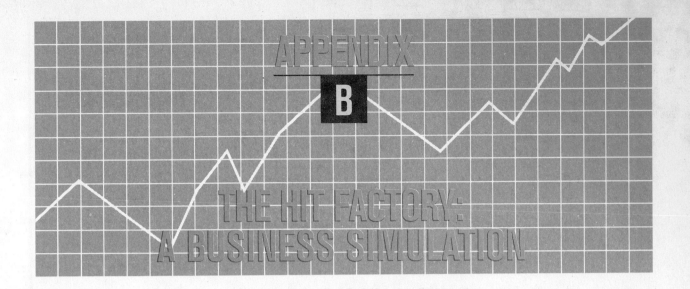

$\mathbf{A}$ssume that you are an employee for The Hit Factory, Inc., a company in the music industry, located in Nashville, Tennessee. Your duties consist of performing various types of business mathematics calculations. The calculations can be performed either manually or with a computer using the spreadsheet available for this simulation. All directions and instructions for calculations you are to make are provided to you on interoffice memos.

THE HIT FACTORY, INC.

3600 MUSIC ROW WEST, NASHVILLE TENNESSEE 37203

October 13, 19--

To Our Newest Employee:

Welcome to The Hit Factory.  I hope that today is the beginning
of a long and mutually beneficial relationship.  Primarily, The
Hit Factory is a professional recording company.  We do,
however, perform three separate activities as described below:

1.  We sign recording artists to contracts.  We record the
    artists in our studio and produce compact discs, cassettes,
    albums, videos, and 45 RPM records for promotion and
    distribution nationwide.  Contemporary, pop, and rock music
    is released on our Cannon label.  Country-western music is
    released on our Nickelodeon label.

2.  We provide custom recording services.  That is, any
    individual or group can rent our recording facilities to
    make their own recordings.  In this case, The Hit Factory
    has no involvement in the making of CDs, records or tapes,
    or in their promotion and distribution.  The recording
    artists handle all of those activities themselves, or they
    hire someone to do it.

3.  We operate a retail music outlet called The Hit Factory
    Music Store.  It's next door to our office and recording
    studio facilities.  The address is 3604 Music Row West.

Your position will be as a Financial Assistant.  Your
activities will primarily involve making mathematical
calculations and analyses.  There is one important aspect I
must impress upon you from your very first day on the job.  The
calculations you perform and the results you obtain must be
absolutely, positively accurate.  Since others within our
company, our recording artists, and our customers will rely on
your calculations, inaccuracies might cause The Hit Factory a
great deal of financial loss and embarrassment.

Lacy Montgomery, the company controller, will be your
supervisor, but you will occasionally be asked by others to
perform calculations as well.

I hope that your association with The Hit Factory will be a
pleasant and beneficial one indeed.

                    Sincerely,

                    T. J. Souther

                    T. J. Souther
                    President

## INTEROFFICE MEMO

THE HIT FACTORY, INC.

TO: Financial Assistant

FROM: Lacy Montgomery, Controller

DATE: October 16, 19--

A summary of each custom recording session sold last month is listed below. Please calculate the total income from each session and supply the data I have requested at the bottom of this memo.

| Session No. | Artist | Type Music | Hours Recording Time | Rate Per Hour | Income |
|---|---|---|---|---|---|
| 1 | Gloria Rhodes | Country | 15 | $125 | _____ |
| 2 | Bamboozle | Pop/Rock | 6 | 125 | _____ |
| 3 | Johnny Dollar | Country | 24 | 125 | _____ |
| 4 | Daydreamer | Pop/Rock | 16 | 125 | _____ |
| 5 | Mickey Silvertone | Country | 28 | 125 | _____ |
| 6 | Caravan | Country | 14 | 125 | _____ |
| 7 | Story Brook Band | Pop/Rock | 20 | 125 | _____ |
| 8 | Waterfall | Pop/Rock | 35 | 125 | _____ |
| 9 | Slow Poke | Country | 12 | 125 | _____ |
| 10 | Mandy | Pop/Rock | 42 | 125 | _____ |
| 11 | Rawhide | Country | 10 | 125 | _____ |
| | | | TOTALS: ___ | | _____ |

Average income per session: _____

Percent of total income from Pop/Rock music groups: _____

Percent of total income from Country music groups: _____

---

## MEMO

FROM THE DESK OF
TJS

*Financial Assistant—
Please make those
calculations for me.
LM*

TO: Lacy Montgomery

DATE: October 17, 19--

We've got a runaway hit on our hands! The Monarchs' "Guitar Pickin' Man" has sold 275,000 copies in its first 8 days. We've got to revise our projections--fast!

Based on new estimated sales of 1,800,000 compact discs, cassettes, and albums at an average of $4.87 each (wholesale), calculate the dollar amount we should budget for the following. All percents should be based on estimated total sales income.

| Item | Percent to Be Budgeted | Budget Amount |
|---|---|---|
| Promotional Tour Sponsorship | 4.0% | _____ |
| Magazine Advertising | 6.5% | _____ |
| Billboard Advertising | 1.4% | _____ |
| Miscellaneous Advertising | 2.5% | _____ |
| Independent Promotions Firm (Hiring) | 3.2% | _____ |
| TOTALS: | ___ | _____ |

# INTEROFFICE MEMO

THE HIT FACTORY, INC.

TO: Financial Assistant

FROM: Lacy Montgomery, Controller

DATE: October 18, 19--

In about three weeks we will begin preparing a mass mailing to promote the J. Lilly Band's new release, "Wild and Wooly." It will be sent to 4,300 AM radio stations, 2,100 FM radio stations, 250 newspapers, and 100 magazines.

Temporary employees will be hired to address and package the materials. I project that a worker can prepare and package one mailing every two minutes. If we hire workers at an hourly rate, we will pay $6.50 per hour. If we pay on a piecework basis, I think 25 cents per mailing would be fair.

Calculate our total cost for preparing the mailing if we pay on the hourly basis--or on the piecework basis.

Cost:  Hourly basis:          _____

Piecework basis:          _____

TO: Lacy Montgomery, Controller

DATE: October 24, 19--

Our current contract with the Monarchs will expire soon.  When
we first signed them nearly three years ago, a 10 percent
royalty rate was appropriate, since they were relatively
unknown.  Now, however, they are <u>hot</u>, and I presume their
manager will want a large cash bonus for signing a new
contract, plus a substantial increase in the royalty rate.
Please prepare some projections at varying cash bonus amounts
and royalty rates for a new three-year contract with the
Monarchs.  Thanks.

Financial Assistant —
    On the attached sheet, I have prepared a projection
for sales of products by the Monarchs for the next
three years. Please calculate the amount of royalty
income we would pay the Monarchs and also
complete the percent analysis indicated.
                                LM

Projected Sales and Royalty Income
3-Year Period
The Monarchs

| | | | | Monarchs' Royalty Income | | |
| Product | Projected Net Sales Volume (Units) | Whole-sale Price Each | Net Product Sales Income | Proposal A 12% Royalty | Proposal B 15% Royalty | Proposal C 18% Royalty |
| --- | --- | --- | --- | --- | --- | --- |
| Albums | 2,250,000 | $3.76 | _____ | _____ | _____ | _____ |
| Cassettes | 3,750,000 | 3.95 | _____ | _____ | _____ | _____ |
| Compact Discs | 4,650,000 | 5.67 | _____ | _____ | _____ | _____ |
| Videos | 650,000 | 4.18 | _____ | _____ | _____ | _____ |
| 45 RPMs | 9,850,000 | .85 | _____ | _____ | _____ | _____ |
| | | TOTALS: | _____ | _____ | _____ | _____ |

Calculate the percent of our total net record sales income that the Monarchs would receive under each of the three proposals if we pay a $3,000,000 cash bonus.

| Proposal | Cash Bonus | + | Royalty Income | = | Monarchs' Income | Percent That Monarchs' Income Is of Total Net Record Sales Income |
| --- | --- | --- | --- | --- | --- | --- |
| A | _____ | + | _____ | = | _____ | _____ |
| B | _____ | + | _____ | = | _____ | _____ |
| C | _____ | + | _____ | = | _____ | _____ |

Make the same calculation, assuming this time that a $1,000,000 cash bonus is paid.

| Proposal | Cash Bonus | + | Royalty Income | = | Monarchs' Income | Percent That Monarchs' Income Is of Total Net Record Sales Income |
| --- | --- | --- | --- | --- | --- | --- |
| A | _____ | + | _____ | = | _____ | _____ |
| B | _____ | + | _____ | = | _____ | _____ |
| C | _____ | + | _____ | = | _____ | _____ |

**MEMO**          FROM THE DESK OF
                        TJS

*Financial Assistant –*
*Please calculate.*
*LM*

TO:  Lacy Montgomery

DATE:  October 25, 19--

In view of the excellent reviews received for Gina McGovern's new release, <u>Genuine Gina</u>, I believe we should revise our promotional budget for the project.  Please calculate the new budgeted amounts based on the percent of increase or decrease I have indicated.

| <u>Promotional Activity</u> | Current <u>Budget</u> | <u>Change</u> | Revised Budget <u>(Amount)</u> |
|---|---|---|---|
| Promotional Tour Sponsorship | $ 30,000 | Increase 250% | _____ |
| Magazine Advertising | 25,500 | Increase 135% | _____ |
| Billboard Advertising | 18,800 | Decrease 15% | _____ |
| Independent Record Promotions Firm | 15,600 | Increase 45% | _____ |
| Telemarketing | 8,250 | Increase 60% | _____ |
| Miscellaneous | <u>16,500</u> | Decrease 40% | _____ |
| TOTALS | _____ | | _____ |

Amount of increase from current budget to revised budget:  _____

Percent of increase from current budget to revised budget:  _____

---

**INTEROFFICE MEMO**

THE HIT FACTORY, INC.

TO:  Financial Assistant

FROM:  Lacy Montgomery

DATE:  October 27, 19--

Shown below are budgeted and actual costs for production of the Sanchez Sisters' video, <u>South by Southwest</u>.  Please calculate the amounts and percents indicated.

| Production <u>Item</u> | Budgeted <u>Amount</u> | Actual <u>Amount</u> | Amount Actual Above (+) or Below (–) Budget | Percent Actual Above (+) or Below (–) Budget |
|---|---|---|---|---|
| Sound Stage Rent | $ 3,500 | $ 3,750 | _____ | _____ |
| Set Design | 4,750 | 4,465 | _____ | _____ |
| Lighting | 800 | 950 | _____ | _____ |
| Special Effects | 1,275 | 1,195 | _____ | _____ |
| Producer's Fee | 15,000 | 16,500 | _____ | _____ |
| Director's Fee | 10,000 | 10,000 | _____ | _____ |
| Camera Operators | 2,500 | 2,675 | _____ | _____ |
| Professional Actors | 8,500 | 9,250 | _____ | _____ |
| Extras (Actors) | 2,000 | 1,850 | _____ | _____ |
| Editing/Sound Mixing | 6,500 | 8,950 | _____ | _____ |
| Film/Supplies | 1,500 | 1,360 | _____ | _____ |
| Costumes | 6,000 | 5,200 | _____ | _____ |
| Transportation | 2,000 | 2,700 | _____ | _____ |
| Food | 1,200 | 1,685 | _____ | _____ |
| Lodging | <u>3,500</u> | <u>3,320</u> | _____ | _____ |
| TOTALS: | _____ | _____ | _____ | _____ |

# INTEROFFICE MEMO

THE HIT FACTORY, INC.

M U S I C · S T O R E

TO: Financial Assistant

FROM: Bill DeLoss, Manager

DATE: October 31, 19--

Lacy Montgomery said I should send this data to you for calculation from now on.

Enclosed is last pay period's payroll register for the staff at The Hit Factory Music Store and federal and state income tax withholding tables. Since you are new to Tennessee, I should explain that the same state income tax withholding table is used for both single and married employees. The table is used exactly like the federal withholding tables.

Each employee's total earnings are subject to the 7.51 percent FICA tax rate.

Be sure to double-check your work--there is <u>nothing</u> that will make an employee madder than a fouled-up paycheck!

Thanks.

---

The Hit Factory
Payroll Register

| NAME | Marital Status | Num. With. | Hours | Hourly Rate | Reg. | Over-Time | Gross Pay | FICA Tax | Fed. Inc. Tax | State Inc. Tax | Other | Total Deduct. | Net Pay |
|------|----------------|------------|-------|-------------|------|-----------|-----------|----------|---------------|----------------|-------|---------------|---------|
| Bandow, A. | S | 1 | 40 | 6.75 | | | | | | | 0.00 | | |
| DeLoss, B. | M | 4 | 43 | 12.25 | | | | | | | 15.00 | | |
| Graham, S. | M | 3 | 45 | 6.25 | | | | | | | 10.00 | | |
| Keir, R. | S | 1 | 45 | 5.75 | | | | | | | 5.00 | | |
| Mikes, P. | S | 0 | 38 | 7.25 | | | | | | | 5.00 | | |
| Peters, J. | S | 1 | 44 | 6.50 | | | | | | | 10.00 | | |
| Wagner, L. | S | 1 | 43 | 8.15 | | | | | | | 20.00 | | |
| Zuccerello, D. | M | 2 | 40 | 7.95 | | | | | | | 15.00 | | |

For Period Ending : Oct. 27, 19--

# FEDERAL INCOME TAX WITHHOLDING TABLE
## WEEKLY PAYROLL PERIOD — SINGLE PERSONS

| And the wages are– | | And the number of withholding allowances claimed is– | | | | | | | | | | |
|---|---|---|---|---|---|---|---|---|---|---|---|---|
| At least | But less than | 0 | 1 | 2 | 3 | 4 | 5 | 6 | 7 | 8 | 9 | 10 |
| | | The amount of income tax to be withheld shall be– | | | | | | | | | | |
| 220 | 230 | 31 | 25 | 20 | 14 | 9 | 3 | | | | | |
| 230 | 240 | 32 | 27 | 21 | 16 | 10 | 5 | | | | | |
| 240 | 250 | 34 | 28 | 23 | 17 | 12 | 6 | 1 | | | | |
| 250 | 260 | 35 | 30 | 24 | 19 | 13 | 8 | 3 | | | | |
| 260 | 270 | 37 | 31 | 26 | 20 | 15 | 9 | 4 | | | | |
| 270 | 280 | 38 | 33 | 27 | 22 | 16 | 11 | 5 | 1 | | | |
| 280 | 290 | 40 | 34 | 29 | 23 | 18 | 12 | 7 | 2 | | | |
| 290 | 300 | 41 | 36 | 30 | 25 | 19 | 14 | 8 | 3 | | | |
| 300 | 310 | 43 | 37 | 32 | 26 | 21 | 15 | 10 | 4 | | | |
| 310 | 320 | 44 | 39 | 33 | 28 | 22 | 17 | 11 | 6 | 1 | | |
| 320 | 330 | 46 | 40 | 35 | 29 | 24 | 18 | 13 | 7 | 2 | | |
| 330 | 340 | 47 | 42 | 36 | 31 | 25 | 20 | 14 | 9 | 3 | | |
| 340 | 350 | 50 | 43 | 38 | 32 | 27 | 21 | 16 | 10 | 5 | | |
| 350 | 360 | 53 | 45 | 39 | 34 | 28 | 23 | 17 | 12 | 6 | 2 | |
| 360 | 370 | 55 | 46 | 41 | 35 | 30 | 24 | 19 | 13 | 8 | 3 | |
| 520 | 530 | 100 | 90 | 80 | 69 | 59 | 49 | 43 | 37 | 32 | 26 | 21 |
| 530 | 540 | 103 | 93 | 83 | 72 | 62 | 52 | 44 | 39 | 33 | 28 | 22 |
| 540 | 550 | 107 | 96 | 85 | 75 | 65 | 55 | 46 | 40 | 35 | 29 | 24 |
| 550 | 560 | 110 | 98 | 88 | 78 | 68 | 57 | 47 | 42 | 36 | 31 | 25 |
| 560 | 570 | 114 | 101 | 91 | 81 | 70 | 60 | 50 | 43 | 38 | 32 | 27 |

# FEDERAL INCOME TAX WITHHOLDING TABLE
## WEEKLY PAYROLL PERIOD — MARRIED PERSONS

| And the wages are– | | And the number of withholding allowances claimed is– | | | | | | | | | | |
|---|---|---|---|---|---|---|---|---|---|---|---|---|
| At least | But less than | 0 | 1 | 2 | 3 | 4 | 5 | 6 | 7 | 8 | 9 | 10 |
| | | The amount of income tax to be withheld shall be– | | | | | | | | | | |
| 240 | 250 | 29 | 24 | 18 | 13 | 7 | 3 | | | | | |
| 250 | 260 | 31 | 25 | 20 | 14 | 9 | 4 | | | | | |
| 260 | 270 | 32 | 27 | 21 | 16 | 10 | 5 | 1 | | | | |
| 270 | 280 | 34 | 28 | 23 | 17 | 12 | 6 | 2 | | | | |
| 280 | 290 | 35 | 30 | 24 | 19 | 13 | 8 | 3 | | | | |
| 290 | 300 | 37 | 31 | 26 | 20 | 15 | 9 | 4 | | | | |
| 300 | 310 | 38 | 33 | 27 | 22 | 16 | 11 | 6 | 1 | | | |
| 310 | 320 | 40 | 34 | 29 | 23 | 18 | 12 | 7 | 3 | | | |
| 320 | 330 | 41 | 36 | 30 | 25 | 19 | 14 | 8 | 4 | | | |
| 330 | 340 | 43 | 37 | 32 | 26 | 21 | 15 | 10 | 5 | 1 | | |
| 340 | 350 | 44 | 39 | 33 | 28 | 22 | 17 | 11 | 6 | 2 | | |
| 350 | 360 | 46 | 40 | 35 | 29 | 24 | 18 | 13 | 7 | 3 | | |
| 360 | 370 | 47 | 42 | 36 | 31 | 25 | 20 | 14 | 9 | 4 | | |
| 370 | 380 | 49 | 43 | 38 | 32 | 27 | 21 | 16 | 10 | 5 | 1 | |
| 380 | 390 | 50 | 45 | 39 | 34 | 28 | 23 | 17 | 12 | 6 | 2 | |
| 540 | 550 | 74 | 69 | 63 | 58 | 52 | 47 | 41 | 36 | 30 | 25 | 19 |
| 550 | 560 | 76 | 70 | 65 | 59 | 54 | 48 | 43 | 37 | 32 | 26 | 21 |
| 560 | 570 | 77 | 72 | 66 | 61 | 55 | 50 | 44 | 39 | 33 | 28 | 22 |
| 570 | 580 | 79 | 73 | 68 | 62 | 57 | 51 | 46 | 40 | 35 | 29 | 24 |
| 580 | 590 | 81 | 75 | 69 | 64 | 58 | 53 | 47 | 42 | 36 | 31 | 25 |

# STATE OF TENNESSEE
Income Tax Withholding Table
Single and Married Persons
Weekly Payroll Period

| And the wages are– | | And the number of withholding allowances claimed is– | | | | | | | | | | |
|---|---|---|---|---|---|---|---|---|---|---|---|---|
| At least | But less than | 0 | 1 | 2 | 3 | 4 | 5 | 6 | 7 | 8 | 9 | 10 |
| | | The amount of income tax to be withheld shall be– | | | | | | | | | | |
| $250 | $260 | $ 4.65 | $ 3.75 | $ 3.00 | $ 2.10 | $ 1.35 | $ .60 | | | | | |
| 260 | 270 | 4.80 | 4.05 | 3.15 | 2.40 | 1.50 | .75 | .15 | | | | |
| 270 | 280 | 5.10 | 4.20 | 3.45 | 2.55 | 1.80 | .90 | .30 | | | | |
| 280 | 290 | 5.25 | 4.50 | 3.60 | 2.85 | 1.95 | 1.20 | .45 | | | | |
| 290 | 300 | 5.55 | 4.65 | 3.90 | 3.00 | 2.25 | 1.35 | .60 | | | | |
| 300 | 310 | 5.70 | 4.95 | 4.05 | 3.30 | 2.40 | 1.65 | .90 | .15 | | | |
| 310 | 320 | 6.00 | 5.10 | 4.35 | 3.45 | 2.70 | 1.80 | 1.05 | .45 | | | |
| 320 | 330 | 6.15 | 5.40 | 4.50 | 3.75 | 2.85 | 2.10 | 1.20 | .60 | | | |
| 350 | 360 | 6.90 | 6.00 | 5.25 | 4.35 | 3.60 | 2.70 | 1.95 | 1.05 | .45 | | |
| 360 | 370 | 7.05 | 6.30 | 5.40 | 4.65 | 3.75 | 3.00 | 2.10 | 1.35 | .60 | | |
| 370 | 380 | 7.35 | 6.45 | 5.70 | 4.80 | 4.05 | 3.15 | 2.40 | 1.50 | .75 | | |
| 530 | 540 | 10.80 | 10.20 | 9.15 | 8.55 | 7.50 | 6.90 | 6.85 | 5.25 | 4.30 | 3.60 | 2.55 |
| 540 | 550 | 11.10 | 10.35 | 9.45 | 8.70 | 7.80 | 7.05 | 6.65 | 5.40 | 4.50 | 3.75 | 2.85 |
| 550 | 560 | 11.40 | 10.75 | 9.75 | 8.85 | 8.10 | 7.20 | 6.45 | 5.55 | 4.80 | 3.90 | 3.15 |

# INTEROFFICE MEMO

## THE HIT FACTORY, INC.

TO: Financial Assistant

FROM: Lacy Montgomery

DATE: November 6, 19--

Our promotions staff consists of eight people who call on radio stations to introduce our new products and contact wholesalers and large retailers to encourage sales and to help arrange special promotions. Currently, each is paid a salary, which varies from person to person depending on their years of experience and size of territory.

Mr. Souther wants us to project how much we would pay each person if we paid them a salary plus a bonus of two percent of all sales made above a quota. Please make the following calculations, based on last year's performance. Round the bonus to the nearest dollar.

| Promotions Staff | Current Annual Salary | Annual Product Sales | Annual Sales Quota | Salary | Bonus Earned | Projected Annual Earnings |
|---|---|---|---|---|---|---|
| S. Dodd | $32,650 | $3,782,065 | $3,500,000 | $25,000 | _____ | _____ |
| T. Lang | 30,075 | 2,147,614 | 2,250,000 | 25,000 | _____ | _____ |
| P. Norris | 28,450 | 2,621,418 | 2,500,000 | 25,000 | _____ | _____ |
| K. Paulson | 28,950 | 2,050,970 | 2,000,000 | 25,000 | _____ | _____ |
| J. Lipka | 44,575 | 5,884,624 | 5,000,000 | 35,000 | _____ | _____ |
| O. Mendoza | 35,690 | 3,056,382 | 2,800,000 | 35,000 | _____ | _____ |
| K. Peta | 38,250 | 4,785,416 | 4,200,000 | 35,000 | _____ | _____ |
| A. Sayers | 34,285 | 2,418,730 | 2,400,000 | 35,000 | _____ | _____ |
| TOTALS | _____ | _____ | _____ | _____ | _____ | _____ |

# INTEROFFICE MEMO

THE HIT FACTORY, INC.
M U S I C · S T O R E

**TO:** Financial Assistant

**FROM:** Bill DeLoss, Manager

**DATE:** November 13, 19--

Enclosed are photocopies of Downtown Sound's trade discount sheet and selected portions of their catalog. Please calculate our purchase price for the items I have indicated and complete the enclosed purchase order. As usual, the goods should be shipped via Carrington Cartage.

---

## DOWNTOWN SOUND CATALOG

*Order 20*

**CASSETTE TAPE CASE** — Holds 24 cassettes. Black vinyl exterior; red velvet interior — delux.
No. TJ-4760          List Price — $18.00

*Order 15*

**ACOUSTIC GUITAR** — Jumbo size, natural spruce top and back, rosewood sides, steel reinforced neck, high-gloss finish, steel strings.
No. GM-3500          List Price — $340.00

*Order 10*

**WESTERN FIDDLE** — Full size, dark finish, precision tuning pegs, solid hardwood top, sides, and back. Includes professional bow and hard shell carrying case.
No. M-1800          List Price — $260.00

# TRADE DISCOUNT SHEET

## DOWNTOWN SOUND
8614 Presley Way
Memphis, TN  38115

Effective September 1, 19--

All terms 2/10, n/30

| Catalog No. | Trade Discount |
|---|---|
| GM-3300 | 40%, 10% |
| GM-3500 | 40%, 10%, 5% |
| GM-4000 | 50% |
| | |
| M-1000 | 40%, 5% |
| M-1800 | 40%, 5% |
| M-2000 | 40%, 10% |
| | |
| TJ-4750 | 30%, 20% |
| TJ-4760 | 30%, 20% |
| TJ-4770 | 40% |

## PURCHASE ORDER

TO:                                              FROM:

DATE:

SHIP VIA:                                        TERMS:

| Quantity | Description | Catalog Number | Price Each | Amount |
|---|---|---|---|---|
| | | | | |
| | | | | |
| | | | | |
| | | | | |

*After trade discounts are deducted.            TOTAL  $ _____

THE HIT FACTORY, INC.

TO: *Financial Assistant*

FROM: *L. M.*

DATE: *November 14, 19—*

*The office worker in charge of the miscellaneous office checking account fell in love about a month ago and hasn't worked since. (He still shows up daily and gets paid, however.)*

*He can't get the Bank Reconciliation for the account to come out. I suspect the balance in the check register is probably wrong — for one thing. Please check the balance and make any necessary corrections.*

| ❤ Mary | | CHECK REGISTER | | | ❤ MARY |
| --- | --- | --- | --- | --- | --- |
| Check Number | Date | Description | Amount of Check | Deposit | Balance |
| | 10-11 | Balance Forwarded | — | — | 121.42 |
| 306 | 10-12 | Music City Stationery | 15.07 | | 106.35 |
| 307 | 10-13 | Flowers By Kathy | 22.15 | | 84.20 |
| ~~308~~ VOID | VOID 10-17 | ~~Mary Moreau~~ VOID | VOID 31.10 | — | 53.10 |
| 309 | 10-17 | Bob's Bake Shop | 31.10 | 35.00 | 49.20 |
| 310 | 10-20 | Music City Stationery | 4.17 | | 45.03 |
| 311 | 10-23 | Boy Scouts | 10.00 | | 35.03 |
| 312 | 10-24 | Speedy Delivery | 8.75 | | 26.28 |
| 313 | 10-27 | Flowers By ~~Mary~~ Kathy | 15.00 | 50.00 | - 8.72 |
| 314 | 11-1 | Red's Office Supply | 7.27 | 50.00 | 34.01 |
| 315 | 11-3 | Speedy Delivery | 12.30 | | 21.71 |
| 316 | 11-8 | Postmaster | 18.00 | | 3.71 |
| 317 | 11-10 | Music City Stationery | 12.08 | 50.00 | 41.63 |

# INTEROFFICE MEMO

THE HIT FACTORY, INC.

M U S I C · S T O R E

TO: Financial Assistant

FROM: Bill DeLoss, Manager

DATE: November 16, 19--

Through comparison shopping, I have determined various prices at which we can sell some of our products and be in line with what our competitors are charging.

From the guidelines furnished to me by Lacy M., I know what percent of markup is needed, based on our selling price.

I have two questions: (1) At what price do we need to buy these products from our suppliers? (2) What percent is markup on cost, rounded to the nearest percent?

| Product | Projected Retail Sales Price | Required Markup Based on Selling Price | Projected Cost | Percent Markup Is of Cost |
|---------|------------------------------|----------------------------------------|----------------|---------------------------|
| Banjo | $ 429.99 | 45% | _____ | _____ |
| Dobro | 459.99 | 48% | _____ | _____ |
| Mandolin | 239.99 | 42% | _____ | _____ |
| Steel Guitar | 2,499.99 | 40% | _____ | _____ |
| Organ (Portable) | 2,874.99 | 46% | _____ | _____ |
| Harmonica | 49.99 | 40% | _____ | _____ |
| Harmonica | 99.99 | 38% | _____ | _____ |
| Acoustic Guitar | 499.99 | 44% | _____ | _____ |
| Tambourine | 47.99 | 37% | _____ | _____ |
| Mikado CD Player | 439.98 | 35% | _____ | _____ |

# INTEROFFICE MEMO

THE HIT FACTORY, INC.

*Financial Assistant—*
*Please prepare a bank*
*reconciliation in rough*
*draft from this data.*
*Use the attached form.*
*LM*

TO:   Lacy Montgomery

FROM:  Ken Kristopherson, Accounting

DATE:  November 20, 19--

I have reviewed the Record Productions Account checkbook and bank statement as you requested.  The information is summarized below:

| | |
|---|---|
| Beginning checkbook balance: | $ 79,535.36 |
| Ending checkbook balance: | 71,665.94 |
| Beginning bank statement balance: | 124,111.55 |
| Ending bank statement balance: | 113,006.13 |

Cancelled checks returned with the bank statement:

| No. | Amount | No. | Amount | No. | Amount |
|---|---|---|---|---|---|
| 714 | $    875.25 | 719 | $144,315.20 | 727 | $ 15,125.14 |
| 715 | 6,415.80 | 720 | 4,610.20 | 728 | 175,016.42 |
| 716 | 37,285.14 | 722 | 987.45 | 729 | 53,247.00 |
| 717 | 74,380.20 | 723 | 287,541.80 | | |
| 718 | 356.82 | 724 | 10,215.20 | | |

Checks written but not presented to the bank for payment:

| No. | Amount | No. | Amount | No. | Amount |
|---|---|---|---|---|---|
| 721 | $10,976.15 | 726 | $37,811.39 | 731 | $9,370.15 |
| 725 | 4,201.16 | 730 | 16,182.40 | | |

Deposits recorded in both checkbook and bank statement: $784,316.20

Deposit recorded in checkbook--not shown on bank statement: $52,100.20

Promissory note collected by bank--shown on bank statement, not on the checkbook:  $15,000.00

Collection fee charged by bank, shown on bank statement, not in checkbook:  $50.00

Interest earned--shown on bank statement, not in checkbook: $399.14

Check No. 720 was recorded in the checkbook as $4,160.20.

There were no errors on the bank statement.

# BANK RECONCILIATION STATEMENT

Depositor _____

_____

| BALANCE SHOWN ON BANK STATEMENT | $_____ |
|---|---|

Plus: Unrecorded Deposits

| Date | Amount | |
|---|---|---|
| | | |
| | | |
| | | |
| Total Unrecorded Deposits | | $ |

SUBTOTAL  $_____

Less: Checks Outstanding

| Number | Amount | |
|---|---|---|
| | | |
| | | |
| | | |
| | | |
| | | |
| | | |
| | | |
| | | |
| Total Checks Outstanding | | $ |

ADJUSTED BANK BALANCE  $_____

| BALANCE SHOWN ON CHECKBOOK STUB | $_____ |
|---|---|

Plus: Additions and Corrections

| Description | Amount | |
|---|---|---|
| | | |
| | | |
| | | |
| Total Additions and Corrections | | $ |

SUBTOTAL  $_____

Less: Charges, Fees, and Corrections

| Description | Amount | |
|---|---|---|
| Service Charge | | |
| | | |
| | | |
| | | |
| | | |
| | | |
| | | |
| | | |
| Total Deductions | | $ |

ADJUSTED CHECKBOOK BALANCE  $_____

## INTEROFFICE MEMO

THE HIT FACTORY, INC.
MUSIC·STORE

TO: Financial Assistant

FROM: Bill DeLoss, Manager

DATE: November 21, 19--

We need to develop some type of standardized pricing system for products sold at the Music Store. It seems a markup of 70 percent to 85 percent on our cost on musical instruments will make us very competitive with other stores in the area and still allow an adequate profit.

Please calculate what the selling price should be for each of the following products, basing markup on our cost. Also determine what percent the markup is of our selling price. Round your percents to the nearest hundredth.

| Product | Inventory Number | Cost | Percent of Markup on Cost | Selling Price | Percent Markup Is of Selling Price |
|---------|------------------|------|---------------------------|---------------|-------------------------------------|
| Banjo | M-1634 | $ 220.49 | 70% | _____ | _____ |
| Ukelele | M-2016 | 69.98 | 80% | _____ | _____ |
| Guitar | M-1305 | 284.65 | 75% | _____ | _____ |
| Harpsichord | M-781 | 70.50 | 70% | _____ | _____ |
| Guitar | M-1308 | 715.99 | 85% | _____ | _____ |
| Mandolin | M-2018 | 160.50 | 82% | _____ | _____ |
| Harmonica | M-6354 | 28.50 | 72% | _____ | _____ |
| Electric Piano | M-4639 | 1,215.25 | 65% | _____ | _____ |
| Electric Piano | M-4643 | 1,565.75 | 68% | _____ | _____ |

## INTEROFFICE MEMO

THE HIT FACTORY, INC.

TO: Financial Assistant

FROM: Lacy M.

DATE: November 22, 19--

We will probably need to borrow $125,000.00 to pay for production costs on the new Donna DeLorean project. This is not unusual for us since there is a considerable time lag from when we must pay the CD manufacturers and record pressers to when we receive payment from the wholesalers we sell to.

I anticipate we will borrow the money on December 4 and will be able to repay on either January 31 or February 14 of next year.

Interest rates have been fluctuating widely, so we cannot say for sure what rate can be negotiated. Therefore, calculate what the amount of interest will be at the various rates shown below, for repayment on each of the projected dates. No doubt, the lender will use a 365-day year and the simple interest formula to calculate interest.

| Interest Rate | Amount of Interest | |
|---------------|--------------------|---|
| | Loan Repaid Jan. 31 | Loan Repaid Feb. 14 |
| 10.75% | _____ | _____ |
| 11.00% | _____ | _____ |
| 11.25% | _____ | _____ |
| 11.50% | _____ | _____ |
| 11.75% | _____ | _____ |
| 12.00% | _____ | _____ |
| 12.25% | _____ | _____ |
| 12.50% | _____ | _____ |
| 12.75% | _____ | _____ |
| 13.00% | _____ | _____ |

# INTEROFFICE MEMO

THE HIT FACTORY, INC.
M U S I C · S T O R E

TO: Financial Assistant

FROM: Bill DeLoss, Manager

DATE: November 23, 19--

Attached is the invoice received from Downtown Sound for merchandise ordered on November 13. Please compare the invoice to the purchase order to determine that all goods ordered were shipped to us and that all prices match. Please furnish the following information:

1. Any discrepancies between the purchase order and the

   invoice: _____

   _____

2. The last date that we can pay and still receive the cash

   discount: _____

3. The amount of cash discount we will receive if we pay

   within the discount period: _____

4. The amount we must submit in full payment if we pay within

   the cash discount period: _____

---

## INVOICE
Downtown Sound
8614 Presley Way
Memphis, TN  38115

SOLD TO: The Hit Factory Music Store
3604 Music Row West
Nashville, TN  37203

DATE: November 21, 19--

TERMS: 2/10, n/30

SHIPPING: Via Carrington Cartage

| Quantity | Description | Catalog Number | Price Each | Total |
|---|---|---|---|---|
| 20 | Cassette tape case | TJ-4760 | 10.08 | 201.60 |
| 12 | Acoustic guitar | GM-3500 | 174.42 | 2,093.04 |
| 10 | Western fiddle | M-1800 | 148.20 | 1,482.00 |
|  |  |  |  |  |

Total for Merchandise: $ 3,776.64

Add Freight: 104.80

TOTAL AMOUNT DUE: $ 3,881.44

# INTEROFFICE MEMO

THE HIT FACTORY, INC.
M U S I C · S T O R E

TO:  Financial Assistant

FROM:  Bill DeLoss, Manager

DATE:  November 30, 19--

On Friday, December 8, from 2:00 p.m. to 5:00 p.m., we will host an autograph party for Jon Elgin to promote his new release, <u>Rock This City</u>. In conjunction with this promotion, we will offer discounts on a number of products. Please calculate the price at which we should mark the items below.

You are invited to a private reception for Jon at 5:30 following the autograph party, to be held in the promotions office area. See you there!

| Product | Retail Price | Percent of Markdown | Sales Price |
|---|---|---|---|
| Compact discs | $ 14.99 | 8% | _____ |
| Compact discs | 16.99 | 12% | _____ |
| Cassettes | 10.49 | 15% | _____ |
| Cassettes | 8.99 | 12% | _____ |
| Albums | 10.98 | 20% | _____ |
| Albums | 9.99 | 18% | _____ |
| 45 RPMs | 1.98 | 15% | _____ |
| Video tapes | 9.99 | 25% | _____ |
| Cassette cases | 12.49 | 16% | _____ |
| Compact disc players | 499.99 | 18% | _____ |

# INTEROFFICE MEMO

THE HIT FACTORY, INC.
M U S I C · S T O R E

TO: Lacy Montgomery

FROM: Bill DeLoss, Mgr.

DATE: December 11, 19--

*Financial Assistant—*
*Please calculate.*
*LM*

We do not currently extend credit at The Hit Factory Music Store, but I would like to begin researching that possibility.

Pertinent data taken from last year's income statement for The Hit Factory Music Store is as follows:

| | |
|---|---|
| Net sales: | $798,455 |
| Cost of goods sold: | 475,240 |
| Total expenses: | 226,852 |
| Net income: | 96,363 |

I have devised several possible credit plans varying from a very strict policy whereby almost no one could qualify to a very loose policy where we would grant credit to virtually anyone who wants it. I anticipate sales, cost of goods, operating expenses, and net income will vary in accordance with the policy we adopt.

Please estimate what net income would be under each of the following projections and determine how much net income would increase or decrease from last year. Base all amounts on last year's figures shown above. Round all amounts to the nearest dollar.

## ANALYSIS OF PROPOSED CREDIT PLANS

### The Hit Factory Music Store

| Credit Plan | Change in Net Sales | Change in Cost of Goods Sold | Change in Operating Expenses | Projected Net Income | Increase or Decrease in Net Income From Last Year |
|---|---|---|---|---|---|
| A | + 5% | + 5% | + 18% | _____ | _____ |
| B | + 10% | + 10% | + 20% | _____ | _____ |
| C | + 15% | + 15% | + 22% | _____ | _____ |
| D | + 20% | + 20% | + 25% | _____ | _____ |
| E | + 25% | + 24% | + 28% | _____ | _____ |
| F | + 30% | + 28% | + 32% | _____ | _____ |
| G | + 35% | + 33% | + 38% | _____ | _____ |
| H | + 40% | + 38% | + 45% | _____ | _____ |
| I | + 45% | + 43% | + 55% | _____ | _____ |
| J | + 50% | + 48% | + 68% | _____ | _____ |
| K | + 55% | + 53% | + 78% | _____ | _____ |
| L | + 60% | + 58% | + 88% | _____ | _____ |
| M | + 65% | + 63% | +100% | _____ | _____ |

*Financial Assistant—
Please take care
of this. LM*

TO: Lacy Montgomery, Controller

DATE: December 18, 19—

For several years, the Elmo Stoddard Band was very popular, and profitable for us to produce and promote. Unfortunately, musical tastes have changed and the Stoddard Band's material no longer sells, as evidenced by their last release.

We are under contract to produce one more release for them, including compact discs, albums, and cassettes. The prospects for a financial success are very slim. Obviously, if the material doesn't sell, the Stoddard Band doesn't receive any royalty income either—everyone loses. Therefore, I am going to propose to buy up their contract and release them. Perhaps a change of label would be beneficial for the Stoddard Band as well.

Please make the following calculations for me, based on the estimates shown below:

1.  The total estimated cost of producing and manufacturing
    the CDs, albums, and cassettes: _____

2.  One-fourth of the total cost amount (calculated in item
    no. 1), which I will offer the Stoddard Band in final
    contract settlement: _____

3.  The amount of interest income we will earn in one year on
    the difference between the total estimated cost (that we
    would have paid out if the project were produced) and the
    contract settlement amount that we will propose to pay.
    Base your calculation on a simple interest rate of 10.25%: _____

4.  Our estimated total net cost of buying up the Stoddard
    Band's contract, after deducting the interest we will
    earn from the contract settlement amount: _____

5.  I estimate we would have gross income of $25,000 from sales
    of CDs, albums, and cassettes. Calculate the amount we will
    save by buying up the contract, rather than producing the
    project. Include the estimated interest income calculated
    in item no. 3 in your calculation. _____

ESTIMATED PRODUCTION COSTS

Elmo Stoddard Band Project

Studio recording time--85 hours @ $175/hour: _____

Engineer's fee--85 hours @ $75/hour: _____

Producer's fee--95 hours @ $100/hour: _____

Sound mixing--40 hours @ $125/hour: _____

Compact disc manufacture--10,000 @ $3.75 each: _____

Cassette tape manufacture--7,500 @ $1.95 each: _____

Album pressing and album jacket fabrication--

  5,000 @ $2.75 each: _____

Promotional costs--$25,000 _____

Miscellaneous--$10,000 _____

  TOTAL ESTIMATED COSTS: ========

TO: Financial Assistant

DATE: December 19, 19--

I am contemplating the sale of the stock listed below.  Please
calculate the net proceeds and the amount of gain or loss from
the sale of each, using the current quotations I just received
from Felby & Co., which are attached.

| Stock Owned | No. of Shares Owned | Price Paid Per Share | Commission Paid |
|---|---|---|---|
| AudioM | 5,000 | 5 1/4 | $316.20 |
| JmElec | 1,000 | 20 | 278.21 |
| MatCrn | 3,000 | 16 3/4 | 602.78 |

---

**WHILE YOU WERE OUT**

To: *TJS*

*John Norris*

*Felby & Co, Stockbrokers*

Stopped In: _____     Telephoned: __X__

Message: *Current stock quotations you requested:*

| Stock | Quotation | Total Sales Commission |
|---|---|---|
| *Audio M* | *9 3/4* | *$599.63* |
| *JM Elec* | *22 1/2* | *294.46* |
| *Mat Crn* | *12 7/8* | *452.03* |

Message taken by: *MJ*

---

| Stock | Net Proceeds | Gain or Loss on Sale |
|---|---|---|
| AudioM | _____ | _____ |
| JmElec | _____ | _____ |
| MatCrn | _____ | _____ |

# INTEROFFICE MEMO

THE HIT FACTORY, INC.

TO: Financial Assistant

FROM: Lacy M.

DATE: December 21, 19--

Every year, we update the insurance coverage on our real estate holdings to be certain we are fully insured. Our buildings are insured under a policy with an 80 percent coinsurance clause. Please calculate the amount of coverage needed on each structure to be fully insured, based on the figures from Rollins & Associates as shown in the attached letter.

Also, calculate the amount of annual premium for each building, using the rates shown below, which I got from our agent yesterday.

| Building | Premium Per $100 Coverage |
|----------|---------------------------|
| Office Building and Studio | $.85 |
| Music Store Building | .80 |
| Warehouse--Waylon Blvd. | .90 |
| Warehouse--Cumberland Street | .75 |

| Building | Replacement Value | Amount of Insurance Required | Annual Premium |
|----------|-------------------|------------------------------|----------------|
| Office Building and Studio: | _____ | _____ | _____ |
| Music Store Building: | _____ | _____ | _____ |
| Warehouse--Waylon Blvd.: | _____ | _____ | _____ |
| Warehouse--Cumberland Street: | _____ | _____ | _____ |
| | | TOTAL: | _____ |

**ROLLINS & ASSOCIATES**
Certified Real Estate Appraisers
4209 Sherman Road
Nashville, TN 37206

December 18, 19--

Ms. Lacy Montgomery, Controller
The Hit Factory
3600 Music Row West
Nashville, TN 37203

Dear Ms. Montgomery:

We have finished our appraisal of your firm's real estate
buildings in Nashville, Tennessee, as per your request. The
values stated below are our estimate of the current replacement
value of the structures and do not include any contents.

| | |
|---|---|
| 3600 Music Row West (Office Bldg. & Recording Studio) | $1,820,000 |
| 3604 Music Row West (Music Store) | 395,000 |
| 1908 Waylon Blvd. (Warehouse) | 180,000 |
| 369 Cumberland Street (Warehouse) | 210,000 |

It should be pointed out that these estimates are based on
current costs of labor and materials. As economic conditions
change, fluctuations will also occur in the replacement value of
your property. Therefore, it is necessary to periodically have
the property appraised to reflect these changes.

We, the appraisers, certify that the estimates of replacement
value are, in our judgment, accurate appraisals of the
property. We do not have, have never had, and do not anticipate
acquiring any financial or other interest in the property herein
appraised.

Cordially,

*Gerald Rollins*

Gerald Rollins
Certified Real Estate Appraiser

# INTEROFFICE MEMO

THE HIT FACTORY, INC.

TO: Financial Assistant

FROM: Lacy M.

DATE: December 22, 19--

The Board of Directors is exploring the possibility of establishing a promotions office in the midwest. Four hundred thousand dollars ($400,000) will be budgeted to build or purchase an office building.

Acceptable locations in several states have been identified. As part of their analysis and decision-making process, the Board wants a report on property taxes in the proposed locations. I have identified each location and the tax rate that applies. Assume that the $400,000 purchase price would be recognized as market value by each county assessor. Calculate the amount of property tax in each location. Thanks.

| Location | Assessment Rate | Tax Rate | Estimated Property Tax |
|---|---|---|---|
| Hennepin Co., Minnesota | 42% | $8.12 per $100 | _____ |
| Cook Co., Illinois | 100% | $41.50 per $1,000 | _____ |
| Jackson Co., Missouri | 33 1/3% | 10.8¢ per $1 | _____ |
| Scott Co., Iowa | 100% | 33 mills per $1 | _____ |
| Douglas Co., Nebraska | 35% | $8.29 per $100 | _____ |

# INTEROFFICE MEMO

THE HIT FACTORY, INC.

TO: Financial Assistant

FROM: Lacy M.

DATE: January 5, 19--

I hope you enjoyed the long holiday weekend. Now, however, comes our busiest time of the year. We must prepare a financial statement analysis of the past year's business.

Attached are financial statements for the year ended December 31. Income statement information is provided for each of the company's three separate activities: record productions, custom recording, and The Music Store. The balance sheet shows combined information for all phases of the company's operation for last year and the previous year. Please make the following calculations.

1.  Prepare a combined income statement, on the column provided on the income statement, by adding the data from the three separate income statements.
2.  Prepare a horizontal analysis of the balance sheet showing the amount of percent of increase or decrease from 19X1 to 19X2.
3.  Make the following analysis:
    a.  Determine what percent each of the company's activities--record productions, custom recording, and The Music Store--contributed to the combined income statement total for the following:

| Activity | Percent of Combined Net Sales | Percent of Combined Gross Profit | Percent of Combined Net Income |
|---|---|---|---|
| Record Productions | _____ | _____ | _____ |
| Custom Recording | _____ | _____ | _____ |
| Music Store | _____ | _____ | _____ |
| TOTAL | _____ | _____ | _____ |

    b.  Percent that last year's combined net income is of combined net sales.
    c.  Percent that last year's current assets are of total assets:
    d.  Percent that last year's fixed assets are of total assets:
    e.  Percent that last year's current liabilities are of total liabilities:
    f.  Percent that last year's long-term liabilities are of total liabilities:
    g.  Percent that last year's total liabilities are of total assets:

    h.  Last year's current ratio:

    i.  Last year's acid test ratio:

    j.  Last year's return on total assets:*

    k.  Last year's return on owner's equity:*

*Note: An amount shown on one year's ending balance sheet becomes the beginning amount for the next year. Use combined net income in your calculations.

```
                              The Hit Factory
                             Income Statement
                    For Year Ended December 31, 19X2
                                                          Combined
                         Record      Custom      Music     Income
                       Productions  Recording    Store    Statement

Revenue From Sales:
  Sales                29,468,250    348,200    897,750   _____

  Less: Sales Ret.& Allow.  892,887       -      22,750   _____

     Net Sales         28,575,363    348,200    875,000   _____

Cost of Goods Sold:
  Beginning Inventory   1,260,173       -       217,000   _____

  Add: Net Purchases    8,841,217       -       558,425   _____

  Cost of Goods for Sale 10,101,390     -       775,425   _____

  Less: Ending Inventory  725,814       -       254,625   _____

     Cost of Goods Sold  9,375,576       -       520,800   _____

Gross Profit on Sales  19,199,787    348,200    354,200   _____


Operating Expenses
  Artist's Royalties     4,744,388       -         -      _____

  Advertising and Promo  3,921,566     21,200     49,875  _____

  Salary Expense         3,556,510    114,340     89,775  _____

  Contract Labor         1,970,983       -         -      _____

  Publisher's Royalties    721,624       -         -      _____

  Depreciation Expense     152,578     14,812     17,226  _____

  Supplies Expense         118,725     13,505      7,175  _____

  Utility Expense           51,470      6,380      6,825  _____

  Insurance Expense         93,225      5,150      6,650  _____

  Interest Expense          60,160      1,280      2,560  _____

  Other Expenses           176,165      4,995     18,090  _____

     Total Expenses     15,567,394    181,662    198,176  _____


Net Income              3,632,393    166,538    156,024   _____
```

The Hit Factory
Comparative Balance Sheet
December 31, 19X1 and 19X2

| Assets | 19X1 | 19X2 | INCREASE OR DECREASE AMOUNT | PERCENT |
|---|---|---|---|---|
| Current Assets: | | | | |
| Cash in Bank | 1,316,290 | 1,609,720 | | |
| Marketable Securities | 286,150 | 346,298 | | |
| Accounts Receivable | 1,609,476 | 2,615,306 | | |
| Inventories | 1,477,173 | 980,439 | | |
| Supplies on Hand | 45,211 | 43,907 | | |
| Other Current Assets | 186,318 | 203,796 | | |
| Total Current Assets | 4,920,618 | 5,799,466 | | |
| Fixed Assets: | | | | |
| Equipment | 486,350 | 455,282 | | |
| Vehicles | 124,960 | 136,871 | | |
| Buildings | 618,347 | 702,838 | | |
| Land | 104,300 | 165,000 | | |
| Total Fixed Assets | 1,333,957 | 1,459,991 | | |
| Total Assets | 6,254,575 | 7,259,457 | | |
| Liabilities and Owners' Equity | | | | |
| Current Liabilities: | | | | |
| Accounts Payable | 224,360 | 276,418 | | |
| Artist's Royalties Payable | 1,079,416 | 1,390,160 | | |
| Notes Payable | 64,381 | 81,060 | | |
| Total Current Liabilities | 1,368,157 | 1,747,638 | | |
| Long-Term Liabilities: | | | | |
| Notes Payable | 324,179 | 367,942 | | |
| Total L-T Liabilities | 324,179 | 367,942 | | |
| Total Liabilities | 1,692,336 | 2,115,580 | | |
| Owners'Equity: | | | | |
| Common Stock | 1,000,000 | 1,000,000 | | |
| Retained Earnings | 3,562,239 | 4,143,877 | | |
| Total Owners' Equity | 4,562,239 | 5,143,877 | | |
| Total Liabilities and Owners' Equity | 6,254,575 | 7,259,457 | | |

TO:  Lacy Montgomery, Controller

DATE:  January 6, 19--

Lacy, we've got some extra cash that I am thinking about
investing in something relatively long-term.  I took the
following information from the bond quotations in this
morning's newspaper.

| Bond | Stated Rate | Years to Maturity | High | Low | Close |
|------|------|------|------|------|------|
| Entco | 4 1/2% | 8 | 48 1/8 | 48 | 48 |
| RecAm | 16 % | 4 | 109 | 108 7/8 | 109 |
| SoNatl | 10 % | 15 | 64 1/8 | 63 3/4 | 64 |

Please calculate the following:  (You can disregard the
brokerage fees in your calculations.)

1.  The total cost to buy 25 of each bond at the closing
    price:

    Entco:  $_____    RecAm:  $_____    SoNatl:  $_____

2.  Interest earned.

| Bond | Interest Earned Per Year Per Bond |
|------|------|
| Entco | $_____ |
| RecAm | $_____ |
| SoNatl | $_____ |

## INTEROFFICE MEMO

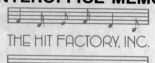

THE HIT FACTORY, INC.

TO: Financial Assistant

FROM: Lacy M.

DATE: February 10, 19--

TJS is still negotiating with the Monarchs on their new contract. He needs a little more data. Listed below are the compact disc releases the Monarchs have had on the Cannon label (rounded to the nearest ten thousand). Please calculate the:

Mean: _____

Median: _____

Mode: _____

| Release | Title | No. of CDs Sold |
|---------|-------|-----------------|
| #1 | Sugar Nights | 80,000 |
| #2 | Crazy Hearts | 110,000 |
| #3 | Comin' Home | 80,000 |
| #4 | Crossin' Over | 240,000 |
| #5 | Sometime Lover | 630,000 |
| #6 | City Life | 950,000 |
| #7 | Heartthrob | 820,000 |
| #8 | Mr. E | 1,160,000 |
| #9 | Guitar Pickin' Man | 1,130,000 |

## INTEROFFICE MEMO

THE HIT FACTORY, INC.

TO: Financial Assistant

FROM: Lacy M.

DATE: February 11, 19--

Somehow, the attached inventory records did not get updated at the end of last year. Please calculate the value of the ending inventory for each asset shown.

```
            INVENTORY RECORD

Asset: Ragga 3" Recording Tape Reels

Inventory Method  FIFO

                Number    Cost
        Date   Purchased  Each
------------------------------------
Beg. Inv          2      $85.00
Jan. 1           16       86.50
Jan. 18          30       84.00
Feb. 28          90       80.00
May 7            30       82.00
June 21          40       88.00
Oct. 12          20       90.00
Nov. 22          10       92.00


Dec. 31, 19---
Ending Inventory:  33

Valuation:  $_____
```

```
            INVENTORY RECORD

Asset: Ragga 1/4" Recording Tape Reels

Inventory Method: FIFO

                Number    Cost
        Date   Purchased  Each
------------------------------------
Beg. Inv         15      $9.50
Jan. 1           10       10.00
Jan. 18          25       10.00
Feb. 28          90        9.00
May 7            30        9.50
June 21          40       10.50
Oct. 12          20       11.00
Nov. 22          10       11.25


Dec. 31, 19--
Ending Inventory: 36

Valuation:  $_____
```

```
            INVENTORY RECORD

Asset: Williams Accoustic Guitars TX

Inventory Method: LIFO

                Number    Cost
        Date   Purchased  Each
------------------------------------
Beg. Inv          1      $75.60
Jan. 1            3       77.00
Feb. 6           5       78.20
Apr. 17          10       80.40
June 4           15       80.40
July 27          15       82.00
Sept. 20         10       84.50
Nov. 30          10       86.00

Dec. 31, 19--
Ending Inventory: 11

Valuation:  $_____
```

```
            INVENTORY RECORD

Asset: Collins Fiddle FJ10

Inventory Method: LIFO

                Number    Cost
        Date   Purchased  Each
------------------------------------
Beg. Inv.
Jan. 1            3      $136.20
Feb. 16          5       139.00
Apr. 7           20      144.50
July 27          5       146.00
Nov. 30          8       148.00

Dec. 31, 19--
Ending Inventory: 12

Valuation: $_____
```

# INTEROFFICE MEMO

THE HIT FACTORY, INC.

TO: Financial Assistant

FROM: Lacy M.

DATE: February 22, 19--

We are now preparing a report of the past year's activities for presentation to the Board of Directors at their next meeting. I want to include some graphs to add some visual attractiveness and to clearly illustrate the progress that the company has made in the past few years.

Please prepare the following graphs. Furnish proper headings and captions. You may need to search your records to locate some of the necessary information.

a. A circle graph showing the proportionate net income earned from each of The Hit Factory's activities: record productions, custom recording, and The Hit Factory music store for last year.

b. A comparative line graph showing net product sales under the Nickelodeon and Cannon labels for the past six years. (Round amounts to the nearest ten thousand.)

| | Net Product Sales | |
|---|---|---|
| Year | Nickelodeon Label | Cannon Label |
| 1 | $ 3,658,025 | $     867,069 |
| 2 | 5,123,950 | 1,114,964 |
| 3 | 7,150,625 | 2,471,078 |
| 4 | 6,032,475 | 1,925,324 |
| 5 | 12,630,950 | 8,710,421 |
| 6 | 13,960,500 | 14,614,863 |

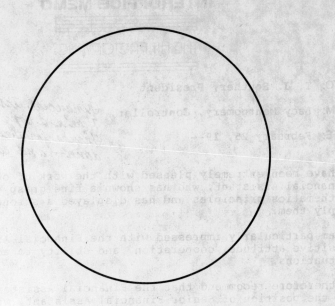

The Hit Factory
NET PRODUCT SALES*
Years 1–6

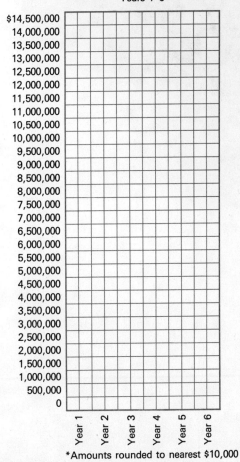

$14,500,000
14,000,000
13,500,000
13,000,000
12,500,000
12,000,000
11,500,000
11,000,000
10,500,000
10,000,000
9,500,000
9,000,000
8,500,000
8,000,000
7,500,000
7,000,000
6,500,000
6,000,000
5,500,000
5,000,000
4,500,000
4,000,000
3,500,000
3,000,000
2,500,000
2,000,000
1,500,000
1,000,000
500,000
0

Year 1　Year 2　Year 3　Year 4　Year 5　Year 6

*Amounts rounded to nearest $10,000

# INTEROFFICE MEMO

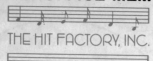

THE HIT FACTORY, INC.

TO: T. J. Souther, President

FROM: Lacy Montgomery, Controller

DATE: February 25, 19--

*Financial Assistant —*
*Here's a copy for your information.*
*Thanks for the fine job you have*
*done — and congratulations!*
*LM*

I have been extremely pleased with the work of our new
Financial Assistant, who has shown a fine grasp of business
mathematics principles and has displayed a strong ability to
apply them.

I am particularly impressed with the Financial Assistant's
positive attitude, cooperation, and ability to adapt to new
situations.

I therefore recommend that the Financial Assistant be promoted
to the position of Senior Financial Assistant, with a
commensurate pay increase.

*Lacy Montgomery, Controller*

*Request approved*
*TJS*

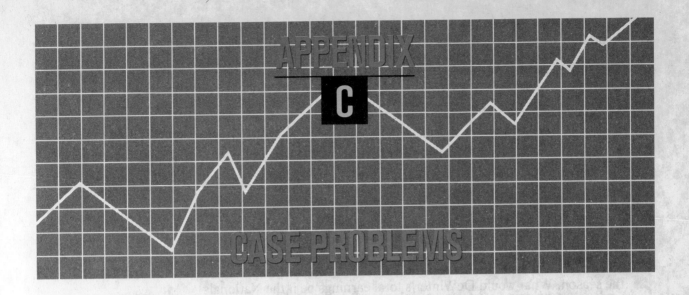

## APPENDIX C

### CASE PROBLEMS

Solve the following case problems.

1. It is estimated that each basketball player runs 40,000 feet in a complete basketball game. The Nationals play 90 games per season. Mike DeWinter has played each game in its entirety for the past three seasons. How many miles has he run in the three seasons? Round off your answer to two decimal places. (Lessons 1-3 and 1-4)

_____

2. Basketball shoes used by the Nationals' players cost $60 per pair. How much will the Nationals spend on basketball shoes if six pairs are purchased for each of the 15 players? (Lesson 1-3)

_____

3. Last season the Nationals' ticket sales were $4,032,000 for 45 home games. How much additional income would be earned from ticket sales if two more home games were added to the schedule? (Assume that each of the two additional games would draw equally as well as last season's home games.) (Lessons 1-3 and 1-4)

_____

**4.** Last year Alex Graham earned a salary of $320,000. He also earned $100,000 from product endorsements, $20,000 from guest appearances, $25,000 for speeches given, and $75,000 as an actor. What were Graham's total earnings for the year? (Lesson 1-1)

_____

**5.** Mike DeWinter's agent proposes that DeWinter receive a salary of $300,000 for next season. In addition, he proposes that DeWinter receive $0.50 for each home game ticket sold in excess of 565,000 for the season. What would DeWinter's total earnings be if the Nationals average 14,000 fans for each of their 45 home games? (Lesson 1-3)

_____

**6.** Last year the Nationals' total income from ticket sales, television rights, and concession sales was $7,032,000. Of this, $4,032,000 was from ticket sales and $850,000 was from concession sales. How much was earned from television rights? (Lessons 1-1 and 1-2)

_____

**7.** It is estimated that 225 square feet will be needed for each parking space in the Nationals' new parking lot. How many vehicles can be parked in a 140,000 square yard lot? (Lessons 1-3 and 1-4)

_____

**8.** Last season Bert Cooper averaged 4.5 fouls in each of the 86 games in which he played. How many fouls did Cooper commit during the season? (Lesson 1-3)

_____

**9.** The Nationals' eight cheerleaders each received $40 per game for each of the 45 home games. What was the total amount paid to the cheerleaders for the season? (Lesson 1-3)

_____

**588** ■ APPENDIX C

10. In the playoffs, the Nationals drew 15,000 fans for each of the four home games. Ticket prices averaged $8.50 each. What was the total ticket sales income for the four games? (Lesson 1-3)

_____

11. It is decided to encourage Rice Real Estate's salespeople to devote 85% of their time to residential property sales next year. Projections are that this, along with several other changes, will increase sales volume by 40% over this year's sales of $3,000,000. What is the projected sales volume for next year? (Lesson 3-3)

_____

12. Next year's office salary expense will be decreased $9,000 from this year's $32,000 expenditure. What percent decrease is this? Round off your answer to two decimal places. (Lesson 3-4)

_____

13. It is projected that a decrease of $9,000 in Rice Real Estate's office salary expense will result in a 35 percent increase in net income from this year's $22,000. What is the projected increase in net income that will result from this decrease in office salary expense? (Lesson 3-2)

_____

14. Rice's advertising expenditures will increase 50% from the $8,300 spent this year in an attempt to spur sales. What amount will be budgeted for advertising next year? (Lesson 3-2)

_____

15. It is projected that an increase of 50% in advertising expenditures will result in a 15 percent increase in commissions earned from this year's $90,000. How much of an increase in commissions earned will this change produce? (Lesson 3-2)

_____

**16.** It is estimated that the new salesperson's incentive program will increase Rice's Real Estate's commissions earned by 35%. This year, commissions earned were $90,000. How much should commissions earned increase next year because of this program? (Lesson 3-2)

_____

**17.** Rice Real Estate's plan for increasing next year's net income includes projections of increasing commissions earned by $129,000 while cutting operating expenses by $50,080. List two factors which might cause actual performance to vary from these projections. (Chapter 3)

_____

_____

_____

_____

_____

_____

_____

_____

_____

_____

_____

Use the tax tables on pages 124 and 125 where necessary.

**18.** Pat Coleman, a single employee, averages $315 in gross earnings a week. How much more or less will her federal income tax withholding be per week if she gets married and still claims one exemption? (Lesson 4-3)

_____

**19.** In 1977 social security taxes were 5.85% on maximum earnings of $15,300. Currently, the rate is 7.51% on maximum earnings of $45,000. Leonard Whiting was taxed on the maximum in each year. How much more did he contribute to social security this year than in 1977? (Lesson 4-3)

_____

**20.** Last year Jane Todd worked 40 hours per week in Summer's retail outlet, earning $4.75 per hour. This year she is on a guaranteed salary of $125 per week plus a commission of 2.5% of all sales she makes. Last week, in 40 hours her sales were $2,900. How much more or less did she earn last week on salary-plus-commission basis, compared to what she would have earned at last year's hourly wage rate? (Lessons 4-1 and 4-2)

_____

**21.** Adam Regency, a married employee claiming two exemptions, averages $378 gross earnings per week. His state income tax withholding averages $14.95 per week. How much of his gross weekly earnings is withheld for combined social security tax, federal income tax, and state income tax? (Lesson 4-3)

_____

**22.** Last year Frank Russell, a married employee claiming two exemptions, switched from his job as a production worker to the sales staff. As a production worker his gross weekly earnings would have averaged $355.77 this year. As a salesperson, he averaged $480.77. Using a 52-week year, calculate how much more or less will be withheld from his earnings for federal income tax this year, compared to what would have been withheld if he had remained a production worker. (Lesson 4-3)

_____

**23.** Summer's management is considering a new salary-plus-commission plan for its retail outlet employees that would pay a base salary of $600 per month and a commission of 10% on all sales made in excess of $5,000 each month. How much would Brian King's earnings be if in a month he sold (a) $8,000 and (b) $9,000? (Lesson 4-2)

**(a)** _____

**(b)** _____

**24.** Emmett Electric Company's checkbook balance on March 31 is $3,164.20. It is now found that Check No. 2161 for $186.50 was recorded on the check stub for $106.50 and that Check No. 2210 for $635.00 was recorded on the check stub as $365.00. What is the correct check stub balance? (Lesson 5-2)

_____

**25.** Dulong Designs offers a dining room set at a list price of $2,500, series trade discounts of 40% and 10%, and terms of 2/10, n/30. Dunne Decorators offers a comparable set with a $2,350 list price, a trade discount of 44%, and terms of n/30. (a) From which supplier can the best buy be obtained? (b) How much lower is the price? (Lesson 6-1)

**(a)** _____

**(b)** _____

**26.** Ebony Furniture Manufacturing Company offers buffets at a list price of $800 with a trade discount of 40% if five or fewer are purchased. If more than five are purchased, the series trade discount is 40% and 10%. What is the buyer's cost per buffet (a) if five or fewer are purchased and (b) if more than five are purchased? (Lesson 6-1)

**(a)** _____

**(b)** _____

**27.** Westfield Furniture Store is contemplating setting up its own factory to manufacture chairs to be sold exclusively in its retail stores. It is estimated that a wooden rocker can be manufactured for $45. Ebony Furniture Manufacturing Company lists a comparable chair at $150 and offers a 45 percent trade discount. (a) How much more or less will the chair cost if it is purchased from Ebony? (b) List two factors Westfield should consider when deciding whether or not to enter manufacturing. (Lesson 6-1)

**(a)** _____

**(b)** _____
_____
_____
_____
_____
_____

**28.** Currently, Westfield Furniture Store buys goods from many manufacturers. Last year, goods with a list price of $3,460,500 were purchased and the average series trade discount was 45% and 10%. It is estimated that an additional 5% series discount can be obtained for buying in quantity if purchases are concentrated in fewer suppliers. How much of a savings would this amount to? (Lesson 6-1)

_____

**29.** A local manufacturer offers Westfield Furniture Store desks at a delivered cost of $100 each. Charleson Manufacturing Company offers similar desks at a list price of $180 and a trade discount of 50%. Freight charges will be $15 per desk. (a) Which supplier offers the lowest price? (b) How much lower is the price per desk? (c) List two factors, other than price, that should be considered in selecting a supplier. (Lesson 6-1)

**(a)** _____

**(b)** _____

**(c)** _____
_____
_____
_____
_____

**30.** Richardson Manufacturing offers Congreaves 300 pairs of skis at $55 per pair and recommends a retail selling price of $100 per pair. Corliss Manufacturing offers 300 pairs at $60 each and recommends a selling price of $115. (a) If each of the manufacturer's recommendations for selling price is followed, which will yield the greatest rate of markup based on selling price? (b) How much greater is it? (c) List three other factors, in addition to price, that Congreaves should consider when selecting a supplier. (Lesson 7-2)

(a) _____

(b) _____

(c) _____

_____

_____

31. Softball gloves are offered by the World Series Corporation for $18 each. Congreaves will mark these at a retail price which yields 70% markup on cost. Congreaves is contemplating buying 200 gloves. It is projected that 150 gloves can be sold at the regular selling price and that the remaining 50 can be sold at a markdown of 30% from regular retail price. If these projections are correct, how much gross profit will be realized from selling the 200 gloves? (Lessons 7-1 and 7-3)

_____

32. Congreaves' research shows that snowshoes will be "hot" items next winter and that sales competition will be stiff. The accounting department estimates operating expenses of $10.50 for each pair of snowshoes sold. The purchasing department determines that snowshoes can be purchased at $30 per pair. Research shows that the price at which the snowshoes are marked will largely determine the number that can be sold. It is projected that if a 40% markup on cost is utilized, 500 pairs will be sold; if a 60% markup on cost is used, 400 pairs will sell. What amount of net income will be earned if (a) 500 pairs sell or (b) 400 pairs sell? (Lesson 7-1)

(a) _____

(b) _____

**33.** Boise City Bank will offer Emmett Electric Company an auto loan for $12,000 at 10% to be amortized over 5 years. Portland Bank will loan the $12,000 at 12% to be amortized over 5 years. (a) How much more or less will the monthly loan payments be at Boise City Bank than at Portland Bank? (b) How much more or less will the total amount of interest paid be at Boise City Bank than at Portland Bank? (Lesson 8-3)

(a) _____

(b) _____

**34.** Exactly 13 years ago, Emmett's purchased a building, borrowing $60,000 at 12% to be amortized over 25 years. Determine (a) the total amount of principal and interest paid on the loan over the past 13 years, (b) the amount of interest paid, and (c) the remaining principal balance. (Lesson 8-3)

(a) _____

(b) _____

(c) _____

**35.** Emmett's is considering the purchase of a utility tractor-mower for $4,000 to be used exclusively for mowing the grounds. The entire amount can be borrowed at 12% to be amortized over 5 years. Randy Wallick, the treasurer's daughter, offers to mow the grounds, using her own mower and paying her own expenses, for $900 for a year. (a) In a year's time, how much more or less would be paid in loan payments for buying the tractor-mower than in hiring Randy? (b) List two other factors Emmett's should consider when deciding whether to buy its own tractor-mower or to hire Randy instead. (Lesson 8-3)

(a) _____

(b) _____

_____

_____

_____

_____

**36.** Emmett Electric Company has the option of leasing a warehouse at $1,200 per month for the next 20 years or of buying it for $120,000. If the building is purchased, a $20,000 down payment will be required and the balance can be borrowed at 11% to be amortized over 20 years. Determine (a) the total cash outlay for rent over 20 years, (b) the total amount that will be paid out to purchase the building over the 20 years, and (c) list two factors Emmett's should consider when deciding whether to lease or purchase. (Lesson 8-3)

(a) _____

(b) _____

(c) _____

_____

_____

_____

_____

_____

**37.** Congreaves has determined that its average installment contract is for $1,000 at 18% APR for 18 months. The actual cost of handling and finance for each such contract is calculated at $178.47. (a) How much does Congreaves make or lose on the finance charge on each account? (b) List two other factors to be considered when analyzing whether it is profitable to sell merchandise on the installment basis. (Lesson 8-2)

(a) _____

(b) _____

_____

_____

_____

_____

_____

**38.** Duchess Cosmetics is planning to purchase a $50,000 whole life insurance policy on Laura Barth, age 43. (a) What will the company's net cost for a whole life insurance policy be if it is kept in force for 10 years (annual premiums paid) and then surrendered for its cash value? (b) What will the company's net cost be if a 10-year term policy for $50,000 is purchased instead? (Lesson 9-1)

(a) _____

(b) _____

**39.** Sites in two different states are under consideration for construction of a new manufacturing plant. It is assumed that the assessor in each county would recognize the building's $2,000,000 cost as its market value. In Cactus County, property is assessed at 25% of market value and the tax rate is $10 per $100. In Winchester County, property is assessed at 60% of market value and the tax rate is 6.5 cents per $1. Calculate the estimated amount of real estate tax in (a) Cactus County and (b) Winchester County. (c) List two other factors, besides the property tax amount, that should be considered in choosing a plant site. (Lesson 9-4)

(a) _____

(b) _____

(c) _____

_____

**40.** Duchess Cosmetics owns a building that is valued at $300,000. Settler's Insurance Company offers a policy with an 80% coinsurance clause at a cost of $0.50 per $100 of coverage. (a) How much will Duchess Cosmetics save in premiums per year if the building is insured for only $150,000 instead of at the 80% coinsurance level? (b) would you recommend that the building be insured at the lower level to save on insurance premiums? Why? (Lesson 9-2)

(a) _____

(b) _____

_____

**41.** Vehicle insurance rates quoted by two different insurance companies for selected coverage are as follows for the same vehicle.

| Insurer | Bodily Injury $1,000,000 | Property Injury $1,000,000 | Collision $500 Deductible | Comprehensive $0 Deductible | Uninsured Motorist |
|---|---|---|---|---|---|
| Settler's Insurance | $315 | $325 | $275 | $275 | $5 |
| Packer Insurance | 290 | 300 | 290 | 210 | 4 |

Calculate the premium for the policy from (a) Settler's Insurance, (b) Parker Insurance. (Lesson 9-3)

(a) _____

(b) _____

**42.** March College has $100,000 to invest for 6 months. Calculate the amount of interest that would be earned from each of the following investment mediums: (a) certificate of deposit paying 5.6536% and (b) money market certificate paying 12.108%. (Lesson 10-1)

(a) _____

(b) _____

**43.** March College invested $20,000 in a 13.078% 6-month money market certificate. Two months before maturity, March redeemed the certificate. A 3-month interest penalty was imposed. (a) What was March's proceeds? (b) Suggest one other investment medium that may have been more appropriate than the money market certificate. (Lesson 10-1)

(a) _____

(b) _____

_____

_____

**44.** March College owns 1,000 shares of Carson Calculators, Inc., stock which cost $25\frac{1}{4}$ per share. The commission was $307. Last year, total dividends of $0.80 per share were received. (a) Calculate last year's annual yield on investment. (b) List one other reason, besides the annual dividend, that March would want to own this stock. (Lesson 11-2)

(a) _____

(b) _____
_____

**45.** Supersound had 25 Model MT100 electric organs available during a quarter, acquired as follows:

| Date | Description | Quantity | Cost Each |
|---|---|---|---|
| July 1 | Beginning inventory | 8 | $2,000 |
| July 27 | Purchase | 5 | 2,200 |
| September 15 | Purchase | 12 | 2,600 |

There are four organs on hand at the end of the period. Supersound wants to show its ending inventory at its lowest possible amount on its financial records. (a) Which method, FIFO, LIFO, or weighted average, should be used? (b) What is the ending inventory valuation? (Lesson 12-2)

(a) _____

(b) _____

**46.** Supersound's goods available for sale for a recent month were $149,500 at cost and $230,000 at retail. Its sales for the month were $120,000. What is the estimated ending inventory at cost? (Lesson 12-2)

_____

**47.** Supersound purchased a new sound system for $15,000. Its estimated life is 5 years and the residual value is $3,000. (a) Which method, declining-balance or sum-of-the-years'-digits, will yield the greatest amount of depreciation expense in the first year? (b) How much greater is it than the other method? (Use twice the straight-line rate for the declining-balance method.) (Lessons 13-3, 13-4)

(a) _____

(b) _____

**48.** For its accounting records, Supersound can depreciate its office equipment over either 4 or 5 years. The cost of the equipment was $50,000 and the salvage value is $5,000. How much more or less depreciation expense can be recognized in the first year if the assets are depreciated over 4 years, using the straight-line method for comparison? (Lesson 13-1)

_____

**49.** Supersound purchased a building for $200,000. It has an estimated life of 40 years and a salvage value of $20,000. How much more or less will Supersound's total expenses be for the first year if the declining-balance method at twice the straight-line rate is used instead of the straight-line rate? (Lessons 13-1, 13-3)

_____

**50.** Supersound installed a new air-conditioning system at a $26,000 cost. It has an estimated life of five years and a salvage value of $2,000. What is the average amount of depreciation expense that would be recognized over the first three years according to the

**600** ■ APPENDIX C

(a) straight-line method and (b) sum-of-the-years'-digits method? (Lessons 13-1, 13-4)

(a) _____

(b) _____

51. Supersound expanded into an adjacent building which it purchased for $110,000. The building has an estimated life of 25 years and a salvage value of $10,000. Supersound's income tax rate amounts to 45% of its net income. (a) Which depreciation method, straight-line or sum-of-the-years'-digits, will allow Supersound to recognize the greatest depreciation expense for the first year the building is owned? (b) How much more income will be taxable than that due by using the other method? (Lessons 13-1, 13-3)

(a) _____

(b) _____

52. In a recent year, First Fashions' budgeted net sales was $980,350 and budgeted cost of goods sold was $588,000. Actual net sales were $910,200 and actual cost of goods sold was $592,000. (a) What percent was budgeted cost of goods sold of budgeted net sales? (b) What percent was actual cost of goods sold of actual net sales? (c) List two reasons why the actual cost of goods sold might vary this much from the budgeted percent. (Lesson 14-1)

(a) _____

(b) _____

(c) _____

_____

_____

**53.** In a recent year, First Fashions had $118,500 current liabilities and $12,800 long-term liabilities. (a) What percent of total liabilities was the current liabilities? (b) List one reason why a business might prefer to have very little long-term debt. (Lesson 14-2)

(a) _____

(b) _____

_____

_____

**54.** In a recent year, First Fashions' total current assets were $560,300 and its total current liabilities were $290,610. A current ratio of 2 to 1 is considered to be desirable. (a) Calculate First's current ratio. (b) What should First's total current assets have been to be in accordance with the desirable ratio? (Lesson 14-3)

(a) _____

(b) _____

**55.** In a recent year, First Fashions' quick assets totaled $220,600 and its total current liabilities were $290,610. (a) Calculate First's acid-test ratio. (b) What should First's quick assets have been to be in accordance with the generally acceptable acid-test ratio level? (Lesson 14-3)

(a) _____

(b) _____

**56.** First Fashions' net income was $42,400 and average owners' equity was $418,700 in a recent year. A return of 12.5% on investment is considered desirable. (a) Calculate the rate of return on First Fashions' total owner's equity. (b) What should net income have been to match the 12.5% rate desired? (Lesson 14-4)

(a) _____

(b) _____

**602** ■ APPENDIX C

**57.** First Fashions' beginning accounts receivable balance was $88,400, its ending accounts receivable balance was $92,500, and its net credit sales were $500,200 in a recent year. First Fashions' considers an accounts receivable turnover ratio of 8 to 1 to be desirable. (a) Calculate the accounts receivable turnover ratio. (b) What should the average accounts receivable have been to be in accordance with the desired ratio? (Lesson 14-3)

**(a)** _____

**(b)** _____

**58.** An analysis of preemployment tests given to all job applicants over the past three years reveals the following statistics: mean, 83.5; median, 87; mode, 85. This month, nine applicants took the preemployment test with the following results: 96, 92, 91, 87, 86, 83, 76, 72, 68. How much above or below the three-year averages were this month's (a) mean, (b) median, and (c) mode? (Lesson 15-1)

**(a)** _____

**(b)** _____

**(c)** _____

**59.** Each employee at Popham's has taken an average of 6 days of sick leave each year for the past 10 years. On the basis of absences so far this year, it is estimated that each employee will average 11 days sick leave for the year. What are some possible reasons for this drastic increase? List at least two. (Lessons 15-1, 15-2)

_____

_____

_____

_____

_____

_____

_____

**60.** Sales made last year by Popham's full-time salespeople are categorized below. Calculate (a) the mean, (b) the median, and (c) the mode. (d) Of what value is this information to management? (Lesson 15-1)

| Number of Employees | Sales | Number of Employees | Sales |
|---|---|---|---|
| 2 | $150,001–$170,000 | 10 | $90,001–$110,000 |
| 0 | 130,001– 150,000 | 5 | 70,001– 90,000 |
| 6 | 110,001– 130,000 | 3 | 50,001– 70,000 |

(a) _____

(b) _____

(c) _____

(d) _____

_____

_____

_____

_____

_____

_____

_____

_____

_____

_____

**61.** The number of persons using Popham's credit cards for each of the past 7 years is as follows: year 1, 800; year 2, 750; year 3, 900; year 4, 1,300; year 5, 1,600; year 6, 1,800; year 7, 2,100. Present this data visually in an appropriate graph. (Lesson 15-2)

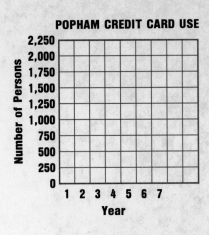

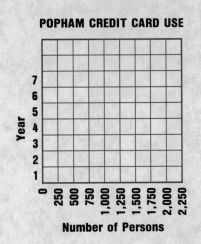

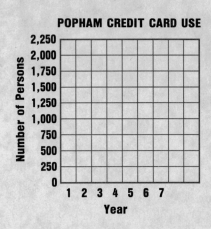